Network-on-Chip Security and Privacy

Prabhat Mishra • Subodha Charles
Editors

Network-on-Chip Security and Privacy

Editors
Prabhat Mishra
University of Florida
Gainesville, FL, USA

Subodha Charles
University of Moratuwa
Colombo, Sri Lanka

ISBN 978-3-030-69133-2 ISBN 978-3-030-69131-8 (eBook)
https://doi.org/10.1007/978-3-030-69131-8

This Springer imprint is published by the registered company Springer Nature Switzerland AG
The registered company address is: Gewerbestrasse 11, 6330 Cham, Switzerland

Preface

System-on-Chip (SoC) integrates a wide variety of hardware components into a single integrated circuit to provide the backbone of modern computing systems ranging from complex navigation systems in airplanes to simple Internet-of-Things (IoT) devices in smart homes. Cars are full of them, as are airplanes, satellites, and advanced military and medical devices. As applications grow increasingly complex, so do the complexities of the SoCs. For example, a typical automotive SoC may include 100–200 diverse components (e.g., processor, memory, controllers, and converters) from multiple third-party vendors. Network-on-Chip (NoC) is a widely used solution for on-chip communication between various components in complex SoCs.

SoCs are designed today using Intellectual Property (IP) components to reduce cost while meeting aggressive time-to-market constraints. Growing reliance on these pre-verified components, often gathered from untrusted third-party vendors, severely affects the security and trustworthiness of SoC computing platforms. These third-party components may come with deliberate malicious implants to incorporate undesired functionality, undocumented test/debug interfaces working as a hidden backdoor, or other integrity issues. Since NoC facilitates communication between various components in an SoC, NoC is the ideal place for any malicious implants (such as hardware Trojans) to hide and launch a wide variety of security attacks. Due to the resource-constrained nature of many embedded and IoT devices, it may not be possible to employ traditional security solutions to protect NoC against malicious attacks. Specifically, there is a need for lightweight countermeasures that can secure NoC without violating any design constraints such as area, power, energy, and performance.

This book provides a comprehensive overview of NoC security attacks and effective countermeasures for designing secure and trustworthy on-chip communication architectures. These techniques are applicable across on-chip communication technologies (e.g., electrical, optical, and wireless) supporting a wide variety of on-chip network topologies (e.g., point-to-point, bus, crossbar, ring, and mesh). Specifically, this book describes state-of-the-art security solutions that satisfy a wide variety of communication (often conflicting) requirements such as securing packets, ensuring route privacy, meeting energy budget and real-time constraints, finding trusted routes in the presence of malicious components, and providing real-time attack detection and mitigation techniques. The presentation of topics has been divided into five categories with each category focusing on a specific aspect of the big picture. A brief outline of the book is provided as follows:

1. *Introduction to NoC Security:* The first part of the book includes three introductory chapters on NoC design and security challenges.

   - Chapter 1 provides an overview of NoC-based SoC design methodology with an emphasis on NoC architectures and security vulnerabilities.
   - Chapter 2 describes accurate modeling and design space exploration of on-chip communication architectures.
   - Chapter 3 presents popular optimization techniques for designing energy-efficient NoC architectures.

2. *Design-for-Security Solutions:* The second part of the book focuses on design-time solutions for securing NoC architectures against attacks.

   - Chapter 4 presents a lightweight encryption scheme using incremental cryptography.
   - Chapter 5 describes a trust-aware routing algorithm that can bypass malicious components.
   - Chapter 6 outlines a lightweight anonymous routing technique.
   - Chapter 7 describes how to efficiently integrate secure cryptography to overcome NoC-based attacks.

3. *Runtime Security Monitoring:* The third part of the book deals with security solutions for runtime detection and mitigation of vulnerabilities.

   - Chapter 8 describes a mechanism for real-time detection and localization of denial-of-service attacks.
   - Chapter 9 utilizes digital watermarking for providing lightweight defense against eavesdropping attacks.
   - Chapter 10 outlines a machine learning framework for detecting attacks on NoC-based SoCs.
   - Chapter 11 presents a routing technique that can provide trusted communication in the presence of hardware Trojans.

4. *NoC Validation and Verification:* The fourth part of the book explores methods for verifying both functional correctness and security guarantees.

   - Chapter 12 describes NoC security and trust validation techniques.
   - Chapter 13 presents post-silicon validation and debug of NoCs.
   - Chapter 14 describes challenges in designing reliable NoC architectures.

5. *Emerging NoC Technologies:* The fifth part of the book surveys security implications in emerging NoC technologies.

   - Chapter 15 describes security solutions for photonic (optical) NoCs.

- Chapter 16 presents security solutions for on-chip wireless networks.
- Chapter 17 provides an overview of securing 3D NoCs from hardware Trojan attacks.

6. *Conclusion and Future Directions:* The last chapter concludes the book with a summary and discussion on future directions.

We hope you enjoy reading this book and find the information on attacks and countermeasures useful in designing secure and trustworthy systems.

Gainesville, FL, USA
Colombo, Sri Lanka
January 1, 2021

Prabhat Mishra
Subodha Charles

Acknowledgments

This book would not be possible without the contributions of many researchers and experts in the field of network-on-chip security and privacy. We would like to gratefully acknowledge the contributions of Prof. Tushar Krishna (Georgia Tech), Dr. Srikant Bharadwaj (AMD), Sumit K. Mandal (Univ. of Wisconsin-Madison), Anish Krishnakumar (Univ. of Wisconsin-Madison), Prof. Umit Ogras (Univ. of Wisconsin-Madison), Dr. Johanna Sepulveda (Airbus), Chamika Sudusinghe (Univ. of Moratuwa), Manju Rajan (IIT Guwahati), Abhijit Das (IIT Guwahati), Prof. John Jose (IIT Guwahati), Aruna Jayasena (Univ. of Florida), Sidhartha Sankar Rout (IIIT Delhi), Mitali Sinha (IIIT Delhi), Prof. Sujay Deb (IIIT Delhi), Prof. Sudeep Pasricha (Colorado State), Ishan Thakkar (Colorado State), Sai Vineel Reddy Chittamuru (Colorado State), Varun Bhatt (Qualcomm), Sairam Sri Vatsavai (Colorado State), Yaswanth Raparti (Colorado State), Prof. Amlan Ganguly (Rochester Institute of Technology), Prof. Sai Manoj PD (George Mason), Abhishek Vashist (Rochester Institute of Technology), Andrew Keats (Rochester Institute of Technology), M. Meraj Ahmed (Rochester Institute of Technology), Noel Daniel Gundi (Utah State), Prabal Basu (Utah State), Prof. Sanghamitra Roy (Utah State), and Koushik Chakraborty (Utah State).

This work was partially supported by the National Science Foundation (NSF) grants SaTC-1936040 and CCF-1908131. Any opinions, findings, conclusions, or recommendations presented in this book are those of the authors and do not necessarily reflect the views of the National Science Foundation.

Contents

Part I
Introduction

Chapter 1
Trustworthy System-on-Chip Design Using Secure on-Chip Communication Architectures

Prabhat Mishra and Subodha Charles

1.1 Introduction

We are living in the era of Internet-of-Things (IoT), an era in which the number of connected smart computing devices exceeds the human population. Various reports suggest that we can expect over 50 billion devices to be deployed and mutually connected by 2025 [66], compared to about 500 million in 2003 [45]. In the past, computing devices like phones with a few custom applications represented the boundary of our imagination. Today, we are developing solutions ranging from smartwatches, smart cars, smart homes, all the way to smart cities. System-on-Chip (SoC) designs are at the heart of these computing devices, which range from simple IoT devices in smart homes to complex navigation systems in airplanes. As applications grow increasingly complex, so do the complexities of the SoCs. For example, a typical automotive SoC may include 100–200 diverse Intellectual Property (IP) blocks designed by multiple vendors. The ITRS (International Technology Roadmap for Semiconductors) 2015 roadmap projected that the increased demand for information processing will drive a 30-fold increase in the number of cores by 2029 [1]. Indeed, one of the most recent many-core processor architectures, Intel "Knights Landing" (KNL), features 64–72 Atom cores and 144 vector processing units [119]. The Intel Xeon Phi processor family, which implements the KNL architecture, is often integrated into workstations to serve machine learning applications. The 256-core CPU—MPPA2, launched by Kalray Corporation [107], is used in many data centers to speed up data processing.

P. Mishra (✉)
University of Florida, Gainesville, FL, USA
e-mail: prabhat@ufl.edu

S. Charles
University of Moratuwa, Colombo, Sri Lanka
e-mail: s.charles@ieee.org

P. Mishra, S. Charles (eds.), *Network-on-Chip Security and Privacy*,
https://doi.org/10.1007/978-3-030-69131-8_1

The increasing number of cores demands the use of a scalable on-chip interconnection architecture, which is also known as Network-on-Chip (NoC). As shown in Fig. 1.1, a typical SoC utilizes NoC to communicate between multiple IP cores including processor, memory, controllers, converters, input/output devices, peripherals, etc. NoC IPs are used in a wide variety of market segments such as mobile phones, tablets, automotive and general purpose processing leading to an exponential growth in NoC IP usage. A survey done by Gartner Inc. has revealed that NoC IP sales of Sonics, a privately-held Silicon Valley IP provider that specializes in NoC and power-management technologies, is ranked number 7 in terms of design IP revenue with a profit growth of 44.8% compared to 2013 [54]. Therefore, it is evident that the NoC has become an increasingly important component in modern SoC designs.

The drastic increase in SoC complexity has led to a significant increase in SoC design and validation complexity [4, 35, 48, 50, 81, 84, 88, 89, 92]. Reusable hardware IP based SoC design has emerged as a pervasive design practice in the industry to dramatically reduce design and verification cost while meeting aggressive time-to-market constraints. Figure 1.2 shows the supply chain of a specific commercial SoC [91]. Growing reliance on these pre-verified hardware IPs, often gathered from untrusted third-party vendors, severely affects the security and trustworthiness of SoC computing platforms. These third-party IPs may come with deliberate malicious implants to incorporate undesired functionality (e.g., hardware Trojan), undocumented test/debug interfaces working as hidden backdoors, or other integrity issues. Based on Common Vulnerability Exposure estimates, if hardware-level vulnerabilities are removed, the overall system vulnerability will reduce by 43% [41, 90].

The security of emerging SoCs is becoming an increasingly important design concern. Beyond the traditional attacks from software on connected devices, attacks originating from or assisted by malicious components in hardware are becoming more common. For example, Quo Vadis Labs has reported backdoors in electronic chips that are used in weapon control systems and nuclear power plants [118], which can allow these chips to be compromised remotely. The well-publicized "Spectre" [73] and "Meltdown" [78] attacks highlight how sensitive data can be stolen from threads executing on multicore processors. It is widely acknowledged that all algorithmically secure cryptographic primitives and protocols rely on a hardware root-of-trust that is resilient to attacks to deliver the expected protections

Fig. 1.1 An example System-on-Chip (SoC) with Network-on-Chip (NoC) based communication fabric to interact with a wide variety of third-party Intellectual Property (IP) cores

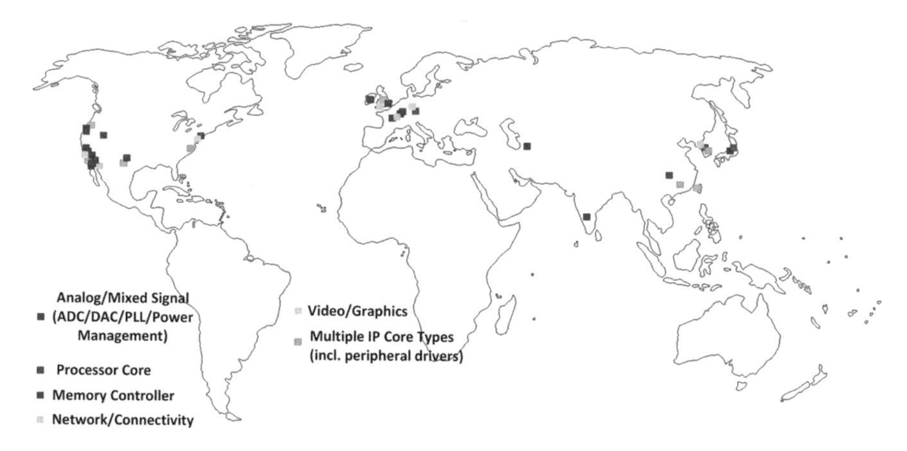

Fig. 1.2 Supply chain of a commercial router SoC with components from multiple third-party companies across the globe [91]

when implemented in software. Clearly, hardware platforms are at an elevated risk for security compromises in today's world.

In order to enable hardware-root-of-trust, we have to ensure that an SoC is trustworthy by ensuring security of computation, communication as well as storage. While the existing efforts have shown promising results in providing computation and storage related security solutions [91], there is limited effort in ensuring on-chip communication security. The ubiquity of devices using NoC-based SoCs has made NoC a focal point for security attacks as well as countermeasures [27–33]. Therefore, in order to secure the cyberspace, it is vital to protect the NoC from potential security threats as well as leverage the advantages given by NoC to minimize security vulnerabilities of other system components.

A fundamental problem of NoC-based SoCs is ensuring security while preserving non-functional requirements such as performance, power, and area. Due to the resource constrained nature of embedded and IoT devices, it may not be possible to implement traditional security measures such as encrypting communication with the AES cipher and using SHA hash functions. Thus, it is evident that considering security alone will not provide conclusive results. A more holistic approach is required that considers security among other non-functional requirements.

This chapter is organized as follows. Section 1.2 provides an overview of NoC architectures. Section 1.3 describes the NoC security landscape. Finally, Sect. 1.4 concludes the chapter.

1.2 Overview of Network-on-Chip (NoC) Architectures

Consider a designer who is responsible for designing the road network of a large city. Roads should be laid out giving easy access to all the offices, schools, houses, parks, etc. If all of the most common places are situated close to each other, it is inevitable that the roads in that area will get congested and other areas will be relatively empty. The designer should make sure that such instances do not occur and the traffic is uniformly distributed as much as possible. Alternatively, the roads should have more lanes and parking lots in such congested areas to cater to the requirement. In addition to accessibility and traffic distribution, the architect should also consider intersections, traffic lights, priority lanes, and potential detours due to occasional road maintenance. Moreover, self-driving cars and drones that deliver various items might come into picture in the future as well. Analogous to this, the designer of an SoC faces a similar set of challenges when designing the communication infrastructure connecting all the cores.

The early SoCs employed bus and crossbar based architectures. Traditional bus architecture has dedicated point-to-point connections, with one wire dedicated to each component. When the number of cores in an SoC is low, buses are cost effective and simple to implement. Buses have been successfully implemented in many complex architectures. ARM's AMBA (Advanced Micro-controller Bus Architecture) bus [8] and IBM's CoreConnect [65] are two popular examples. Figure 1.3 shows an overview of the ARM AMBA bus architecture [8]. Buses do not classify activities depending on their characteristics. For example, the general classification as transaction, transport, and physical layer behavior are not distinguished by buses. This is one of the main reasons why they cannot adapt to changes in architecture or make use of advances in silicon process technology. Due to increasing SoC complexity coupled with increasing number of cores, buses often become the performance bottleneck in complex SoCs. This coupled with other drawbacks, such as non-scalability, increased power consumption, non-reusability, variable wire delay, and increased verification cost, motivated researchers to search for alternative solutions.

The inspiration for network-on-chip (NoC) came from traditional networking solutions, more specifically, the Internet. The NoC, a miniature version of the wide area network with routers, packets, and links, was proposed as the solution for on-

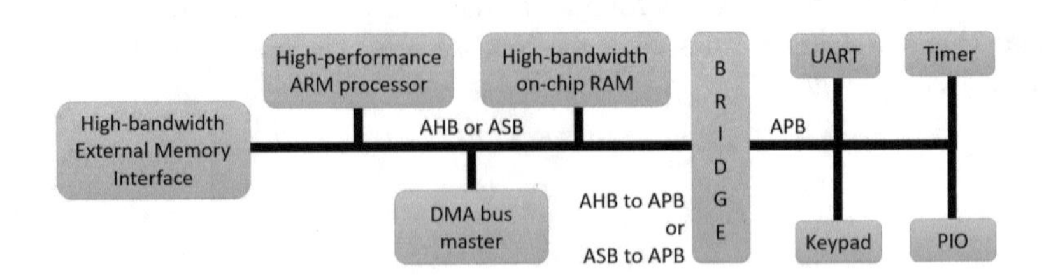

Fig. 1.3 Overview of the ARM AMBA bus architecture

chip communication [12, 40]. The new paradigm described a way of communicating between IPs including features such as routing protocols, flow control, switching, arbitration, and buffering. With increased scalability, resource reuse, improved performance, and reduced costs, NoC became the solution for the complex SoCs that required a scalable interconnection architecture. The remainder of this section covers various aspects of NoC architectures.

1.2.1 Network-on-Chip Architecture and Communication Protocol

Figure 1.4 shows an example NoC interconnection architecture consisting of several processing elements connected together via routers and regular sized wires (links). A processing element can be any component such as a microprocessor, an ASIC (application specific integrated circuit), or an intellectual property block that performs a dedicated task as shown in Fig. 1.1. We refer these processing elements as IPs. IPs are connected to the routers via a network interface (NI). We call the combination of an IP, an NI and a router as a "node" in the NoC. It can be observed that words node and "tile" are used interchangeably in existing literature to refer to NoC components connected to one router [119, 129].

NoC interconnection architecture uses a packet-based communication approach. A request or response that goes to a cache or to off-chip memory is divided into packets, and subsequently to "flits", and injected to the network. A flit is the smallest unit of flow control in an NoC. A packet may consist of one or more flits. For example, assume S is a processor IP, whereas node D is connected to an off-chip memory interface (memory controller). When a load instruction is executed at S, it first checks the private cache located in the same node and if it is a cache miss, the required data has to be fetched from the memory. Therefore, a memory fetch request message is created and sent on the appropriate virtual network to the NI. The network interface then converts it into network packets according to the packet format, divides each packet into flits, and sends the flits into the network via the

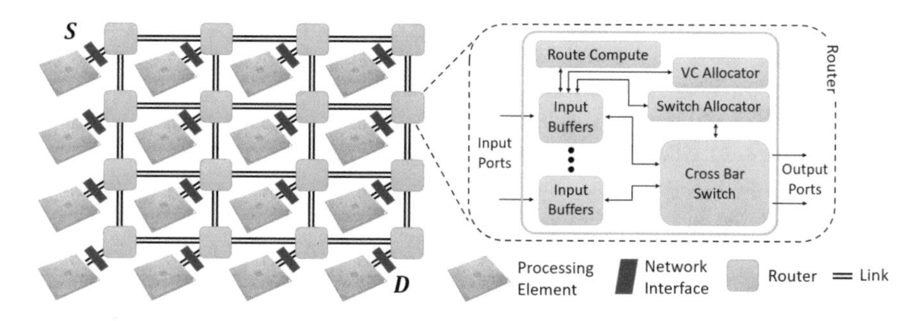

Fig. 1.4 Example of an NoC connecting 16 IPs

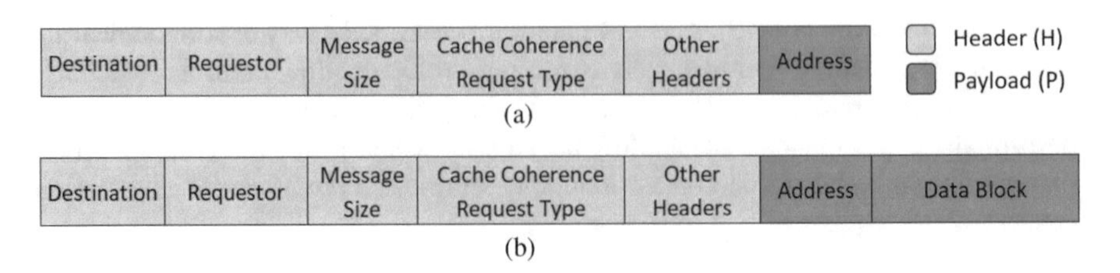

Fig. 1.5 NoC control (memory request) and data (response data) packet formats used in the gem5 simulator. (**a**) Memory request packet. (**b**) Response data packet from memory

local router. The network is then responsible to route the flits to the destination, D. Flits are routed either along the same path or different paths depending on the routing protocol. The NI at D creates the packet from the received flits and forwards the request to D, which then initiates the memory fetch request. The response message from the memory that contains the data block follows a similar process. Similarly, all IPs integrated in the SoC leverage the resources provided by the NoC to communicate with each other. Figure 1.5 shows the format of a memory request packet and a response data packet used in the gem5 architectural simulator [15].

Previous works have proposed several NoC architectures such as Nostrum [76], SOCBUS [130], Proteo [117], Xpipes [39], Æthereal [57], etc. based on different requirements. The choice of the parameters in the architecture depends on the design requirements such as performance/power/area budgets, reliability, quality-of-service guarantees, scalability, and implementation cost. Some of the existing NoC architectures have been surveyed in literature [2, 17]. NoC architecture design needs to consider two important factors—network topology and routing protocol. The next two subsections describe these aspects in detail.

1.2.1.1 Network Topology

The topology defines the physical organization of IPs, routers, and links of an interconnect. The organization in Fig. 1.4 shows a mesh topology. Crossbar, point-to-point, tree, 3-D mesh are few other commonly used topologies. Figure 1.6 shows some examples of them. The topology is chosen depending on the cost and performance requirements of an SoC. The topology directly impacts the communication latency when two IPs are communicating, since it affects the number of links and routers a flit has to traverse through to reach a given destination. A major trade-off when deciding the topology for a given requirement is between connectivity and cost. Higher connectivity (e.g., point-to-point) allows increased performance, but has higher area and power overhead. The 2-D mesh is the most common topology in NoC designs [119, 129]. Each link in a mesh has the same length leading to ease of design, and the area occupied by the mesh grows linearly with the number of nodes.

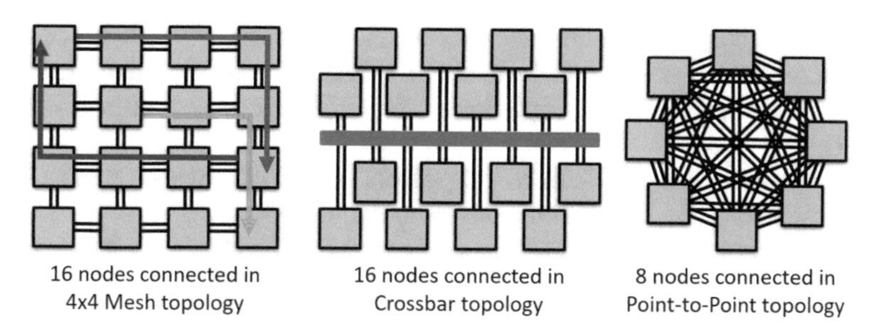

Fig. 1.6 NoC topologies and an example of X-Y routing in a mesh NoC

1.2.1.2 Router and Routing Protocol

The routers comprise input buffers that accept packets from the local IP via the NI or from other routers connected to it. For example, in the mesh topology, except for the routers in the border, each router is connected to the local IP and four other routers. Based on the addresses in the packet header and the routing protocol, the crossbar switch routes data from the input buffers to the appropriate output port. Buffers are allocated for virtual channels which helps avoid deadlock. The switch allocator handles input port arbitration for output ports [37].

The routing protocol defines the path a flit should take in a given topology. Routing protocols can be broadly classified as deterministic and adaptive. In deterministic routing, each packet traversing from S to D follows the same path. X-Y routing is one common example of deterministic routing. In X-Y routing, packets use X-directional links first, before using Y-directional links [42]. An example including three paths taken by X-Y routing in a mesh NoC is shown in Fig. 1.6. Adaptive routing takes network states such as congestion, security, and reliability into account, and sends the flits through different paths based on the current state of the network [136].

1.2.2 Emerging NoC Technologies

When NoC was first introduced, the focus was on electrical (copper) wires connecting NoC components together, referred to as "electrical NoC." However, recent advancements have demanded exploration of alternatives. With the advancement of manufacturing technologies, the computational power of IPs have increased significantly. As a result, the communication between SoC components have become the bottleneck. Irrespective of the architectural optimizations, electrical NoC exhibits inherent limitations due to the physical characteristics of electrical wires [97].

- The resistance of wires, and as a result, the resistance of NoC, is increasing significantly under combined effects of enhanced grain boundary scattering, surface scattering, and the presence of a highly resistive diffusion barrier layer [122, 123].
- Electrical NoC can contribute to a significant portion of the on-chip capacitance. In some cases, about 70% of the total capacitance [93].
- The electrical NoC is a major source of power dissipation due to the above two factors.

Therefore, it is becoming increasingly difficult for electrical NoC to keep up with the delay, power, bandwidth, reliability, and delay uncertainty requirements of state-of-the-art SoC architectures [34, 121]. These challenges can only intensify in future giga and tera-scale architectures. In fact, the International Technology Roadmap for Semiconductors (ITRS) has mentioned optical and wireless based on-chip interconnect innovation to be key to addressing these challenges [109].

There are various emerging NoC technologies such as "wireless NoC" [43] and "optical NoC" [96]. While the focus of this book is on security attacks and countermeasures in electrical NoCs, a majority of these security solutions are also applicable for wireless and optical NoCs. This is primarily due to the fact that they have inherent similarities in terms of network topology and routing protocols. For example, both electrical and optical NoCs represent similar topologies using wired connectivity. Similarly, wireless NoC always use one-hop routing, while optical and electrical NoCs utilize one-hop or multi-hop communication depending on the source and destination. Figure 1.7 shows an overview of how different NoC technologies can be used to connect heterogeneous SoC components.

1.2.2.1 Wireless NoC

Wireless NoC was proposed as a solution to the latency experienced by electrical NoCs, which are based on metal interconnects and multi-hop communication. Wireless NoC integrates on-chip antennas and suitable transceivers that enable communication between two IPs without a wired medium. Silicon integrated

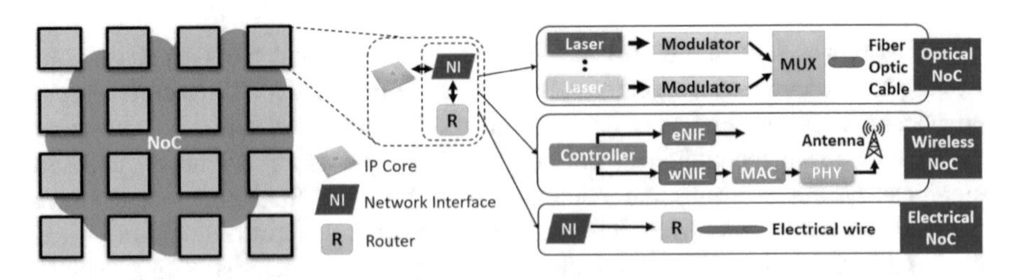

Fig. 1.7 NoC enables communication between IPs. The network interface (NI), router (R), and links can be implemented using optical, wireless, or electrical communication technologies

antennas communicating using the millimeter wave range is shown to be a viable technology for on-chip communication [43].

1.2.2.2 Optical NoC

On the other hand, optical NoC, also known as photonic NoC, uses photo emitters, optical wave guides, and transceivers for communication [135]. The major advantage over electrical NoC is that it is possible to physically intersect light beams with minimal crosstalk. This enables simplified routing and together with other properties, optical NoC can achieve bandwidths in the range of Gbps.

1.3 Security Landscape in NoC-Based System-on-Chip

The widespread adaptation of NoCs has made it a focal point for security attacks as well as countermeasures. There is a growing interest in the industry to use the NoC to secure the SoC as evident from NoC-Lock [120] and FlexNoC resilience package [7]. On the other hand, the NoC itself can be a threat when different IP blocks come from different vendors. A compromised NoC IP can corrupt data, degrade performance, or even steal sensitive information. NoC security is crucial for three related reasons: (1) NoC has access to all system data, (2) NoC spans across the entire SoC, and (3) NoC elements are repetitive in a way that any modification can be easily replicated. In the following subsections, we discuss how SoCs can become vulnerable to security threats (Sect. 1.3.1), why securing NoC-based SoCs has become a hard problem (Sect. 1.3.2) and different threat models in existing literature related to NoC security (Sect. 1.3.3).

1.3.1 Security Vulnerabilities in SoCs

SoC complexity and tight time-to-market deadlines have shifted the in-house SoC manufacturing process to a global supply chain. SoC manufacturers outsource parts of the manufacturing process to third-party IP vendors. This globally distributed mechanism of design, validation, and fabrication of IPs can lead to security vulnerabilities. Adversaries have the ability to implant malicious hardware/software components in the IPs. Existing literature has discussed three forms of vulnerabilities: (1) malicious implants, (2) backdoor using test/debug interfaces, and (3) unintentional vulnerabilities [47]. An adversary can utilize the malicious implants (hardware Trojans) to cause malfunction or facilitate information leakage [91]. An adversary can also exploit legitimate test and debug interfaces as a backdoor for information leakage [118]. Many security vulnerabilities can be created unintentionally by design automation/computer-aided design (CAD) tools or by designers' mistakes [131]. These vulnerabilities can lead to untrusted (potentially malicious) IPs.

Attacks based on malicious implants, such as hardware Trojans, rely on Trojans being integrated in the SoC without being detected at the post-silicon verification stage or during runtime [91]. Hardware Trojans can be inserted into the design in several places such as by an untrusted CAD tool or designer or at the foundry via reverse engineering [13]. Even if all the IPs are tested before integration, hardware Trojans can still go undetected because of the complexity of designs with billions of transistors which make physical inspection or 100% coverage in design verification/validation a costly or even impossible target [124]. Furthermore, Trojans can mask their behavior as transient errors and can be activated only when a specific condition or a combination of conditions are satisfied [14]. A smart attacker can carefully craft the Trojan activation method so that it becomes difficult to detect. Previous work has discussed external/internal Trojan activation modes [124], software-hardware coalition [10], and triggers based on time, input sequence, traffic pattern, and even thermal conditions [14].

The usage of third-party NoC IPs has grown rapidly over the years. Due to the widespread use of NoC IPs, outsourcing NoC IP fabrication has become a common practise. iSuppli, an independent market research firm, has concluded from their research that the FlexNoC on-chip interconnection architecture [7] is used by four out of the top five Chinese fabless semiconductor OEM (original equipment manufacturer) companies [116]. This has led to Arteris, the company that developed FlexNoC, achieve a sales growth of 1002% over a 3 year time period through IP licensing [9]. Therefore, there is ample opportunity for adversaries to attack the SoC through malicious implants in NoC IPs. Furthermore, due to the complexity of the design, NoC IPs are ideal candidates to insert hardware Trojans [101].

1.3.2 Unique Challenges in Securing NoC-Based SoCs

The general problem of securing the interconnect has been well studied in the computer networks domain and other related areas [24, 72, 134]. However, implementation of security features introduces area, power, and performance overhead. While complex security countermeasures are practical in computer networks domain, the resource constrained nature of embedded and IoT devices pose additional unique challenges as outlined below.

1.3.2.1 Conflicting Requirements

While enabling communication between IPs, NoCs need to satisfy a wide variety of requirements including security, privacy, energy efficiency, domain-specific requirements, and real-time constraints. While security is the primary focus of this book, we cannot ignore other NoC design constraints. Designers employ a wide variety of techniques to improve energy efficiency in NoC-based SoCs [5, 25, 26, 59, 62, 126]. It is difficult to satisfy conflicting requirements such as security and energy efficiency. For example, it may not be possible to implement traditional security measures such as encrypting text with the AES cipher and using SHA hash functions in resource-constrained IoT devices. Similarly, security and domain-specific requirements may not be compatible. For example, in an automotive network, when a potential security breach is detected, pausing all systems to check the malfunction is not an option since the car is moving, and stopping it abruptly can lead to catastrophic consequences. Thus, there is a need for innovative solutions to secure NoCs with lightweight security mechanisms customized for application domains.

1.3.2.2 Increased Complexity

The complexity of SoC designs have made exhaustive security validation an impossible task. Most IPs come as black boxes from vendors that do not reveal design details in order to maintain the competitive advantage in a niche market. As a result, the complete design is not visible to verification engineers. Modern verification tools often try to detect missing or erroneous functionality, whereas security vulnerabilities can be hidden in dormant functions in large and complex designs that get triggered only by a specific set of inputs as discussed in Sect. 1.3.1. Therefore, it is not feasible to capture all security vulnerabilities using security validation tools during design time [3, 46, 47, 49, 86, 87, 91, 94].

1.3.2.3 Diverse Technologies

While electrical communication is widely used in designing NoC-based SoCs, emerging NoCs can also support chip-scale photonics (optical NoC) as well as wireless communication (wireless NoC) as shown in Fig. 1.7. Security solutions for NoCs thus need to not only address security over electrical wires but also consider the emerging challenges from data transfers over photonic waveguides and wireless channels. While broadcast may be preferred for wireless NoCs, optical and electrical NoCs need to consider a wide variety of network topologies as well.

1.3.3 Threat Models

The intention of a hardware Trojan can vary from design to design. Commonly discussed threats include information leakage, denial-of-service, and data corruption. A recent occurrence of a hardware Trojan (spying on data) raised concerns across top US companies and authorities including Apple, Amazon and CIA [18].

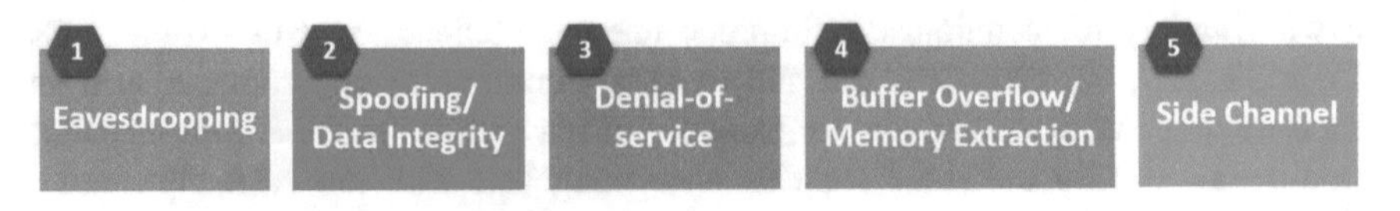

Fig. 1.8 Five classes of security attacks discussed in existing literature

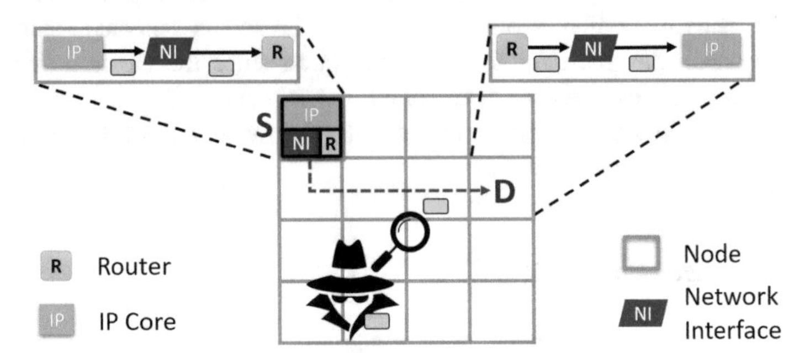

Fig. 1.9 An example of eavesdropping attack

In this section, we provide an overview of five classes of attacks on NoC-based SoCs (Fig. 1.8).

These classes of attacks have been well studied in the computer networks domain and other related areas. However, implementation of security features introduces area, power, and performance overhead. To address this issue, we need lightweight security countermeasures that can provide the desired security with tolerable impact on area, power, and performance. In the remainder of this section, we provide an overview of attacks explored in NoC-based SoCs.

1.3.3.1 Eavesdropping Attacks

Eavesdropping attack, also known as snooping/sniffing, refers to an attacker passively listening to on-chip communication in an attempt to steal sensitive information as shown in Fig. 1.9. The intention of the attacker is to leak information over long time periods without being detected. Recent occurrences of hardware security breaches where hard-to-detect hardware components, that were not a part of the original design, integrated into the original design leaking information have attracted more attention to eavesdropping attacks [18].

As discussed in Chap. 1.1, IPs integrated on the same SoC use the NoC to communicate between each other using message passing as well as shared memory. Therefore, eavesdropping on the NoC allows an attacker to extract secret information without relying on memory access (either through on-chip cache or off-chip memory) or hacking into individual IPs. Bus-based communication (e.g., broadcast in wireless NoCs) is inherently vulnerable to eavesdropping attacks. Existing literature on NoC security has explored several variations of the eavesdropping attack.

One commonly explored threat model is where the malicious NoC IP colludes with an accompanying malicious application running on another IP to launch an eavesdropping attack. It includes a Trojan infected router copying packets passing through it and sending the duplicated packets to another IP running a malicious application in an attempt to steal confidential information. This threat model has been extensively used to study eavesdropping attacks specially since the attack is hard to detect [10, 21, 64, 70, 114]. Trojans can also directly eavesdrop on the NoC communication without relying on re-routing duplicated packets to an accomplice application. This can be facilitated by external I/O pins attached to the NoC [55]. However, NoCs are generally more resistant against bus-probing attacks compared to the traditional bus-based architectures.

Similar to the malicious router and application colluding to launch the attack, a Trojan infected network interface and an application can work together to launch an eavesdropping attack [101]. In the threat model presented in [101], the hardware Trojan embedded in the NI can tamper with the flits in the circular flit queue, which is used to store flits before sending them to the corresponding router. When a flit is sent to the router, it waits in the queue until the next flit overwrites it. The Trojan keeps track of such outstanding flits, modifies the header flit with a new destination address, and updates the header pointer so that it gets re-sent to the router. The duplicated flits are received by the malicious application. The area overhead of the Trojan is shown to be 1.3% [101].

Common countermeasures against eavesdropping attacks include packet encryption, authentication, additional validation checks during NoC traversal and information obfuscation. Encryption ensures that the plaintext of the secure information is not leaked and authentication detects any tampering with the packet including header information. Several prior studies have tried to develop lightweight encryption and authentication schemes for on-chip data communication. Ancajas et al. [10] proposed a simple XoR cipher together with a packet certification technique that calculates a tag and validates at the receiver. A configurable packet validation and authentication scheme was proposed by merging two robust error detection schemes, namely algebraic manipulation detection and cyclic redundancy check, in [21]. Intel's TinyCrypt—a cryptographic library with a small footprint is built for constrained IoT devices [125]. It provides basic functionality to build a secure system with very little overhead. It gives SHA-256 hash functions, message authentication, a pseudo-random number generator which can run using minimal memory, digital signatures, and encryption. It also has the basic cryptographic building blocks such as entropy sources, key exchange, and the ability to create nonces and challenges. The duplicated packets in router-application combination as well as NI-application combination can be detected by additional validation checks. In [101], the authors implemented a snooping invalidator module (SIM) at the NI output queue to discard duplicate packets. On the other hand, information obfuscation can make the attack harder to initiate. For example, hiding the source and destination information of NoC packets can ensure that the malicious agents in the NoC are unable to select the target application to eavesdrop. Onion routing, a well-known mechanism in the computer networks domain, can hide the origin and target of a network packet [56]. However, implementing such complex security mechanisms is not feasible in resource-constrained SoCs. Several previous studies tried to propose lightweight solutions that are compatible with the NoC context [10, 27].

1.3.3.2 Spoofing and Data Integrity Attacks

SoC relies on the integrity of data communicated through the NoC for correct execution of tasks. If a malicious agent corrupts data intentionally, it can lead to erroneous execution of programs as well as system failures. On the other hand, spoofing is the act of disguising a communication from an unknown source as being from a known (trusted) source. Therefore, a malicious agent pretending to be a trusted source can inject new packets to the network causing system to malfunction as shown in Fig. 1.10. Spoofing can be used to bypass memory access protection by impersonating a core that has permission to read from (or write in) prohibited regions to steal sensitive information or disrupt execution. Spoofing may also be leveraged to respond to legitimate requests with wrong information to cause system failure. Spoofing can be achieved by an attacker replacing the source address of a packet by an address of a trusted IP.

Spoofing and data integrity attacks intentionally corrupt data transferred on the NoC to cause malfunction. Sepúlveda et al. presented "MalNoC," a Trojan infected NoC that can perform multiple attacks on NoC packets [114]. The infected MalNoC router copies packets arriving at a router, replaces the packet data with the content in a malicious register, modifies source and/or destination address in the header to the desired IP, and injects it back into the NoC. A control register within the router controls the Trojan operation. A similar threat model that discussed eavesdropping, DoS, and illegal packet forwarding, all of which utilized packet corruption at a router was presented in Sect. [64]. Kumar et al. [70] discussed a Trojan that corrupts flits arriving at the input buffers of a router.

Trojans can also be inserted in links to corrupt NoC packets. To avoid being detected, the Trojans change only the header flits causing deadlock, livelock, and packet loss situations [132]. Even if hardware Trojans are not present, bit flipping can happen when packets are transferred through the links due to other reasons.

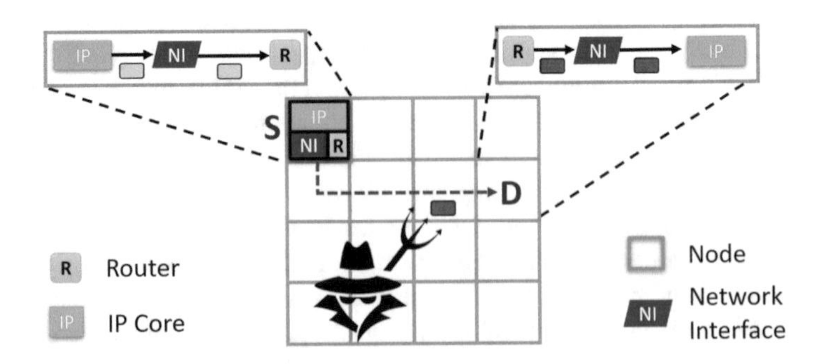

Fig. 1.10 An example of data integrity attack

Error correction codes are used to correct such bit flips. The Trojan in the link attempts to mask its malicious behavior as an error rather than a security attack to avoid being detected. The authors have explored the impact of Trojans embedded in different links (boundary links versus center links) in a 5×5 Mesh NoC [132].

Authenticated encryption schemes provide data confidentiality through encryption and data integrity through authentication [71, 108, 114]. If the authentication tag is calculated using the entire packet (header as well as payload), any packet corruption can be detected at the receiver's side when the packet is validated using authentication. Hussain et al. [64] argued that since the Trojan is rarely activated to avoid detection, authenticating each packet can lead to reduction in energy efficiency. In their work, they proposed an efficient Trojan detection design where the authentication gets activated only when the hardware Trojan has been triggered in the system. A combination of security modules placed at the IPs as well as at the routers provided attack detection as well as Trojan localization capabilities [64].

Error correcting codes (ECC) are widely used in the telecommunications domain [63]. ECCs have been used in NoCs to correct bit errors due to particle strikes, crosstalk, and spurious voltage fluctuation in NoCs. Yu et al. introduced a method to detect Trojan induced errors using ECCs in [132]. Their method consisted of two main components. (1) Link reshuffling: to avoid the Trojan from affecting the same bit in an attempt to create deadlocks/livelocks, the odd and even bits are switched in the retransmitted flit in case of an error detected by the ECC. This is effective for scenarios where the Trojan is triggered by specific flits. If the Trojan gets activated by a certain input, reshuffling the bits during the retransmission can make the Trojan inactive again. (2) Link isolation: an algorithm to isolate links that are suspected to have Trojans. Trojans that are triggered by external signals can remain active for a long time. In such cases, wire isolation is used to reduce the number of retransmissions.

1.3.3.3 Denial-of-Service Attacks

Denial-of-service (DoS) in a network is an attack that prevents legitimate users from accessing services and information. The most common example is an attacker flooding a network with information as shown in Fig. 1.11. When a user is trying to access a website, the request is sent to that web server to view the page. The server has a certain bandwidth and can only serve a limited number of requests at a time. If the attacker overloads the server with requests, it will not be able to process the user's legitimate request. This is "denial-of-service." In the context of an NoC, several threat models have been explored. In general, DoS in NoC-based SoCs are attacks that overwhelm the network resources in an attempt to cause performance degradation, real-time guarantee violations, and reduction of battery lifetime.

Several threat models related to DoS attacks have been studied in prior work. One common threat model is where malicious IPs manipulate the availability of on-chip resources by flooding the NoC with packets. The performance of an SoC can heavily depend on few components. For example, a memory intensive application is likely

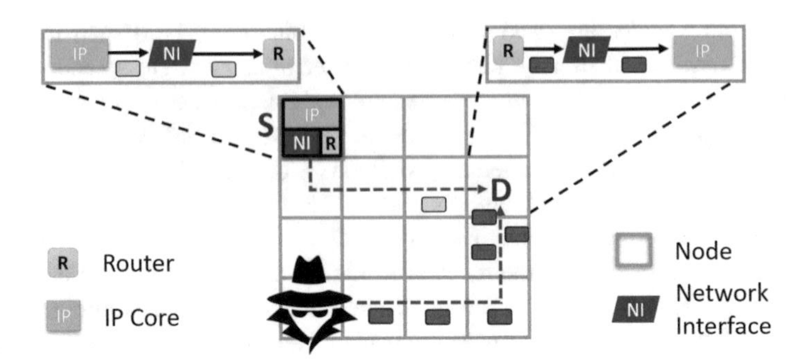

Fig. 1.11 An example of denial-of-service attack

to send many requests to memory controllers, and as a result, routers connected to them will experience heavy traffic. If a malicious IP targets the same node, the SoC performance will suffer significant degradation [28, 29, 52, 99]. This is known as a flooding-type DoS attack.

Continuous corruption of packets can also lead to a DoS attack [53, 70]. In [70], hardware Trojans tamper flits arriving at the input buffer of a router causing performance degradation. Performance degradation is caused by dropped packets, wastage of NoC resources such as buffer space, response delays, and retransmissions. Boraten et al. [20] discussed a similar threat model where hardware Trojans influenced resource allocations and corrupted data to degrade performance. The same authors further explored possible DoS attacks in [22]. Compared to router-based packet corruption, they discussed a Trojan that performs deep packet inspection on links and inject faults when the target is identified. The injected faults trigger retransmissions from the error correcting mechanism. Therefore, repeated injection of faults causes repeated retransmission to starve network resources and create deadlocks capable of rendering single application to full chip failures.

Rajesh et al. [100] discussed a threat model where the packets are unfairly treated at the router to cause a DoS attack. The malicious NoC IP, once integrated on the SoC, picks a victim IP that is an important SoC component and manipulates the traffic flow to/from the victim IP. The traffic flow is manipulated by denying fair access to the allocator and arbiter units in the router. The allocator is responsible for granting flits access to the crossbar. DoS is achieved by the allocator delaying packets to/from the victim IP. At the arbiter, the Trojan infected router gives least priority to the flits that have the victim IP as the source/destination. Both these scenarios lead to flits to/from one IP getting significantly delayed.

To address these different threat models, researchers proposed several solutions. As a countermeasure to denial-of-service through packet corruption, Kumar et al. proposed a bit shuffling method that makes flits less sensitive to the attack [70]. The authors proposed to shuffle the critical bit fields of the flits among themselves and others so that the Trojan is attacking on randomly shuffled data and not on the critical fields within the packets such as flit indication bits, source and destination addresses. While fuzzing can make the attack difficult, it does not guarantee prevention. Furthermore, the attack is not detected, and as a result, future attacks are not prevented either. Boraten et al.'s work was motivated by this, where they coupled switch-to-switch scrambling, inverting, shuffling, and flit reordering with a heuristic-based fault detection model [22]. Their solution addresses the challenge of differentiating fault injections from transient and permanent faults. Another technique that exhibits similar defense characteristics as fuzzing—partitioning, tries to reduce interference of communication between different applications/packet types. As a result, overwhelming the NoC with DoS attacks becomes difficult [128].

Monitoring the traffic flow to detect abnormalities is another common defense against DoS attacks. Rajesh et al. [100] proposed a defense against their traffic flow manipulation threat model that is based on identifying the latency elongation of packets caused by the DoS attack. Their method relied on injecting additional packets to the network and observing their latencies. SoC firmware then examines the latencies of the injected packets. If two packets are injected at the same time and traverse paths with significant overlap, they are expected to exhibit comparable latencies. If not, it will be flagged as a potential threat. Similar methods that profiled normal behavior of traffic during design time and monitored NoC traffic to detect deviations from normal behavior were proposed in [16, 52]. Exploring another orthogonal direction, work in [20, 53, 99, 110] proposed additional formal verification and runtime checks integrated in to the NoC to prevent and detect DoS attacks.

1.3.3.4 Buffer Overflow and Memory Extraction Attacks

The goal of a buffer overflow attack is to alter the function of a privileged program so that the attacker can gain access and execute his own code. A program with high privileges (root programs) typically becomes the target of buffer overflow attacks. To accomplish this, the adversary has to insert malicious code and make the program execute it. "Code injection" is the first step to accomplish this where the malicious code is inserted into the privileged program's address space. This can be achieved by providing a string as input to the program which will be stored in the program buffer. The string will contain some root level instructions which the adversary wants the program to execute [38]. Then, the adversary creates an overflow in the program buffer to alter states of the program. For example, it can alter a return address of a function so that the program will jump to that location and start executing the malicious code [79]. This can be accomplished when buffers have weak or no bound checking. Buffer overflow attacks can also be used to read privileged memory locations from the address space. In an NoC context, the threat gets aggravated due to memory spaces being shared between multiple cores.

Similar to the buffer overflow attacks in the computer networks domain, execution of malicious code can launch a buffer overflow attack in NoC-based SoCs. If a malicious IP writes on the stack and modifies the return address of a function to point at the malicious code, the malicious code will be executed. Return address modification in the stack is done by writing more data to a buffer located on the stack than what is actually allocated for that buffer. This is known as "smashing the stack" [77]. Even if the stack memory is made non-executable, or kept separate, it is possible to overwrite both the return address as well as the saved registers. Work done in [79] explored this threat model. Buffer overflow attacks pose a significant threat in NoC-based SoCs where the memory is shared among multiple cores.

Kapoor et al. in their work considered some IPs on the SoC to contain confidential information (secure/trusted IP cores) and some untrusted IPs which can potentially carry hardware Trojans (non-secure/untrusted IP cores) [71]. The information inside secure IP cores should be protected from non-secure IP cores. Since all IPs are integrated on the same NoC, non-secure cores can communicate with secure cores. Non-secure cores can try to install Trojans in the secure cores and try to extract information. The confidential information in registers in the secure cores such as cryptographic keys, configuration register information, and other secure data can be compromised in such an attack [71]. This threat model of non-secure IP cores trying to access secure IP cores has been used in several other work as well [44, 51, 52, 106, 108].

Lukovic et al. proposed two methods to counter buffer overflow attacks. The first method focused on protecting the processing cores by embedding additional security in the network interface (NI) [79]. In their work, a data protection unit, which is similar to a firewall sits on the NI attached to the shared memory block. It secures the memory by filtering unauthorized memory access requests. A stack protection unit (SPU) is developed which protects the stack from attacks that targets the return addresses. The SPU is developed as a part of the processor protection system which combines software and hardware units that replicate return addresses stored in the stack and protects it against code injection attacks. These countermeasures also stopped the attack from getting propagated to other parts of the NoC. Their second method extends the solutions proposed in [79] to a hierarchical security architecture [80]. The authors introduced four levels of security working at system level, NoC cluster level, per core, and in a layer specific to the attack (e.g., code injection). Similar to software protection mechanisms and the data protection unit in [79], many existing works provide access control by monitoring the incoming requests [51, 52, 106]. For example, Saeed et al. introduced a method to mitigate buffer overflow attacks in an NoC-based shared memory architecture by deploying an ID and address verification unit (IAV) [106]. This minimizes the threats caused by malicious IPs in the NoC because the IAV verifies each incoming packet by its ID and address.

Adding an extra layer of security to access authorization, commercial products such as Sonic SMART Interconnect [120] and ARM TrustZone [6] divide memory blocks into different protection regions and isolate secure and normal execution environments from each other. If the non-secure cores access secure cores, requests are validated by access authorization techniques [71, 108]. It is possible that security zones have to be modified due to task migration, new applications starting and ending. Therefore, security zones have to be created, modified and eliminated during runtime. Sepúlveda et al. [111] achieved this by using a partitioning method that used a lightweight Diffie–Hellman key-exchange protocol. The same authors proposed a method to create dynamic firewalls at the network interface to monitor and filter the NoC traffic [113]. The dynamic firewalls create "elastic security zones" by wrapping a desired set of components in a 3D NoC according to a trust policy. Porquet et al. [98] presented a method to co-host several secure applications running in parallel using the same shared memory space. Secure hardware implemented at the NI of the NoC enables secure and flexible partitioning of the shared memory space between multiple applications. Their approach is similar to the operation of a virtualization hypervisor that protects code, data, exclusive peripheral device usage, etc., when multiple virtual machines are running on the same host machine [23].

1.3.3.5 Side-Channel Attacks

Side-channel attacks exploit non-functional behavior such as time, power, electromagnetic radiation, heat and acoustic waveforms to attack a secure system [133]. The switching behavior of the CMOS (complementary metal oxide semiconductor) transistors can be analyzed to infer the underlying circuit functionality. Therefore, even a flawless implementation of a security mechanism can be vulnerable against side-channel attacks as shown in Fig. 1.12. For example, Zhen et al. presented a method to implement a timing attack on Nvidia Kepler K40 GPU and successfully recovered the complete 128-bit AES encryption key [69]. In contrast, a paper published in 2012 showed that a brute-force attack on AES using a super computer can take 149 trillion years [11]. Even though computing resources have significantly improved since then, a brute-force attack on AES-128 is still not possible. Possibility of side-channel attacks escalated, since in a realistic scenario, more constraints are imposed on the system such as performance and power. Even for systems with theoretically proven security bounds, revealing the secrets through these non-functional physical properties is a likely scenario.

Due to the difference in computation requirements, secure systems often take different times to perform different operations. By carefully measuring these time

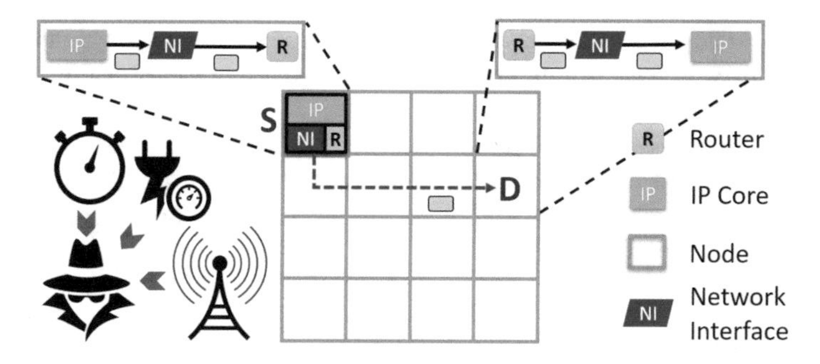

Fig. 1.12 An example of side-channel attack

differences, it is possible to extract secret information from vulnerable systems. Reinbrecht et al. demonstrated a practical "Prime+Probe" timing attack on an NoC-based SoC [103]. The target of their attack was the communication between an ARM Cortex-A9 core and a shared cache memory. Other studies carried out on timing attacks also used similar concepts on timing analysis of network traffic for attacks [67, 68, 102, 127]. The threat model in [67] included four cores. Two of which are carrying out a secure communication and the other two, which lies on the secure communication path will be infected by the adversary. The two infected cores inject traffic to the network. Adversary is then able to observe latencies of maliciously injected traffic to infer information about timing, frequency and volume of the secure communication.

Wang et al. [127] in their work showed that the number of ones in the RSA [105] key can be inferred with a timing side-channel attack on NoC, which can then be used to infer the entire key. A major part of the RSA algorithm is to do the modulo multiplication of two large (1024 or 2048-bit) numbers. The modulo multiplication is shown to be vulnerable to timing side-channel attacks [75], mainly because the algorithm examines each bit in the RSA key and multiplies only if it is one. Wang et al.'s attack is based on observing the additional network traffic caused due to multiplications [127]. Similar to recovering the RSA key through timing attacks, existing work used the AES cipher as case studies as well. In 2010, Bogdanov et al. [19] proposed a differential cache collision attack on embedded systems. While their work did not consider an NoC-based setup, in 2018, Reinbrecht et al. [104] showed that combining their previous work on NoC timing attacks [102] with Bogdanov et al's cache collision attack [19] can significantly enhance the AES key recovery effort.

Measuring the power consumption will give information about the process that is occurring inside the system. For example, if the processor is performing a simple addition versus executing an encryption instruction (Intel chips come with "AESENC" instruction that performs one round of AES encryption on a given plaintext), observing their power consumption can give reasonable information to differentiate the two operations. Similarly, many data encryption standard (DES) implementations have visible differences within permutations and shifts which can be utilized to break the security scheme [36]. Differential power analysis is a powerful attack technique based not only on power observations, but also on statistical analysis and noise filtering methods to gain more information about the underlying security scheme [74].

In addition to timing and power, existing work has explored thermal side channels. Similar to power, the SoC thermal characteristics are highly correlated to the SoC operation. Guo et al. [58] discussed two main thermal characteristics:

1. Spatial distribution: by observing the heatmap, attackers can identify active cores in the SoC.
2. Temporal variation: different instructions have different thermal profiles when executed. The temperature trace over time allows attackers to infer the executed instructions with a certain probability.

As a countermeasure to the "Prime+Probe" attack, the authors proposed "Gossip NoC" [102, 103]—a two stage security mechanism which first detects the attack and then protects the SoC. The detection process monitors the bandwidth and sends an alert message in case of a potential security breach. The protection mechanism gets triggered by this alert message which then alters the routing protocols to route packets avoiding the sensitive path that contains the malicious IP. The same route randomization concept was used as a mitigation technique in [67, 68]. Sepúlveda et al. combined random arbitration with adaptive routing to dynamically allocate NoC resources, and as a result, minimized interference between secure packets and packets injected by the attacker [115].

As a solution to the thermal side-channel attacks discussed in [58], the authors presented a task mapping scheme that minimized the thermal information leakage. In their work, a mathematical model was developed to quantify the security cost corresponding to a certain application mapping. A greedy optimization algorithm was then used to map application threads to cores such that the leakage is minimized. The optimization algorithm is implemented in the operating system and it receives SoC status from a hardware monitor. The security cost is then calculated according to the model for each core and a new application mapping is generated if required.

To avoid timing side-channel attacks similar to the one introduced in [127], the same authors proposed to partition network traffic based on its security level. The basic idea is to make sure packets from applications running on secure IPs do not interfere with the packets from applications running on non-secure IPs. As a result, the communication latency and throughput of non-secure applications become independent of the dynamic behavior of secure application traffic. An obvious way to achieve this goal is to statically partition NoC resources (link bandwidth, buffers, etc.) spatially or temporally. However, it can lead to sub-optimal results causing performance degradation. Wang et al. introduced a priority-based arbitration technique for resources such as the router crossbar along with static allocation of virtual channels [127]. A similar principal was used in the "Secure Enhanced Router" architecture proposed by Sepúlveda et al. [112]. In their work, the router architecture included a shared buffer space and the number of virtual channels per input port was decided during runtime according to communication and security requirements. Similar to [127], the goal was to make the non-secure traffic flow oblivious of the secure traffic flow. Recent efforts try to combine the advantages of logic testing and side-channel analysis for effective Trojan detection in integrated circuits [60, 61, 82, 83, 85, 95].

1.4 Summary

This chapter provided an overview of Network-on-Chip (NoC) based System-on-Chip (SoC) design methodology. It also introduced the security vulnerabilities in electrical, optical as well as wireless NoCs. We have considered existing literature covering state-of-the-art attacks and defense mechanisms in NoC-based SoCs. In particular, we have discussed the research efforts under five classes of attacks highlighting their threat models and respective countermeasures.

Acknowledgement This work was partially supported by the National Science Foundation (NSF) grant SaTC-1936040.

References

1. 2015 International Technology Roadmap for Semiconductors (ITRS). www.semiconductors.org/main/2015_international_technology_roadmap_for_semiconductors_itrs/
2. A. Agarwal, C. Iskander, R. Shankar, Survey of network on chip (NoC) architectures & contributions. J. Eng. Comput. Archit. **3**(1), 21–27 (2009)
3. A. Ahmed, F. Farahmandi, Y. Iskander, P. Mishra, Scalable hardware trojan activation by interleaving concrete simulation and symbolic execution, in *2018 IEEE International Test Conference (ITC)* (IEEE, Piscataway, 2018), pp. 1–10
4. A. Ahmed, F. Farahmandi, P. Mishra, Directed test generation using concolic testing on RTL models, in *2018 Design, Automation & Test in Europe Conference & Exhibition (DATE)*, 2018, pp. 1538–1543
5. A. Ahmed, Y. Huang, P. Mishra, Cache reconfiguration using machine learning for vulnerability-aware energy optimization. ACM Trans. Embed. Comput. Syst. **18**(2), 1–24 (2019)
6. T. Alves, Trustzone: integrated hardware and software security. White paper, 2004
7. Alteris FlexNoC Resilience Package. www.arteris.com/flexnoc-resilience-package-functional-safety
8. ARM: 'Amba specification', Technical report, ARM, Revision 2.0, 1999. developer.arm.com/products/architecture/amba-protocol
9. Arteris makes big gains on inc. 500 list of America's fastest-growing private companies (2013). www.arteris.com/Inc-500-Arteris-pr-2013-august-20
10. D.M. Ancajas, K. Chakraborty, S. Roy, Fort-NOCs: Mitigating the threat of a compromised NoC, in *2014 51st ACM/EDAC/IEEE Design Automation Conference (DAC)*, 2014 pp. 1–6
11. M Arora, How secure is AES against brute force attacks? in *Freescale Semiconductor*, USA, 2012
12. L. Benini, G. De Micheli, Networks on chips: a new SoC paradigm. Computer **35**(1), 70–78 (2002)
13. S. Bhunia, M.S. Hsiao, M. Banga, S. Narasimhan, Hardware trojan attacks: threat analysis and countermeasures. Proc. IEEE **102**(8), 1229–1247 (2014)
14. S. Bhunia, M. Tehranipoor, *The Hardware Trojan War* (Springer, Berlin, 2018)
15. N. Binkert, B. Beckmann, G. Black, S.K. Reinhardt, A. Saidi, A. Basu, J. Hestness, D.R. Hower, T. Krishna, S. Sardashti, R. Sen, K. Sewell, M. Shoaib, N. Vaish, M.D. Hill, D.A. Wood, The gem5 simulator. SIGARCH Comput. Archit. News **39**(2), 1–7 (2011)
16. A.K. Biswas, S.K. Nandy, R. Narayan, Router attack toward NoC-enabled MPSoC and monitoring countermeasures against such threat. Circuits Systems Signal Process. **4**(10), 3241–3290 (2015)
17. T. Bjerregaard, S. Mahadevan, A survey of research and practices of network-on-chip. ACM Comput. Surv. **38**(1), 1-es (2006)
18. Bloomberg, *The Big Hack: How China Used a Tiny Chip to Infiltrate U.S. Companies*. https://www.bloomberg.com/news/features/2018-10-04/the-big-hack-how-china-used-a-tiny-chip-to-infiltrate-america-s-top-companies

19. A. Bogdanov, T. Eisenbarth, C. Paar, M. Wienecke, Differential cache-collision timing attacks on AES with applications to embedded CPUs, in *Cryptographers' Track at the RSA Conference* (Springer, Berlin, 2010), pp. 235–251
20. T. Boraten, D. DiTomaso, A.K. Kodi, Secure model checkers for network-on-chip (NoC) architectures, in *2016 International Great Lakes Symposium on VLSI (GLSVLSI)*, 2016, pp. 45–50
21. T. Boraten, A.K. Kodi, Packet security with path sensitization for NoCs, in *2016 Design, Automation Test in Europe Conference Exhibition (DATE)*, 2016, pp. 1136–1139
22. T. Boraten, A. Karanth Kodi, Mitigation of denial of service attack with hardware Trojans in NoC architectures, in *IPDPS*, 2016, pp. 1091–1100
23. T.C. Bressoud, F.B. Schneider, Hypervisor-based fault tolerance. ACM Trans. Comput. Syst. **14**(1), 80–107 (1996)
24. T. Bui, S. Prakash Rao, M. Antikainen, V.M. Bojan, T. Aura, Man-in-the-machine: exploiting ill-secured communication inside the computer, in *27th USENIX Security Symposium (USENIX Security 18)*, 2018, pp. 1511–1525
25. S. Charles, A. Ahmed, U.Y. Ogras, P. Mishra, Efficient cache reconfiguration using machine learning in NoC-based many-core CMPs. ACM Trans. Des. Autom. Electron. Syst. **24**(6), 1–23 (2019)
26. S. Charles, H. Hajimiri, P. Mishra, Proactive thermal management using memory-based computing in multicore architectures, in *International Green and Sustainable Computing Conference (IGSC)*, 2018, pp. 1–8
27. S. Charles, M. Logan, P. Mishra, Lightweight Anonymous Routing in NoC based SoCs, in *Design Automation & Test in Europe (DATE)*, 2020
28. S. Charles, Y. Lyu, P. Mishra, Real-time detection and localization of dos attacks in NOC based socs, in *Design Automation & Test in Europe (DATE)*, 2019 pp. 1160–1165
29. S. Charles, Y. Lyu, P. Mishra, Real-time detection and localization of distributed DoS attacks in NoC based SoCs. IEEE Trans. Comput. Aided Des. Integr. Circuits Syst. **39**, 4510–4523 (2020)
30. S. Charles, P. Mishra, Lightweight and trust-aware routing in NoC based SoCs, in *IEEE Computer Society Annual Symposium on VLSI (ISVLSI)*, 2020
31. S. Charles, P. Mishra, Reconfigurable network-on-chip security architecture. ACM Trans. Des. Autom. Electron. Syst. **25**(6), 1–25 (2020)
32. S. Charles, P. Mishra, Securing network-on-chip using incremental cryptography, in *IEEE Computer Society Annual Symposium on VLSI (ISVLSI)*, 2020
33. S. Charles, C. Arvind Patil, U.Y. Ogras, P. Mishra, Exploration of memory and cluster modes in directory-based many-core CMPs, in *IEEE/ACM International Symposium on Networks-on-Chip (NOCS)*, 2018, pp. 1–8
34. G. Chen, H. Chen, M. Haurylau, N.A. Nelson, D.H. Albonesi, P.M. Fauchet, E.G. Friedman, On-chip copper-based vs. optical interconnects: delay uncertainty, latency, power, and bandwidth density comparative predictions, in *2006 International Interconnect Technology Conference* (IEEE, Piscataway, 2006), pp. 39–41
35. M. Chen, X. Qin, H.-M. Koo, P. Mishra, *System-Level Validation: High-Level Modeling and Directed Test Generation Techniques* (Springer, Berlin, 2012)
36. O. Choudary et al., *Breaking Smartcards Using Power Analysis* (University of Cambridge, Cambridge, 2005)
37. É. Cota, A. de Morais Amory, M.S. Lubaszewski, NoC basics, in *Reliability, Availability and Serviceability of Networks-on-Chip* (Springer, Berlin, 2012), pp. 11–24
38. C. Cowan, P. Wagle, C. Pu, S. Beattie, J. Walpole, Buffer overflows: attacks and defenses for the vulnerability of the decade, in *DARPA Information Survivability Conference and Exposition,*, vol. 3. Los Alamitos, CA, USA (IEEE Computer Society, Silver Spring, 2000), p. 1119
39. M. Dall'Osso, G. Biccari, L. Giovannini, D. Bertozzi, L. Benini, Xpipes: a latency insensitive parameterized network-on-chip architecture for multi-processor SoCs, in *2012 IEEE 30th International Conference on Computer Design (ICCD)* (IEEE, Piscataway, 2012), pp. 45–48

40. W.J. Dally, B. Towles, Route packets, not wires: on-chip interconnection networks, in *Proceedings of the 38th Design Automation Conference (IEEE Cat. No.01CH37232)*, 2001, pp. 684–689
41. DARPA System Security Integrated Through Hardware and Firmware (SSITH) 2017. https://www.fbo.gov
42. A.V. de Mello, L.C. Ost, F.G. Moraes, N.L.V. Calazans, Evaluation of routing algorithms on mesh based NoCs. *PUCRS, Av. Ipiranga*, 2004, p. 22
43. S. Deb, A. Ganguly, P.P. Pande, B. Belzer, D. Heo, Wireless NoC as interconnection backbone for multicore chips: promises and challenges. IEEE J. Emerging Sel. Top. Circuits Syst. **2**(2), 228–239 (2012)
44. J. Diguet, S. Evain, R. Vaslin, G. Gogniat, E. Juin, NoC-centric security of reconfigurable SoC, In *First International Symposium on Networks-on-Chip (NOCS'07)*, 2007, pp. 223–232
45. D. Evans, The internet of things: How the next evolution of the internet is changing everything. CISCO White Paper **1**, 1–11 (2011)
46. F. Farahmandi, Y. Huang, P. Mishra, Trojan localization using symbolic algebra, in *2017 22nd Asia and South Pacific Design Automation Conference (ASPDAC)* (IEEE, Piscataway, 2017), pp. 591–597
47. F. Farahmandi, Y. Huang, P. Mishra, *System-on-Chip Security: Validation and Verification.* (Springer Nature, 2019)
48. F. Farahmandi, P. Mishra, Automated debugging of arithmetic circuits using incremental gröbner basis reduction, in *2017 IEEE International Conference on Computer Design (ICCD)* (IEEE, Piscataway, 2017), pp. 193–200
49. F. Farahmandi, P. Mishra, FSM anomaly detection using formal analysis, in *2017 IEEE International Conference on Computer Design (ICCD)* (IEEE, Piscataway, 2017), pp. 313–320
50. F. Farahmandi, P. Mishra, Automated test generation for debugging multiple bugs in arithmetic circuits. IEEE Trans. Comput. **68**(2), 182–197 (2018)
51. L. Fiorin, G. Palermo, S. Lukovic, V. Catalano, C. Silvano, Secure memory accesses on networks-on-chip. IEEE Trans. Comput. **57**(9), 1216–1229 (2008)
52. L. Fiorin, G. Palermo, C. Silvano, *A Security Monitoring Service for NoCs* (Association for Computing Machinery, New York, 2008), pp. 197–202.
53. J. Frey, Q. Yu, A hardened network-on-chip design using runtime hardware trojan mitigation methods. Integr. VLSI J. **56**(C), 15–31 (2017)
54. Gartner Inc., Semiconductor design IP revenue, chip infrastructure, worldwide, 2012 and 2013, in *Gartner Research*, 2014, pp. 70–78
55. C.H. Gebotys, R.J. Gebotys, A framework for security on NOC technologies, in *Proceedings of the IEEE Computer Society Annual Symposium on VLSI (ISVLSI'03)*, ISVLSI '03 (IEEE Computer Society, Silver Spring 2003), p. 113
56. D. Goldschlag, M. Reed, P. Syverson, Onion routing. Commun. ACM **42**(2), 39–41 (1999)
57. K. Goossens, J. Dielissen, A. Radulescu, Æthereal network on chip: concepts, architectures, and implementations. IEEE Design Test Comput. **22**(5), 414–421 (2005)
58. S. Guo, J. Wang, Z. Chen, Z. Lu, J. Guo, L. Yang, Security-aware task mapping reducing thermal side channel leakage in CMPs. IEEE Trans. Ind. Inf. **15**(10), 5435–5443 (2019)
59. U. Gupta, C.A. Patil, G. Bhat, P. Mishra, U.Y. Ogras, DyPO: Dynamic pareto-optimal configuration selection for heterogeneous MpSoCs. ACM Trans. Embed. Comput. Syst. **16**(5s), 1–20 (2017)
60. Y. Huang, S. Bhunia, P. Mishra, MERS: statistical test generation for side-channel analysis based trojan detection, in *ACM SIGSAC Conference on Computer and Communications Security (CCS)*, 2016, pp. 130–141
61. Y. Huang, S. Bhunia, P. Mishra, Scalable test generation for trojan detection using side channel analysis. IEEE Trans. Inf. Forensics Secur. **13**(11), 2746–2760 (2018)
62. Y. Huang, P. Mishra, Vulnerability-aware energy optimization for reconfigurable caches in multitasking systems. IEEE Trans. Comput. Aided Des. Integr. Circuits Syst. **38**(5), 809–821 (2019)

63. W.C. Huffman, V. Pless, *Fundamentals of Error-Correcting Codes* (Cambridge University Press, Cambridge, 2010)
64. M. Hussain, A. Malekpour, H. Guo, S. Parameswaran, EETD: an energy efficient design for runtime hardware trojan detection in untrusted network-on-chip, in *2018 IEEE Computer Society Annual Symposium on VLSI (ISVLSI)*, 2018, pp. 345–350
65. IBM Core Connect. http://www.chips.ibm.com/product/coreconnect/docs/crconwp.pdf
66. IDC, The growth in connected IoT devices is expected to generate 79.4 ZB of data in 2025, according to a new IDC forecast, 2019
67. L.S. Indrusiak, J. Harbin, M.J. Sepulveda, Side-channel attack resilience through route randomisation in secure real-time networks-on-chip, in *2017 12th International Symposium on Reconfigurable Communication-centric Systems-on-Chip (ReCoSoC)*, 2017, pp. 1–8
68. L.S. Indrusiak, J. Harbin, C. Reinbrecht, J. Sepúlveda, Side-channel protected MPSoC through secure real-time networks-on-chip. Microsc. Microanal. **68**, 34–46 (2019)
69. Z. H. Jiang, Y. Fei, D. Kaeli, A novel side-channel timing attack on GPUs, in *Proceedings of the on Great Lakes Symposium on VLSI 2017*, 2017, pp. 167–172
70. J.Y.V. Manoj Kumar, A.K. Swain, S. Kumar, S.R. Sahoo, K. Mahapatra, Run time mitigation of performance degradation hardware trojan attacks in network on chip, in *2018 IEEE Computer Society Annual Symposium on VLSI (ISVLSI)*, 2018, pp. 738–743
71. H.K. Kapoor, G. Bhoopal Rao, S. Arshi, G. Trivedi, A security framework for NoC using authenticated encryption and session keys. Circuits Systems Signal Process. **32**(6), 2605–2622 (2013)
72. S.T. King, J. Tucek, A. Cozzie, C. Grier, W. Jiang, Y. Zhou, Designing and implementing malicious hardware. Leet **8**, 1–8 (2008)
73. P. Kocher, J. Horn, A. Fogh, D. Genkin, D. Gruss, W. Haas, M. Hamburg, M. Lipp, S. Mangard, T. Prescher, M. Schwarz, Y. Yarom, Spectre attacks: exploiting speculative execution, in *2019 IEEE Symposium on Security and Privacy (SP)*, 2019 pp. 1–19
74. P. Kocher, J. Jaffe, B. Jun, P. Rohatgi, Introduction to differential power analysis. J. Cryptogr. Eng. **1**(1), 5–27 (2011)
75. P.C. Kocher, Timing attacks on implementations of Diffie–Hellman, RSA, DSS, and other systems, in *Annual International Cryptology Conference* (Springer, Berlin, 1996), pp. 104–113
76. S. Kumar, A. Jantsch, J.-P. Soininen, M. Forsell, M. Millberg, J. Oberg, K. Tiensyrja, A. Hemani, A network on chip architecture and design methodology, in *Proceedings IEEE Computer Society Annual Symposium on VLSI. New Paradigms for VLSI Systems Design. ISVLSI 2002* (IEEE, Piscataway, 2002), pp. 117–124
77. E. Levi, Smashing the stack for fun and profit. Phrack Mag. **7**(49) (1996)
78. M. Lipp, M. Schwarz, D. Gruss, T. Prescher, W. Haas, A. Fogh, J. Horn, S. Mangard, P. Kocher, D. Genkin, Y. Yarom, M. Hamburg, Meltdown: reading kernel memory from user space, in *27th USENIX Security Symposium (USENIX Security 18)* (2018)
79. S. Lukovic, N. Christianos, Enhancing network-on-chip components to support security of processing elements, in *Proceedings of the 5th Workshop on Embedded Systems Security, WESS '10* (Association for Computing Machinery, New York, 2010)
80. S. Lukovic, N. Christianos, Hierarchical multi-agent protection system for NoC based MPSoCs, in *Proceedings of the International Workshop on Security and Dependability for Resource Constrained Embedded Systems* (Association for Computing Machinery, New York, 2010)
81. Y. Lyu, A. Ahmed, P. Mishra, Automated activation of multiple targets in RTL models using concolic testing, in *2019 Design, Automation & Test in Europe Conference & Exhibition (DATE)* (IEEE, Piscataway, 2019), pp. 354–359
82. Y. Lyu, P. Mishra, A survey of side-channel attacks on caches and countermeasures. J. Hardw. Syst. Secur. **2**(1), 33–50 (2018)
83. Y. Lyu, P. Mishra, Efficient test generation for trojan detection using side channel analysis, in *Design Automation & Test in Europe Conference (DATE)*, 2019, pp. 408–413

84. Y Lyu, P. Mishra, Automated test generation for activation of assertions in RTL models, in *2020 25th Asia and South Pacific Design Automation Conference (ASPDAC)* (IEEE, Piscataway, 2020), pp. 223–228
85. Y. Lyu, P. Mishra, Automated test generation for trojan detection using delay-based side channel analysis, in *2020 Design, Automation & Test in Europe Conference & Exhibition (DATE)*, 2020, pp. 1031–1036
86. Y. Lyu, P. Mishra, Automated trigger activation by repeated maximal clique sampling, in *Asia and South Pacific Design Automation Conference (ASPDAC)*, 2020, pp. 482–487
87. Y. Lyu, P. Mishra, Scalable activation of rare triggers in hardware trojans by repeated maximal clique sampling. *IEEE Transactions on Computer-Aided Design of Integrated Circuits and Systems* (2020)
88. Y. Lyu, P. Mishra, Scalable concolic testing of RTL models. IEEE Trans. Comput. (2020). https://ieeexplore.ieee.org/document/9099620
89. Y. Lyu, X. Qin, M. Chen, P. Mishra, Directed test generation for validation of cache coherence protocols. IEEE Trans. Comput. Aided Des. Integr. Circuits Syst. **38**(1), 163–176 (2018)
90. R.A. Martin, *Common Weakness Enumeration* (MITRE Corporation, McLean, 2007)
91. P. Mishra, S. Bhunia, M. Tehranipoor, *Hardware IP security and Trust* (Springer, Berlin, 2017)
92. P. Mishra, F. Farahmandi, *Post-Silicon Validation and Debug* (Springer, Berlin, 2019)
93. A. Naeemi, R. Sarvari, J.D. Meindl, On-chip interconnect networks at the end of the roadmap: limits and nanotechnology opportunities, in *2006 International Interconnect Technology Conference* (IEEE, Piscataway, 2006), pp. 201–203
94. Z. Pan, P. Mishra, Automated test generation for hardware trojan detection using reinforcement learning, in *Asia and South Pacific Design Automation Conference (ASPDAC)* (2021)
95. Z. Pan, J. Sheldon, P. Mishra, Test generation using reinforcement learning for delay-based side channel analysis, in *IEEE/ACM International Conference on Computer-Aided Design (ICCAD)* (2020)
96. S. Pasricha, N. Dutt, Orb: An on-chip optical ring bus communication architecture for multi-processor systems-on-chip, in *2008 Asia and South Pacific Design Automation Conference* (IEEE, Piscataway, 2008), pp. 789–794
97. S. Pasricha, N. Dutt, *On-Chip Communication Architectures: System on Chip Interconnect* (Morgan Kaufmann, Los Altos, 2010)
98. J. Porquet, A. Greiner, C. Schwarz, NoC-MPU: a secure architecture for flexible co-hosting on shared memory MPSoCs, in *2011 Design, Automation Test in Europe*, 2011, pp. 1–4
99. N. Prasad, R. Karmakar, S. Chattopadhyay, I. Chakrabarti, Runtime mitigation of illegal packet request attacks in networks-on-chip, in *2017 IEEE International Symposium on Circuits and Systems (ISCAS)* (IEEE, Piscataway, 2017), pp. 1–4
100. J.S. Rajesh, D.M. Ancajas, K. Chakraborty, S. Roy, Runtime detection of a bandwidth denial attack from a rogue network-on-chip, in *Proceedings of the 9th International Symposium on Networks-on-Chip, NOCS '15* (Association for Computing Machinery, New York, 2015)
101. V.Y. Raparti, S. Pasricha, Lightweight mitigation of hardware trojan attacks in NoC-based manycore computing, in *Proceedings of the 56th Annual Design Automation Conference 2019* (ACM, New York, 2019), p. 48
102. C. Reinbrecht, A. Susin, L. Bossuet, J. Sepúlveda, Gossip NoC—avoiding timing side-channel attacks through traffic management, in *2016 IEEE Computer Society Annual Symposium on VLSI (ISVLSI)*, 2016, pp. 601–606
103. C. Reinbrecht, A. Susin, L. Bossuet, G. Sigl, J. Sepúlveda, Side channel attack on NoC-based MPSoCs are practical: NoC prime+probe attack, In *2016 29th Symposium on Integrated Circuits and Systems Design (SBCCI)*, 2016, pp. 1–6
104. C. Reinbrecht, B. Forlin, A. Zankl, J. Sepúlveda, Earthquake—a NoC-based optimized differential cache-collision attack for MPSoCs, in *2018 Design, Automation & Test in Europe Conference & Exhibition (DATE)* (IEEE, Piscataway, 2018), pp. 648–653
105. R.L. Rivest, A. Shamir, L. Adleman, A method for obtaining digital signatures and public-key cryptosystems. Commun. ACM **26**(1), 96–99 (1983)

106. A. Saeed, A. Ahmadinia, M. Just, C. Bobda, An ID and address protection unit for NoC based communication architectures, in *Proceedings of the 7th International Conference on Security of Information and Networks* (ACM, New York, 2014), p. 288
107. S. Saidi, R. Ernst, S. Uhrig, H. Theiling, B. D. de Dinechin, The shift to multicores in real-time and safety-critical systems, in *2015 International Conference on Hardware/Software Codesign and System Synthesis (CODES+ ISSS)* (IEEE, Piscataway, 2015), pp. 220–229
108. K. Sajeesh, H.K Kapoor, An authenticated encryption based security framework for NoC architectures, in *2011 International Symposium on Electronic System Design* (IEEE, Piscataway, 2011), pp. 134–139
109. Semiconductor Industry Association et al. International technology roadmap for semiconductors (ITRS), 2003 edition. Incheon, Korea, Dec, 2011
110. J. Sepúlveda, D. Aboul-Hassan, G. Sigl, B. Becker, M. Sauer, Towards the formal verification of security properties of a network-on-chip router, in *2018 IEEE 23rd European Test Symposium (ETS)* (IEEE, Piscataway, 2018), pp. 1–6
111. J. Sepúlveda, D. Flórez, G. Gogniat, Reconfigurable security architecture for disrupted protection zones in NoC-based MPSoCs, In *2015 10th International Symposium on Reconfigurable Communication-centric Systems-on-Chip (ReCoSoC)* (IEEE, Piscataway, 2015), pp. 1–8
112. J. Sepúlveda, D. Flórez, M. Soeken, J.-P. Diguet, G. Gogniat, Dynamic NoC buffer allocation for MPSoC timing side channel attack protection, in *2016 IEEE 7th Latin American Symposium on Circuits & Systems (LASCAS)* (IEEE, Piscataway, 2016), pp. 91–94
113. J. Sepúlveda, G. Gogniat, D. Florez, J.-P. Diguet, C. Zeferino, M. Strum, Elastic security zones for NoC-based 3d-MPSoCs, in *2014 21st IEEE International Conference on Electronics, Circuits and Systems (ICECS)* (IEEE, Piscataway, 2014), pp. 506–509
114. J. Sepúlveda, A. Zankl, D. Flórez, G. Sigl, Towards protected MPSoC communication for information protection against a malicious NoC. Proc. Comput. Sci. **108**, 1103–1112 (2017)
115. M.J. Sepulveda, J.-P. Diguet, M. Strum, G. Gogniat, NoC-based protection for SoC time-driven attacks. IEEE Embed. Syst. Lett. **7**(1), 7–10 (2014)
116. K. Shuler, Majority of leading China semiconductor companies Rely on Arteris network-on-chip interconnect IP (2013)
117. D. Sigüenza-Tortosa, T. Ahonen, J. Nurmi, Issues in the development of a practical NoC: the Proteo concept. Integration **38**(1), 95–105 (2004)
118. S. Skorobogatov, C. Woods, Breakthrough silicon scanning discovers backdoor in military chip, in *Cryptographic Hardware and Embedded Systems—CHES 2012*, ed. by E. Prouff, P. Schaumont (Springer, Berlin, 2012), pp. 23–40
119. A. Sodani, R. Gramunt, J. Corbal, H. Kim, K. Vinod, S. Chinthamani, S. Hutsell, R. Agarwal, Y. Liu, Knights landing: Second-generation Intel Xeon Phi product. IEEE Micro **36**(2), 34–46 (2016)
120. SONICS NoCk-Lock Security. www.sonicsinc.com/wp-content/uploads/NoC-Lock.pdf
121. N. Srivastava, K. Banerjee, A comparative scaling analysis of metallic and carbon nanotube interconnections for nanometer scale VLSI technologies, in *Proceedings of the 21st International VLSI Multilevel Interconnect Conference*, 2004, pp. 393–398
122. N. Srivastava, K. Banerjee, Performance analysis of carbon nanotube interconnects for VLSI applications, in *ICCAD-2005. IEEE/ACM International Conference on Computer-Aided Design, 2005* (IEEE, Piscataway, 2005), pp. 383–390
123. W. Steinhögl, G. Schindler, G. Steinlesberger, M. Engelhardt, Size-dependent resistivity of metallic wires in the mesoscopic range. Phys. Rev. B **66**(7), 075414 (2002)
124. M. Tehranipoor, F. Koushanfar, A survey of hardware trojan taxonomy and detection. IEEE Design Test Comput. **27**(1), 10–25 (2010)
125. Using TinyCrypt Library, Intel Developer Zone, Intel, 2016. https://software.intel.com/en-us/node/734330
126. W. Wang, P. Mishra, S. Ranka, *Dynamic Reconfiguration in Real-Time Systems* (Springer, 2012)
127. Y. Wang, G.E. Suh, Efficient timing channel protection for on-chip networks, in *2012 IEEE/ACM Sixth International Symposium on Networks-on-Chip*, 2012, pp. 142–151

128. H.M.G. Wassel, Y. Gao, J.K. Oberg, T. Huffmire, R. Kastner, F.T. Chong, T. Sherwood, SurfNoC: a low latency and provably non-interfering approach to secure networks-on-chip, in *Proceedings of the 40th Annual International Symposium on Computer Architecture, ISCA '13* (Association for Computing Machinery, New York, 2013), pp. 583–594
129. D. Wentzlaff, P. Griffin, H. Hoffmann, L. Bao, B. Edwards, C. Ramey, M. Mattina, C. Miao, J.F. Brown III, A. Agarwal, On-chip interconnection architecture of the tile processor. IEEE Micro **27**(5), 15–31 (2007)
130. P.T. Wolkotte, G.J.M. Smit, G.K. Rauwerda, L.T. Smit, An energy-efficient reconfigurable circuit-switched network-on-chip, in *19th IEEE International Parallel and Distributed Processing Symposium* (IEEE, Piscataway, 2005), p. 8
131. K. Xiao, A. Nahiyan, M. Tehranipoor, Security rule checking in IC design. Computer **49**(8), 54–61 (2016)
132. Q. Yu, J. Frey, Exploiting error control approaches for hardware trojans on network-on-chip links, in *2013 IEEE International Symposium on Defect and Fault Tolerance in VLSI and Nanotechnology Systems (DFTS)* (IEEE, Piscataway, 2013), pp. 266–271
133. W. Yu, O.A. Uzun, S. Köse, Leveraging on-chip voltage regulators as a countermeasure against side-channel attacks, in *Proceedings of the 52nd Annual Design Automation Conference*, 2015, pp. 1–6
134. S. Zander, G. Armitage, P. Branch, A survey of covert channels and countermeasures in computer network protocols. IEEE Commun. Surv. Tutorials **9**(3), 44–57 (2007)
135. L. Zhang, Optical network-on-chip architectures and designs (2011)
136. H. Zhu, P.P. Pande, C. Grecu, Performance evaluation of adaptive routing algorithms for achieving fault tolerance in NoC fabrics, in *2007 IEEE International Conf. on Application-Specific Systems, Architectures and Processors (ASAP)* (IEEE, Piscataway, 2007), pp. 42–47

Chapter 2
Interconnect Modeling for Homogeneous and Heterogeneous Multiprocessors

Tushar Krishna and Srikant Bharadwaj

With the end of Dennard scaling and slowing of Moore's law, computer systems have become increasingly complex, with multiple processing units (multicore CPUs, GPUs, and other accelerators) integrated with an interconnection system. The high-demand for increase in the number of cores has led to a need for greater integration of units across a chip. Thus, the need for a scalable and high-bandwidth communication fabric to connect the units has become critically important.

Modeling of such complex interconnect systems is needed for design-space explorations as well as for evaluating novel optimizations. Several interconnect modeling tools have been developed over the past decades with various levels of detailing. The interconnect models are often combined with different types of traffic patterns originating from various IPs (such as CPU, GPU, FPGA, and accelerators). In addition to application traffic, performance models allow stress testing the network with a myriad of synthetic traffic patterns to characterize the latency and throughput characteristics.

In this chapter, we discuss the various approaches to modeling interconnection systems that have been adopted for modern systems. We start with describing the type of traffic patterns that are experienced by these interconnection models. This would include the cache coherence and memory traffic observed in modern systems as well as the synthetic traffic patterns used for evaluating networks. We then describe the typical modeling approaches taken by performance simulators and power-area models.

T. Krishna (✉)
Georgia Institute of Technology, Atlanta, GA, USA
e-mail: tushar@ece.gatech.edu

S. Bharadwaj
AMD Research, Bellevue, WA, USA
e-mail: srikant.bharadwaj@amd.com

P. Mishra, S. Charles (eds.), *Network-on-Chip Security and Privacy*,
https://doi.org/10.1007/978-3-030-69131-8_2

2.1 Interconnects in Modern Systems

Modern systems require the interconnect system (or data fabric) for several types of communications across the system. In shared memory systems, the on-chip network is a key component to connect the different units of the memory subsystem hierarchy (L1, L2, directory, memory controller, and so on). In addition to the memory traffic, modern processors also demand inter-core communication often referred to as the cache coherence traffic. The recent advances to the systems in the form of heterogeneous integration has further led to the demand for advanced interconnection systems. These systems require heterogeneity in the form of clock domain crossings, 2.5D and 3D topologies, and serialization-deserialization. In this section, we describe some key characteristics of modern cache-coherent heterogeneous systems.

2.1.1 Cache Coherence

One of the most important utilities of an interconnect system is to support cache coherence protocols. In modern shared memory systems, communication occurs primarily through loading and storing of data to the memory. Practically, these shared memory systems utilize hierarchical cache structures to improve performance of the systems. However, these cache structures create complexity to the unified shared memory paradigm. The role of the cache coherence protocol in modern systems is to maintain the semantics of one writer or many readers in parallel programs. Cache coherence protocols are thus required to maintain coherency across the multiple copies of data that could be present in a system. These coherence protocols thus require strong communication semantics for efficient governance of coherency.

Cache coherency protocols often adopt one of two key types of mechanisms. Directory-based protocols rely on a logical structure which houses the information on the location of cache blocks across the system. Alternative to directory-based protocols is the broadcast-based system, where requests are broadcasted to all sharers over the communication fabric. We can classify cache coherence protocols into four broad categories.

- **Full-state Directory**. In this design, the directory has a bit-vector to track all sharers and the owner (if any) for any line. The storage requirement for this design goes up as O(N) where N is the number of cores. Figure 2.1 shows a transaction in a full-state directory protocol, and the corresponding messages in the network. L2 #8 sends a write miss (*WRITE A*) to its home node L2 #6 which houses a part of the distributed directory. L2 #6 forwards (*FWD A*) this request to the sharer (S) of this data L2#0 and the owner (O) of this data[1] L2# 14. L2

[1]If there is no owner on-chip, L2 #6 also forwards this request to the memory controller since DRAM is the owner of the line.

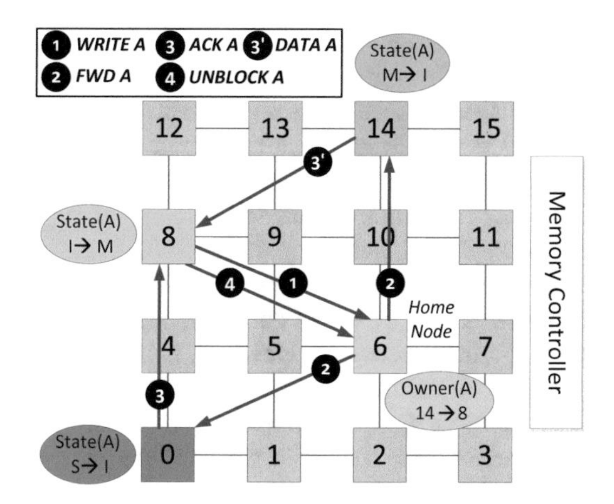

Fig. 2.1 Cache Coherence transactions with full-state directory. (1) The requester (#8) sends a write miss to the home node (#6). (2) The home node forwards it to the owner (#14) and the sharer (#0). (3) The owner invalidates its copy and sends the data to the requester. The sharer invalidates its copy and sends an ACK to the requester. (4) The requester then unblocks the home node

#14 responds with data (*DATA A*) to L2 #8, while L2#0 sends an invalidation acknowledgement (ACK) to L2 #8. On receiving all ACKs and data, L2 #8 then unblocks its home node (*UNBLOCK A*). At the end of the transaction, L2 #8 holds line A in modified (M) state, while L2# 14 and L2# 0 hold the line in invalid (I) state.

- **Partial-state Directory**. In these designs, the directory only tracks the owner [12] or a subset of sharers [22] or tracks lines at a coarser granularity [15], to reduce the storage requirement. To provide coverage over the complete chip, it resorts to occasional broadcasts.
- **No-state Directory**. In such protocols (e.g., AMD's HyperTransport™ [19], used in its early Opteron chips), the directory does not have any state, but simply acts as an ordering point among different requests to the same line, to help maintain memory consistency semantics [2].
- **No Directory**. In these designs, there is no home node for data. The requester broadcasts all its requests, all other nodes snoop, and the owner (cache or memory) responds. A key requirement for these snoopy protocols is global ordering. On a bus-based design, where these protocols are highly prevalent, the central bus arbiter serves as the ordering point. On a distributed mesh, on the other hand, other techniques are required to guarantee race-free correct functionality in case of competing requests. Token Coherence [25] and INSO [4] are two techniques to run snoopy protocols on a mesh topology.

2.1.1.1 Message Classes and Virtual Networks (Vnets)

The series of messages sent by a coherence protocol as part of a coherence transaction fall within different **message classes**. For instance, most directory protocols (full-state, partial-state, and no-state) use 3–4 message classes: *request*, *forward*, *response*, and *unblock*.

A potential deadlock can occur in the protocol if a request for a line from a L2 is unable to enter the network because the L2 is waiting for a response for a previous request, while the response is unable to reach the L2 since all queues in the network are full of such waiting requests. To avoid such deadlocks, protocols require messages from different message classes to use different set of queues within the network. This is implemented by using **virtual networks (vnets)** within the physical network. Virtual Networks are identical to Virtual Channels(VC) in terms of their implementation: all vnets have separate buffers but multiplex over the same physical links. In fact many works on coherence protocols use the term virtual channels to refer to virtual networks. However, in this chapter, we will strictly adhere to using the term virtual networks or vnets to refer to protocol level message classes. The number of vnets is thus fixed by the protocol. Each vnet, on the other hand, can have one or more VCs within each router, to avoid head-of-line blocking or avoid routing deadlocks.

2.1.1.2 Message Sizes

The size of messages generated by the cache coherency controllers vary depending on the type of message and system requirements. These messages could include control messages (requests/forwards/unblocks) which fit within one single flit, or data responses which span multiple flits. For example, if a physical link or router supports 128-bit flits, a 64B cache line would have to broken down into 5 flits (including header flit) for transmission through the link or router. Thus VCs within the request, forward or unblock vnets could be 1-flit deep, while VCs within the response vnet could be more than 1-flit deep.

2.1.1.3 Point-to-Point Ordering

Certain message classes (and thus their vnets) require **point-to-point ordering** in the network for functional correctness. This means that two messages injected from the same source, for the same destination, should be delivered in the order of their issue. Most interconnect systems implement point-to-point ordering for flits within ordered vnets by (1) using deterministic routing, and (2) using FIFO/queuing arbiters at each router. The first condition guarantees that two messages from the same source do not use alternate paths to the same destination as that could result in the older message getting delivered after the newer one if the former's path has more congestion. The second condition guarantees that flits at a router's input port leave in the order in which they came in.

2.1.2 Heterogeneous Interconnect Systems

With the end of Dennard scaling and slowing of Moore's law, computer systems have become increasingly complex, with multiple processing units (multicore CPUs, GPUs, and other accelerators) integrated with an interconnection system. These multiprocessor architectures are typically composed of complex memory hierarchy operating in multiple independently-clocked voltage/frequency islands (VFIs) connected to each other. Modern architectures consisting of multi-chip System-On-Chip (SoC) design, where multiple small chips are assembled to form a large system, are being produced commercially.

Further, heterogeneous architectures that consist of CPUs, GPUs, and other accelerators often involve tight integration through diverse connectivity. NoCs have emerged as the scalable solution for these interconnections. NoCs have also been extended to GPUs, heterogeneous architectures, and domain-specific architectures, such as machine learning and FPGAs. NoCs have, thus, become an integral part of modern architectures. This has led to a large amount of research in improving various aspects of the NoC such as topology, flow control, deadlock prevention and detection, and router microarchitecture.

Emerging systems are diverse in many dimensions. First, their traffic sources can be diverse, where various computation and memory components with different characteristics are integrated into the same system. Second, their physical interconnects can be diverse, where interconnect materials (e.g., on-chip wire, TSV, micro bump) and widths can be different across the whole system. Finally, their system structure can be diverse, where multiple small systems operating at different voltages/frequencies are connected to form a larger system on a chip, package (NoP) [32], or interposer (NoI) [8].

To support this diversity, modern commercial NoCs typically consist of a comprehensive set of features. These features not only include functionalities such as supporting the memory hierarchy and different message types through VCs, arbitration at various memory levels to ensure performance and quality of service, but also include supporting multiple voltage/frequency domains, transmitting messages in irregular networks and more. The heterogeneous nature of the networks in modern systems directly affects overall system performance and energy.

2.1.2.1 Multi-Domain Interconnect Systems

Modern systems are diverse, with multiple small systems operating at different voltages/frequencies connected to form a larger system on a chip or package. Such systems need multiple voltage-frequency domains to support the inherent heterogeneity of the system. The interconnect system which connects these domains in turn needs to support communication across these domains for functionality. These requirements in modern heterogeneous systems lead to integration of units such as clock domain crossings (CDCs) within the system.The CDCs within the interconnect system need to support packets and flits travelling across the system.

2.1.2.2 Serializer-Deserializer Units

Emerging technology of building larger systems using small *chiplets* embedded on a substrate such as an interposer has introduced the idea of heterogeneously sized interconnect system. Unlike classical systems where the data fabric was designed to support flits and packets of a fixed size, modern system allows conversion of flits from one size to another on the go. Modeling such complex interconnect systems requires modeling links of different widths as well as units which can serialize and deserialize different sized flits. These Serializer-Deserializer (SerDes) units are introduced at domain boundaries and are used to allow long-distance low-latency communication to other chiplets.

2.2 Traffic Models

2.2.1 Synthetic Traffic

Synthetic traffic patterns are used to stress the interconnect systems. Several types of synthetic traffic patterns are generally used to evaluate interconnect topologies and designs. Table 2.1 lists some common synthetic traffic patterns used for studying a mesh network, along with their average hop-counts and theoretical throughput with XY routing. The theoretical throughput or capacity is the injection rate at which some link(s) in the mesh is (are) sending 1-flit every cycle.[2] This is the best a topology can do, with perfect routing, flow control, and microarchitecture.

2.2.2 Application Traffic

2.2.2.1 Trace-Based Simulation

Researchers often use traces of network injections from applications running on a real system or a full-system simulator. Network traces provide a fairly realistic way of exploring the effectiveness of proposed on-chip network designs, but clearly, it should be noted that their characteristics depend heavily on the simulated many-

[2]Table 2.1 shows that uniform random traffic offers the highest throughput, since it saturates when the bisection links of the mesh are fully occupied. For traffic patterns that saturate other links, throughput is lower.

Table 2.1 Synthetic Traffic Patterns for $k \times k$ Mesh **Source** (binary coordinates): $(y_{k-1}, y_{k-2}, \ldots, y_1, y_0, x_{k-1}, x_{k-2}, \ldots, x_1, x_0)$

Traffic pattern	Destination (binary coordinates)	Avg hops (for $k = 8$)	Throughput (for $k = 8$) (flits/nodes/cycle)
Bit-complement	$(\bar{y}_{k-1}, \bar{y}_{k-2}, \ldots, \bar{y}_1, \bar{y}_0, \bar{x}_{k-1}, \bar{x}_{k-2}, \ldots, \bar{x}_1, \bar{x}_0)$	8	0.25
Bit-reverse	$(x_0, x_1, \ldots, x_{k-2}, x_{k-1}, y_0, y_1, \ldots, y_{k-2}, y_{k-1})$	5.25	0.14
Shuffle	$(y_{k-2}, y_{k-3}, \ldots, y_0, x_{k-1}, x_{k-2}, x_{k-3}, \ldots, x_0, y_{k-1})$	4	0.25
Tornado	$(y_{k-1}, y_{k-2}, \ldots, y_1, y_0, x_{k-1+\lceil \frac{k}{2} \rceil -1}, \ldots, x_{\lceil \frac{k}{2} \rceil -1})$	3.75	0.33
Transpose	$(x_{k-1}, x_{k-2}, \ldots, x_1, x_0, y_{k-1}, y_{k-2}, \ldots, y_1, y_0)$	5.25	0.14
Uniform random	$random()$	5.25	0.5

core platform. The number of cores/IP blocks, the memory hierarchy, the number of memory controllers, etc. significantly influence the network trace. The lack of feedback effects when using network traces also impacts the accuracy. For instance, faster on-chip networks than the ones on which traces were collected could lead to pathological scenarios such as responses getting delivered before their requests were injected. Tracking or inferring dependencies between packet is important to being able to replay the trace correctly [29].

2.2.2.2 Full-System Simulation

In addition to the use trace-based simulation, researchers often integrate detailed interconnect models to full-system simulation. A full-systems simulation involves booting an OS on the detailed model of the hardware and launching the actual software application. Full-system simulations guarantee functionality by offloading the actual instructions from the simulator on the host CPU, while detailed timing for the hardware being modeled is computed separately. The timing of different aspects of hardware (e.g., compute, memory-system, NoC) can be modeled at various levels of fidelity. In the context of NoCs, the timing for the various messages traversing the NoC are computed as part of the delay incurred by the overall memory subsystem that comprises the coherence protocol, cache hierarchy, NoC, memory controller, and DRAM.

Simulators such as HeteroGarnet [8], BookSim2 [28], Garnet [3] have been integrated with full-system simulators such as gem5 [9, 24] and GPGPU-sim [20] for higher accuracy of evaluations. Such a tight integration is often done by integrating the interconnect simulator to the memory subsystem of the simulator such that the coherency and memory traffic of the system is redirected to the network system. We will describe this process in detail in later sections.

It is important to note though that detailed network simulation often comes at the cost of wall-clock time for real workloads. In the next two sections, we describe how NoCs can be modeled at increasing levels of fidelity.

2.3 Analytical Modeling

Modeling of interconnect topologies for performance, power, and area are key for developing novel optimized designs and architectures. Analytical modeling of interconnect topologies has provided quick insight as design-time metric for comparing topologies. Figure 2.2 shows some of the common network-on-chip topologies considered in this section.

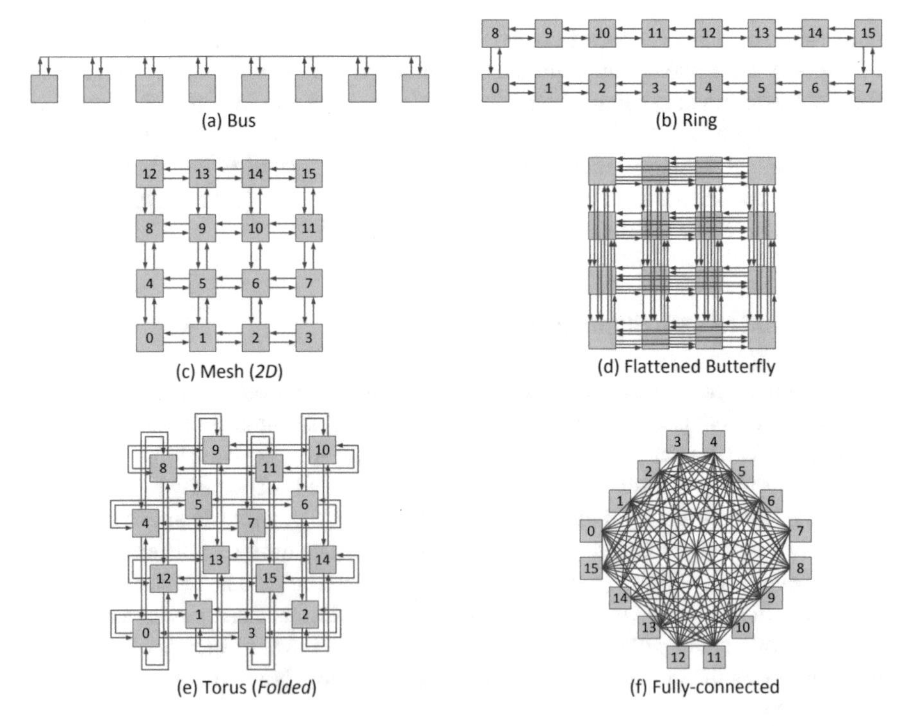

Fig. 2.2 Common network-on-chip topologies. (**a**) Bus. (**b**) Ring. (**c**) Mesh (2*D*). (**d**) Flattened Butterfly. (**e**) Torus (*Folded*). (**f**) Fully connected

2.3.1 Latency

Hop Count The number of hops [16] a message takes from source to destination, or the number of links it traverses, defines hop count. This is a very simple and useful proxy for network latency, since every node and link incurs some propagation delay, even when there is no contention. The maximum hop count is given by the *diameter* of the network. In addition to the maximum hop count, average hop count is very useful as a proxy for network latency. It is given by the average hops over all possible source-destination pairs in the network.

For the same number of nodes, and assuming uniform random traffic where every node has an equal probability of sending to every other node, a ring will lead to higher hop count than a mesh or a torus [13]. For instance, assuming bidirectional links and shortest-path routing, the maximum hop count of an 8-core ring is four, that of a 3x3 mesh is also four, while a torus improves the hop count to two. Looking at average hop count, we see that the torus again has the lowest average hop count ($1\frac{1}{3}$). The mesh has a higher average hop count of $1\frac{7}{9}$. Finally, the ring has the worst average hop count of the three topologies with an average of $2\frac{2}{9}$.

Effective Hop Count The average hop count defined for homogeneous systems cannot be accurately extended to heterogeneous interconnect systems. The catch in average hop count is that the operating frequency f (time duration of each hop) is often not taken into account. For example, a topology could use long express links to provide single hop transmission, but could be restricted by the maximum possible f. Hence the wire and router latencies would go up, making it worse than a topology with higher average hop count but higher. This is often not an issue in homogeneous NoCs since the (low) operating frequency of the NoC is limited by CPU/GPU cores and caches, which have plateaued since Dennard's scaling stopped. However, heterogeneous interconnect systems such as NoP [32] or NoI [8] need not be coupled to core frequencies, and could thus operate at a different and (multi-GHz) operating frequency. This necessitates co-optimizing average hop latency and the operating frequency. Recent work has defined a new metric called effective hop count as the proxy and takes into account the frequency trade off involved in heterogeneous systems. The effective hop count is defined as

$$Effective_{HopCount}{=}Average\ Hop\ Count\ (H_{avg})/Operating\ Frequency\ (f)$$

Latency Estimation The latency of every packet in an on-chip network can be described by the following equation:

$$\begin{aligned} T_{Network} &= T_{wire} + T_{router} + T_{contention} \\ &= H \cdot t_{wire} + (H+1) \cdot t_{router} + \sum_{h=1}^{H+1} t_{contention}(h) \end{aligned}$$

where H is the average hop count through the topology, t_{router} is the pipeline delay through a single router, t_{wire} is the wire delay between two routers, and $t_{contention}(h)$ is the delay due to contention between multiple messages competing for the same network resources at a router h-hops from the start. A factor of $H+1$ is considered for router power and contention since a packet traverses the input router prior to the first hop through the network. t_{router} accounts for the time each packet spends in various stages at each router as the router coordinates between multiple packets; depending on the implementation, this can consist of one to several pipeline stages as discussed later in Sect. 2.4.4. t_{router} and t_{wire} are design-time metrics. They can be used to determine a lower-bound on the latency of any packet. H and $t_{contention}(h)$ are runtime metrics that depend on traffic.

2.3.2 Throughput

Bisection Bandwidth The bisection bandwidth is the bandwidth across a cut that partitions the network into two equal parts.[3] For example, in Fig. 2.2, two links cross the bisection for the ring, three for the mesh, and six for the torus. This bandwidth is often useful in defining worst-case performance of a particular network, since it limits the total data that can be moved from one side of the system to the other. It also serves as a proxy for cost since it represents the amount of global wiring that will be necessary to implement the network. As a metric, bisection bandwidth is less useful for on-chip networks as opposed to off-chip networks, since global on-chip wiring is considered abundant relative to off-chip pin bandwidth.

Effective Bandwidth The concept of calculating the effective value of a metric can also been extended to bandwidth for measuring the bisection bandwidth of a heterogeneous interconnect system.

$$Effective_{BisectionBW} = BisectionBW.Operating\ Frequency\ (f)$$

Throughput Estimation The bisection bandwidth is a design-time metric for the throughput of any network. It is defined as the inverse of the maximum load across the bisection channels of any topology. Ideal throughput assumes perfect flow control and perfect load balancing from the routing algorithm. The actual throughput at saturation, however, might vary heavily, depending on how routing and flow control interact with runtime traffic. Throughput higher than the bisection bandwidth can be achieved if traffic does not go from one end of the network to the other over the bisection links. However, often times, the achieved saturation throughput is lower than the bisection bandwidth. A deterministic routing algorithm, such as XY, might be unable to balance traffic across all available links in the topology in response to network load. Heavily used paths will saturate quickly, reducing the rate of accepted traffic. On the other hand, an adaptive routing algorithm using local

[3] If there are multiple such cuts possible, it is the minimum among all the cuts.

congestion metrics could lead to more congestion in downstream links. The inability of the arbitration schemes inside the router to make perfect matching between requests and available resources can also degrade throughput. Likewise, limited number of buffers and buffer turnaround latency can drive down the throughput of the network.

2.3.3 Energy

The energy consumed by each flit during its network traversal is given by

$$\begin{aligned} E_{Network} &= H \cdot E_{wire} + (H+1) \cdot E_{router} \\ &= H \cdot E_{wire} + (H+1) \cdot (E_{ST} + E_{BW} + E_{BR} + E_{RC}) \\ &\quad + \sum_{h=1}^{H+1} t_{contention}(h) \cdot (E_{VA} + E_{SA}) \end{aligned}$$

where E_{BW}, E_{RC}, E_{VA}, E_{SA}, E_{BR}, and E_{ST} are the energy consumption for buffer write, route computation, VC arbitration, switch arbitration, buffer read, and switch traversal, respectively. E_{RC} and E_{VA} are only consumed by the head-flit. The relative contribution of these parameters is topology and flow control specific. For instance, a high-radix router might have a larger E_{ST} and E_{wire}, but lower H. Similarly, a wormhole router will not consume E_{VA}. Contention at every router determines the number of times a flit may need to perform VA and SA before winning both and getting access to the switch. E_{VA} and E_{SA} depend on the specific allocator implementation.

2.3.4 Area

The area footprint of an on-chip network depends on the area of routers.

$$\begin{aligned} A_{Network} &= N \cdot (A_{router}) \\ &= N \cdot (p \cdot v \cdot A_{VC} + p \cdot A_{RouteUnit} + p \cdot A_{Arbiter inport} \\ &\quad + p \cdot A_{Arbiter outport} + A_{Crossbar}) \end{aligned}$$

where N is the number of routers (assuming all of them are homogeneous input buffered designs), p is the number of ports, and v is the number of VCs per input port. A_{VC} is the area consumed by the buffers and control for each VC, which in turn depends on its implementation. This equation assumes a separable switch allocator design; $A_{Arbiter inport}$ represents the area of all the arbiters at each input port, and $A_{Arbiter outport}$ represents the area of all the arbiters at each output port.

Wires do not directly contribute to the area footprint as they are often routed on higher metal layers above logic; the link drivers are embedded within the crossbar while the link receiver is within the input VC.

2.4 Cycle-Level Software Simulators

Software modeling of the various aspects of modern interconnect system is key to unlocking the understandings of the micro-architectural behavior of traffic patterns. Several modeling tools have been developed over the years to support varied levels of interconnect modeling. Cycle-level simulators have been the most commonly utilized performance simulators for both research and industrial explorations. Power and area modeling tools have been developed for different types of interconnect systems. These models generally utilize RTL and openly available process design kits (PDK) to estimate power and area consumption of various units within the interconnect system.

Cycle-level simulators are widely used in computer architecture research. Both industry and academia heavily rely on cycle-level simulators that model the processing units and memory systems in various levels of detail. Network-on-chip simulators such as Garnet [3], BookSim2 [28], Topaz [1], HeteroGarnet [8] are used to evaluate the different network metrics. These simulators have also been integrated into larger simulators such as gem5 and GPGPU-sim for simulating real applications. In particular, many studies have used Garnet to evaluate both the execution of real programs (combined with gem5) and isolated network performance (using synthetic traffic patterns). Table 2.2 shows some of the widely used interconnect modeling tools and their major features.

Most of these tools were designed mainly to support interconnection network studies such as topology, flow control, and deadlock prevention. The increasingly heterogeneous nature of interconnect systems has propelled the development of

Table 2.2 State of the Art NoC simulators

Simulator	Language	Environment	Target interconnect
Garnet [3]	C++	Standalone + full-system (gem5 [9])	Silicon and optical interconnects
Booksim2 [28]	C++	Standalone + full-system (gpgpusim [6])	Silicon interconnects
Topaz [1]	C++	Standalone	Silicon interconnects
SuperSim [26]	C++	Standalone	Silicon large networks
HeteroGarnet [8]	C++	Standalone + full-system (gem5 [9])	Chip and package-level silicon, optical, and wireless interconnects
PhoenixSim [11]	C++	Standalone	Photonic interconnect
NoxSim [10]	C++	Standalone	Wireless interconnect

modern network simulators such as SuperSim [26] and HeteroGarnet [8]. These new simulators enable detailed exploration of heterogeneous interconnects with support for multi-clock domain and serializer-deserializer units in addition to the typical features. HeteroGarnet, which was released as Garnet 3.0 in gem5 [24], extensively builds over the Garnet and includes several other features.

We discuss the general software modeling approach taken by such cycle-level simulators by taking Garnet 3.0 (which includes HeteroGarnet) as an example in the upcoming sections.

2.4.1 Topology

Most simulators enable configuration of the entire network through a configuration file. All kinds of end points (e.g., Cores, Caches, DMA nodes) connected to the network can be configured to form a topology. This allows complete configuring of complex and irregular network-topologies. A typical example of a topology is shown in Fig. 2.3.

Garnet allows users to configure complex topologies using a python configuration file as the topology. The overall topology configuration could include the complete interconnect definition of the system including any heterogeneous components. The general flow of defining a topology involves the following steps:

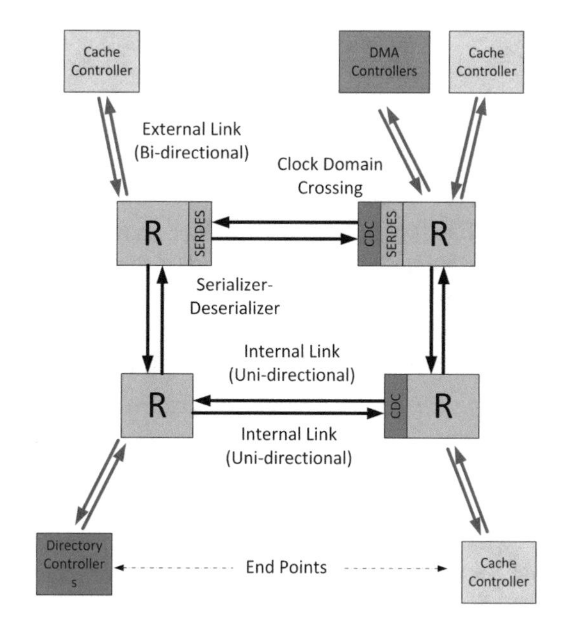

Fig. 2.3 Typical form of a network topology in modern systems. The end points connect to the network through external links. The internal links are then used to connect routers and other units within the system

1. Determine the total number of routers in the system and instantiate them.
   - Use the *Router* class to instantiate individual routers.
   - Configure properties of each router, such as clock domain, supported flit width, depending on the requirements.

   ```
   routers = Router(id, latency, clock_domain,
                    flit_width, supported_vnets,
                    vcs_per_vnet)
   ```

2. Determine the total end points of the interconnect system and instantiate network interface for each of them.
   - Use the *NetworkInterface* class to instantiate individual network interfaces.

   ```
   network_interfaces = NetworkInterface(id)
   ```

3. Connect the routers which connect to the end points (e.g., Cores, Caches, Directories) using external physical interconnects.
   - Use *ExternalLink* class to instantiate the links connecting the end points.
   - Configure properties of each external link, such as clock domain, link width, depending on the requirements.
   - Enable clock domain crossings(CDC) and Serializer-Deserializer(SerDes) units at either depending on the interconnect topology.

   ```
   external_link = ExternalLink(id, latency, clock_domain,
                                flit_width, supported_vnets,
                                serdes_enable, cdc_enable)
   ```

4. Connect the individual routers within the network depending upon the topology.
   - Use *InternalLink* class to instantiate the links connecting the end points.
   - Configure properties of each external link, such as clock domain, link width, depending on the requirements.
   - Enable clock domain crossings and Serializer-Deserializer units at either depending on the interconnect topology.

   ```
   internal_link = InternalLink(id, latency, clock_domain,
                                flit_width, supported_vnets,
                                serdes_enable, cdc_enable)
   ```

Garnet 3.0 also provides several pre-configuration scripts which automatically do some of the steps, such as instantiating network interfaces, domain crossings, and SerDes units. The several types of units used to configure the topologies are discussed below.

2.4.1.1 Physical Links

The physical link model in Garnet represents the interconnect wire itself. A link is a single entity which has its own latency, width, and the types of flit it can transmit. The links also support a credit-based backpressuring mechanism. Similar to the upgraded Garnet 3.0 router, each Garnet 3.0 link can be configured to an operating frequency and width using appropriate parameters. This allows links and routers operating at different frequencies to be connected to each other.

2.4.1.2 Network Interface

The network interface controller (NIC) is an object which sits between the network end points (e.g., Caches, DMA nodes) and the interconnection system. The NIC receives messages from the controllers and converts them into fixed-length flits, short for flow control units. These flits are sized appropriately according to the outgoing physical links. The network interface also governs the flow control and buffer management for the outgoing and incoming flits. Garnet 3.0 allows multiple ports to be attached to a single end point. Thus, the NIC decides where a certain message/flit must be scheduled.

2.4.1.3 Clock Domain Crossing Units

To support multiple clock domains, Garnet 3.0 introduces Clock Domain Crossing (CDC) unit as shown in Fig. 2.4a, which consists of first-In-First-Out (FIFO) buffers and can be instantiated anywhere within the network model. The CDC unit enables architectures with different clock domains across the system. The delay of each CDC unit is configurable. The latency can also be calculated dynamically depending on the clock domains connected to it. This enables accurate modeling of DVFS techniques as CDC latencies are generally a function of the operating frequency of producer and consumer.

2.4.1.4 Serializer-Deserializer Units

Another critical feature necessary in modeling SoCs and heterogeneous architectures is supporting various interconnect widths across the system. Consider a link between two routers within a GPU and a link between a memory controller and on-chip memory. These two links might be of different widths. To enable such configuration, Garnet 3.0 introduces the Serializer-Deserializer unit (Fig. 2.4b), which converts flits into appropriate widths at bit-width boundaries. These SerDes units can be instantiated anywhere in the Garnet 3.0 topology similar to the CDC unit described in the previous subsection.

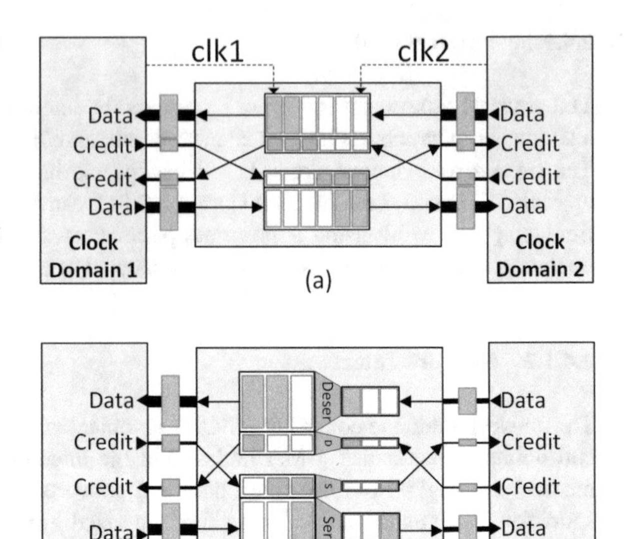

Fig. 2.4 Data and credit flow through a pair of (**a**) Clock Domain Crossing (CDC) units between two different clock domains, and (**b**) Serializer-Deserializer (SerDes) units instantiated between a link width crossing

2.4.2 *Routing*

The routing algorithm decides how the flits travel through the topology. The objective of a routing policy is to minimize contention while maximizing the bandwidth offered by the interconnect. Garnet 3.0 provides several standard routing policies that the user can select from.

Routing Policies There are several generic routing policies that have been proposed for deadlock free routing of flits through the interconnect network.

Table Based Routing Garnet also features table based routing policy which users can select to set custom routing policies using a weight-age based system. Lower weighted links are preferred over links which are configured to have higher weights.

2.4.3 *Flow Control and Buffer Management*

Flow control mechanisms determine the buffer allocation in interconnect systems. The aim of a good flow control system is minimizing the impact of buffer allocation to the overall latency of a message in the system. Implementation of these mechanisms often involve micro-management of physical packets within the interconnect system.

Coherence messages generated by cache controllers are often broken down into fixed-length flits (flow control units). A set of flits carrying a message is often termed as a packet. A packet could have a head-flit, body-flit, and a tail-flit to carry the contents of the message along with any additional meta data of the packet itself. Several flow control techniques have been proposed and implemented at various granularities of resource allocation.

Garnet 3.0 implements a credit-based flit-level flow control mechanism with support for virtual channels.

Virtual Channels Virtual Channels (VCs) in a network act as separate queues which can share physical wires (physical links) between two routers or arbiters. Virtual channels are mainly used to alleviate head-of-line blocking. However, they are also used as a means for deadlock-avoidance.

Buffer Backpressure Most implementations of interconnection networks do not tolerate dropping of packets or flits during traversal. Thus, there is a need to strictly manage the flits using backpressuring mechanisms. Credit-based backpressuring mechanism is often used for low-latency implementation of flit-stalling. Credits track the number of buffers available at the next intermediate destination by decrementing the overall buffers every time a flit is sent. A credit is then sent back by the destination when it is vacated.

2.4.4 Router Microarchitecture

Routers in interconnect systems perform arbitration, allocation of buffers, and flow control within the network. The objective of the router microarchitecture is to minimize the contention within the router while offering minimal per-hop latency for the flits. The complexity of the router microarchitecture also affects the overall energy and area consumption of the interconnect system.

The microarchitecture of the router comprises the logic and state blocks that implement the components described so far. The microarchitecture is similar to the design of a traffic intersection, such as the different lanes (left-only, right-only, etc.), the algorithm running inside the signal to decide when to switch from Red to Green, and so on.

Figure 2.5 shows the microarchitecture of a state-of-the-art NoC router. Each input port has buffers that are organized into separate VCs. Buffers are FIFO queues that can be implemented using Flip Flops or register files or SRAM.

Each input port connects to a crossbar switch which provides cycle-by-cycle non-blocking connectivity from any input port to any output port. A crossbar is fundamentally a mux at every output port. Mux-based crossbars are actually implemented by synthesizing muxes at every output port, while matrix crossbars layout the crossbar as a grid with switching elements at cross-points.

Each input port also houses a route compute unit, an arbiter for the crossbar's input port, and a table tracking the state of each VC. Each output port has an

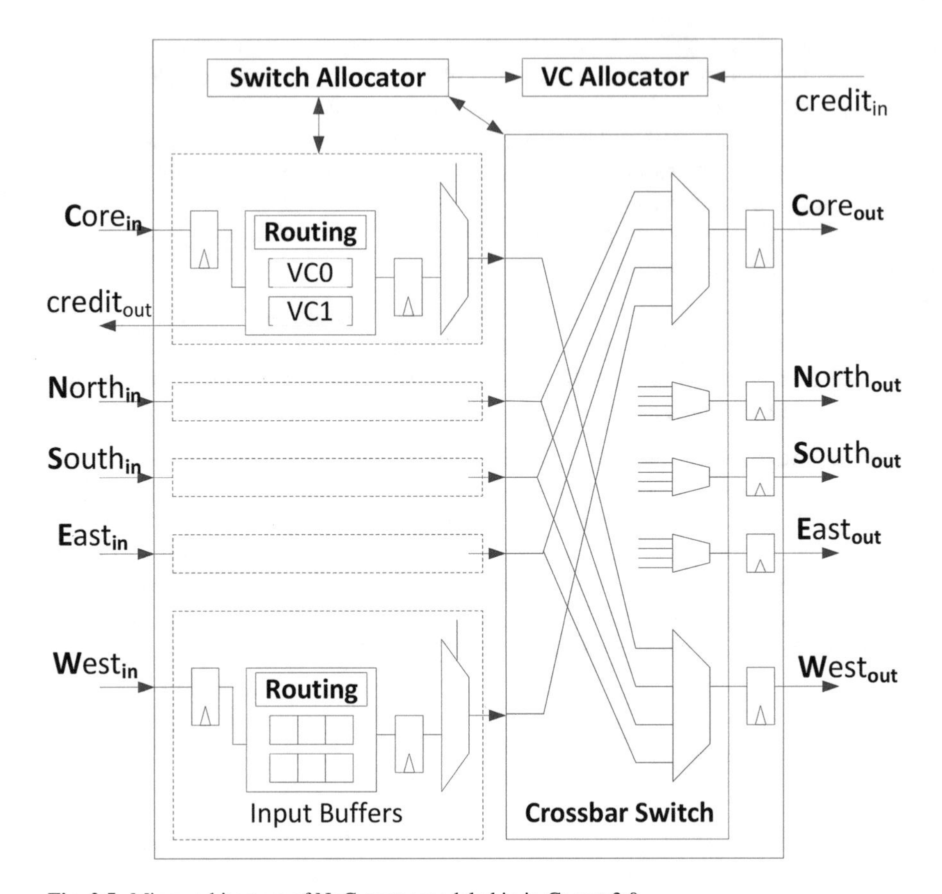

Fig. 2.5 Microarchitecture of NoC router modeled in in Garnet 3.0

arbiter for the crossbar's output port, and also tracks the free VCs and credits at the neighboring router's input port. A $n : 1$ arbiter allows up to n requests for a resource, and grants it to one of them. Each flit that goes through a router needs to perform the following actions on its *control-path*:

- **Route Compute (RC)**. All *head* and *head_tail* flits need to compute their output ports, before they can arbitrate for the crossbar. RC can be performed either by a table lookup or simply by combinational logic. The former is used for complex routing algorithms, while the latter is used for simpler routing schemes like XY which we assume in most of this thesis. To remove RC from the critical path, many NoC routers implement lookahead routing [16] where each flit computes the output port at the next router, instead of the current one so that its output port request is ready as soon as it arrives.
- **Switch Allocation (SA)**. All flits arbitrate for access to the crossbar's input and output ports. For a $n \times n$ router with v VCs per input port, Switch Allocation is fundamentally a matching problem between n resources (output links) and $n \times v$

contenders (total VCs in the router). To simplify the allocator design in order for it to be realizable at a reasonable clock frequency, we often use a separable allocator [31]. The idea is to first arbitrate among the input VCs at each input port using a v : 1 arbiter at every input port, and then arbitrate among the input ports using a n : 1 arbiter at every output port.[4]

- **VC Allocation (VA)**. All flits need a guaranteed VC at the next router before proceeding. VC Allocation is only performed by *head_tail* and *head* flits, while *body* and *tail* flits use the same VC as their *head*. VC Allocation can also be performed in a separable manner [31] like SA. Garnet uses a simpler VA scheme proposed by Kumar *et al.* [21] which we refer to as **VC Select (VS)**. Each output port maintains a queue of VC ids corresponding to the free VCs at the neighbor's input port. The SA winner for that output port gets assigned the VCid at the head of the queue, and the VCid is dequeued. When a VC becomes free at the next router and it sends back a credit, the VCid is enqueued into the queue. If the free VC queue is empty, then flits are not allowed to perform SA.

Once a flit completes RC, SA, and VA, it can proceed to its *data-path*:

- **Switch Traversal (ST)**. Winners of SA traverse the crossbar in this stage. The select lines of the crossbar are set by the grant signals of SA.
- **Link Traversal (LT)**. Flits coming out of the crossbar traverse the link to the next router.
- **Buffer Write (BW)**. Incoming flits are buffered in their VC. While the flit remains buffered, its control-path (RC, SA and VA) is active.
- **Buffer Read (BR)**. Winners of SA are read out of their buffers and sent to the crossbar.

The input and output units of the router model are equipped with a credit system which enables back pressure on their respective producer and consumer routers/NICs as shown in Fig. 2.5. Although the Garnet3.0 router still uses the same four logical stages described above to accurately model contention for buffers and links, it allows configuring the minimum number of cycles spent by a flit inside the router to any value $>= 1$ by setting the router latency parameter. This allows Garnet3.0 to easily simulate simple 1-cycle routers to multi-cycle high-radix routers in the same framework. Moreover, unlike earlier simulators, different routers in a heterogeneous network can be given different latencies by individually setting each router latency in the topology file. This flexibility allows modeling modern router designs and reach higher accuracy. The router can be configured using the parameters in the configuration file.

[4]If u-turns are disallowed, for instance, in minimal routing schemes, the arbiter at the output ports can be $n - 1 : 1$.

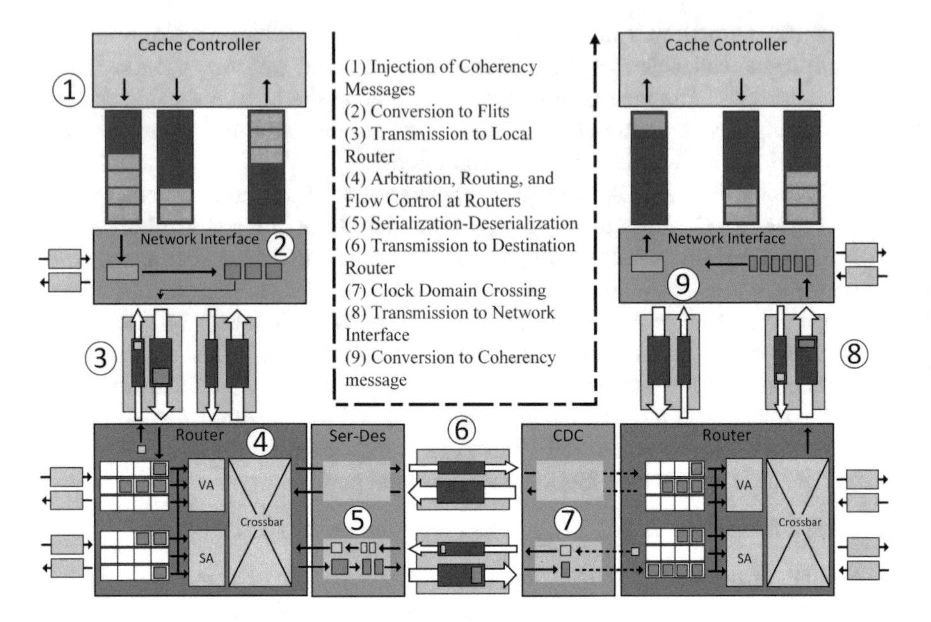

Fig. 2.6 Life of coherency message in Garnet 3.0 simulations

2.4.5 Life of a Message in Garnet 3.0

In this section we describe the life of a message in the NoC after it is generated by a cache controller unit. We take the case of Garnet 3.0 for describing the process, but the general modeling principles can be extended to other software simulation/modeling tools as well.

The overall flow of the system is shown in detail in Fig. 2.6. We take a simple example scenario where a message is generated by a cache controller destined for another cache controller which is connected through routers via physical links, serializer-deserializer units, and clock domain crossings.

Injection of Message The source cache controller creates a message and assigns one or more cache controllers as the destination. This message is then injected into message queues. A cache controller often has several outgoing and incoming message buffers for different kinds of messages.

Conversion to Flits A network interface controller unit (NIC) is attached to each cache controller. This NIC wakes up and consumes the messages from the message queues. Each message is then converted to unicast messages before being broken down into fixed-length flits according to the size supported by the outgoing physical links. These flits are then scheduled for transmission depending on the availability of buffers at the next hop through one of the output links. The outgoing link is chosen depending on the destination, routing policy, and the type of message.

Transmission to Local Router Each network interface is connected to one or more "local" routers which is could be connected through an "External" link. Once a flit is scheduled, it is transmitted over these external links which deliver the flit to the router after a period of defined latency.

Router Arbitration The flit wakes up the router which is a multi-stage unit. The router houses the input buffers, VC allocation, switch arbitration, and crossbar units. On arrival the flit is first placed in an input buffer queue. There are several input buffer queues in a router which contend for an output link and a VC for the next hop. This is done using the VC allocation and switch arbitration stages. Once a flit is selected for transmission, the crossbar stage directs the flit to the output link. A credit is then sent back to the NIC as the input buffer space is vacated for the next flit to arrive.

Serialization-Deserialization The serialization-deserialization (SerDes) is an optional unit that can be enabled depending on the design requirements. The SerDes units consumes the flits and appropriately converts it into outgoing flit size. In addition to manipulating the data packets, the SerDes also handles the credit system, by serializing or deserializing the credit units.

The flit eventually reaches the destination router where a similar router arbitration is performed before being transmitted to the destination network interface. The destination network interface finally forms the message from the flits before queuing it to the destination cache controller.

2.4.6 Area, Power and Energy Model

Frameworks like Orion2.0 [18] and DSENT [27, 33] provide models for the area and power for the various building blocks of a NoC router and links. Garnet integrates DSENT as an external tool to report area, power and energy (which depends on activity) at the end of the simulation.

2.5 NoC RTL Generators

In addition to C++ simulators like Garnet, described above, there also exist various NoC RTL generators provided by commercial vendors and academic researchers for plug-and-play into CMPs and/or MPSoCs. These generators use a library of modularized components to build routers with varying number of ports, data widths, and buffer depths. Some of these provide application-specific synthesis for heterogeneous SoCs, while some generate homogeneous NoCs for multicores with different topology and routing algorithms. Modeling NoC details in RTL and using Cadence/Synopsys/MentorGraphics tools for RTL simulation gives the most accurate network implementation, and the most cycle-accurate timing information,

Table 2.3 State of the art network-on-chip RTL generators

Generator	Language	Target
FlexNoC [5]	Verilog	ASIC
Basejump [36]	Verilog	ASIC
OpenPiton [7]	Verilog	ASIC
PyOCN [34]	PyMTL	ASIC
OpenSMART [23]	BSV	ASIC
CONNECT [30]	BSV	FPGA
OpenSoC [14]	Chisel	ASIC
Ratatoskr [17]	VHDL	ASIC
NoCTegra [35]	VHDL	FPGA

and area/power estimate. The NoC RTL serves as both a model as well as the final design and is a common design-space exploration methodology in industry. Table 2.3 lists some of the state-of-the-art NoC RTL generators in use today.

2.6 Conclusion

In this chapter, we presented details on modeling and simulating modern networks-on-chip—both analytically and in a cycle-accurate manner. Specifically, we described details of the NoC modeling within Garnet, the NoC simulator within gem5, that can model both homogeneous and heterogeneous systems. While the focus of this chapter was on electrical interconnects, the simulation can be extended to also model emerging interconnect technologies like photonics and wireless by changing the link latencies, bandwidths, and arbitration policies.

References

1. P. Abad, P. Prieto, L.G. Menezo, A. Colaso, V. Puente, J. Gregorio, Topaz: an open-source interconnection network simulator for chip multiprocessors and supercomputers, in *2012 IEEE/ACM Sixth International Symposium on Networks-on-Chip*, 2012, pp. 99–106
2. S.V. Adve, K. Gharachorloo, Shared memory consistency models: a tutorial. Computer **29**(12), 66–76 (1996)
3. N. Agarwal, T. Krishna, L. Peh, N.K. Jha, Garnet: a detailed on-chip network model inside a full-system simulator, in *2009 IEEE International Symposium on Performance Analysis of Systems and Software*, 2009, pp. 33–42
4. N. Agarwal, L. Peh, N.K. Jha, In-network snoop ordering (INSO): snoopy coherence on unordered interconnects, in *2009 IEEE 15th International Symposium on High Performance Computer Architecture*, 2009, pp. 67–78.
5. Arteris, Flexnoc. http://www.arteris.com/flexnoc
6. A. Bakhoda, G. Yuan, Wilson W. L. Fung, H. Wong, Tor M. Aamodt, Analyzing CUDA Workloads Using a Detailed GPU Simulator, in IEEE International Symposium on Performance Analysis of Systems and Software (ISPASS), Boston, MA, April 19–21, 2009

7. J. Balkind, M. McKeown, Y. Fu, T.M. Nguyen, Y. Zhou, A. Lavrov, M. Shahrad, A. Fuchs, S. Payne, X. Liang, M. Matl, D. Wentzlaff, Openpiton: an open source manycore research framework, in *Proceedings of the Twenty-First International Conference on Architectural Support for Programming Languages and Operating Systems, ASPLOS '16, Atlanta, GA, USA, April 2–6, 2016* (ACM, New York, 2016), pp. 217–232. https://doi.org/10.1145/2872362.2872414
8. S. Bharadwaj, J. Yin, B. Beckmann, T. Krishna, Kite: a family of heterogeneous interposer topologies enabled via accurate interconnect modeling, in *2020 57th ACM/IEEE Design Automation Conference (DAC)*, 2020, pp. 1–6
9. N. Binkert, B. Beckmann, G. Black, S.K. Reinhardt, A. Saidi, A. Basu, J. Hestness, D.R. Hower, T. Krishna, S. Sardashti, R. Sen, K. Sewell, M. Shoaib, N. Vaish, M.D. Hill, D.A. Wood, The gem5 simulator. SIGARCH Comput. Archit. News **39**(2), 1–7 (2011). https://doi.org/10.1145/2024716.2024718
10. V. Catania, A. Mineo, S. Monteleone, M. Palesi, D. Patti, Noxim: an open, extensible and cycle-accurate network on chip simulator, in *2015 IEEE 26th International Conference on Application-specific Systems, Architectures and Processors (ASAP)*, 2015, pp. 162–163
11. J. Chan, G. Hendry, A. Biberman, K. Bergman, L.P. Carloni, Phoenixsim: a simulator for physical-layer analysis of chip-scale photonic interconnection networks, in *2010 Design, Automation Test in Europe Conference Exhibition (DATE 2010)*, 2010, pp. 691–696
12. P. Conway, N. Kalyanasundharam, G. Donley, K. Lepak, B. Hughes, Cache hierarchy and memory subsystem of the AMD opteron processor, IEEE Micro **30**(2), 16–29 (2010)
13. W.J. Dally, C.L. Seitz, The torus routing chip. Distrib. Comput. **1**(4), 187–196 (1986). https://doi.org/10.1007/BF01660031
14. F. Fatollahi-Fard, D. Donofrio, G. Michelogiannakis, J. Shalf, OpenSoC fabric: on-chip network generator, in *IEEE International Symposium on Performance Analysis of Systems and Software, ISPASS*, 2016, pp. 194–203
15. A. Gupta, W.-D. Weber, T. Mowry, Reducing memory and traffic requirements for scalable directory-based cache coherence schemes, in *Scalable Shared Memory Multiprocessors.* (Springer, Berlin, 1992), pp. 167–192
16. N.D.E. Jerger, T. Krishna, L. Peh, *On-Chip Networks*, 2nd edn., ser. Synthesis Lectures on Computer Architecture (Morgan & Claypool Publishers, 2017)
17. J.M. Joseph, L. Bamberg, I. Hajjar, A. Drewes, B.R. Perjikolaei, A. García-Ortiz, T. Pionteck, in *Ratatoskr: An open-source framework for in-depth power, performance and area analysis in 3d NoCs* (2020)
18. A. Kahng, B. Li, L.-S. Peh, K. Samadi, Orion 2.0: a fast and accurate NoC power and area model for early-stage design space exploration, in *Proceedings of the Conference on Design, Automation and Test in Europe*, April 2009
19. C.N. Keltcher, K.J. McGrath, A. Ahmed, P. Conway, The AMD opteron processor for multiprocessor servers. IEEE Micro **23**(2), 66–76 (2003)
20. M. Khairy, Z. Shen, T.M. Aamodt, T.G. Rogers, Accel-sim: an extensible simulation framework for validated GPU modeling, in *2020 ACM/IEEE 47th Annual International Symposium on Computer Architecture (ISCA)*, 2020, pp. 473–486
21. A. Kumar, P. Kundu, A.P. Singh, L.-S. Peh, N.K. Jha, A 4.6 tbits/s 3.6 GHz single-cycle NoC router with a novel switch allocator in 65 nm CMOs, in *2007 25th International Conference on Computer Design* (IEEE, Piscataway, 2007), pp. 63–70
22. G. Kurian, J.E. Miller, J. Psota, J. Eastep, J. Liu, J. Michel, L.C. Kimerling, A. Agarwal, Atac: a 1000-core cache-coherent processor with on-chip optical network, in *Proceedings of the 19th International Conference on Parallel Architectures and Compilation Techniques, ser. PACT '10* (Association for Computing Machinery, New York, 2010), p. 477–488. https://doi.org/10.1145/1854273.1854332
23. H. Kwon, T. Krishna, OpenSMART: single-cycle multi-hop NoC generator in BSV and Chisel, in *IEEE International Symposium on Performance Analysis of Systems and Software, ISPASS*, 2017

24. J. Lowe-Power, A.M. Ahmad, A. Akram, M. Alian, R. Amslinger, M. Andreozzi, A. Armejach, N. Asmussen, B. Beckmann, S. Bharadwaj, G. Black, G. Bloom, B.R. Bruce, D.R. Carvalho, J. Castrillon, L. Chen, N. Derumigny, S. Diestelhorst, W. Elsasser, C. Escuin, M. Fariborz, A. Farmahini-Farahani, P. Fotouhi, R. Gambord, J. Gandhi, D. Gope, T. Grass, A. Gutierrez, B. Hanindhito, A. Hansson, S. Haria, A. Harris, T. Hayes, A. Herrera, M. Horsnell, S.A.R. Jafri, R. Jagtap, H. Jang, R. Jeyapaul, T.M. Jones, M. Jung, S. Kannoth, H. Khaleghzadeh, Y. Kodama, T. Krishna, T. Marinelli, C. Menard, A. Mondelli, M. Moreto, T. Mück, O. Naji, K. Nathella, H. Nguyen, N. Nikoleris, L.E. Olson, M. Orr, B. Pham, P. Prieto, T. Reddy, A. Roelke, M. Samani, A. Sandberg, J. Setoain, B. Shingarov, M.D. Sinclair, T. Ta, R. Thakur, G. Travaglini, M. Upton, N. Vaish, I. Vougioukas, W. Wang, Z. Wang, N. Wehn, C. Weis, D.A. Wood, H. Yoon, Éder F. Zulian, The gem5 simulator: version 20.0+ (2020)
25. M.M.K. Martin, M.D. Hill, D.A. Wood, Token coherence: decoupling performance and correctness. SIGARCH Comput. Archit. News **31**(2), 182–193 (2003). https://doi.org/10.1145/871656.859640
26. N. McDonald, A. Flores, A. Davis, M. Isaev, J. Kim, D. Gibson, Supersim: extensible flit-level simulation of large-scale interconnection networks, in *2018 IEEE International Symposium on Performance Analysis of Systems and Software (ISPASS)* (IEEE Computer Society, Los Alamitos, 2018), pp. 87–98. https://doi.ieeecomputersociety.org/10.1109/ISPASS.2018.00017
27. MIT-DSENT, MIT-DSENT. https://sites.google.com/site/mitdsent/
28. J. Nan, D.U. Becker, G. Michelogiannakis, J. Balfour, B. Towles, D.E. Shaw, J. Kim, W.J. Dally, A detailed and flexible cycle-accurate network-on-chip simulator, in *2013 IEEE International Symposium on Performance Analysis of Systems and Software (ISPASS)*, 2013, pp. 86–96
29. C. Nitta, K. Macdonald, M.K. Farrens, V. Akella, Inferring packet dependencies to improve trace based simulation of on-chip networks, in *NOCS 2011, Fifth ACM/IEEE International Symposium on Networks-on-Chip, Pittsburgh, Pennsylvania, USA, May 1–4, 2011*, ed. by R. Marculescu, M. Kishinevsky, R. Ginosar, K.S. Chatha (ACM/IEEE Computer Society, 2011), pp. 153–160
30. M.K. Papamichael, J.C. Hoe, CONNECT: re-examining conventional wisdom for designing NoCs in the context of FPGAs, in *Proceedings of the ACM/SIGDA international symposium on Field Programmable Gate Arrays*, 2012, pp. 37–46
31. L.-S. Peh, W.J. Dally, A delay model and speculative architecture for pipelined routers, in *Proceedings HPCA Seventh International Symposium on High-Performance Computer Architecture* (IEEE, Piscataway, 2001), pp. 255–266
32. Y.S. Shao, J. Clemons, R. Venkatesan, B. Zimmer, M. Fojtik, N. Jiang, B. Keller, A. Klinefelter, N.R. Pinckney, P. Raina, S.G. Tell, Y. Zhang, W.J. Dally, J.S. Emer, C.T. Gray, B. Khailany, S.W. Keckler, Simba: scaling deep-learning inference with multi-chip-module-based architecture, in *Proceedings of the 52nd Annual IEEE/ACM International Symposium on Microarchitecture, MICRO 2019, Columbus, OH, USA, October 12–16, 2019* (ACM, New York, 2019), pp. 14–27. https://doi.org/10.1145/3352460.3358302
33. C. Sun, C.-H.O. Chen, G. Kurian, L. Wei, J. Miller, A. Agarwal, L.-S. Peh, V. Stojanovic, DSENT—a tool connecting emerging photonics with electronics for opto-electronic networks-on-chip modeling, in *Proceedings of International Symposium on Networks-on-Chip*, 2012, pp. 201–210
34. C. Tan, Y. Ou, S. Jiang, P. Pan, C. Torng, S. Agwa, C. Batten, Pyocn: a unified framework for modeling, testing, and evaluating on-chip networks, in *2019 IEEE 37th International Conference on Computer Design (ICCD)* (IEEE, Piscataway, 2019), pp. 437–445
35. The NoC system generator. https://noctegra.github.io
36. S. Xie, M.B. Taylor, The basejump manycore accelerator network. abs/1808.00650, 2018. http://arxiv.org/abs/1808.00650

Chapter 3
Energy-Efficient Networks-on-Chip Architectures: Design and Run-Time Optimization

Sumit K. Mandal, Anish Krishnakumar, and Umit Y. Ogras

3.1 Introduction

Diminishing instruction-level parallelism (ILP) and power wall led to the introduction of multicore NoC architectures more than a decade ago and fueled their growth to date [20, 44]. In turn, the growing number of cores has continuously increased the importance of efficient data movement between cores and memory. One can view the communication cost as a "necessary overhead" incurred to move data both within a chip or across multiple chips. Thus, we can evaluate the efficiency of communication architectures as a function of their performance, measured in terms of latency and throughput, versus their cost, measured by their contribution to power and energy consumption. Besides facilitating the design process through reuse and regularity, networks-on-chip architectures have proven their effectiveness with respect to this efficiency metric and become the mainstream communication fabric choice for multicore architectures [81, 93].

The transition from single core to multicore architectures played a pivotal role in satisfying the continuous demand for higher processing power. However, homogeneous multicore architectures have also reached their limits to exploit thread- and data-level parallelism. General-purpose cores facilitate programmability, but their flexibility comes at the expense of energy-efficiency, which is orders of magnitude lower than special-purpose processors, such as video processors and hardware accelerators. Therefore, the heterogeneity of cores increases together with their number to drive the processing power and energy-efficiency [52]. In turn, the growing heterogeneity continues to increase the requirements on the communication architectures. On the one hand, communication latency should be

S. K. Mandal · A. Krishnakumar · U. Y. Ogras (✉)
Department of Electrical and Computer Engineering, University of Wisconsin-Madison, Madison, WI, USA
e-mail: skmandal@wisc.edu; anish.n.krishnakumar@wisc.edu; uogras@wisc.edu

P. Mishra, S. Charles (eds.), *Network-on-Chip Security and Privacy*,
https://doi.org/10.1007/978-3-030-69131-8_3

in the same order as, or ideally lower than, the processing times of special-purpose processors, which could be in the order of nanoseconds. On the other hand, the power consumption overhead should scale down to match those of the special-purpose processors. Otherwise, the communication cost can undermine the benefits of heterogeneous architectures. Hence, the NoCs continue to play a pivotal role in developing processing systems with higher performance and energy-efficiency.

NoC architectures are designed to meet the performance requirements, such as latency, throughput, and quality-of-service (QoS), while minimizing their area, power, and energy overhead. NoCs achieve a higher energy-efficiency than bus and point-to-point communication architectures [56]. The exact contribution of the NoC architecture to the total power consumption is a function of many design parameters, such as the number and types of processing cores, communication bandwidth, and technology. While there is no public data for commercial designs, published academic work shows that NoC power consumption amounts to 18%–36% of the total chip power, as shown in Fig. 3.1. Ghosh et al. [34], Huang et al. [43], and Li et al. [58] report absolute values, while Taylor et al. [94], Kim et al. [49], Mukherjee et al. [69], Hoskote et al. [39], Salihundam et al. [89], Fallin et al. [29], Sodani et al. [93], and Adhinaryanan et al. [3] report the percentage power consumption, which can be useful references for future work.

This chapter focuses on the design and run-time approaches developed to optimize the energy-efficiency of NoC architectures. Design-time approaches of NoCs include optimization in router architecture, routing technique, switching technique, and NoC architecture optimization, as detailed in Sect. 3.2. For example, three-dimensional integrated circuits (3D ICs) have proved several advantages over planar ICs in terms of flexibility in floorplanning, improved packaging density, and reduced power consumption [30]. 3D NoCs were proposed to provide a scalable interconnect solution to 3D ICs. Several energy-efficient design methodologies for 3D NoCs have been proposed. Long-range communications are usually a bottleneck for electrical NoCs. Wireless NoCs address the issue by providing

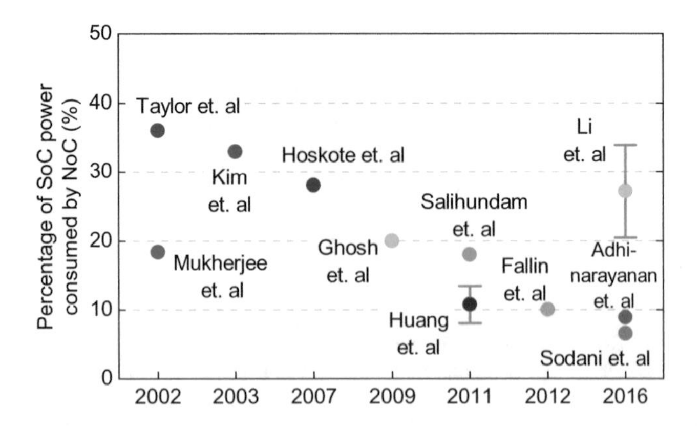

Fig. 3.1 Percentage of the total power consumed by NoC over the past two decades

energy-efficient long-range communications. Multiple researchers explored various energy-efficient design techniques for wireless NoCs. Some of the techniques utilize the advantages of both wired and wireless communication [1, 25]. Optical NoCs provide high bandwidth density and the power consumption of these NoCs do not depend on the distance between the source and destination [38]. Therefore, optical NoCs have emerged as a promising alternative to electrical NoCs. Recently, researchers also proposed optical NoCs for systems with GPUs [6, 111]. Unlike design-time approaches, run-time techniques are applied dynamically in real-time, as discussed in Sect. 3.3. Traditionally, these techniques were limited to congestion- and workload-aware routing as opposed to static routing choices. With the advances in dynamic voltage and frequency scaling, researchers proposed techniques to manage the power states of the NoCs components. For example, the voltage of the whole NoC, as well as last-level cache, can be controlled to save power when performance targets are relaxed [15].

Both academic and industrial work demonstrated the effectiveness of NoCs in chip multiprocessors, embedded systems, and high-performance computing. Thriving deep neural network (DNN) and heterogeneous computing domains indicate that the need for faster and more energy-efficient on-chip communication will continue to grow. Therefore, we conclude this chapter by reviewing the recent use of NoCs in DNN hardware implementations. We note that a complete review of energy-efficient NoC design is beyond the scope of a single chapter. For example, a class of techniques also considered the mapping of tasks to resources to design NoC, both using static and dynamic techniques [41, 87]. This chapter focuses only on NoC architecture design and run-time management techniques. Interested readers can refer to NoC books and comprehensive surveys for other aspects [21, 27, 45, 64], and the remaining chapters of this book for NoC security review.

The rest of this chapter is organized as follows. Section 3.2 reviews the design techniques proposed for energy-efficient NoCs. This section discusses traditional NoCs with wired interconnects, as well as more recent developments on 3D, wireless, and optical NoCs. Section 3.3 presents the run-time techniques that manage the routing decisions and power states of the major NoC components. Section 3.4 analyzes the emerging workload trends and their implications to NoC design. Finally, Sect. 3.5 concludes this chapter and discusses new directions.

3.2 Design Strategies for Energy-Efficient NoCs

The most direct and effective approach towards energy-efficient NoCs is designing an energy-aware architecture that can enable further savings through dynamic management. This section presents static design-time techniques, while Sect. 3.3 discusses run-time techniques that work on top of the static architecture. Energy-aware design starts with the fundamental building blocks, i.e., router architecture, reviewed in Sect. 3.2.1. Section 3.2.2 presents energy-efficient NoC architectures and routing techniques that are built by these routers. As the NoC sizes continue to increase, the performance of traditional 2D designs scale poorly. Therefore, 3D NoC designs have emerged to enable continued scaling, as detailed in Sect. 3.2.3. Another promising and complementary approach to energy-efficient long-range communication is wireless NoCs. Section 3.2.4 reviews recent techniques that combine wired and on-chip wireless communication. Similarly, Sect. 3.2.5 discusses the feasibility and potential of optical on-chip communication to provide throughput and latency significantly larger than electrical networks. Finally, researchers recognized the critical impact of application mapping and scheduling on the system performance and energy-efficiency in both homogeneous and heterogeneous SoCs.

3.2.1 NoC Router Design

Router design has a crucial impact on power consumption and performance since they are the fundamental building blocks of NoCs. On the one hand, fast and simple router architectures are preferred to achieve low latency (in terms of clock cycles) and small area overhead. On the other hand, deeper pipelines and more buffering space are needed for higher clock frequencies and throughput [21, 27]. Similarly, simpler crossbar and arbitration, such as round-robin, reduce the area-power overhead, but they can also severely limit the performance. As a result, the optimal router design must provide an intricate balance between power, performance, energy, and area.

Many techniques have been proposed to improve the router energy-efficiency since the introduction of NoCs [2, 64]. For example, Wang et al. explore the bottlenecks that contribute to power dissipation in interconnection networks [98]. The analysis prompts the authors to devise power-efficient microarchitecture techniques such as a cut-through crossbar, a segmented crossbar, and a write-through buffer. These techniques achieve an NoC power reduction of up to 44.9% and 37.9% with no performance overheads for both synthetic and real application traces, respectively. One of the early key design insights is exploiting application- or domain-specific information to customize the router architecture [42]. Hu et al. use this idea to allocate a given amount of buffer area non-uniformly across the NoC to maximize the performance [42]. Similarly, router architecture with shared buffers can improve both the area utilization and throughput in [84]. Decreasing the router pipeline delay has been an important design consideration. The basic idea is to bypass certain router pipeline stages or the whole router whenever possible, i.e., when there is no contention and packets already waiting in the queue [51, 73]. Towards achieving a single-cycle router latency, Kumar et al. [54] present a non-speculative single-cycle NoC router pipeline targeting 3.6 GHz frequency. A customized mesh NoC that exploits application-specific behavior to introduce single-cycle links between a source and destination is introduced in [14]. These clockless repeater links traverse up to 8 mm in a single cycle at a clock frequency of 2 GHz in a 45 nm process node. The application-specific link configuration introduces complexities in reconfiguring the routers when applications are frequently preempted during execution by the system scheduler. A 2-dimensional mesh NoC coprocessor is designed for data movement in an Epiphany 64-core superscalar processor architecture [96]. Each router uses three communication channels for read requests, on-chip writes, and off-chip write transactions to improve the NoC throughput. Recently, Arm has introduced a scalable low latency, one cycle per-hop, high bandwidth network-on-chip architecture that supports coherency [81]. The NoC can scale from 1×2 mesh to 8×8 mesh to support edge devices and high-performance computing systems. More recent studies focus on non-conventional technologies discussed in Sect. 3.2.3–3.2.5.

3.2.2 NoC Architecture and Packet Routing

The interconnection of the routers defines the NoC architecture. Most of the early work and industrial solutions employ a 2D mesh topology due to its regularity [67, 81, 93]. A regular layout simplifies the floorplan and wiring, while providing predictable delays between the routers. Consequently, regular NoC topologies facilitate energy-efficient design as well as simplified testing and validation [44, 64]. However, both the average and worst-case latency scale poorly with the network size. Since the NoC plays a crucial role in determining the latency, energy, and power, numerous studies evaluate other topologies with smaller diameter, such as hypercube, folded-torus, and fat tree [2, 27]. For example, an extensive power-performance evaluation of mesh and torus topologies on power and performance is presented in [67]. NoC topology can be optimized for a given application domain by either synthesizing a custom topology [72, 82] or altering a regular topology, such as a 2D mesh [73]. Kilo-NoC, a heterogeneous NoC architecture in which only a selected number of routers support QoS requirements is presented in [37]. The heterogeneity in router architecture enables 45% and 29% savings in area and power when compared to a homogeneous NoC that supports QoS in all routers.

Switching, or flow control, techniques determine how the packets are transmitted from their sources to destinations [21, 27, 44]. Due to the large buffer requirements of the packet and virtual cut-through switching techniques, wormhole routing has been a popular choice for NoCs [64]. Virtual channels have commonly been added to address head-of-line blocking and handling multiple traffic classes while avoiding deadlocks. Circuit switching have also been employed to take advantage of high-volume data transfer over persistent connections [60]. While the early work focused on buffered NoCs, bufferless solutions are later employed to minimize the buffer requirements, hence the NoC area and static power [24, 65]. Bufferless NoCs do not store the packets in the intermediate routers. Routers try to forward the packets towards their destinations and then deflect them if their preferred direction is not available [44]. Studies show that bufferless NoC can perform better at low traffic loads, but they suffer from traffic congestion as the traffic load increases [65]. The advantages of buffered and bufferless techniques are combined in [29] to achieve as much as 16% higher energy-efficiency than the state-of-the-art buffered routers.

Intel Xeon-Phi processors [93] use priority-aware bufferless NoCs, which can provide predictable latency within the network [63]. The NoC uses fewer buffers, which in turn helps to reduce energy consumption.

Routing strategies play a crucial role in determining the area, performance, power, and energy of an NoC [21, 27]. For a given NoC topology and switching technique, routing algorithms determine the path taken by each packet while traveling to their destination. The choice of the routing algorithm is important since the length of the path and traffic congestion have a significant impact on both power and performance, while their complexity impacts the area [64]. Specifically, routing techniques influence: (1) end-to-end latency, (2) selection of routing paths, (3) livelock and deadlock avoidance, (4) fault-tolerance, and (5) starvation avoidance. Routing techniques are broadly classified into deterministic and adaptive routing techniques. Deterministic algorithms require fewer resources and are simpler than their adaptive counterpart. Therefore, dimension-ordered routing, such as XY routing, and other deterministic algorithms are used more commonly in industrial and academic designs [9, 93]. At the same time, run-time adaptation to the workload can provide significant benefits [40], as discussed in Sect. 3.3.1. Since the traditional routing techniques are covered in detail in the literature, we refer the reader to existing surveys and books on this topic for a complete taxonomy and review [2, 21, 27, 45, 75].

3.2.3 3D NoC Architectures

As the number of transistors on a single chip increases, three-dimensional integrated circuits (3D ICs) provide more floorplanning flexibility than traditional planar designs [30]. 3D ICs also provide more packaging density since they are not limited to two dimensions [23]. Moreover, 3D ICs can reduce power consumption since they use shorter wire lengths than planar ICs [83]. Since they have emerged as a new technology paradigm due to these advantages, 3D NoCs are employed to interconnect the cores in 3D IC.

Feero et al. [30] present a detailed performance evaluation of 3D NoCs and compare them to traditional 2D NoCs. This work considers 3D mesh, 3D stacked mesh, ciliated 3D mesh, 3D butterfly tree (BFT), and 3D fat-tree topologies. The experimental evaluations show that 3D mesh, ciliated 3D mesh, and stacked mesh consume 42%, 47%, 33% less energy per packet than 2D mesh, respectively, Authors also show that both 3D fat-tree and 3D BFT consume 49% less energy per packet than their 2D counterpart. Similar to the dimension-ordered routing techniques in 2D NoCs, an X–Y–Z static routing can be applied to 3D NoC. For example, Ahmed et al. [4] present a 3D NoC (OASIS-NoC) with wormhole switching and 3-stage router pipeline: routing calculation, switch allocation, and switch traversal. Evaluations using synthetic traffic show that a 2×2×4 3D OASIS-NoC reduces 22% delay on average compared to a 4×4 2D OASIS-NOC. Since a fixed design may not be suitable for different applications, authors in [70] propose a synthesis approach to construct a power-efficient 3D NoC for a given application. Experiments on real applications show that the NoC topologies synthesized by the proposed methodology reduce power by 38% on average. Similarly, a floorplan-aware application-specific 3D NoC synthesis algorithm is proposed in [109]. The authors construct an irregular 3D NoC architecture that meets specific objectives for a given application in this work. The proposed algorithm's input is a directed graph, where each node represents a core, and the edges represent the traffic flow between the cores. A multi-commodity flow (MCF) problem is formulated to optimize multiple objectives. The objectives include network power, average network latency and number of through silicon vias (TSV). Authors incorporate simulated allocation (SAL) to solve the MCF problem. The proposed methodology enables 22% power saving with respect to [70] for a set of synthetic benchmarks [101]. A summary of different 3D NoC technologies, their advantages, and drawbacks can be found in [83].

Several research teams have recently applied machine-learning techniques to design 3D NoCs [22, 46]. For example, Das et al. [22] present a monolithic 3D-enabled energy-efficient NoC. Smaller dimensions of monolithic inter-tier vias provide the scope of high-density integration with reduced wire length compared to TSVs. Experimental evaluations show that the proposed methodology enables 32% lower energy-delay-product (EDP) than mesh-based interconnect and 28% lower EDP than TSV-based interconnect. TSV-based 3D NoCs are prone to failure. Therefore, a Near Field Inductive Coupling (NFIC)-based 3D NoC is proposed in [36]. In this work, convex optimization is used to co-optimize latency, power, and area of the 3D NoC. Experimental evaluation on real benchmarks shows that the proposed NFIC-based 3D NoC is 34.5% more energy-efficient than TSV-based 3D NoC.

Joardar et al. [46] propose a 3D NoC for heterogeneous many-core systems. The authors employ a machine learning-based multi-objective optimization (MOO) to construct 3D NoC, which jointly optimizes latency, throughput, temperature, and energy. The proposed algorithm determines the optimal placement of the CPUs, GPUs, LLCs, and planar links. The authors show that the proposed technique enables 9.6% better EDP than a thermally-optimized 3D NoC design.

3.2.4 Wireless NoC Architectures

The delay and energy-consumption of global on-chip interconnects do not scale down with the gate delay. Wireless NoCs (WiNoCs) emerged to provide energy-efficient long-range communication and extra bandwidth [31]. WiNoCs consist of small on-chip antennas and transceivers that enable efficient communication between remote core within the same chip or package. For example, Ganguly et al. present design methodologies and technology required for scalable WiNoC architectures [32]. The authors perform an extensive evaluation of wireless NoC for various traffic patterns to show the superiority of WiNoC over the existing alternatives. WiNoC consumes 133× less energy per packet than a wired mesh architecture for a 512-core system. Moreover, the authors demonstrate that WiNoC also significantly improves energy consumption compared to 3D NoCs.

A mix of wired links (for short-range) and wireless links (for long-range communication) can combine the benefits of wired and wireless NoCs [1, 25]. Deb et al. [25] divide the target NoC into smaller networks (subnets) with a relatively smaller number of cores connected through wired NoC architectures. Each subnet is connected through wired or wireless links via a centrally located hub. The authors optimize the placement of the wireless interface to maximize performance. The proposed methodology provides up to 8× improvement in energy per bandwidth for a system size of 512 compared to conventional wired mesh NoC. A broadcast-oriented dual-plane wireless NoC architecture, OrthoNoC, is proposed in [1]. Unlike existing hybrid NoCs, OrthoNoC uses two independent network planes: wired and wireless. The wireless plane is customized to perform efficient broadcast, whereas the wired plane is customized to achieve efficient unicast communication. Communication between two planes is performed through network interfaces instead of routers. Experimental evaluation with different traffic patterns shows that OrthoNoC helps to reduce energy consumption significantly compared to other alternatives.

Given the wide range of target applications, wireless NoC need to adapt their performance as a function of the workload. DiTomaso et al. [26] propose an adaptive wireless NoC for chip multiprocessors. The proposed design increases the network throughput by dynamically reassigning the channels depending on bandwidth requirements from different cores. The authors show that the proposed methodology is scalable to a system with 256 cores and provides 21% energy improvement compared to a state-of-the-art baseline. A more detailed discussion about various wireless NoC design methodologies can be found in [97].

3.2.5 Optical NoC Architectures

Optical on-chip communication is a promising alternative to wired interconnects since it can provide high bandwidth density using wavelength division multiplexing. Furthermore, the power consumption of optical NoCs is distance independent [38]. Hence, optical NoCs based on nanophotonics have emerged as an energy-efficient alternative to traditional NoCs.

Authors in [90] show that the integration of photonic NoC can decrease the power consumption significantly compared to electrical NoCs. An optical ring bus NoC architecture proposed in [79] provides 10× improvement in power consumption than an electrical bus-based communication architecture. Vantrease et al. [95] present a 3D many-core architecture, called Corona, which uses nanophotonic communication for both inter-core and off-stack communication. Experimental evaluations show that the power consumption of the proposed NoC can be significantly lower than that of electrical mesh NoC for applications with higher memory demands.

It is important to note that static power contributes a significant portion of the total power in nanophotonics communication [8]. To address this issue, a hybrid NoC architecture with electrical and nanophotonics communication is proposed in [78]. The authors use electrical NoCs for short, local communications and nanophotonics for long communications. Experimental evaluation on synthetic workload shows that the proposed NoC reduces energy consumption over 34% over a purely optical NoC similar to Corona [95]. Another hybrid photonic NoC is proposed in [5]. The authors use a photonic ring to communicate between processor located far apart and memory cores in this work. Other communications are performed through electrical links. The proposed technique results in 13× reduction in power consumption with respect to electrical 2D mesh and torus NoC architectures.

The aforementioned optical NoCs mainly focus on the topology and routing algorithm of the NoC. A bufferless photonic NoC (BLOCON) is proposed in [47]. BLOCON incorporates wormhole routing and does not require virtual channels at the output. The authors also present a scheduling algorithm to resolve contention at the output. BLOCON consumes up to 4× less energy compared to 2D mesh for synthetic traffic.

Recent work has also combined multiple novel technologies to design NoCs. A 3D mesh-based optical NoC proposed in [104] takes advantage of both 3D NoC and optical NoC. In this NoC, optical routers are placed in a single layer. The authors show that the proposed NoC reduces 52% energy consumption compared to 2D mesh-based electrical NoC. Sikder et al. proposed an NoC architecture named Optical-Wireless Network-on-Chip (OWN) in [92]. This proposed architecture combines the advantages of wireless and photonics technologies. Photonics NoC is used within a cluster and one-hop wireless interconnect is used beyond a cluster. The authors show that OWN scales up to 1024-core chip multiprocessors. It consumes 30.6% less energy than wireless architecture.

The techniques mentioned above mainly target an SoC consisting of only CPUs. There exist few techniques which construct optical NoC targeting GPUs [6, 111]. Ziabari et al. proposed a hybrid photonic NoC to improve the performance of the GPUs. The proposed approach replaces *L1–to–L1* transaction into a *L1–to–L2* and a *L2–to–L1* transaction. Photonic components are used to perform these transactions. Authors show that the proposed optical NoC consumes 29% less power than a 2D electrical mesh. Recently, an optical NoC for GPUs has been proposed in [6]. In this work, the authors proposed a laser modulation scheme which reduces static power consumption of the optical NoC. The proposed optical NoC reduces laser power consumption by 67% compared to [111]. Overall, the photonic-silicon integration provides the ability to design high-speed optical NoCs, which are also highly suitable for datacenters [105]. Interested readers can find more research on optical NoC in [7, 80].

3.3 Run-Time Power and Energy Management Techniques

The abundance of computing resources in multicore architectures enable executing many complex applications in parallel. NoC traffic exhibits widely varying behavior both as a function of these applications and nontrivial interactions among them. For instance, motion tracking applications are characterized by periodic processing of several pixels representing each image frame. As an essential shared resource among all applications, NoCs can become the possible cause of their performance degradation. Design-time approaches rely excessively on the expected behavior, which can deviate significantly from the actual state observed during execution. First, it is infeasible to predict the launch and exit times of different applications in a concurrent system. Second, each application itself may exhibit time-varying behavior and uncertainties in its inputs. Therefore, run-time techniques are crucial to allow adaptive and dynamic management and improve the energy-efficiency of the NoC. This section reviews run-time techniques that address adaptive routing, congestion and flow control, and dynamic voltage-frequency scaling in electrical, 3D, wireless, and optical NoCs.

3.3.1 Adaptive Routing Approaches

Routing algorithms can be predominantly classified into oblivious and adaptive techniques. In oblivious routing algorithms, the routing path does not depend on any run-time condition, such as congestion. Deterministic algorithms lead to simpler and faster logic, but they result in poor performance under heavy traffic load, as discussed in Sect. 3.2.2. Adaptive algorithms address this limitation by routing packets through different paths as a function of the dynamic traffic conditions. The probability of congestion is alleviated by using performance counters to provide alternate routing paths [21, 27, 75].

NoCs have traditionally employed partial adaptive routing techniques since fully adaptive routing can cause deadlock and out-of-order packet delivery, besides increasing the hardware cost significantly [21, 45, 64]. The seminal work by Glass et al. [35] lay the foundation of partial adaptive routing by defining how certain turns can be avoided to break cyclic dependencies and avoid deadlocks. The degree of adaptivity is improved by reducing the locations at which turns are restricted, thereby improving the average communication latency for the same congestion scenarios [18]. This foundation enabled a myriad of deadlock-free partial adaptive routing algorithms tailored for NoCs [2]. However, the complexity of partial adaptive routing algorithms is still higher than widely used deterministic algorithms, such as XY routing. Hence, they can have a lower performance at low traffic loads than deterministic algorithms. Adaptive algorithms, such as DyAD [40], combine the advantages of two approaches by using deterministic routing during light loads and switching to adaptive routing when the NoC experiences congestion.

Another class of hybrid approaches combines circuit switching and packet switching techniques. While circuit switching offers lower latency, packet switching is much more efficient in utilizing the resources in the NoC. Typically, NoC traffic is generated by read/write requests from applications, which is harder to predict at run-time. But a request is always followed by a response. This behavior is exploited in [61], where the technique uses packet switching for the request traffic and uses circuit switching by setting up the route path in anticipation of the response traffic to provide lower latency. This technique improves the performance of a 64-core chip multiprocessor by 15% compared to conventional packet-switched mesh NoCs. Similarly, Liu et al. [60] present an interesting analysis to determine the cross-over point in the traffic flow below which packet switching offers lower latency. The key challenges in hybrid architectures can be classified into design-time and run-time challenges. Design-time challenges include identifying the optimal distribution of NoC activity between the two techniques, and run-time challenges include determining the best point to switch between the different techniques.

3D NoCs also benefit from run-time routing techniques to avoid degradation under congestion. For example, DyXYZ [28] algorithm uses information from input buffers of neighboring routers as a measure of congestion and performs fully adaptive routing for 3D NoCs. This technique uses virtual channels for all three dimensions, eliminates deadlocks, and achieves better latency under high traffic injection. Considering routers beyond the neighbors can provide better congestion estimation and higher performance, but it also increases the algorithm and implementation complexity. Technological restrictions limit the bandwidth on the vertical links of 3D NoCs, which are constructed by through silicon vias (TSVs). TSVs involve a larger area overhead, higher design cost and are also prone to defects and failures at run-time. By limiting non-minimal routing in the vertical direction, a deadlock-aware routing strategy that achieves 50% higher throughputs than adaptive-XYZ routing is presented in [110]. The area overhead of TSVs inspired the design of partially connected 3D NoCs. To address the reliability concerns, a resilient routing algorithm that guarantees packet delivery with at least one healthy vertical link on the east-most column is presented in [88]. Rout3D relaxes the above requirement by requiring at least one functional vertical link anywhere in the network[13]. Adaptive routing techniques play a larger role in 3D NoCs since they address the crucial reliability aspect, apart from improving latency and throughput.

Wireless NoCs leverage wireless links to facilitate long-range communication in the NoC. However, such architectures face congestion threats as several packets can swamp the wireless router. Therefore, adaptive routing mechanisms play a crucial role in wireless NoCs in alleviating congestion, which is more probable due to wireless links. A routing technique that considers the input buffer utilization to calculate the routing path is presented in [99]. The authors also ensure that the routing algorithm is deadlock-free by including virtual channels in the architecture. The CPCA algorithm partitions the wireless NoC into sub-networks and transmits congestion information of the wireless router along paths within the sub-networks [77]. The traffic information enables a near-optimal selection of wired and wireless networks for packet routing. CPCA achieves up to 25% improvement in throughput over existing routing techniques for different traffic patterns.

Silicon photonic technologies, hence Optical NoCs, are highly sensitive to temperature since thermo-optic coupling causes shifts in optical wavelengths and leads to signal loss. Hence, maintaining the temperature within the allowable limits is of paramount importance in optical interconnection networks. Since adaptive routing techniques can acclimatize to dynamic variations, they are highly suitable for developing thermal-aware routing algorithms. Aurora enables a reliable optical NoC by utilizing a cross-layer approach that spans across architecture, operating system, and device [59]. Specifically, it routes packets through colder regions of the chip towards the destination. The multi-layer approach improves the bit error rate by 96% and achieves 37% better power efficiency over traditional optical NoCs. A Q-learning technique that incorporates the temperature gradient of the chip presents a thermal-aware adaptive routing algorithm for optical NoCs [106]. The approach leverages reinforcement learning to predict route paths that cause minimal thermally-induced optical power loss. The authors report up to 36% and 19% reduction in optical power loss for synthetic and real applications, respectively. Hybrid photonic-electric NoC is composed of both electrical and photonic interconnect components. The availability of two modalities of packet routing and on-chip thermal conditions are exploited by TAFT, a thermal-aware fault-tolerating adaptive routing technique for hybrid NoCs [102]. TAFT achieves a bit error rate of 10^{-11} and an improved power efficiency of 30% compared to traditionally designed hybrid NoCs. Adaptive routing plays a pivotal role in designing reliable and power-efficient optical NoCs, as comprehensively discussed in this section.

3.3.2 Run-Time Flow Control and Source Throttling Techniques

Adaptive routing techniques use non-minimal routes to alleviate congestion by efficiently distributing the packets in the network. However, routing alone does not address the root cause of the problem if the traffic sources can continue injecting packets regardless of the congestion level. In fact, even a subset of active cores can congest the entire network leading to increased latency and power consumption. Therefore, congestion control mechanisms (a.k.a. flow control) manage the traffic sources and traffic congestion by regulating packet injection into the network [74]. For example, all of a subset of the cores can be throttled until the congestion clears. Even if the packets are delayed temporarily during this time, the overall performance and energy-efficiency improves since the messages experience much shorter queuing delays overall [11].

In the source throttling mechanism, a perennial challenge is to effectively evaluate the congestion status (local congestion vs. global congestion) in the NoC and communicate it to all routers with low latency and energy. A predictive closed-loop flow control technique for 2D mesh NoCs is presented in [74]. The proposed approach uses router and traffic source models to control packet injection into the network and avoid congestion. This source throttling mechanism incurs low overhead since it utilizes information only from the neighboring routers and achieves better average latency for synthetic and real benchmarks. However, excessive source throttling leads to starvation in some sources and thereby, reduced throughput. Additionally, throttling latency-sensitive applications can lead to adverse effects. Chang et al. [11] present an application-aware throttling mechanism, called HAT, that adjusts the throttling rate based on the load in the network. The energy-efficiency improves by up to 14.7% as packets move through the NoC with lower latency as a function of reduced congestion. In [100], the authors present a throttling technique with bounded wait times for packets in the source queues for the Hoplite NoC, making it highly suitable for real-time applications. Implementing the proposed NoC on FPGA demonstrates that it conforms to the upper bounds for synthetic and real applications. A reinforcement learning technique enables each router to learn its throttling rate by using global starvation indicators and achieve fairness across all sources [24].

The smaller footprint and diversified thermal conductance of components of 3D NoCs exacerbate the need for run-time thermal management techniques. To this end, Chen et al. [16] proposed a proactive thermal management technique that throttles the sources using temperature sensing and prediction models. Aggressive throttling adversely impacts the throughput of NoC architectures. The power density in 3D NoCs is higher than their 2D counterparts due to the vertical stacking of dies, making them more prone to temperature variations. Considering the heat transfer characteristics in a 3D NoC, the throttling scheme proposed in [12] reduces average throttling time by 70% and the throttling ratio by around 9–15% under network congestion.

3.3.3 Voltage-Frequency Scaling

As the size and complexity of NoC architectures increase to handle large-sized multicore architectures, they contribute to a substantial portion of the total chip power and energy, as shown in Fig. 3.1. In some cases, the energy to move data in the chip has even exceeded the energy spent for computations [15, 58]. Dynamic voltage and frequency scaling (DVFS) is a popular and effective technique to reduce the dynamic power consumption. DVFS has shown extensive benefits in improving the scalability of NoC architectures, as we comprehensively discuss in this section.

The benefits of global DVFS diminish with growing NoC sizes, and hence fine-grained power management becomes crucial and gains importance [33]. The globally asynchronous locally synchronous (GALS) design paradigm enables local clock generation for smaller logic in the design, making it highly suitable to be integrated with fine-grained DVFS control [10]. A larger number of voltage-frequency islands (VFI) enable a higher degree of freedom, but it also increases the design cost and area overhead. Therefore, it is essential to determine the optimal number of voltage-frequency domains and minimize energy consumption [76]. The authors show that optimal VFI design followed by a run-time energy management approach can minimize energy consumption while meeting the performance goals. On the contrary, a bottom-up approach is proposed in [103], where communicating threads seek voltage and frequency requirements based on their run-time characteristics by voting. The votes are sent to a local DVFS controller that controls the voltage and frequency levels for a subset of routers in the network. As asserted in prior sections, the NoC traffic is a function of the applications. The traffic also depends on the type of computing resources. For instance, CPU traffic typically benefits from low latency in the NoC, and GPU traffic prefers high bandwidth. An adaptive and reconfigurable design for an integrated CPU-GPU system that optimizes for both types of traffic is presented in BiNoCHS [66]. It provides a high-frequency, low hop-count mode suitable for CPU traffic and a higher virtual channel count with more routers and links suitable for GPU traffic. The CPU performance improves by 28% with predominantly CPU-type traffic, whereas the CPU and GPU performance improve by 57% and 34% under congested conditions.

Growing NoC sizes also lead to an increased number of parameters to be monitored for optimal DVFS performance. This challenge inspired the development of a reinforcement learning (RL) based DVFS technique, in which a router learns to choose the optimal voltage and frequency state based on interactions with system parameters [107]. Compared to a baseline implementation, developing optimal router control policies using RL achieve 26% lower power consumption along with 7% improved performance for applications in the PARSEC benchmark suite. Reinforcement learning suffers from excessive time overheads to learn the optimal decisions [107], rendering the convergence time problem a key challenge for future work.

The emergence of 3D NoCs has enabled smaller chip footprints compared to conventional 2D mesh NoCs. Moreover, compact chips and newer technology increase the power density and lead to thermal hotspots. A VFI technique that uses performance and thermal profiles to alleviate thermal hotspots and reduce performance degradation is presented in [57]. The worst-case temperature is reduced by 15.2% with the deployment of VFI based partitioning. Aging in multicore circuits leads to degradation in supply voltage, thereby causing gradual circuit slowdown. ARTEMIS presents an aging-aware voltage scaling technique for 3D NoC based multiprocessor systems to reduce the stress on the power delivery network [85]. Such techniques play a crucial role in improving circuit reliability and lifetime.

The DVFS strategies also extend to wireless NoCs in [71], in which link utilization and bandwidth are estimated to scale up and down the voltage and frequency levels. In [50], a machine learning-based technique exploits the variations in the workloads to determine the optimal voltage-frequency clusters and levels. This approach achieves 38.9% lower energy-delay product (EDP) than non-VFI based wireless NoCs for the evaluated benchmarks. By utilizing a model to estimate the router utilization, the technique presented in [68] scales the supply voltage of routers and wireless interfaces to improve the power consumption. With negligible performance and area overheads, the voltage scaling approach reduces the power consumption by 56% in the network and up to 62.5% in the wireless interfaces.

The study of energy-efficient algorithms for optical NoCs is in nascent stages [80]. However, without loss of generality, the core ideas of the techniques proposed for electrical, 3D, and wireless NoCs apply to optical NoCs. PROBE, proposed in [108], reduces the power of lasers in an optical NoC by reducing the voltage as a function of expected bandwidth. The proposed approach saves 68% optical power with 12% performance overheads. While the power overhead of voltage regulators is within 3%, the performance overheads are significant, opening up opportunities for energy-efficient techniques.

3.4 NoCs for Deep Neural Networks

Deep neural networks (DNNs) have recently shown strong potential in many fields ranging from autonomous driving to edge computing [53, 62, 86]. Figure 3.2 shows that different DNNs proposed in recent years consist of more than 50 million trainable parameters. The training and inference processes of DNNs consist of operations distributed across multiple layers. These operations involve a significant amount of data communication between processing element and memory. Communication overhead itself contributes up to 90% of total latency and 20–40% of total energy for DNN hardware for a wide range of DNNs [53]. Therefore, an energy-efficient communication strategy is needed to ensure peak performance for the DNN hardware.

Stringent performance and energy-efficiency requirements of target applications motivate hardware implementations for DNNs and CNNs [53, 91]. Therefore, energy-efficient NoCs for DNN/CNN hardware have attracted attention. For instance, Shafiee et al. [91] present an NoC with c-mesh topology for DNN hardware. The authors consider in-memory computing (IMC)-based DNN, where the computation is distributed in multiple tiles connected by an NoC with c-mesh

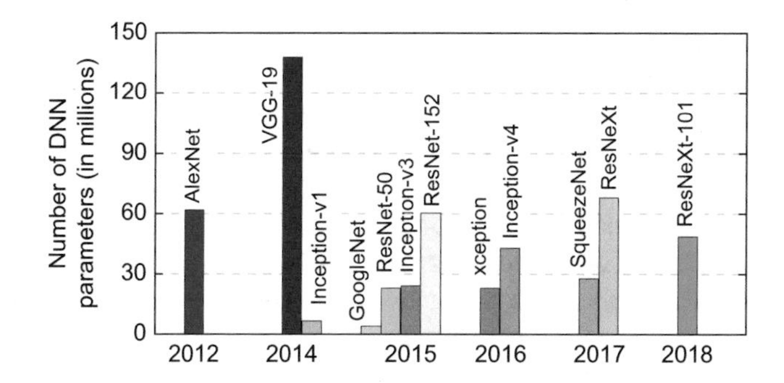

Fig. 3.2 Number of parameters for different DNNs [48, 53]

topology. A recent work proposes a programmable NoC with fat-tree topology to provide flexibility in NoC architecture [55]. The programmability helps to incorporate variations in convolutions, recurrent layers, irregular filter sizes, and sparsity of different DNNs. Chen et al. [17] also propose a flexible hierarchical NoC with mesh topology for DNN hardware. The proposed architecture efficiently reuses data for different data-flow patterns. However, these architectures do not target to reduce NoC energy specifically. An energy-efficient mesh NoC for DNN hardware is presented in [53]. The authors first propose a multi-objective optimization-based technique to determine the size of the mesh NoC for a given DNN. Then, they design a scheduling technique to optimally schedule the output activations for different layers of a given DNN. These two optimizations together result in a 74% reduction in NoC energy on average compared to other DNN hardware previously proposed. A reconfigurable NoC architecture is proposed in [62] for DNN hardware to minimize the communication energy. In this work, the links between the routers are customized for different DNNs. As a result, the technique achieves minimum possible communication latency and reduces power consumption by up to 47% compared to a regular mesh interconnect.

While the studies discussed above target hardware accelerators for DNNs, a different line of work considers efficient NoCs for manycore systems to facilitate the computations related to DNNs [19]. In this work, the authors propose a heterogeneous system with hybrid NoC to accelerate the DNN training time. The hybrid NoC is a combination of wired and wireless links which facilitates communication between CPU and GPU. Authors show that the proposed architecture provides up to 30% improvement in EDP compared to mesh NoC with long-range links.

3.5 Conclusion

This chapter presented an overview of energy-efficient NoC architectures, as well as design and run-time optimization techniques. As the computing systems and applications become more complex, efficient data transfer between processing elements and memory becomes extremely important. On-chip communication architecture must provide low latency and high throughput while minimizing the power and energy consumption overheads. We broadly analyzed the energy-efficient techniques into three groups: design-time strategies, run-time techniques, and techniques incorporated in modern workload trends. We discussed various design-time and run-time energy-efficient techniques for electrical, wireless, 3D, and optical NoCs. Finally, this chapter discussed the communication needs of artificial intelligence (AI) applications. The trends in network and parameter sizes indicate that NoC architectures will continue playing an essential role in energy-efficient communication for AI hardware.

References

1. S. Abadal, J. Torrellas, E. Alarcón, A. Cabellos-Aparicio, OrthoNoC: a broadcast-oriented dual-plane wireless network-on-chip architecture. IEEE Trans. Parall. Distrib. Syst. **29**(3), 628–641 (2017)
2. A. Abbas et al., A survey on energy-efficient methodologies and architectures of network-on-chip. Comput. Electr. Eng. **40**(8), 333–347 (2014)
3. V. Adhinarayanan et al., Measuring and modeling on-chip interconnect power on real hardware, in *IEEE International Symposium on Workload Characterization*, pp. 1–11
4. A.B. Ahmed, A.B. Abdallah, K. Kuroda, Architecture and design of efficient 3D network-on-chip (3D NoC) for custom multicore SoC, in *International Conference on Broadband, Wireless Computing, Communication and Applications* (2010), pp. 67–73
5. S. Bahirat, S. Pasricha, Exploring hybrid photonic networks-on-chip for emerging chip multiprocessors, in *Proceedings of International Conference on Hardware/Software Codesign and System Synthesis* (2009), pp. 129–136
6. J. Bashir, S.R. Sarangi, GPUOPT: power-efficient photonic network-on-chip for a scalable GPU. ACM J. Emerg. Technol. Comput. Syst. **17**(1), 1–26 (2020)
7. J. Bashir, E. Peter, S.R. Sarangi, A survey of on-chip optical interconnects. ACM Comput. Surv. **51**(6), 1–34 (2019)
8. C. Batten et al., Building many-core processor-to-DRAM networks with monolithic CMOS silicon photonics. IEEE Micro **29**(4), 8–21 (2009)
9. S. Bell et al., Tile64-processor: a 64-core SoC with mesh interconnect, in *International Solid-State Circuits Conference-Digest of Technical Papers* (2008), pp. 88–598
10. P. Bogdan, R. Marculescu, S. Jain, R.T. Gavila, An optimal control approach to power management for multi-voltage and frequency islands multiprocessor platforms under highly variable workloads, in *International Symposium on Networks-on-Chip* (2012), pp. 35–42
11. K.K.-W. Chang, R. Ausavarungnirun, C. Fallin, O. Mutlu, HAT: heterogeneous adaptive throttling for on-chip networks, in *International Symposium on Computer Architecture and High Performance Computing* (2012), pp. 9–18
12. C.-H. Chao et al., Traffic-and thermal-aware run-time thermal management scheme for 3D NoC systems, in *Proceedings of International Symposium. on Networks-on-Chip* (2010), pp. 223–230
13. A. Charif, N.-E. Zergainoh, A. Coelho, M. Nicolaidis, Rout3d: a light-weight adaptive routing algorithm for tolerating faulty vertical links in 3D-NoCs, in *IEEE European Test Symposium* (2017), pp. 1–6
14. C.-H. O. Chen et al., SMART: a single-cycle reconfigurable NoC for SoC applications, in *Design, Automation & Test in Europe Conference & Exhibition* (2013), pp. 338–343
15. X. Chen et al., In-network monitoring and control policy for DVFS of CMP networks-on-chip and last level caches. ACM Trans. Des. Autom. Electron. Syst. **18**(4), 1–21 (2013)
16. K.-C. Chen, E.-J. Chang, H.-T Li, A.-Y A. Wu, RC-based temperature prediction scheme for proactive dynamic thermal management in throttle-based 3D NoCs. IEEE Trans. Parall. Distrib. Syst. **26**(1), 206–218 (2014)
17. Y.-H. Chen, T.-J. Yang, J. Emer, V. Sze, Eyeriss v2: a flexible accelerator for emerging deep neural networks on mobile devices. IEEE J. Emerg. Select. Top. Circ. Syst. **9**(2), 292–308 (2019)
18. G.-M. Chiu, The odd-even turn model for adaptive routing. IEEE Trans. Parall. Distrib. Syst. **11**(7), 729–738 (2000)
19. W. Choi et al., On-chip communication network for efficient training of deep convolutional networks on heterogeneous manycore systems. IEEE Trans. Comput. **67**(5), 672–686 (2017)
20. W.J. Dally, B. Towles, Route packets, not wires: on-chip interconnection networks, in *Proceedings of the Design Automation Conference* (2001), pp. 684–689
21. W.J. Dally, B.P. Towles, *Principles and Practices of Interconnection Networks* (Elsevier, Amsterdam, 2004)

22. S. Das, J.R. Doppa, P.P. Pande, K. Chakrabarty, Monolithic 3D-enabled high performance and energy efficient network-on-chip, in *Proceedings of the International Conference on Computer Design* (2017), pp. 233–240
23. W.R. Davis et al., Demystifying 3D ICs: the Pros and Cons of going vertical. IEEE Des. Test Comput. **22**(6), 498–510 (2005)
24. B.K. Daya, L.-S. Peh, A.P. Chandrakasan, Quest for high-performance bufferless NOCs with single-cycle express paths and self-learning throttling, in *Proceedings of Design Automation Conference* (2016), pp. 1–6
25. S. Deb et al., Wireless NoC as interconnection backbone for multicore chips: promises and challenges. IEEE J. Emerg. Select. Top. Circ. Syst. **2**(2), 228–239 (2012)
26. D. DiTomaso et al., AWiNoC: adaptive wireless network-on-chip architecture for chip multiprocessors. IEEE Trans. Parall. Distrib. Syst. **26**(12), 3289–3302 (2014)
27. J. Duato S. Yalamanchili, L. Ni, *Interconnection Networks: An Engineering Approach* (M. Kaufmann Publishers Inc., Burlington, 2002)
28. M. Ebrahimi et al., DyXYZ: fully adaptive routing algorithm for 3D NoCs, in *Euromicro International Conference on Parallel, Distributed, and Network-Based Processing* (2013), pp. 499–503
29. C. Fallin et al., MinBD: minimally-buffered deflection routing for energy-efficient interconnect, in *International Symposium on Networks-on-Chip* (2012), pp. 1–10
30. B.S. Feero, P.P. Pande, Networks-on-chip in a three-dimensional environment: a performance evaluation. IEEE Trans. Comput. **58**(1), 32–45 (2008)
31. B.A. Floyd, C.-M. Hung, K.K. O, Intra-chip wireless interconnect for clock distribution implemented with integrated antennas, receivers, and transmitters. IEEE J. Solid-State Circ. **37**(5), 543–552 (2002)
32. A. Ganguly et al., Scalable hybrid wireless network-on-chip architectures for multicore systems. IEEE Trans. Comput. **60**(10), 1485–1502 (2010)
33. S. Garg, D. Marculescu, R. Marculescu, U. Ogras, Technology-driven limits on DVFS controllability of multiple voltage-frequency island designs: a system-level perspective, in *Proceedings of Design Automation Conference* (2009), pp. 818–821
34. P. Ghosh, A. Sen, A. Hall, Energy efficient application mapping to NoC processing elements operating at multiple voltage levels, in *International Symposium on Networks-on-Chip* (2009), pp. 80–85
35. C.J. Glass, L.M. Ni, The turn model for adaptive routing. ACM SIGARCH Comput. Archit. News **20**(2), 278–287 (1992)
36. S. Gopal, S. Das, D. Heo, P.P. Pande, Energy and area efficient near field inductive coupling: a case study on 3D NoC, in *Proceedings of IEEE/ACM International Symposium on Networks-on-Chip* (2017), pp. 1–8
37. B. Grot, J. Hestness, S.W. Keckler, O. Mutlu, Kilo-NOC: a heterogeneous network-on-chip architecture for scalability and service guarantees, in *International Symposium on Computer Architecture* (2011), pp. 401–412
38. M. Haurylau et al., On-chip optical interconnect roadmap: challenges and critical directions. IEEE J. Select. Top. Quant. Electron. **12**(6), 1699–1705 (2006)
39. Y. Hoskote et al., A 5-GHz mesh interconnect for a teraflops processor. IEEE Micro **27**(5), 51–61 (2007)
40. J. Hu, R. Marculescu, DyAD: smart routing for networks-on-chip, in *Proceedings of Design Automation Conference* (2004), pp. 260–263
41. J. Hu, R. Marculescu, Energy-and performance-aware mapping for regular NoC architectures. IEEE Trans. Comput.-Aided Des. Integr. Circ. Syst. **24**(4), 551–562 (2005)
42. J. Hu, U.Y. Ogras, R. Marculescu, System-level buffer allocation for application-specific networks-on-chip router design. IEEE Trans. Comput.-Aided Des. Integr. Circ. Syst. **25**(12), 2919–2933 (2006)
43. J. Huang, C. Buckl, A. Raabe, A. Knoll, Energy-aware task allocation for network-on-chip based heterogeneous multiprocessor systems, in *International Euromicro Conference on Parallel, Distributed and Network-Based Processing* (2011), pp. 447–454

44. A. Jantsch, H. Tenhunen, *Networks on Chip*, vol. 396 (Springer, New York, 2003)
45. N.E. Jerger, T. Krishna, L.-S. Peh, On-chip networks. Synth. Lect. Comput. Archit. **12**(3), 1–210 (2017)
46. B.K. Joardar et al., Learning-based application-agnostic 3D NoC design for heterogeneous manycore systems. IEEE Trans. Comput. **68**(6), 852–866 (2018)
47. Y.H. Kao, H.J. Chao, BLOCON: a bufferless photonic Clos network-on-chip architecture, in *Proceedings of International Symposium on Networks-on-Chip* (2011), pp. 81–88
48. A. Khan, A. Sohail, U. Zahoora, A.S. Qureshi, A survey of the recent architectures of deep convolutional neural networks. Artif. Intell. Rev. **53**(8), 5455–5516 (2020)
49. J.S. Kim, M.B. Taylor, J. Miller, D. Wentzlaff, Energy characterization of a tiled architecture processor with on-chip networks, in *Proceedings of International Symposium on Low Power Electronics and Design* (2003), pp. 424–427
50. R.G. Kim et al., Wireless NoC and dynamic VFI codesign: energy efficiency without performance penalty. IEEE Trans. Very Large Scale Integr. Syst. **24**(7), 2488–2501 (2016)
51. T. Krishna et al., NoC with near-ideal express virtual channels using global-line communication, in *IEEE Symposium on High Performance Interconnects* (2008), pp. 11–20
52. A. Krishnakumar et al., Runtime task scheduling using imitation learning for heterogeneous many-core systems. IEEE Trans. Comput.-Aided Des. Integr. Circ. Syst. **39**(11), 4064–4077 (2020)
53. G. Krishnan et al., Interconnect-aware area and energy optimization for in-memory acceleration of DNNs, in *IEEE Design & Test* (2020)
54. A. Kumary et al., A 4.6 Tbits/s 3.6 GHz single-cycle NoC Router with a novel switch allocator in 65 nm CMOS, in *International Conference on Computer Design* (2007), pp. 63–70
55. H. Kwon, A. Samajdar, T. Krishna, MAERI: enabling flexible dataflow mapping over DNN accelerators via reconfigurable interconnects. ACM SIGPLAN Not. **53**(2), 461–475 (2018)
56. H.G. Lee, N. Chang, U.Y. Ogras, R. Marculescu, On-chip communication architecture exploration: a quantitative evaluation of point-to-point, bus, and network-on-chip approaches. ACM Trans. Des. Autom. Electron. Syst. **12**(3), 1–20 (2008)
57. D. Lee, S. Das, P.P. Pande, Analyzing power-thermal-performance trade-offs in a high-performance 3D NoC architecture. Integration **65**, 282–292 (2019)
58. D. Li, J. Wu, Energy-efficient contention-aware application mapping and scheduling on NoC-based MPSoCs. J. Parall. Distrib. Comput. **96**, 1–11 (2016)
59. Z. Li et al., Aurora: a cross-layer solution for thermally resilient photonic network-on-chip. IEEE Trans. Very Large Scale Integr. Syst. **23**(1), 170–183 (2014)
60. S. Liu, A. Jantsch, Z. Lu, Analysis and evaluation of design trade-offs between circuit switched NoC and packet switched NoC, in *Euromicro Conference on Digital System Design* (2013), pp. 21–28
61. P. Lotfi-Kamran, M. Modarressi, H. Sarbazi-Azad, An efficient hybrid-switched network-on-chip for chip multiprocessors. IEEE Trans. Comput. **65**(5), 1656–1662 (2015)
62. S.K. Mandal et al., A latency-optimized reconfigurable NoC for in-memory acceleration of DNNs. IEEE J. Emerg. Select. Top. Circ. Syst. **10**(3), 362–375 (2020)
63. S.K. Mandal et al., Analytical performance modeling of NoCs under priority arbitration and bursty traffic, in *IEEE Embedded Systems Letters* (2020)
64. R. Marculescu et al., Outstanding research problems in NoC design: system, microarchitecture and circuit perspectives. IEEE Trans. Comput.-Aided Des. Integr. Circ. Syst. **28**(1), 3–21 (2008)
65. G. Michelogiannakis, D. Sanchez, W.J. Dally, C. Kozyrakis, Evaluating bufferless flow control for on-chip networks, in *International Symposium on Networks-on-Chip* (2010), pp. 9–16
66. A. Mirhosseini et al., BiNoCHS: bimodal network-on-chip for CPU-GPU heterogeneous systems, in *Proceedings of International Symposium on Networks-on-Chip* (2017), pp. 1–8
67. M. Mirza-Aghatabar, S. Koohi, S. Hessabi, M. Pedram, An empirical investigation of mesh and torus NoC topologies under different routing algorithms and traffic models, in *Euromicro Conference on Digital System Design Architectures, Methods and Tools* (2007), pp. 19–26

68. H.K. Mondal, S.H. Gade, S. Kaushik, S. Deb, Adaptive multi-voltage scaling with utilization prediction for energy-efficient wireless NoC. IEEE Trans. Sustain. Comput. **2**(4), 382–395 (2017)
69. S.S. Mukherjee et al., The alpha 21364 network architecture. IEEE Micro **22**(1), 26–35 (2002)
70. S. Murali, C. Seiculescu, L. Benini, G. De Micheli, Synthesis of networks on chips for 3D systems on chips, in *Proceedings of Asia and South Pacific Design Automation Conference* (2009), pp. 242–247
71. J. Murray, P.P. Pande, B. Shirazi, DVFS-enabled sustainable wireless NOC architecture, in *IEEE International SOC Conference* (2012), pp. 301–306
72. U.Y. Ogras, R. Marculescu, Energy-and performance-driven NoC communication architecture synthesis using a decomposition approach, in *Proceedings of Design, Automation and Test in Europe* (2005), pp. 352–357
73. U.Y. Ogras, R. Marculescu, It's a small world after all: NoC performance optimization via long-range link insertion. IEEE Trans. Very Large Scale Integr. Syst. **14**(7), 693–706 (2006)
74. U.Y. Ogras, R. Marculescu, Analysis and optimization of prediction-based flow control in networks-on-chip. ACM Trans. Des. Autom. Electron. Syst. **13**(1), 1–28 (2008)
75. U.Y. Ogras, R. Marculescu, *Modeling, Analysis and Optimization of Network-on-Chip Communication Architectures*, vol. 184 (Springer Science & Business Media, Berlin, 2013)
76. U.Y. Ogras, R. Marculescu, D. Marculescu, E.G. Jung, Design and management of voltage-frequency island partitioned networks-on-chip. IEEE Trans. Very Large Scale Integr. Syst. **17**(3), 330–341 (2009)
77. Y. Ouyang et al., CPCA: an efficient wireless routing algorithm in WiNoC for cross path congestion awareness. Integration **69**, 75–84 (2019)
78. Y. Pan et al., Firefly: illuminating future network-on-chip with nanophotonics, in *Proceedings of International Symposium on Computer Architecture* (2009), pp. 429–440
79. S. Pasricha, N. Dutt, ORB: an on-chip optical ring bus communication architecture for multi-processor systems-on-chip, in *Proceedings of Asia and South Pacific Design Automation Conference* (2008), pp. 789–794
80. S. Pasricha et al., A survey on energy management for mobile and IoT devices, in *IEEE Design & Test* (2020)
81. A. Pellegrini et al., The arm neoverse N1 platform: building blocks for the next-gen cloud-to-edge infrastructure SoC. IEEE Micro **40**(2), 53–62 (2020)
82. A. Pinto, L.P. Carloni, A.L. Sangiovanni-Vincentelli, Efficient synthesis of networks on chip, in *Proceedings of International Conference on Computer Design* (2003), pp. 146–150
83. A.-M. Rahmani et al., Research and practices on 3D networks-on-chip architectures, in *NORCHIP* (2010), pp. 1–6
84. R.S. Ramanujam, V. Soteriou, B. Lin, L.-S. Peh, Design of a high- throughput distributed shared-buffer NoC router, in *ACM/IEEE International Symposium on Networks-on-Chip* (2010), pp. 69–78
85. V.Y. Raparti, N. Kapadia, S. Pasricha, ARTEMIS: an aging-aware run-time application mapping framework for 3D NoC-based chip multiprocessors. IEEE Trans. Multi-Scale Comput. Syst. **3**(2), 72–85 (2017)
86. A. Rasouli, J.K. Tsotsos, Autonomous vehicles that interact with pedestrians: a survey of theory and practice. IEEE Trans. Intell. Transport. Syst. **21**(3), 900–918 (2019)
87. P.K. Sahu, S. Chattopadhyay, A survey on application mapping strategies for network-on-chip design. J. Syst. Archit. **59**(1), 60–76 (2013)
88. R. Salamat, M. Khayambashi, M. Ebrahimi, N. Bagherzadeh, A resilient routing algorithm with formal reliability analysis for partially connected 3D-NoCs. IEEE Trans. Comput. **65**(11), 3265–3279 (2016)
89. P. Salihundam et al., A 2 Tb/s 6×4 mesh network for a single-chip cloud computer with DVFS in 45 nm CMOS. IEEE J. Solid-State Circuits **46**(4), 757–766 (2011)
90. A. Shacham, K. Bergman, L.P. Carloni, Photonic networks-on-chip for future generations of chip multiprocessors. IEEE Trans. Comput. **57**(9), 1246–1260 (2008)

91. A. Shafiee et al., ISAAC: a convolutional neural network accelerator with in-situ analog arithmetic in crossbars. ACM SIGARCH Comput. Archit. News **44**(3), 14–26 (2016)
92. M.A.I. Sikder et al., OWN: optical and wireless network-on-chip for kilo-core architectures, in *Proceedings. of IEEE Symposium on High-Performance Interconnects* (2015), pp. 44–51
93. A. Sodani et al., Knights landing: second-generation Intel Xeon Phi product. IEEE Micro **36**(2), 34–46 (2016)
94. M.B. Taylor et al., The raw microprocessor: a computational fabric for software circuits and general-purpose programs. IEEE Micro **22**(2), 25–35 (2002)
95. D. Vantrease et al., Corona: system implications of emerging nanophotonic technology. ACM SIGARCH Comput. Archit. News **36**(3), 153–164 (2008)
96. A. Varghese B. Edwards, G. Mitra, A.P. Rendell, Programming the Adapteva Epiphany 64-core network-on-chip coprocessor. Intl. J. High Perform. Comput. Appl. **31**(4), 285–302 (2017)
97. S. Wang, T. Jin, Wireless network-on-chip: a survey. J. Eng. **2014**(3), 98–104 (2014)
98. H. Wang, L.-S. Peh, S. Malik, Power-driven design of router microarchitectures in on-chip networks, in *Proceedings of International Symposium on Microarchitecture* (2003), pp. 105–116
99. C. Wang, W.-H. Hu, N. Bagherzadeh, A wireless network-on-chip design for multicore platforms, in *International Euromicro Conference on Parallel, Distributed and Network-Based Processing* (2011), pp. 409–416
100. S. Wasly, R. Pellizzoni, N. Kapre, HopliteRT: an efficient FPGA NoC for real-time applications, in *International of Conference on Field Programmable Technology* (2017), pp. 64–71
101. S. Yan, B. Lin, Design of application-specific 3D networks-on-chip architectures, in *3D Integration for NoC-based SoC Architectures* (2011), pp. 167–191
102. M. Yang, P. Ampadu, Thermal-aware adaptive fault-tolerant routing for hybrid photonic-electronic NoC. *Proceedings of International Workshop on Network on Chip Architectures* (2016), pp. 33–38
103. Y. Yao, Z. Lu, DVFS for NoCs in CMPs: a thread voting approach, in *IEEE International Symposium on High Performance Computer Architecture* (2016), pp. 309–320
104. Y. Ye et al., 3-D besh-based optical network-on-chip for multiprocessor system-on-chip. IEEE Trans. Comput.-Aided Des. Integr. Circ. Syst. **32**(4), 584–596 (2013)
105. C. Zhang, J.E. Bowers, Silicon photonic terabit/s network-on-chip for datacenter interconnection. Opt. Fiber Technol. **44**, 2–12 (2018)
106. W. Zhang, Y. Ye, A table-free approximate Q-learning based thermal-aware adaptive routing for optical NoCs, in *IEEE Transactions on Computer-Aided Design of Integrated Circuits and Systems* (2020)
107. H. Zheng, A. Louri, An energy-efficient network-on-chip design using reinforcement learning, in *Proceedings of Design Automation Conference* (2019), pp. 1–6
108. L. Zhou, A.K. Kodi, Probe: prediction-based optical bandwidth scaling for energy-efficient NoCs, in *Proceedings of International Symposium on Networks-on-Chip* (2013), pp. 1–8
109. P. Zhou, P.-H. Yuh, S.S. Sapatnekar, Application-specific 3D network-on-chip design using simulated allocation, in *Proceedings of Asia and South Pacific Design Automation Conference* (2010), pp. 517–522
110. M. Zhu, J. Lee, K. Choi, An adaptive routing algorithm for 3D mesh NoC with limited vertical bandwidth, in *International Conference on VLSI and System-on-Chip* (2012), pp. 18–23
111. A.K.K. Ziabari et al., Leveraging silicon-photonic NoC for designing scalable GPUs, in *Proceedings of International Conference on Supercomputing* (2015), pp. 273–282

Part II
Design-for-Security Solutions

Chapter 4
Lightweight Encryption Using Incremental Cryptography

Subodha Charles and Prabhat Mishra

4.1 Introduction

With the growing demand for high-performance and low-power designs, multi-core architectures are widely used in general purpose chip multiprocessors as well as special purpose system-on-chip (SoC) designs [2, 12, 14, 25, 26, 44]. The desired performance improvement of multi-core architectures cannot be fully achieved by parallelizing the applications unless an efficient interconnect is used to connect all the heterogeneous components on the chip. Network-on-chip (NoC) has become the standard interconnect solution [13]. Due to increasing SoC complexity, it is crucial to develop efficient NoC fabrics [14]. The importance of the information passing through the NoC has made it one of the focal points of security attacks. Diguet et al. have classified the major NoC security vulnerabilities as denial-of-service attack, extraction of secret information, and hijacking [19]. Typically, SoCs contain several assets (e.g., encryption and authentication keys, random numbers, configuration keys, and sensitive data) that reside in different Intellectual Property (IP) cores [7, 21, 37]. Protecting communications between IPs, which involve asset propagation, is a major challenge and requires additional hardware implementing security such as on-chip encryption and authentication units. However, implementation of security features introduces area, power, and performance overhead. While designers employ a wide variety of techniques to improve energy efficiency in NoC-based SoCs [2, 12, 14, 25, 26, 44], security engineers have to take into account these

S. Charles (✉)
University of Moratuwa, Colombo, Sri Lanka
e-mail: s.charles@ieee.org

P. Mishra
University of Florida, Gainesville, FL, USA
e-mail: prabhat@ufl.edu

P. Mishra, S. Charles (eds.), *Network-on-Chip Security and Privacy*,
https://doi.org/10.1007/978-3-030-69131-8_4

non-functional and real-time constraints while designing secure architectures to address various threats [11, 16]. The threat model used in this chapter is as follows:

Threat Model Figure 4.1 shows a typical NoC-based many-core architecture which encrypts packets transferred between IP cores. When packets are sent through the NoC, a router infected by a hardware Trojan can copy or re-route packets and send to a malicious IP sitting on the same NoC to leak sensitive information. Therefore, the threat model assumes that some of the IPs, as well as the routers, can be malicious. The IPs that can be trusted to be non-malicious are referred to as *secure IPs*. The goal is to ensure secure communication between these secure IPs. The network interfaces (NI) that connect IPs with routers are assumed to be secure. This assumption is valid since the NIs are used to integrate components of an SoC and are typically built in house. A similar threat model and assumptions have been used in previous work on NoC security, proving the validity of the model [3, 41].

Prior research on security architectures have explored trust-zones [10, 45], lightweight encryption [43], DoS attack detection [15, 17], and side-channel analysis [27, 28, 31–33, 39]. The method outlined in this chapter utilizes *incremental encryption* to encrypt packets in NoC. The proposed solution takes advantage of the unique characteristics of NoC traffic, and as a result, it has the ability to construct a "lighter-weight" encryption scheme without compromising the security. Incremental cryptography has been explored in areas such as software virus protection [6] and code obfuscation [23]. The goal of using incremental encryption is to design cryptographic algorithms that can reduce the effort of encryption/decryption by reusing the previously encrypted/decrypted memory fetch requests/responses rather than re-computing them from the scratch. In this framework, data is encrypted at the NI of each secure IP core. The NI is chosen to accommodate the encryption framework so that each packet can be secured before injecting into the NoC. Prior research on NoC security have proposed similar architectures where the security

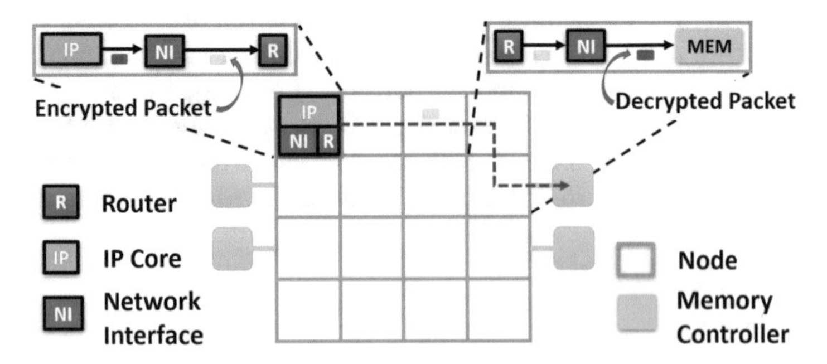

Fig. 4.1 NoC-based many-core architecture connecting IPs on a single SoC using a 4 × 4 Mesh topology. Each node contains an IP that connects to a router via a network interface. Communication between two IPs (in this case, a processor IP and a memory controller) is encrypted so that an eavesdropper cannot extract the packet content

framework was implemented at the NI [22, 41]. Major contributions of this chapter are as follows:

- This chapter shows that consecutive NoC packets that contain memory fetch requests/responses differ only by a few bits while communicating between IP cores and memory controllers in an SoC.
- A lightweight encryption scheme based on incremental cryptography is proposed that exploits the unique NoC traffic characteristics observed above.
- The proposed encryption scheme is shown to be resilient against existing NoC attacks, and it significantly improves the performance compared to state-of-the-art NoC encryption methods.

The rest of the chapter is organized as follows. Section 4.2 provides a background on related concepts. Section 4.3 presents prior related research efforts. Section 4.4 motivates the need for this work. Section 4.5 describes the lightweight encryption scheme. Section 4.6 presents the experimental results and finally, Sect. 4.7 concludes the chapter.

4.2 Background

4.2.1 Symmetric Encryption Schemes

A symmetric encryption scheme $S = (\mathcal{K}, \mathcal{E}, \mathcal{D})$ consists of three algorithms defined as follows:

- The key generation algorithm is written as $K \leftarrow \mathcal{K}$. This denotes the execution of the randomized key generation algorithm $\mathcal{K}$ and storing the return string as K where β is the length of the key.
- The *encryption* algorithm $\mathcal{E}$ produces the *ciphertext* $C \in \{0, 1\}^l$ by taking the key K and a *plaintext* $M \in \{0, 1\}^l$ as inputs, where l is the length of the plaintext. This is denoted by $C \leftarrow \mathcal{E}_K(M)$.
- Similarly, the *decryption* algorithm $\mathcal{D}$ denoted by $M \leftarrow \mathcal{D}_K(C)$ takes a key K and a ciphertext $C \in \{0, 1\}^l$ and returns the corresponding $M \in \{0, 1\}^l$.

4.2.2 Block Ciphers

A *block cipher* typically acts as the fundamental building block of the encryption algorithm ($\mathcal{E}$). Formally, it is a function (E) that takes a β-bit key (K) and an n-bit plaintext (m) and outputs an n-bit long ciphertext (c). The values of β and n depend on the design and are fixed for a given block cipher. For every $c \in \{0, 1\}^n$, there is exactly one $m \in \{0, 1\}^n$ such that $E_K(m) = c$. Accordingly, E_K has an inverse

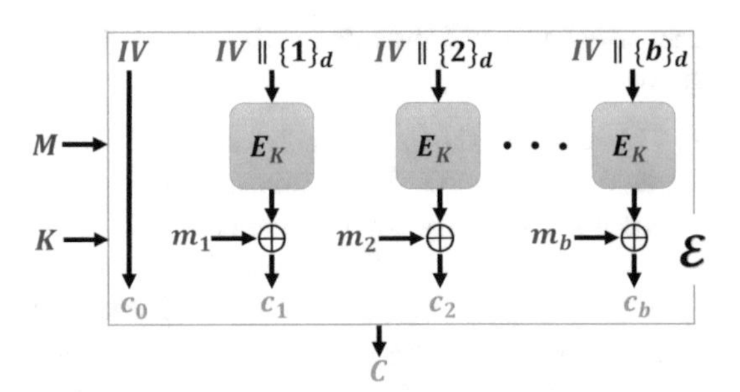

Fig. 4.2 A block cipher-based encryption scheme using counter mode. Each block cipher (E_K) encrypts an n-bit block (m_q) and b block ciphers together encrypt the entire message M and outputs ciphertext C. This constructs $\mathcal{E}$ of the encryption scheme S

block cipher denoted by E_K^{-1} such that $E_K^{-1}(E_K(m)) = m$ and $E_K(E_K^{-1}(c)) = c$ for all $m, c \in \{0, 1\}^n$.

When using block ciphers to encrypt long messages, the plaintext (M) of a given length l is divided into b substrings (m_q) where each substring is $n(= \frac{l}{b})$ bits long and n is called the *block size*. Block ciphers are used in operation modes where one or more block ciphers work together to encrypt n-bit blocks and concatenate the outputs at the end to create the ciphertext of l bits. Figure 4.2 shows the *counter mode (CM)* which is a popular operation mode. CM also uses an *initialization vector (IV)* which is concatenated with a d-bit value counter (e.g., if $d = 4$, $\{1\}_d = 0001$) before inputting to the block cipher. This is done to create domain separation by giving per message and per block variability. The decryption process is shown in Algorithm 1. In fact, the decryption process is the inverse of the encryption scheme shown in Fig. 4.2.

4.2.3 *Incremental Cryptography*

Consider a scenario that involves encrypting sensitive files/documents. Once a file is encrypted initially, there may be minor changes in the original file. In such a scenario, if typical encryption is used, the previous encrypted file will be discarded

Algorithm 1: Decryption process of Counter Mode

```
1:  Inputs: ciphertext to decrypt C
2:  Output: plaintext corresponding to the ciphertext M
3:  procedure D_K
4:      for all q = 1, ..., b do
5:          r_q ← E_K(IV||{q}_d)
6:          m_q ← r_q ⊕ c_q
7:      end for
8:      M ← m_1 || m_2 || ... || m_b
9:      return M
10: end procedure
```

and a new encryption will be performed on the modified file. However, since these changes are very small in comparison to the size of the file, encrypting the entire file again is clearly inefficient. Incremental encryption can give significant advantages in such a setup [5]. Updating an obfuscated code to accommodate patches and video transmission of images when there are minor changes between frames are two similar scenarios [23]. Incremental encryption allows to find the cryptographic transformation of a modified input not from scratch, but as a function of the encrypted version of the input from which the modified input was derived. When the changes are small, the incremental method gives considerable improvements in efficiency.

4.3 Related Work

4.3.1 Packet Security and Integrity

The most commonly proposed solution against eavesdropping attacks is to use an authenticated encryption scheme [24, 41]. While encryptions ensure packet security, authentication preserves data integrity. An overview of how an authenticated encryption scheme can be integrated in the NoC is shown in Fig. 4.3. The block diagram closely resembles the Galois Counter Mode (GCM) based encryption and authentication [35]. Packets originating from each IP are encrypted (ciphertext denoted by C) and an authentication tag (T) is appended to each packet at the NI before injecting it to the NoC. The entire packet, which consists of $H \parallel C \parallel T$, traverses the NoC and arrives at the destination. The header H is sent as plaintext

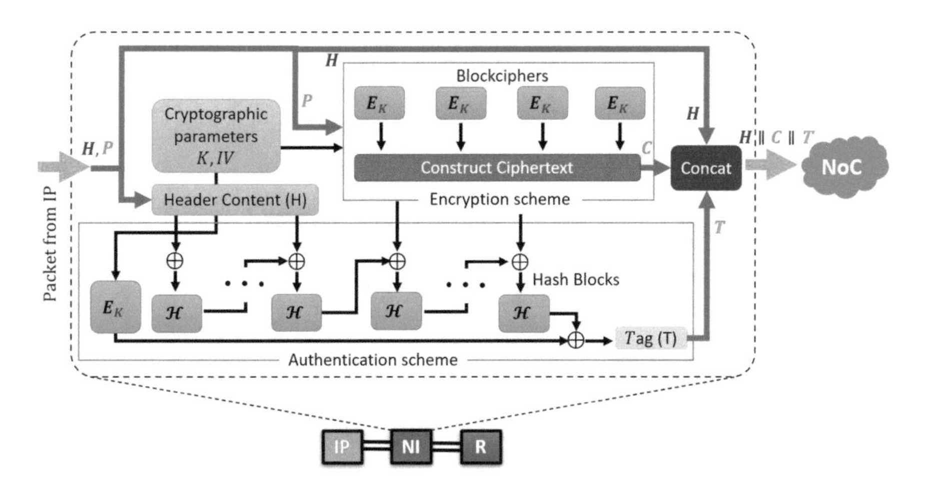

Fig. 4.3 Overview of an authenticated encryption scheme implemented to provide security to NoC

so that intermediate routers can use the header information, such as source and destination addresses, for routing. At the destination NI, the inverse process takes place. The tag T is validated and if valid, the ciphertext C is decrypted to send the plaintext to the desired IP. Encryption ensures that the plaintext of the secure information is not leaked and authentication detects any tampering with the packet including header information.

Several prior studies have tried to develop lightweight encryption and authentication schemes for on-chip data communication. Sepúlveda et al. [41] proposed a variation of authenticated encryption where only the destination is sent as plaintext and the source and data is encrypted using AES Counter mode [18]. The hash of the entire packet is calculated using SipHash [4] to ensure data integrity. An overview of the architecture is shown in Fig. 4.4. The authors introduce their solution as a tunnel-based communication mechanism to isolate sensitive information transferred between IPs. Tunnels are created by encapsulating packets in a way such that the communication is isolated by the compromised NoC. AES Counter mode provides high parallelizability at the expense of power and area. SipHash is chosen as the hash function since it provides a fast and lightweight message authentication code. SipHash is especially suited to provide a secure and fast MAC function for short inputs, which is an ideal fit for NoC packets. AES Counter mode and SipHash also provide reconfigurability, in terms of number of AES blocks and SipHash rounds, depending on the performance requirement of the SoC.

Ancajas et al. [3] proposed a simple XoR cipher together with a packet certification technique that calculates a tag and validates the tag at the receiver. Boraten et al. proposed a configurable packet validation and authentication scheme by merging two robust error detection schemes, namely algebraic manipulation detection and cyclic redundancy check, in [9]. Intel's TinyCrypt—a cryptographic library with a small footprint is built for constrained IoT devices [43]. It provides basic functionality to build a secure system with very little overhead. It gives SHA-256 hash functions, message authentication, a pseudo-random number generator

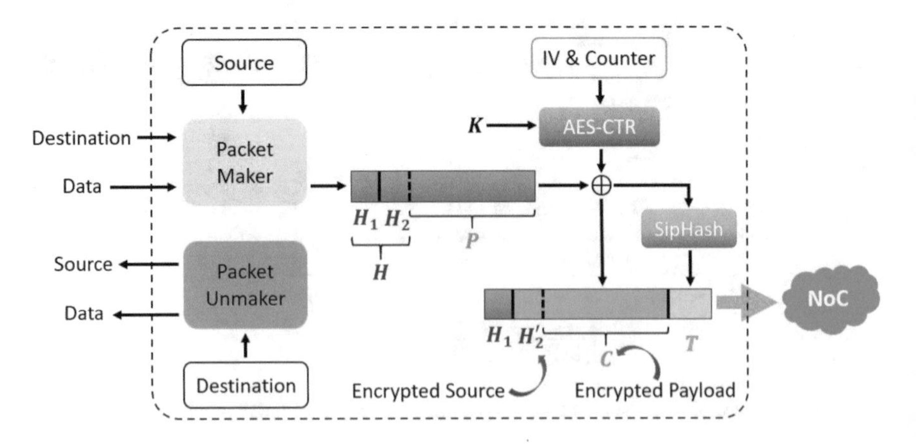

Fig. 4.4 Overview of the tunnel-based communication mechanism proposed in [41]

which can run using minimal memory, digital signatures, and encryption. It also has the basic cryptographic building blocks such as entropy sources, key exchange, and the ability to create nonces and challenges.

Several researchers have proposed other lightweight encryption solutions in the IoT domain [38]. Exploiting the unique characteristics of RFID communication, Engels et al. have proposed a low-cost encryption algorithm and a protocol [20]. The approach presented in this chapter utilizes incremental encryption to create a lightweight NoC security framework that minimizes performance overhead with minor impact on area and power.

4.3.2 *Incremental Cryptography*

Incremental encryption was first introduced by Bellare et al. [6]. In their work, they encrypt documents undergoing minor changes. Rather than encrypting every document from scratch after each change, they propose to encrypt only the change(s) and send it together with the previous encryption such that the encryption of the modified version can be constructed. Each document D is treated as a sequence of symbols $D = D[1]...D[l]$ where $D[i]$ denotes the ith symbol. The encrypted version contains two sequences of encrypted values, denoted E_1 and E_2. The first sequence, E_1, is obtained by encrypting the original document D block-by-block. The other sequence, E_2, encodes only the sequence of modifications, denoted $M = M[1]...M[t]$, which caused the current document to change from the original document D. The algorithm appends the encryption of the modifications to E_2. In every l steps, the algorithm takes the modified un-encrypted document and re-encrypts it using traditional block-by-block encryption and assigns the result to E_1 while setting E_2 to be empty. The amortized complexity of this algorithm amounts to two block encryptions per each modification.

There are fundamental challenges when using incremental encryption to encrypt packets in the NoC. In the file setup, when a file undergoes some number of modifications, every previous modification becomes redundant. In other words, intermediate steps lose their values as long as the latest version is available. However, when encrypting packets in the NoC, we cannot drop certain packets and encrypt after some modifications because each packet is important for the correct functionality of the SoC. Ideas from incremental cryptography have been adopted in other areas such as hashing and signing [5], program obfuscation [23], and cloud computing [29]. This chapter presents a technique that is able to use incremental encryption in the domain of NoC and can increase the efficiency of secure NoCs.

4.4 Motivation

The IPs use the capabilities given by the NoC to communicate with each other and to request/store data from/in memory. The packets injected into the network can be classified into two main categories—(1) control packets and (2) data packets. For example, a cache miss at an IP will cause a control packet to be injected into the network requesting for that data from the memory. The memory controller upon receiving the request will reply back with a data packet containing the cache block corresponding to the requested address. The formats of these packets are shown in Fig. 4.5. The NI divides the packet into flits ("fliticization") before injecting into the network. Flits are the basic building blocks of information transfer between routers. Sensitive data of each flit is encrypted by the NI and injected into the network through the local router. Encryption process of a packet consumes time as each block has to be encrypted and concatenated to create the encrypted packet. Depending on the parameters used for the block cipher (block size, key size, number of encryption rounds, etc.), the time complexity of the process differs. If each packet is encrypted independently, it takes $z \times T$ time to encrypt all of them, where z is the number of packets and T is the average time needed to encrypt one packet.

As discussed in Sect. 4.2, the idea of incremental encryption is to develop a scheme where the time taken to encrypt an incoming packet should not be dependent on the packet size, but rather on the amount of modifications done compared to the previous packet. To explore how to use this idea in the context of NoC, the number of bit changes between consecutive packets generated by a particular IP was profiled. Figure 4.6 shows the number of bit differences as a percentage of memory fetch requests (control packets) when running five benchmarks (FFT, FMM, LU, RADIX, OCEAN) from the SPLASH-2 benchmark suite on the gem5 full-system simulator [8]. More details about the experimental setup are given in Sect. 4.6.1. Out of the 64 bits of data to be encrypted, according to the default gem5 packet size, the maximum number of bit difference between consecutive packets was 13 bits in all benchmarks. On average, 30% of the packets differed by only one bit. This

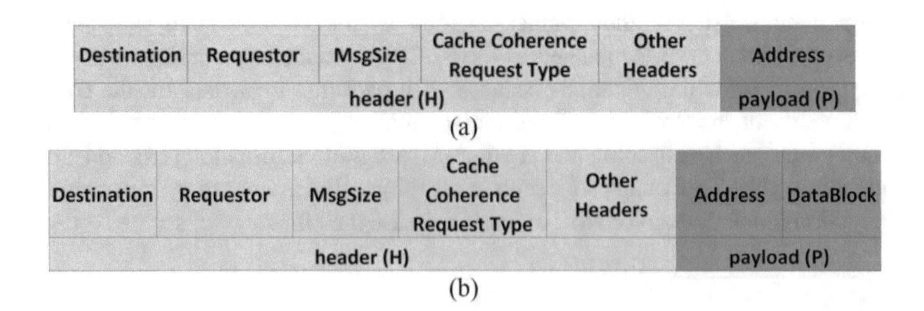

Fig. 4.5 Packet formats for (**a**) control and (**b**) data packets. Blue shows header (H) which is sent as plaintext. Red shows the payload (P) with sensitive data encrypted

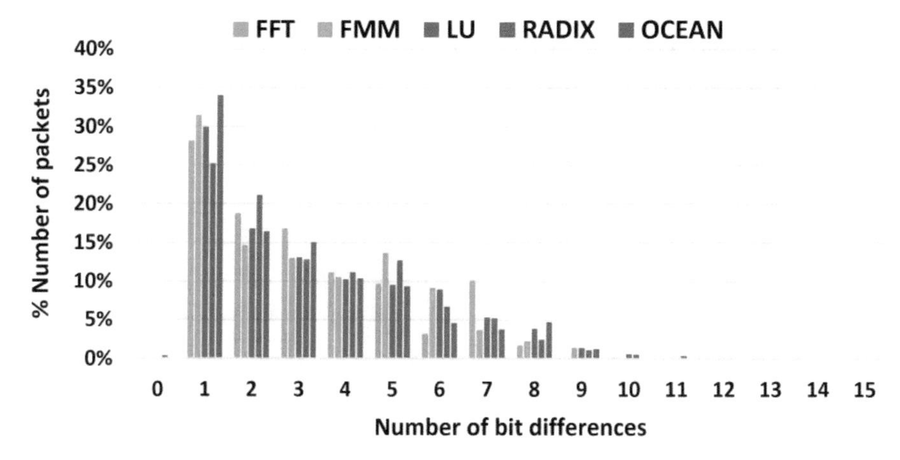

Fig. 4.6 Number of bit differences between consecutive memory fetch requests in SPLASH-2 benchmarks

is expected since an application running on a core most likely accesses memory locations within the same memory page which differs by only a few bits.

Since encryption is done in blocks, the data was profiled assuming a block size of 16 bits [20]. In this case, up to 16 consecutive bit differences can be considered for each block, and the maximum number of blocks for 64 bits of secure data is 4. The results showed that on average, 80% of the packets differ by only one block and the other 20% differ by two blocks for the benchmarks that was used. This provides a significant opportunity for optimizing the encryption process with incremental encryption. Similar to memory fetch requests, the response memory data packets were profiled as well. Since the response contains a whole cache block consisting of data modified by calculations, the optimization opportunity was less compared to the memory fetch requests. However, it still shows that 15% of consecutive packets are identical. These observations show that the encryption process can be significantly optimized using incremental encryption.

4.5 Incremental Encryption

This section describes the incremental encryption scheme in detail. First, an illustrative example is given to demonstrate the merit of exploiting unique traffic characteristics using incremental encryption. Then the major components in the proposed framework are elaborated.

Illustrative Example Figure 4.7 shows an example on how incremental encryption can improve the performance of an NoC. It shows the encryption process of three consecutive NoC packets (each with 16 bits) using two methods (1) traditional encryption, (2) incremental encryption. In traditional encryption, both packets are

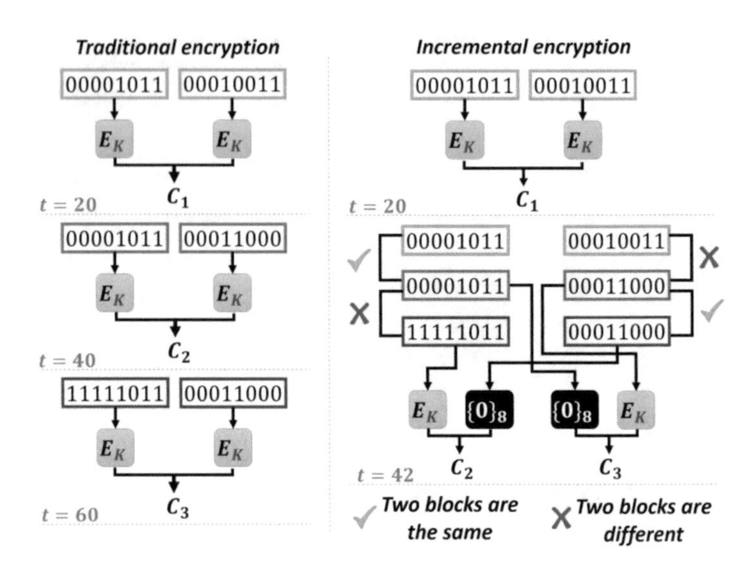

Fig. 4.7 Illustrative example of using incremental encryption. *Assumptions: encryption takes 20 cycles for each block cipher, comparing two bit strings to identify different blocks take 1 cycle each*

encrypted sequentially using the two 8-bit block ciphers. In incremental encryption, each packet is compared with the previous packet and only the different blocks are encrypted. Identical blocks are filled with zeros and header bits are added to indicate the changed blocks. The decryption process uses previously received packets and header information to reconstruct the new packets. Only the first packet has to be fully encrypted since there is no prior packet for comparison. This example shows a speedup of 1.43 times. However, when many packets are encrypted, the time spent to encrypt the first packet becomes negligible and as a result, a significant performance improvement can be achieved as shown in Sect. 4.6.2. A detailed description of the methodology is given in the next four subsections.

4.5.1 Overview

Figure 4.8 shows an overview of the proposed NoC security framework. It consists of two main components: (1) incremental crypto engine, and (2) encryption scheme which includes the block ciphers. Each packet sent from an IP core has two main parts: (1) packet header (H) which is sent as plaintext across the network, and (2) payload (P) which should be encrypted before sending to the network. Both header and payload are sent to the incremental crypto engine to start the incremental encryption process. The payload is assumed to be divided into b blocks. For example, the 64-bit payload of a control packet will contain four 16-bit blocks

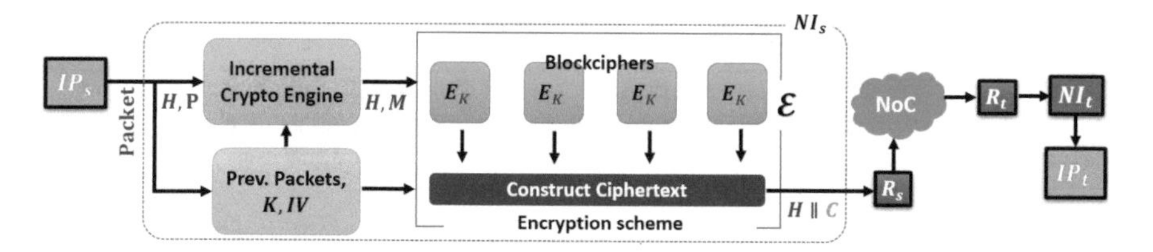

Fig. 4.8 Overview of the proposed security framework. The packet sent from source (IP_s) goes through the encryption process implemented in the network interface (NI_s). It traverses the NoC, and NI_t of the target IP (IP_t) decrypts before forwarding the packet to IP_t

($b = 4$) numbered 1 through 4 starting from the least significant byte. The proposed encryption scheme uses block ciphers arranged in counter mode [34]. A detailed explanation of parameters used in the experiments is given in Sect. 4.6.1.

Algorithm 2 describes the incremental encryption process. When a packet is sent from the IP core, the incremental crypto engine first identifies which blocks are different compared to the previous packet (line 6). This is done by comparing with the previous packet payload (P_{i-1}) which is stored in a register inside the NI. In this model, only two packets are required to be stored for the two different packet types (control and data) at the sender's end. Similarly, the receiver's side also stores the most recent packet for each packet type. In addition to that, the key (K) and initialization vector (IV) for the encryption scheme are also stored by both sender and receiver IPs. Once block differences are computed, it is then sent to the encryption scheme which encrypts only the different blocks (line 7). The final ciphertext is derived from the encrypted blocks and block comparison results (line 8). Additional header bits are also computed in this step to be used by the decryption process. Finally, the header and encrypted payload are concatenated to create the final packet and injected into the network (line 9). At the destination node, the inverse process takes place. It also stores the previous packet for each packet type, and therefore, can construct the next packet using the stored packet and the incoming packet data. Since the previous packets are stored in special registers, there is no need to encrypt/decrypt the full packet. Only the changed blocks are sent and the receiver replaces the changed blocks with its modifications to construct the new packet.

The remainder of this section elaborates the major components of the NoC security framework. Section 4.5.2 explains the $compareBlocks$ function which is implemented in the incremental crypto engine. Section 4.5.3 presents the encryption scheme $\mathcal{E}$ and $constructCipherText$ function in Algorithms 4 and 5, respectively. Section 4.5.4 explains how keys and IVs are generated and managed throughout the communication process.

Algorithm 2: Encryption process

1: **Inputs**: current packet $packet_i$, previous payload P_{i-1}, key K, initialization vector IV
2: **Output**: encrypted packet consisting of header H_i and encrypted payload C_i
3: **procedure** $encryptPackets$
4: $P_i \leftarrow packet_i.payload$
5: $H_i \leftarrow packet_i.header$
6: $M_i, \delta_i \leftarrow compareBlocks(P_i, P_{i-1})$
7: $C' \leftarrow \mathcal{E}(IV, K, M_i)$
8: $C_i \leftarrow constructCipherText(C', \delta_i)$
9: **return** $H_i \parallel C_i$
10: **end procedure**

4.5.2 Incremental Crypto Engine

The operation of the incremental crypto engine is outlined in Algorithm 3. The payload (P_i) sent from the IP core is compared with the previous payload of that type (P_{i-1}) to identify the blocks that are different (M_i). This can be implemented with a simple XOR operation in hardware (line 4). Once the bitwise differences are obtained, the payload is split into blocks (line 5) to see which blocks are different (lines 6–9). Only different blocks are sent for encryption. The incremental crypto engine also sends the different block numbers (δ_i) to build the complete ciphertext as well as to set the header bits indicating the different blocks to be used by the decryption algorithm.

As discussed before, the performance improvement is gained by encrypting multiple blocks in parallel. For example, if two consecutive control packets have differences in two blocks each, this method can achieve twice the speedup by encrypting both at the same time compared to the traditional (non-incremental) approach where all four block ciphers will be used to encrypt each packet.

Algorithm 3: Finding block-wise packet differences

1: **Inputs**: current payload P_i, previous payload P_{i-1}
2: **Output**: different blocks M_i, different block indices δ_i
3: **procedure** $compareBlocks$
4: $bitDiff \leftarrow P_i \oplus P_{i-1}$
5: $B[1], ..., B[k] \leftarrow split(bitDiff, blockSize)$
6: **for all** $x = 1, ..., size(B)$ **do**
7: **if** $B[x] > 0$ **then**
8: $M_i.append(B[x])$
9: $\delta_i[x] = 1$
10: **end if**
11: **end for**
12: **return** M_i, δ_i
13: **end procedure**

4.5.3 Encryption Scheme

The security architecture uses the counter mode for encryption which uses an initialization vector (IV), a key and the message to be encrypted as inputs and produces the ciphertext. The $IV \parallel \{q\}_d$ string, which is the standard format of the input nonce to counter mode, is used to give per message and per block variability. In this framework, it is calculated using the sequence number of the packet (let seq_j be the sequence number of packet P_j), a counter, and the IV as $IV \parallel seq_j \parallel q$ to identify different blocks. The block cipher ID ($q \in \{1, 2, 3, 4\}$) changes with each block cipher and the sequence number seq_j varies from packet to packet. Algorithm 4 shows the major steps of the encryption scheme. The structure of the encryption scheme is the same as shown in Fig. 4.2 with the number of block ciphers (b) set to four.

Algorithm 4: Encrypt selected blocks

1: **Inputs**: initialization vector IV, key K, different blocks M_i
2: **Output**: encrypted blocks C'
3: **procedure** $\mathcal{E}$
4:     **for all** $q = 1, ..., 4$ **do**
5:         $seq_j \leftarrow getSequenceNumber(P_j)$
6:         $r_q \leftarrow E_K(IV \parallel seq_j \parallel q)$
7:         $C'.append(r_q \oplus M_i[q])$
8:     **end for**
9:     **return** C'
10: **end procedure**

C' is stored in a buffer. The final ciphertext is constructed using δ_i and C' as shown in Algorithm 5. Algorithm 5 takes the encrypted value from the buffer for the changed blocks (lines 5–6) and appends n (block size) zeros to identical blocks compared to the previous packet (lines 7–8). It ensures the construction of the same packet size, and as a result, every other functionality from fliticization to NoC traversal remains the same.

4.5.4 Initialization and Parameter Refresh

The generation and management of keys and nonces have been addressed in several ways [30, 42]. One possible method is to use a key distribution center (KDC). The KDC is a specialized IP in the SoC. The KDC is also responsible for generating the IVs. The keys and IVs are distributed among the cores and stored in special registers. This architecture also provides the flexibility to refresh the parameters depending on the application. For example, if the number of bits allocated for the sequence number in a different packet format is small, the sequence number will be

Algorithm 5: Construct the encrypted payload

```
1: Inputs: encrypted blocks C', different block indices δ_i
2: Output: Encrypted payload C_i
3: procedure constructCipherText
4:     for all x = 1, ..., size(δ) do
5:         if δ_i[x] > 0 then
6:             C_i.append(C'[x])
7:         else
8:             C_i.append({0}_n)
9:         end if
10:    end for
11:    return C_i
12: end procedure
```

re-initialized more often. In such a setup, using a different IV is mandatory to make sure the counter string in CM is a nonce.

One way to generate an IV is to utilize hash functions, such as SHA-256. During initialization, KDC picks a random number as salt S. When it receives a request for an IV, KDC utilizes the hash function and the salt to generate a few random bytes, i.e., SHA-256(S), then increments S. As SHA-256 is non-invertible and collision resistant, the output of SHA-256 is a good source of nonce and randomness. Since it is computationally impossible to compute the value of S by inspecting the plaintext IV (output of SHA-256), malicious IPs cannot infer the next IV generated by KDC. As the required bits of the IV is less than the output of SHA-256, an IV buffer is used to store all the unused bits. After KDC performs SHA-256, it appends the 256 bits output to the end of the IV buffer. Since the IV used in this chapter is less than 16 bits, one SHA-256 operation can serve more than 16 IV requests.

4.6 Experiments

This section first describes the experimental setup used to evaluate the approach. Then, results are presented to show the performance gain achieved through incremental encryption by comparing it with traditional encryption. Next, the security of the proposed framework and associated overhead are discussed.

4.6.1 Experimental Setup

The framework was validated using five benchmarks chosen from the SPLASH-2 benchmark suite. Traffic traces were generated by the cycle-accurate full-system simulator—gem5 [8]. The 4×4 Mesh NoC was built on top of "GARNET2.0" model that is integrated with gem5 [1]. The network interface (NI) was modified to simulate the proposed security framework. The following options were chosen to simulate architectural choices in a resource-constrained NoC.

Packet Format For control and data packet formats, the experimental setup used the default GARNET2.0 implementations which allocates 128 bits for a flit. This value results in control messages fitting in 1 flit, and data packets, in 5 flits. Out of the 128 bits, 64 bits are allocated for the payload (address) in a control packet and data packets have a payload of 576 bits (64-bit address and 512-bit data). This motivated the use of 16-bit blocks to evaluate the performance of the proposed incremental encryption scheme.

Block Cipher This approach uses an ultra-lightweight block cipher—"Hummingbird-2" as the block cipher of the encryption scheme [20]. Hummingbird-2 was chosen in the experiments mainly because it is lightweight and also, with the block size being 16, other encryption schemes can be broken using brute-force attacks in such small block sizes. However, it has been shown in [20] that Hummingbird-2 is resilient against attacks that try to recover the plaintext from ciphertext. It uses a 128-bit key and a 128-bit internal state which provides adequate security for on-chip communication. Considering the payload and block sizes, four block ciphers were used in counter mode for the encryption scheme. Each block cipher is assumed to take 20 cycles to encrypt a 16-bit block and each comparison of two bit strings incurs a 1-cycle delay [20]. This framework is flexible to accommodate different packet formats, packet sizes and block ciphers depending on the design requirements. For example, if a certain architecture requires 128-bit blocks, AES can be used while keeping the incremental encryption approach intact.

4.6.2 Performance Evaluation

The performance improvement achieved by this approach can be presented in two main steps: (1) time taken for encryption (Fig. 4.9) and (2) execution time (Fig. 4.10). The experiments that were carried out measured the cycles spent for encryption alone (encryption time) and total cycles executed to run the benchmark (execution time) including encryption time, using this approach (incremental

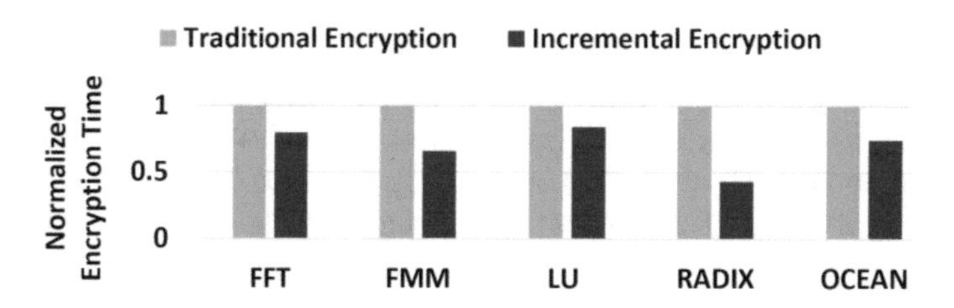

Fig. 4.9 Encryption time comparison using traditional encryption and incremental encryption

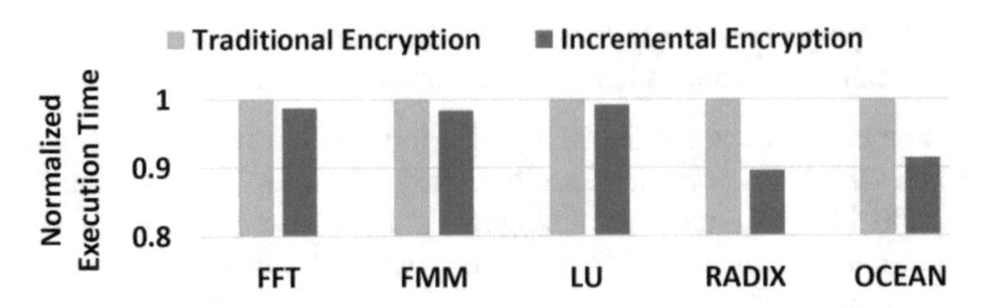

Fig. 4.10 Execution time comparison using traditional encryption and incremental encryption

encryption) as well as traditional encryption. Figure 4.9 shows the encryption time comparison. This approach improves the performance of encryption by 57% (30% on average) compared to the traditional encryption schemes. The locality in data and the differences in operand values affect the number of changed blocks between consecutive packets. This is reflected in the encryption time. For example, if an application is doing an image processing operation on an image stored in memory, accessing pixel data stored in consecutive memory locations provides an opportunity for performance gain using this approach.

The total execution time was also compared using traditional encryption as well as incremental encryption. Figure 4.10 presents these results. When the overall system including CPU cycles, memory load/store delays, and delays traversing the NoC is considered, the total execution time improves up to 10% (5% on average). Benchmarks that have significant NoC traversals such as RADIX and OCEAN show higher performance improvement (10%).

4.6.3 Security Analysis

When discussing the security of this approach, three main components have to be considered: (1) incremental encryption, (2) encryption scheme that uses counter mode, and (3) block cipher.

Incremental Encryption Due to the inherent characteristics of incremental encryption, this approach reveals the amount of differences between consecutive packets. Studies on incremental encryption have shown that even though hiding the amount of differences is not possible, it is possible to hide "everything else" by using secure block ciphers and secure operation modes [6]. Attacks on incremental encryption using this vulnerability relies on the adversary having many capabilities in addition to the ones defined in the threat model. When using incremental encryption to encrypt documents undergoing frequent, small modifications as explained in Sect. 4.2, it is reasonable to assume that the adversary not only has availability to the previously encrypted versions of documents but is also able to modify documents and obtain encrypted versions of the modified ones. This attack model allows the adversary to launch chosen plaintext attacks [6]. Discussing security of this approach for known plaintext, chosen plaintext and chosen ciphertext attacks are irrelevant since the adversary does not have access to separate hardware that implements the design, nor access to known plaintext/ciphertext pairs. In other words, as long as the block cipher and operation mode are secure, incremental encryption does not allow recovering of plaintext from the ciphertext. The same argument has been proven to hold true in previous work on incremental encryption [6, 36].

Counter Mode Encryption Using this approach, each block is treated independently while encrypting, and blocks belonging to multiple packets can be encrypted in parallel. In such a setup, using the same $IV \| \{q\}_d$ string with the same key K can cause the "two time pad" situation. This is solved by setting the string to $IV \| seq_j \| q$ as shown in Algorithm 4. It gives per message and per block variability and ensures that the value is a nonce. The proposed usage of counter mode adheres to the security recommendations outlined in [34].

Block Cipher As discussed above, the security of the proposed framework depends on the security of the block cipher. The security of the block cipher used in this framework, Hummingbird-2, has been discussed extensively in [20]. The first version of the Hummingbird scheme was shown to be insecure [40] and Hummingbird-2 was developed to address the security flaws. After thousands of hours of cryptanalysis, no significant flaws or sub-exhaustive attacks against Hummingbird-2 have been found [20]. Hummingbird-2 approach has been shown to be resilient against birthday attacks on the initialization, differential cryptanalysis, linear cryptanalysis, and algebraic attacks. Zhang et al. presented a related-key chosen-IV attack against Hummingbird-2 that recovered the 128-bit secret key [46]. However, the attack requires 2^{28} pairs of plaintext to recover the first 4 bits of the key adding up to a data complexity of $O(2^{32.6})$ [46]. As discussed before, launching such chosen plaintext attacks is not possible in the NoC setting. A brute-force key recovery takes 2^{128} attempts which is not computationally feasible according to modern computing standards as well as for computing power in the foreseeable future.

The proposed approach allows easy plug-and-play of security primitives. Any block size/key size/block cipher can be combined with the proposed incremental encryption approach. Note that stronger security comes at the expense of performance. Therefore, security parameters can be decided depending on the desired security and performance requirements.

4.6.4 Overhead Analysis

The proposed incremental encryption approach was implemented using Verilog to show the area overhead in comparison with the original Hummingbird-2 implementation. Figure 4.11 provides an overview of the hardware implementation of

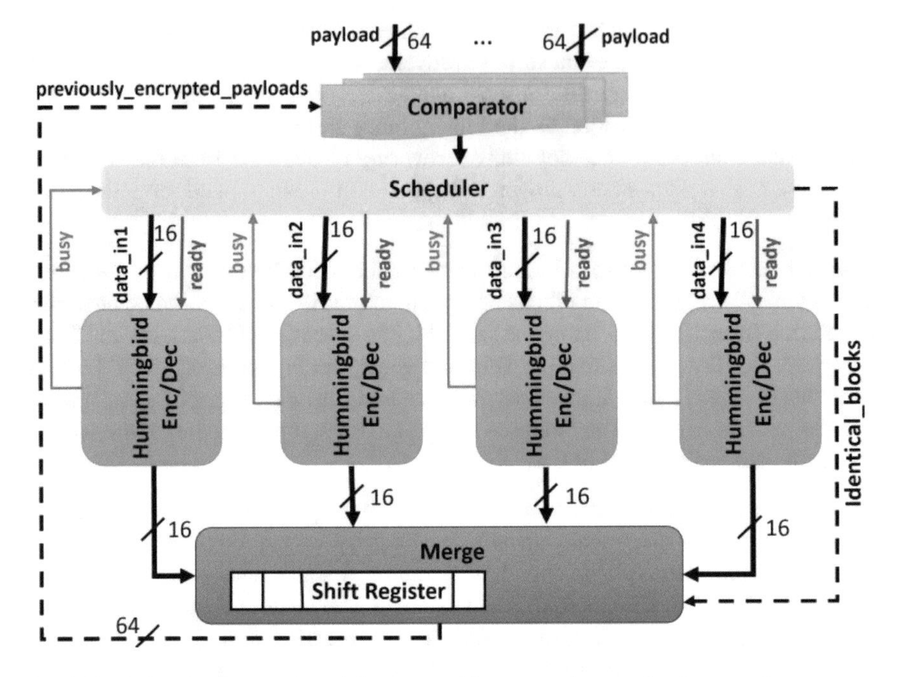

Fig. 4.11 Verilog implementation of the proposed incremental encryption framework

Table 4.1 Area overhead of the proposed approach

	Incremental encryption	Traditional encryption	Overhead (encryption)	Overhead (entire NoC)
Area	175,649 μm^2	152,424 μ m^2	15%	2%

incremental encryption. This implementation is capable of assigning blocks to idle block ciphers and encrypting up to four payloads in parallel. Basically, if an upcoming payload needs to be encrypted/decrypted for the number of blocks which are less than or equal to the number of idle block ciphers, this method can encrypt/decrypt it immediately. Merger and scheduler units were implemented to ensure the correctness of final encrypted/decrypted payloads. The proposed framework was compared with an implementation of traditional encryption, modeled as four Hummingbird-2 block ciphers without the incremental crypto engine. The experiments were conducted using the Synopsys Design Compiler with 90 nm Synopsis library (saed90nm).

Based on the results shown in Table 4.1, the proposed approach introduces less than 2% overall area overhead with respect to the entire NoC. When only the encryption unit is considered, the overhead is 15%. This overhead is caused due to components responsible for buffering and scheduling of modified blocks to idle block cipher units as well as computations related to the construction of the final result.

Therefore, the proposed encryption approach has a negligible area overhead and it can be efficiently implemented as a lightweight security mechanism for NoCs.

While there is a minor increase in power overhead due to the additional components, there is no penalty on overall energy consumption due to the reduction in execution time.

4.7 Summary

In this chapter, we presented a lightweight security mechanism that improves the performance of traditional encryption schemes used in NoC while incurring negligible area and power overhead. The security framework consists of an encryption/decryption scheme that provides secure communication on the NoC. The proposed approach uses incremental encryption to improve performance by utilizing the unique traffic characteristics of packets observed in an NoC. The framework was validated in terms of security to prove that the performance gain is not achieved at the expense of security. Experimental results show a performance improvement of up to 57% (30% on average) in encryption time and up to 10% (5% on average) in total execution time compared to traditional encryption while introducing less than 2% overall area overhead.

Acknowledgments This work was partially supported by the National Science Foundation (NSF) grant SaTC-1936040. We would like to acknowledge the contributions of Prof. Farimah Farahmandi and Dr. Yangdi Lyu for providing helpful comments.

References

1. N. Agarwal, T. Krishna, L. Peh, N.K. Jha, Garnet: a detailed on-chip network model inside a full-system simulator, in *2009 IEEE International Symposium on Performance Analysis of Systems and Software* (2009), pp. 33–42
2. Ahmed, A., Huang, Y., Mishra, P., Cache reconfiguration using machine learning for vulnerability-aware energy optimization. ACM Trans. Embedded Comput. Syst. **18**(2), 1–24 (2019)
3. D.M. Ancajas, K. Chakraborty, S. Roy, Fort-NoCs: mitigating the threat of a compromised NoC, in *2014 51st ACM/EDAC/IEEE Design Automation Conference (DAC)* (2014), pp. 1–6
4. J.-P. Aumasson, D.J. Bernstein, Siphash: a fast short-input PRF, in *International Conference on Cryptology in India* (Springer, New York, 2012), pp. 489–508
5. M. Bellare, O. Goldreich, S. Goldwasser, Incremental cryptography: the case of hashing and signing, in *Advances in Cryptology — CRYPTO '94*, ed. by Y.G. Desmedt (Springer, Berlin, Heidelberg, 1994), pp. 216–233
6. M. Bellare, O. Goldreich, S. Goldwasser, Incremental cryptography and application to virus protection, in *Proceedings of the Twenty-Seventh Annual ACM Symposium on Theory of Computing, STOC '95*, pp. 45–56 (Association for Computing Machinery, New York, NY, 1995)
7. S. Bhunia, M. Tehranipoor, *The Hardware Trojan War* (Springer, New York, 2018)
8. N. Binkert, B. Beckmann, G. Black, S.K. Reinhardt, A. Saidi, A. Basu, J. Hestness, D.R. Hower, T. Krishna, S. Sardashti, R. Sen, K. Sewell, M. Shoaib, N. Vaish, M.D. Hill, D.A. Wood, The gem5 simulator. SIGARCH Comput. Archit. News **39**(2), 1–7 (2011)

9. T. Boraten, A.K. Kodi, Packet security with path sensitization for NoCs, in *2016 Design, Automation Test in Europe Conference Exhibition (DATE)* (2016), pp. 1136–1139
10. S. Charles, P. Mishra, Lightweight and trust-aware routing in NoC-based SoCs, in *2020 IEEE Computer Society Annual Symposium on VLSI (ISVLSI)* (2020), pp. 160–167
11. S. Charles, P. Mishra, Reconfigurable network-on-chip security architecture. ACM Trans. Des. Autom. Electron. Syst. **25**(6), 1–25 (2020)
12. S. Charles, H. Hajimiri, P. Mishra, Proactive thermal management using memory-based computing in multicore architectures, in *International Green and Sustainable Computing Conference (IGSC)*, pp. 1–8 (2018)
13. S. Charles, C.A. Patil, U.Y. Ogras, P. Mishra, Exploration of memory and cluster modes in directory-based many-core CMPs, in *IEEE/ACM International Symposium on Networks-on-Chip (NOCS)* (2018), pp. 1–8
14. S. Charles, A. Ahmed, U.Y. Ogras, P. Mishra, Efficient cache reconfiguration using machine learning in NoC-based many-core CMPs. ACM Trans. Des. Autom. Electron. Syst. **24**(6), 1–23 (2019)
15. S. Charles, Y. Lyu, P. Mishra, Real-time detection and localization of dos attacks in NoC based SoCs, in *Design Automation & Test in Europe (DATE)* (2019), pp. 1160–1165
16. S. Charles, M. Logan, P. Mishra, Lightweight anonymous routing in NoC based SoCs, in *Design Automation & Test in Europe (DATE)* (2020)
17. S. Charles, Y. Lyu, P. Mishra, Real-time detection and localization of distributed DoS attacks in NoC based SoCs. IEEE Transactions on Computer-Aided Design of Integrated Circuits and Systems (2020). https://doi.org/10.23919/DATE.2019.8715009
18. J. Daemen, V. Rijmen, *The Design of Rijndael: AES-the Advanced Encryption Standard* (Springer Science & Business Media, Cham, 2013)
19. J. Diguet, S. Evain, R. Vaslin, G. Gogniat, E. Juin, NOC-centric security of reconfigurable SoC, in *First International Symposium on Networks-on-Chip (NOCS'07)* (2007), pp. 223–232
20. D. Engels, M.-J.O. Saarinen, P. Schweitzer, E.M. Smith, The hummingbird-2 lightweight authenticated encryption algorithm, in *RFID. Security and Privacy*, ed. by A. Juels, C. Paar (Springer, Berlin, Heidelberg, 2012), pp. 19–31
21. F. Farahmandi, Y. Huang, P. Mishra, *System-on-Chip Security: Validation and Verification.* (Springer, Nature, 2019)
22. L. Fiorin, G. Palermo, C. Silvano, *A Security Monitoring Service for NoCs* (Association for Computing Machinery, New York, NY, 2008), pp. 197–202
23. S. Garg, O. Pandey, Incremental program obfuscation, in *Annual International Cryptology Conference* (Springer, New York, 2017), pp. 193–223
24. C.H. Gebotys, R.J. Gebotys, A framework for security on NOC technologies, in *Proceedings of the IEEE Computer Society Annual Symposium on VLSI (ISVLSI'03), ISVLSI '03* (IEEE Computer Society, Washington, DC, 2003), pp. 113
25. U. Gupta, C.A. Patil, G. Bhat, P. Mishra, U.Y. Ogras, DyPo: Dynamic pareto-optimal configuration selection for heterogeneous MpSoCs. ACM Trans. Embedded Comput. Syst. **16**(5s), 1–20 (2017)
26. Y. Huang, P. Mishra, Vulnerability-aware energy optimization for reconfigurable caches in multitasking systems. IEEE Trans. Comput.-Aided Des. Integr. Circ. Syst. **38**(5), 809–821 (2019)
27. Y. Huang, S. Bhunia, P. Mishra, MERS: statistical test generation for side-channel analysis based trojan detection, in *ACM SIGSAC Conference on Computer and Communications Security (CCS)* (2016), pp. 130–141
28. Y. Huang, S. Bhunia, P. Mishra, Scalable test generation for trojan detection using side channel analysis. IEEE Trans. Inf. Forensics Secur. **13**(11), 2746–2760 (2018)
29. Itani, W., Kayssi, A., Chehab, A., Energy-efficient incremental integrity for securing storage in mobile cloud computing, in *2010 International Conference on Energy Aware Computing* (2010), pp. 1–2
30. B. Lebiednik, S. Abadal, H. Kwon, T. Krishna, Architecting a secure wireless network-on-chip, in *2018 Twelfth IEEE/ACM International Symposium on Networks-on-Chip (NOCS)* (2018), pp. 1–8

31. Y. Lyu, P. Mishra, A survey of side-channel attacks on caches and countermeasures. J. Hardw. Syst. Secur. **2**(1), 33–50 (2018)
32. Y. Lyu, P. Mishra, Efficient test generation for trojan detection using side channel analysis, in *Design Automation & Test in Europe Conference (DATE)* (2019), pp. 408–413
33. Y. Lyu, P. Mishra, Automated test generation for trojan detection using delay-based side channel analysis, in *2020 Design, Automation & Test in Europe Conference & Exhibition (DATE)* (2020), pp. 1031–1036
34. D.A. McGrew, *Counter Mode Security: Analysis and Recommendations*, vol. 2(4) (Cisco Systems, San Jose, 2002)
35. D. McGrew, J. Viega, The galois/counter mode of operation (GCM). Submission to NIST Modes of Operation Process **20**, 0278–0070 (2004)
36. I. Mironov, O. Pandey, O. Reingold, G. Segev, Incremental deterministic public-key encryption, in *Proceedings of the 31st Annual international conference on Theory and Applications of Cryptographic Techniques*, pp. 628–644 (Springer, New York, 2012)
37. P. Mishra, S. Bhunia, M. Tehranipoor, *Hardware IP security and Trust*. (Springer, New York, 2017)
38. E.R. Naru, H. Saini, M. Sharma, A recent review on lightweight cryptography in IoT, in *2017 International Conference on I-SMAC (IoT in Social, Mobile, Analytics and Cloud) (I-SMAC)* (2017), pp. 887–890
39. Z. Pan, J. Sheldon, P. Mishra, Test generation using reinforcement learning for delay-based side channel analysis, in *IEEE/ACM International Conference on Computer-Aided Design (ICCAD)* (2020)
40. M.-J.O. Saarinen, Cryptanalysis of hummingbird-1, in *Fast Software Encryption*, ed. by A. Joux (Springer, Berlin, Heidelberg, 2011), pp. 328–341
41. J. Sepúlveda, A. Zankl, D. Flórez, G. Sigl, Towards protected MpSoC communication for information protection against a malicious NoC. Proc. Comput. Sci. **108**, 1103–1112 (2017). *International Conference on Computational Science, ICCS 2017*, 12–14 June 2017, Zurich
42. J. Sepulveda, D. Flórez, V. Immler, G. Gogniat, G. Sigl, Efficient security zones implementation through hierarchical group key management at NoC-based MpSoCs. Microprocess. Microsyst. **50**, 164–174 (2017)
43. Using TinyCrypt Library, Intel Developer Zone, Intel, 2016. https://software.intel.com/en-us/node/734330. [Online]
44. W. Wang, P. Mishra, S. Ranka, *Dynamic Reconfiguration in Real-Time Systems* (Springer, New York, 2012)
45. J. Winter, Trusted computing building blocks for embedded Linux-based arm trustzone platforms, in *Proceedings of the 3rd ACM Workshop on Scalable Trusted Computing, STC '08*, New York, NY (Association for Computing Machinery, New York, 2008), pp. 21–30
46. K. Zhang, L. Ding, J. Guan, Cryptanalysis of hummingbird-2. Technical report, Cryptology ePrint Archive, Report 2012/207 (2012)

Chapter 5
Trust-Aware Routing in NoC-Based SoCs

Subodha Charles and Prabhat Mishra

5.1 Introduction

Reusable hardware Intellectual Property (IP) based System-on-Chip (SoC) design has emerged as a pervasive design practice in the industry to dramatically reduce design/verification cost while meeting aggressive time-to-market constraints [3, 9, 18, 21, 23, 36, 40–42, 45]. Growing reliance on these pre-verified hardware IPs, often gathered from untrusted third-party vendors, severely affects the security and trustworthiness of SoC computing platforms [2, 22, 24, 25, 38, 39, 46, 49]. Since the malicious third-party IPs share the same Network-on-Chip (NoC) with secure IPs, malicious IPs can adversely affect the communication between the secure IPs. Figure 5.1 shows an NoC-based SoC divided into secure and non-secure zones similar to the architecture proposed in the ARM TrustZone architecture [52]. An IP in one secure zone (top left) communicates secure information with a secure IP in the other zones (bottom right). Since the packets traverse through the non-secure zone, a malicious IP can tamper the packets.

Consider a scenario where the integrity of exchanged data is ensured using a message authentication code (MAC) . The sender IP sends a packet together with an authentication tag, and the receiver re-computes the tag to check for data integrity. If it does not match, the packet has been tampered during communication, and a re-transmission is required. This method of error correction is widely employed in NoC-based SoCs [48]. However, re-transmissions due to corrupt packets can lead to several problems:

S. Charles (✉)
University of Moratuwa, Colombo, Sri Lanka
e-mail: s.charles@ieee.org

P. Mishra
University of Florida, Gainesville, FL, USA
e-mail: prabhat@ufl.edu

P. Mishra, S. Charles (eds.), *Network-on-Chip Security and Privacy*,
https://doi.org/10.1007/978-3-030-69131-8_5

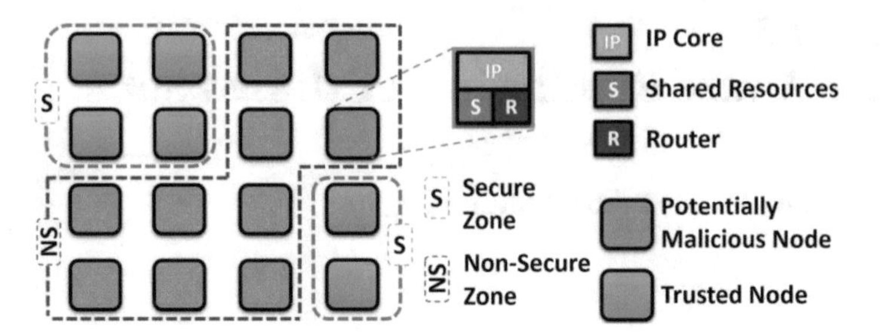

Fig. 5.1 Overview of a typical SoC architecture with secure and non-secure zones

- Increased latency because of re-transmission as well as additional stall cycles introduced by the IP cores while waiting for the requested data.
- This can increase the number of packets traversing the network and, as a result, increase energy consumption and performance penalty.
- In MAC-then-encrypt protocols [19],[1] authentication tag is computed on the plaintext, appended to the data, and then tag and plaintext are encrypted together. When MAC is computed in this way, the receiver IP has no way of knowing whether the message is indeed authentic or tampered until the message is decrypted. Therefore, the resources spent to decrypt a tampered packet are wasted.

Systematic exploitation of error correction protocols, such as the one explained above, can lead to denial-of-service (DoS) attacks. For example, a malicious IP can corrupt data on purpose and cause continuous re-transmissions leading to a DoS attack [7]. Specifically, the threat model is as follows.

Threat Model Figure 5.1 shows a standard NoC-based many-core architecture with IPs connected in a mesh topology. Each IP connects to a router via a network interface. The network interface accommodates the authentication scheme that implements MAC-based authentication [32]. A packet originating from a source IP (*src*) in a secure zone has to traverse through the non-secure zone in order to reach the destination IP (*dest*) in another secure zone. The IPs in the non-secure zone are potentially malicious. In reality, out of all the potentially malicious IPs, only a small fraction is actually malicious. We call them malicious IPs (MIP) in this chapter. If the packet traverses through such an MIP, it can tamper with the packet and, therefore, at *dest*, the authentication tag computation will not match and the packet will be dropped. The *src* will re-transmit the packet since a response is not received from the *dest* within the time-out period. The problem of minimizing this impact gets aggravated due to two challenges. (1) The MIP will not always behave maliciously. In other words, it will tamper packets only in sporadic intervals. (2). Since the *src* depends on the response from the *dest* to know whether the packet was

[1]MAC-then-encrypt is the standard method used in TLS [19].

received or not, the MIP can tamper the packet between *src* and *dest* or tamper the response packet between *dest* and *src*, and both of these scenarios lead to the same outcome from the *src*'s point of view. The security countermeasure outlined in this chapter considers both of these challenges.

In this chapter, we discuss denial-of-service attacks caused by packet corruption and highlight a trust-aware routing protocol that avoids MIPs when two secure IPs are communicating with each other. This approach leads to less re-transmissions and, as a result, improved performance and energy efficiency. Trust-aware routing can complement the existing NoC attack detection and mitigation techniques by allowing on-chip communication even in the presence of an adversary while minimizing the energy and performance overhead. Designers can employ orthogonal techniques to improve energy efficiency in NoC-based SoCs [4, 12, 15, 26, 27, 55].

Major contributions of this chapter can be summarized as follows:

1. We discuss performance degradation type denial-of-service attacks in detail together with the corresponding countermeasures proposed in the existing literature.
2. We outline a mechanism to quantify trust and a routing protocol that uses the trust values between routers to make routing decisions such that the MIPs are avoided by packets when routing from source to destination.
3. The effectiveness of the approach is evaluated using both real benchmarks and synthetic traffic patterns to demonstrate that it leads to significant improvement in both performance and energy efficiency.

The chapter outlines a routing protocol that avoids MIPs in the NoC rather than detecting them. Therefore, this approach can be used together with any MIP detection mechanism while increasing the overall performance and energy efficiency. The remainder of this chapter is organized as follows. Section 5.2 demonstrates how the attack can lead to significant overhead. Section 5.3 discusses other related research efforts in mitigating denial-of-service attacks caused by packet corruption. Section 5.4 presents the NoC trust model. Section 5.5 describes the trust-aware routing protocol that utilizes the NoC trust model. Section 5.6 presents the experimental results. Finally, Sect. 5.7 summarizes the chapter.

5.2 Motivation

Lightweight authentication schemes implemented on NoC-based SoCs try to provide desired security while consuming minimum number of cycles. However, if the MAC fails to match at the receiver's end, the src has to re-transmit again, leading to wasted effort in repeated NoC traversal and MAC calculation [48]. This is the goal of packet corruption by the attacker. The challenge is aggravated in MAC-then-encrypt protocols because MAC can only be calculated and matched after decryption is done. If the packet is tampered, time and energy spent on decryption are wasted. To analyze these overheads, FFT, RADIX (RDX), FMM, and LU benchmarks from the SPLASH-2 benchmark suite were run on an 8 × 8 Mesh NoC-based SoC with 64 cores that implements a MAC-then-encrypt security protocol and a XY routing protocol. The behavior of an MIP was simulated by one of the IPs along the routing path dropping n consecutive packets after every p (period) packets. NoC delay (total NoC traversal delay for all packets) including encryption/decryption and MAC calculation time, execution time, and number of packets injected were recorded with and without the presence of an MIP. The encryption/decryption and authentication process is assumed to take 20 cycles per transmission [20]. Results are shown in Fig. 5.2a, b, and c, respectively. The results show 67.2% increase in NoC delay and a 4.7% increase in execution time on average across all benchmarks. The number of packets injected increased by 60.1%. The combination of execution time and number of injected packets directly affects the energy consumption since both time spent to execute the task and dynamic power are increased [15].

Therefore, it is evident that in addition to checking data integrity, a mechanism to avoid MIPs when routing through the non-secure zone can lead to less re-transmissions and, as a result, increased performance and energy efficiency.

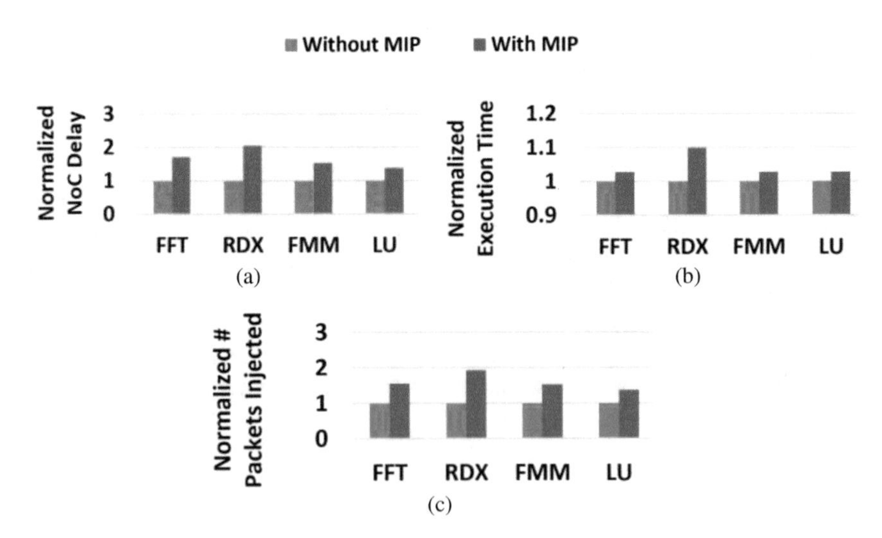

Fig. 5.2 NoC delay, execution time, and number of packets injected in comparison with and without the presence of an MIP when $p = 20$ and $n = 14$. (**a**) NoC delay. (**b**) Execution time. (**c**) No. of packets injected

5.3 Related Work

Previous research on securing the NoC has proposed lightweight security schemes [11, 16], DoS attack mitigation techniques [14, 17, 51], and methods to prevent side-channel attacks [28, 29, 31, 34, 35, 37, 50]. Most of the methods try to exploit the unique characteristics offered by the structure of the NoC and traffic transferred through the NoC when developing security schemes. In this chapter, we discuss denial-of-service attacks caused by packet corruption [10].

In [44], the authors introduced a threat model where hardware Trojans tamper flits arriving at the input buffer of a router causing performance degradation. Performance degradation is caused by dropped packets, wastage of NoC resources such as buffer space, response delays, and re-transmissions. In their work, four Trojans were introduced that differ based on the field in the flit they tamper with. The format of the flits considered is shown in Fig. 5.3.

1. Quan Trojan: Modifies the flit quantity indication field. The destination detects that more or less flits are there than that indicated by the flit quantity in head flit and drops the flits.
2. Address Trojan: Modifies the destination address field. This can lead to both information leakage and performance degradation since legitimate packets sent to the destination can be blocked.
3. Head hardware Trojan: Modifies the head bit of the head flit. As a result, the route computation module will not access the address field and the entire packet will be dropped requesting a re-transmission.
4. Tail hardware Trojan: Modifies the tail indication bit of the tail flit. The destination waits for the tail to arrive causing packet mixing and packet loss.

As a countermeasure for this attack, the same paper proposed a bit shuffling method that makes flits less sensitive to the attack [44]. The attacker can target the critical fields since the attacker is aware of the packet structure. An example of the attack scenario is shown in Fig. 5.4. The defense mechanism is proposed to shuffle the critical bit fields of the flits among themselves and others so that the Trojan is attacking on randomly shuffled data and not on the critical fields within the packets. The shuffling patterns are changed frequently. Therefore, the Trojan,

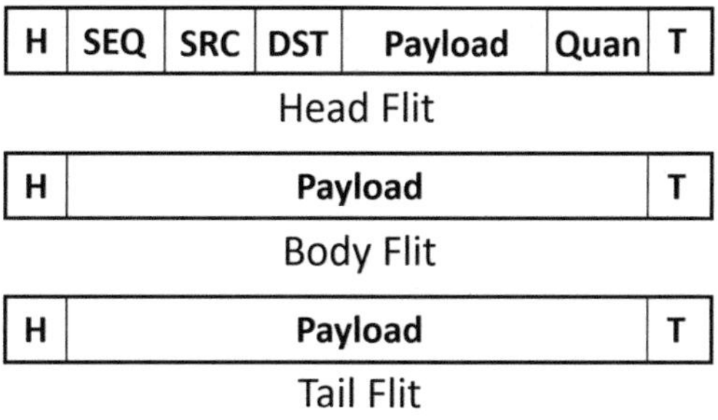

Fig. 5.3 Flit format. H: Head bit (1 for head flit), T: Tail bit (1 for Tail flit), SEQ: Packet sequence number, Quan: Flit quantity of packets, SRC: Source tile's $x - y$ positions, DST: Destination tile's $x - y$ positions

H	SEQ	SRC	DST	Payload	Quan	T

Head Flit

H	Payload	T

Body Flit

H	Payload	T

Tail Flit

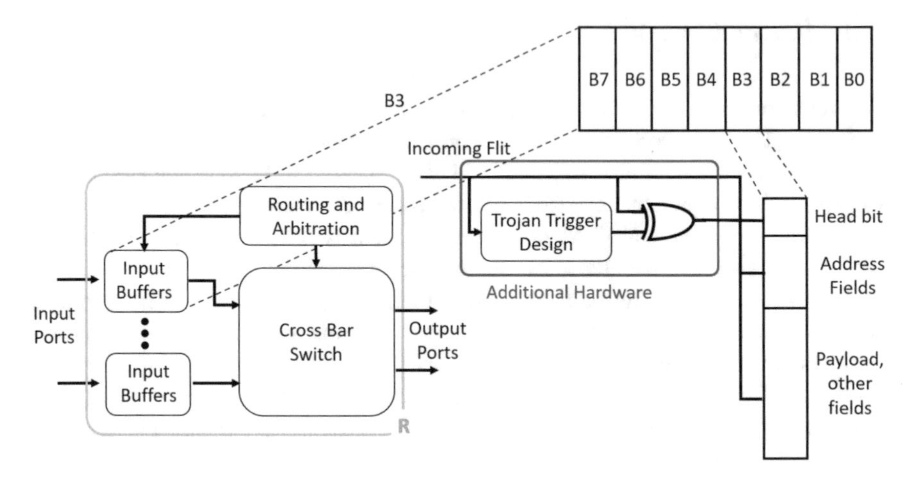

Fig. 5.4 Trojan design and attack scenario

which is unaware of the shuffling patterns, cannot target particular fields to launch a meaningful attack. Since the flit indication fields are typically one bit long, an error correcting code based on a 1-bit Hamming code was also proposed that can complement the bit shuffling mechanism.

Boraten et al. [7] discussed a similar threat model where hardware Trojans influenced resource allocations and corrupted data to degrade performance. The same authors further explored possible DoS attacks in [6]. Compared to router-based packet corruption, they discussed a Trojan that performs deep packet inspection on links and inject faults when the target is identified. The injected faults trigger re-transmissions from the error correcting mechanism. Therefore, repeated injection of faults causes repeated re-transmission to starve network resources and create deadlocks capable of rendering single application to full chip failures. The Trojan tries to prevent infected links from being detected by altering the locations of faults to disguise them as transient faults. While fuzzing can make the attack difficult, it does not guarantee prevention. Furthermore, the attack is not detected, and as a result, future attacks are not prevented either. The security countermeasure proposed in [6] was motivated by this, where the method coupled switch-to-switch scrambling, inverting, shuffling, and flit reordering with a heuristic-based fault detection model [6]. Their solution addressed the challenge of differentiating fault injections from transient and permanent faults.

Sepúlveda et al. presented *MalNoC*, a Trojan infected NoC that can perform multiple attacks on NoC packets [53]. MalNoC launches data integrity attacks by replacing the packet content with content in a malicious register. A similar threat model that discussed eavesdropping, DoS, and illegal packet forwarding, all of which utilized packet corruption at a router, was presented in [30].

The concept of "trust" between inter-connected entities has been studied before in the networking domain [54]. It tries to enhance the security of distributed networks such as ad-hoc networks by identifying attacks against trust evaluation systems and building defense techniques based on trust models. Concepts such as "web-of-trust" and "pretty-good-privacy" (PGP), which are widely used in internet communication, establish a similar notion that discusses the authenticity of binding a public key to the owner [56]. The OpenPGP email protocol is one such example [8].

5.4 NoC Trust Model

This section describes the trust model to quantitatively measure the trust between two nodes. Trust is established between two nodes to handle packets without tampering with the data. In particular, one node trusts the other node to perform the intended action on the received packet (in the case of routing, forward the packet to the next hop). In this chapter, the first node is referred to as the "producer" (α) and the second node as the "consumer" (β). The notation $\{producer \rightarrow consumer\}$ ($\alpha \rightarrow \beta$) is used to denote a trust relationship.[2] Trust can be established in two ways—(1) delegated trust and (2) direct trust. Direct trust is established when a node calculates trust about one of its neighbors. Trust is said to be delegated when one node recommends a consumer node to another producer node that is not directly connected to the consumer. The recommending node is referred to as "recommender." Figure 5.5a shows such an example. In this three node setup, direct trust can be established between B and C, and A and B. But, trust between A and C can only be established via B's recommendation. Therefore, $A \rightarrow C$ has a delegated trust relationship.

To quantify trust between two entities, a measure of trust is required. Keeping a binary value per node (either trusted or not) does not capture the entire trust model due to several reasons: (i) trust can be delegated (in the example in Fig. 5.5a, the

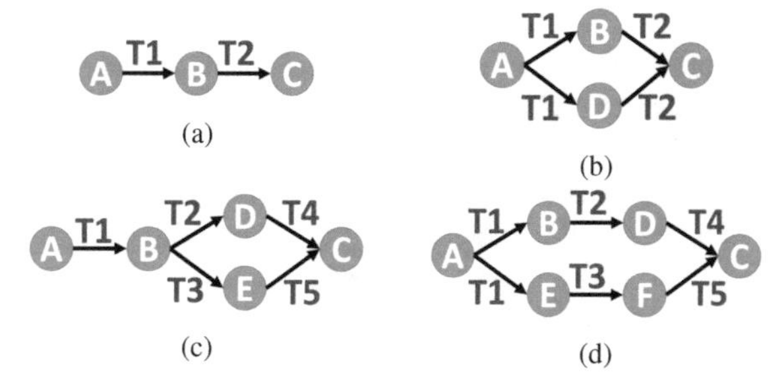

Fig. 5.5 Trust delegation across NoC. The values on the arrows represent the trust. For example, $T1$ in (a) denotes $T^{(a)}_{A \rightarrow B}$, where the superscript (a) corresponds to (a)

[2]The producer and consumer notations are different from src and dest since any two routers along the routing path can be producer/consumer, whereas src and dest refer to the origin of the packet and its destination, respectively.

amount of trust A places on C depends on how much A trusts B) and (ii) a malicious node might not launch an attack at first but do so after a while or periodically. Therefore, a value (denoted as $T_{\alpha \to \beta}$) between -1 and 1 is assigned for each trust relationship ($-1 \leq T_{\alpha \to \beta} \leq 1$) to indicate a trust value in the "potentially malicious" spectrum. The two bounds are defined as follows:

- when the producer is confident that the consumer will always function correctly: $T_{\alpha \to \beta} = 1$;
- when the producer is confident that the consumer is definitely malicious: $T_{\alpha \to \beta} = -1$.

In addition to the two bounds, $T_{\alpha \to \beta} = 0$ implies that the producer has no idea whether the consumer is malicious or not. Therefore, at the beginning of network packet transmission, all trust relationships are initialized to the value of zero. During operation, with information received from nodes, trust values are calculated. It is important to note that when B recommends C to A (delegated trust), $T^{(a)}_{A \to C}$ can be established only if $T^{(a)}_{A \to B} \geq 0$. In other words, A should not trust its enemy to recommend someone as trustworthy. Once this condition is met, three axioms are presented such that the trust delegation calculation adheres to those. The remainder of this section describes these axioms (Sect. 5.4.1) and elaborates how delegated trust (Sect. 5.4.2) and direct trust (Sect. 5.4.3) are calculated.

5.4.1 Axioms for Trust Delegation

Axiom 1 In delegated trust, trust value between producer and consumer should not be higher than the trust between producer and recommender as well as the trust between recommender and consumer. This can be formalized using Fig. 5.5a

$$\left| T^{(a)}_{A \to C} \right| \leq min(T^{(a)}_{A \to B}, T^{(a)}_{B \to C}). \tag{5.1}$$

Axiom 2 Producer receiving the same recommendation about the same consumer via multiple different recommenders should not reduce the trust between producer and consumer. In other words, the producer will be more certain about the consumer or at least maintain the same level of certainty if the producer obtains an extra recommendation that agrees with the producer's current opinion. For example, Fig. 5.5a and b shows two scenarios where A in first figure establishes trust with C via only one path and in the second scenario, trust with C is established through two same-trust paths.

$$T^{(b)}_{A \to C} \geq T^{(a)}_{A \to C} \geq 0, \text{ for } T1 > 0 \text{ and } T2 \geq 0, \tag{5.2}$$

$$T^{(b)}_{A \to C} \leq T^{(a)}_{A \to C} \leq 0, \text{ for } T1 > 0 \text{ and } T2 < 0. \tag{5.3}$$

This holds only if the multiple paths give the same recommendations.

Axiom 3 In a setup similar to Fig. 5.5c, it is possible to receive multiple recommendations from a single node (B). Compared to that, recommendations from independent nodes such as the ones shown in Fig. 5.5d (B and E) should always be trusted more. In other words, recommendations from independent nodes can reduce uncertainty more effectively than the recommendations from correlated nodes. Formally,

$$T^{(d)}_{A\rightarrow C} \geq T^{(c)}_{A\rightarrow C} \geq 0, \text{ if } T^{(c)}_{A\rightarrow C} \geq 0, \tag{5.4}$$

$$T^{(d)}_{A\rightarrow C} \leq T^{(c)}_{A\rightarrow C} \leq 0, \text{ if } T^{(c)}_{A\rightarrow C} < 0. \tag{5.5}$$

5.4.2 Delegated Trust Calculation

The calculation of trust from the point of view of any given node should adhere to the above axioms. For the example shown in Fig. 5.5a, the necessary condition is to satisfy Axiom 1. To achieve this, trust can be calculated by concatenation as $T^{(a)}_{A\rightarrow C} = T^{(a)}_{A\rightarrow B} \cdot T^{(a)}_{B\rightarrow C}$. In general,

$$T_{\alpha\rightarrow\beta} = T_{\alpha\rightarrow\gamma} \cdot T_{\gamma\rightarrow\beta}, \tag{5.6}$$

where γ is the recommender. As mentioned before, this can only be calculated if $T_{\alpha\rightarrow\gamma} \geq 0$. It can be noticed that if α has no idea about the trustworthiness of γ ($T_{\alpha\rightarrow\gamma} = 0$), no matter how much γ trusts β, α will not trust β ($T_{\alpha\rightarrow\beta} = 0$).

In case of multi-path trust delegation such as the example in Fig. 5.5b, Axioms 2 and 3 have to be satisfied in addition to Axiom 1. When α can establish trust with β via two paths, one via δ and another via ϵ ($\alpha - \delta - \beta$ and $\alpha - \epsilon - \beta$), the ratios of trust concatenation can be combined.

$$T_{\alpha\rightarrow\beta} = z_1 \cdot (T_{\alpha\rightarrow\delta} \cdot T_{\delta\rightarrow\beta}) + z_2 \cdot (T_{\alpha\rightarrow\epsilon} \cdot T_{\epsilon\rightarrow\beta}), \tag{5.7}$$

where

$$z_1 = \frac{T_{\alpha\rightarrow\delta}}{T_{\alpha\rightarrow\delta} + T_{\alpha\rightarrow\epsilon}}, \text{ and } z_2 = \frac{T_{\alpha\rightarrow\epsilon}}{T_{\alpha\rightarrow\delta} + T_{\alpha\rightarrow\epsilon}}. \tag{5.8}$$

5.4.3 Direct Trust Calculation

The direct trust is calculated based on the "sigmoid function" ($\frac{1}{1+e^{-x}}$), where x keeps track of the number of successful transmissions at a given router. Since the

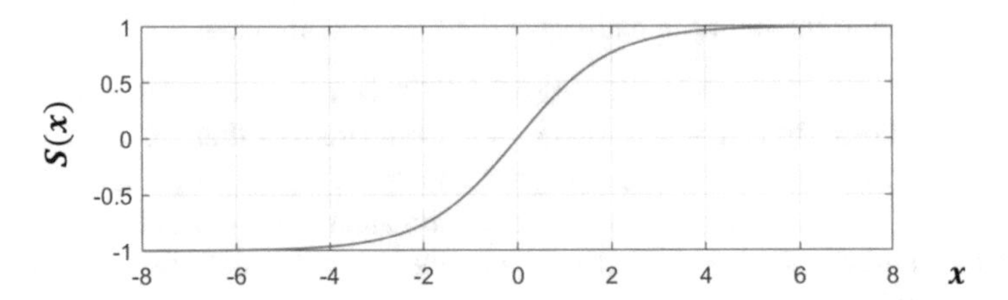

Fig. 5.6 Sigmoid function $S(x)$ variation with input x

sigmoid function ranges between 0 and 1, it is scaled to range between -1 and 1 (Fig. 5.6).

$$S(x) = 2 \cdot \frac{1}{1 + e^{-x}} - 1. \tag{5.9}$$

Assume that α and β are neighbors. Initially, α has no trust information about β. Therefore, $x = 0$, and as a result, $S(x) = 0$. When α learns about β's behavior, it changes the value x and re-calculates $S(x)$. For example, if α gets a positive feedback about β's trust, direct trust is calculated as $T_{\alpha \to \beta} = S(x + \delta)$, where δ is a small positive number. Since $S(x)$ is an increasing function as shown in Fig. 5.6, α's trust about β is now increased. Similarly, to reduce trust, $T_{\alpha \to \beta} = S(x - \delta)$. Therefore, direct trust is calculated as

$$x = x \pm \delta, \qquad T_{\alpha \to \beta} = S(x). \tag{5.10}$$

5.5 Trust-Aware Routing

Once the trust values are established, they are used by the routing protocol. The basic idea is to route packets through highly trusted nodes so that MIPs are avoided. It is important to note that trust values have to be dynamically updated during SoC execution since MIPs shift between malicious and non-malicious behaviors according to the threat model. The following subsections explain in detail how direct trust and delegated trust are updated at each router (Sects. 5.5.1 and 5.5.2, respectively) and how those trust values are used in routing (Sect. 5.5.3).

5.5.1 Updating Trust

According to the threat model used in this chapter, if a src IP does not receive a response to the packet sent, it can be because of two reasons:

- **Message was lost between src and dest**: In this case, a response is never received. The src times out after a while and re-transmits the packet. The routers along the routing path observe that this is a re-transmission and reduces the direct trust of their next hop neighbors. Direct trust is reduced since a packet took that path before and it was tampered. Direct trust re-calculation is done every time a re-transmission is observed. Once the trust values go down compared to the other possible paths, the packet takes an alternate path avoiding the MIP according to the routing protocol and is received at the dest.
- **Response was lost between dest and src**: This means that the packet was received at dest, but the response was not received by src. Again, src sends a re-transmission that is received by dest. The dest observes that this is an address that was previously served and sends the response again. Again, routers along the path observe that this is a re-transmission and reduces direct trust. This process is repeated until the response is received by src. This causes the routers between src and dest to reduce trust unnecessarily (false negative). However, this is not corrected because to do that, src has to keep track of all the paths the re-transmitted packets took to reset trust values. Furthermore, the routers should also maintain previous trust values. Therefore, false negatives are allowed to happen. With several ongoing communications overlapped between routers, the false negatives will regain trust over time.

Considering these scenarios, this method uses an event-driven approach to update trust. The overview of the algorithm is shown in Algorithm 1. To keep track of the re-transmissions and to increase/decrease direct trust according to that, a separate data structure is implemented at each router "Communication Table" (ComTable). It stores each pending communication using src, dest, address of corresponding memory location (addr), timestamp to indicate when the entry was added to the table, and a re-transmission flag (rtx flag). When a new packet arrives at a router, it checks to see if there is a pending communication between the same src and dest by matching src and dest fields in the packet header to entries in the ComTable (line 1). If yes, it can either be for the same address (line 3) or for a different address. If it is for a different address, it means that the previous communication has completed successfully. If it is for the same address, then it is identified as a re-transmission. The rtx flag is set to indicate this (line 4), and direct trust with the next hop ($getNextHop$ routine elaborated in Section 5.5.3) is reduced (line 6). If it is a new communication, the rtx flag is checked to see whether the previous communication between the same src and dest has not been flagged as a re-transmission before (line 8). If it has not been flagged before, the path can be trusted. Then, the direct trust with next hop router is increased (line 10) and the trust is delegated (line 11) to other neighbors as explained in Sect. 5.5.2. If it has already been flagged as a re-transmission, no further action is taken since it has already been penalized and as a result, direct trust has been reduced in a previous iteration (lines 4–6). In both cases, when it is a new communication, the ComTable is updated by removing the old entry and adding the new one (line 13). If it is the first communication that is passing through that router for that src and dest pair, a new entry is added in the

ComTable (line 16). The ComTable also records a timestamp for each entry. The timestamp is used to stop the exponential growth of the ComTable by removing old entries after a certain time threshold.

One limitation of this model is that it assumes that an IP will only send a second request to the same destination once the first one is served. For architectures that support multiple pending requests, this scheme can be easily extended. The sender maintains a list of pending requests and adds a header bit in the next packet to indicate that this is another request with the same src and dest but has a different address. Then, the routers check this bit before removing the previous entry, and trust is increased only if this bit is not set. The rest of the methodology remains the same.

5.5.2 Delegating Trust in the NoC

Once a communication is successfully completed, trust about the next hop ($T_{\alpha\rightarrow\beta}$) is delegated to nearby routers by each router ($delegateTrust$ routine in Algorithm 1). This is done by broadcasting a packet that contains $T_{\alpha\rightarrow\beta}$ with a pre-defined time-to-live (τ) value in the header in all directions except for the direction of the next hop router. During experiments, $\tau = 1$ is used. This causes the trust about the next hop router to be delegated to all other neighboring routers. An illustrative example of this mechanism is shown in Fig. 5.7. Once router α completes a communication where according to the routing protocol, the next hop router is β, it sends the direct trust value ($T_{\alpha\rightarrow\beta}$) to B, D, and E. These three routers now calculate $T_{B\rightarrow\beta}$, $T_{D\rightarrow\beta}$ and $T_{E\rightarrow\beta}$, which are delegated trust values, according to the trust model in Sect. 5.4.2 (Eq. (5.6)). As a result, B, D, and E learn about the trustworthiness of a router (β) two hops away from them.

It is possible that this delegated trust packet itself is tampered, and in that case, delegated trust will not be updated. This has no impact since a delegated trust packet being dropped means an MIP is on that path and its trust value will be negative. Delegated trust is updated only when it comes from a trusted source with a positive trust value according to Eq. (5.6).

5.5.3 Routing Protocol

The goal of the routing protocol is to avoid MIPs in the non-secure zone while routing through the most trusted routers. Each router stores the trust values of routers that are one (direct trust) and two hops away from it (delegated trust). When a router receives a packet, it first updates the trust values according to Algorithm 1. Next, the packet is forwarded to the next hop. Both forwarding and Algorithm 1 use the $getNextHop$ routine, which works as follows:

Algorithm 1: Updating direct and delegated trust

```
    This routine is called by each router every time a packet arrives.
    Input: packet
    Current node is assumed to be α
 1: entry ← checkComTable(packet)
 2: if entry ≠ NULL then
 3:     if entry.addr = packet.addr then
 4:         entry.rtxFlag ← 1
 5:         β ← getNextHop(packet)
 6:         T_{α→β} ← S(x − δ)
 7:     else
 8:         if entry.rtxFlag ≠ 1 then
 9:             β ← getNextHop(packet)
10:             T_{α→β} ← S(x + δ)
11:             delegateTrust()
12:         end if
13:         updateComTable(packet)
14:     end if
15: else
16:     updateComTable(packet)
17: end if

    Routine:checkComTable
    Input: packet
18: for entry ∈ comTable do
19:     if entry.src = packet.src & entry.dest = packet.dest then
20:         return entry
21:     end if
22: end for
23: return NULL

    Routine:updateComTable
    Input: packet
24: for entry ∈ comTable do
25:     if entry.src = packet.src & entry.dest = packet.dest then
26:         comTable.delete(entry)
27:     end if
28: end for
29: newEntry.src ← packet.src, newEntry.dest ← packet.dest
30: newEntry.addr ← packet.addr, newEntry.rtxFlag ← 0
31: newEntry.timestamp ← 0
32: comTable.add(newEntry)
```

- Read the dest ID of the packet.
- Compare dest and current router IDs.
  - If dest is located in the same row or column as the current router, the next hop is the neighboring router along that row or column toward dest.
  - Else, check the sum of trust values of routers one and two hops toward the dest, and select the neighbor along the path that has the largest trust value as

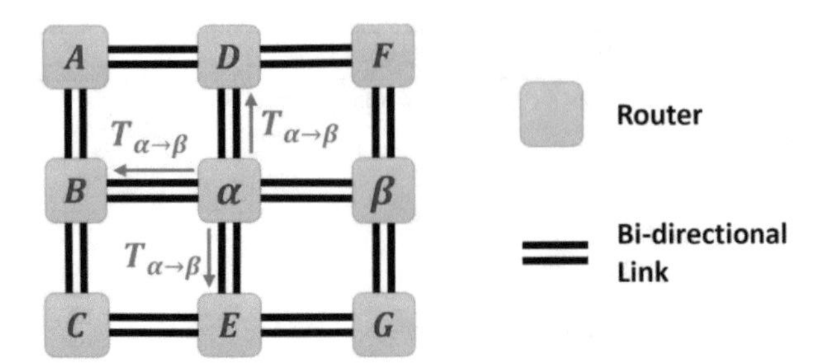

Fig. 5.7 Illustrative example showing that once a communication completes, the direct trust between α and β ($T_{\alpha\to\beta}$) is delegated to nodes one hop away from α

the next hop. If two paths have the same largest trust value, randomly pick one.

For example, in Fig. 5.7, assume a packet arrives at router B with the destination G. Since B is not in the same row or column as G, next hop is selected based on trust values. When considering routers that are one and two hops away from B in the direction of G, there are three possible paths: $B-\alpha-\beta$, $B-\alpha-E$, and $B-C-E$. Therefore, B calculates $max(T_{B\to\alpha}+T_{B\to\beta}, T_{B\to\alpha}+T_{B\to E}, T_{B\to C}+T_{B\to E})$, and if $T_{B\to C}+T_{B\to E}$ gives the maximum trust value, next hop is C. Considering nodes that are always toward the destination (in the example, B only considers α and C as next hops) ensures that the packet traverses the network following only one of the shortest paths. This together with the use of bi-directional links ensures the deadlock- and livelock-free nature of the routing algorithm. This routing protocol is identical to the congestion-aware routing protocol presented in [43] except that this method uses trust values instead of congestion values. Therefore, it can be shown that the routing protocol is also deadlock-free and livelock-free using the same arguments from [43].

It is important to note that this trust-aware routing protocol works even if all the IPs in the non-secure zone are malicious or, MIPs isolate the untrusted zone into several disconnected subzones of secure IPs. If all the neighbors of a router have a trust value of -1 (all routers are malicious), it will still be routed through that path since -1 is the largest value. Therefore, the packet is guaranteed to reach the destination but might be corrupted. If there is a path from source to destination that does not contain an MIP, this approach is guaranteed to find that path and deliver the packets without being corrupted.

5.6 Experiments

This section explores the feasibility and effectiveness of the approach by presenting experimental results and discussing the overheads associated with it.

5.6.1 Experimental Setup

An 8×8 Mesh NoC-based SoC was modeled with 64 cores using the gem5 cycle-accurate full-system simulator [5, 13]. The interconnection network was built on top of "GARNET2.0" model that is integrated with gem5 [1]. Each router in the mesh topology connects to four neighbors and a local IP via bi-directional links. Each IP connects to the local router through a network interface, which implements the MAC-then-encrypt protocol. The default XY routing protocol was modified to implement the trust-aware routing protocol. During experiments, $\delta = 0.5$ (Eq. (5.10)) was used when increasing/reducing direct trust. The value 0.5 was chosen experimentally such that the algorithm chooses alternative paths as quickly as possible while minimizing the impact of false negatives.

The system was tested using 4 real benchmarks (FFT, RADIX, FMM, LU) from the SPLASH-2 benchmark suite and 7 synthetic traffic patterns (uniform random (URD), tornado (TRD), bit complement (BCT), bit reverse (BRS), bit rotation (BRT), shuffle (SHF), transpose (TPS)). When running both real benchmarks and synthetic traffic patterns, each IP in the top (first) row of the Mesh NoC instantiated an instance of the task. Real benchmarks used 8 memory controllers that provide the interface to off-chip memory, which were connected to the bottom eight IPs. As synthetic traffic patterns do not use memory controllers, the destination of injected packets was selected based on the traffic pattern. For example, uniform random selected the destination from the IPs at the bottom row with equal probability. Source and destination modelling was done this way to mimic the secure and non-secure zones. Four MIPs were modeled and assigned at random to IPs in the other six rows. To simulate the sporadic behavior of the MIPs as discussed in the threat model, each MIP corrupted n consecutive packets after every p (period) packets. According to the architecture model, the IPs in the top row (secure zone) communicate with the IPs in the bottom row (secure zone) through the other 6 rows (non-secure zone) of IPs out of which 4 are malicious. This approach will work the same for any other secure, non-secure zone selection and MIP placement. The output of the gem5 simulation statistics was fed to the McPAT power modelling framework to obtain power consumption [33].

5.6.2 Performance Improvement

Figure 5.8 shows results related to the performance improvement when running real benchmarks. The figure compares performance results without the presence of MIPs (without MIP), with the presence of MIPs when default XY routing is used (with MIP-default), and when this approach is used with the presence of MIPs (with MIP-trust routing). We can observe that this approach reduces NoC delay by 53% (43.6% on average) compared to the default XY routing protocol. Execution time and number of packets injected are reduced by 9% (4.7% on average) and 71.8%

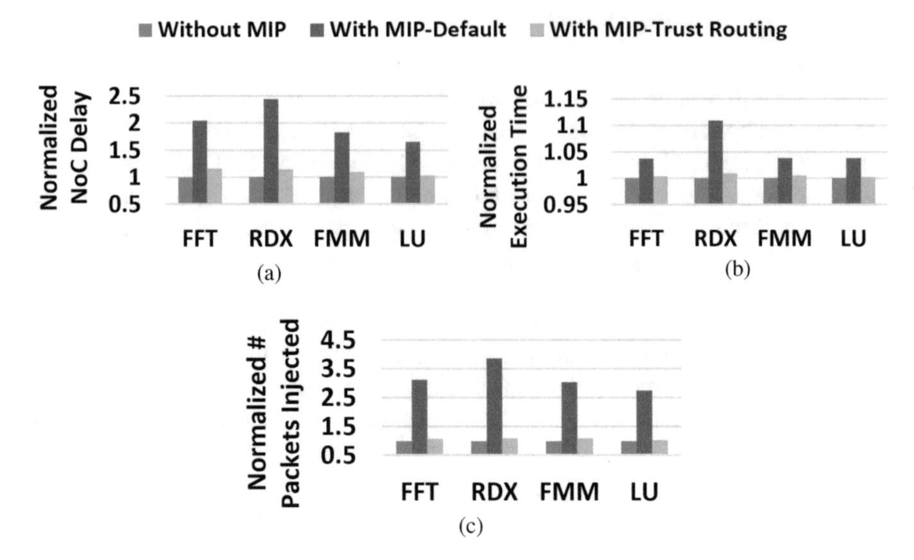

(a) (b) (c)

Fig. 5.8 NoC delay, execution time, and number of packets injected with and without the trust-aware routing model when running real benchmarks. $p = 20$ and $n = 14$. This figure is an extension of Fig. 5.2. (**a**) NoC delay. (**b**) Execution time. (**c**) No. of packets injected

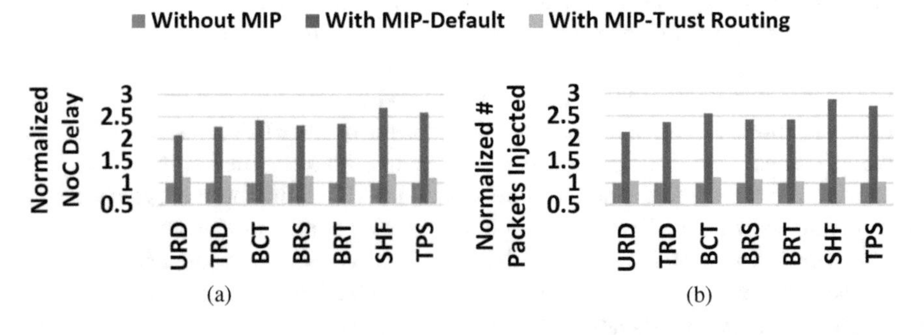

(a) (b)

Fig. 5.9 Execution time and number of packets injected with and without the trust-aware routing model when running synthetic traffic patterns. $p = 20$, $n = 14$. (**a**) NoC delay. (**b**) No. of packets injected

(66% on average), respectively. When the MIPs corrupt packets, re-transmissions are caused and its trust is reduced. As a result, alternative paths are chosen. The performance improvement depends on how quickly the algorithm chooses an alternative path once an attack is initiated.

In addition to real benchmarks, experiments were conducted with synthetic traffic traces as well. Results related to synthetic traffic patterns are shown in Fig. 5.9. The comparison is the same as that of Fig. 5.8. It shows that NoC delay and number of packets injected on the NoC are reduced by 57.1% (51.2% on average) and 56.7% (50.1% on average), respectively.

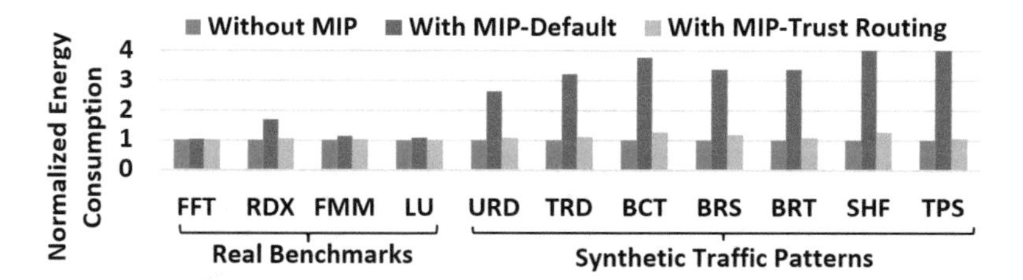

Fig. 5.10 Energy consumption with and without the trust-aware routing model when running real benchmarks and synthetic traffic patterns. $p = 20, n = 14$

5.6.3 *Energy Efficiency Improvement*

As a result of reduced execution time and reduced number of re-transmissions, the energy consumption of the SoC also reduces. Figure 5.10 shows the energy consumption comparison. Note that 47.4% (28.3% on average) of less energy is consumed by real benchmarks when routing using the approach presented in this chapter compared to the default XY routing in the presence of MIPs. Synthetic traffic demonstrates energy savings of up to 75.6% (67.6% on average). Compared to real benchmarks, synthetic traffic patterns show more energy reduction since synthetic traffic focuses only on network traversals unlike real benchmarks that go through the entire processor pipeline including instruction execution, NoC traversal, and memory operations.

5.6.4 *Overhead Analysis*

To implement this routing protocol, additional hardware is required at each router. This includes extra memory to store trust values and hardware to calculate, update, and propagate trust. To accommodate a row in the ComTable, 10 bytes of memory is required (6-bit src, 6-bit dest, 32-bit addr, 1-bit rtx flag, 32-bit timestamp). The maximum size of the ComTable during the experiments was 24. This leads to 240 bytes of extra memory requirement per router.

The default 5-stage router pipeline (buffer write, virtual channel allocation, switch allocation, switch traversal, and link traversal) was implemented in gem5. Once separate hardware is implemented, computations related to trust can be carried out in a pipelined fashion in parallel to the computations in the router pipeline. To evaluate the area overhead, the RTL design of an open source NoC router [47] was modified and synthesized the design with 180nm GSCLib library from Cadence using Synopsis Design Compiler. This resulted in an area overhead of 6% compared to the default router. This shows that the trust-aware routing protocol is lightweight and can be effectively implemented at routers in an NoC-based SoC.

5.7 Summary

In this chapter, we discussed denial-of-service attacks caused by packet corruption in NoC-based SoCs. We outlined several work in the existing literature that addressed the challenge and outlined a trust-aware routing protocol that is capable of routing packets by avoiding malicious IPs. The routing protocol is implemented based on a trust model that calculates how much a neighboring node can be trusted to route packets through that router. The experiments conducted by using both real benchmarks and synthetic traffic patterns demonstrated significant performance and energy efficiency improvements compared to traditional XY routing in the presence of a MAC-then-encrypt security protocol. Overhead analysis has revealed that the area overhead to implement the routing protocol is only 6%. This approach can be integrated with any existing authentication scheme as well as other threat mitigation techniques, to secure the SoC while minimizing the performance and energy efficiency degradation caused by malicious IP tampering packets.

Acknowledgments This work was partially supported by the National Science Foundation (NSF) grant SaTC-1936040.

References

1. N. Agarwal, T. Krishna, L. Peh, N.K. Jha, Garnet: A detailed on-chip network model inside a full-system simulator, in *2009 IEEE International Symposium on Performance Analysis of Systems and Software* (2009), pp. 33–42
2. A. Ahmed, F. Farahmandi, Y. Iskander, P. Mishra, Scalable hardware Trojan activation by interleaving concrete simulation and symbolic execution, in *2018 IEEE International Test Conference (ITC)* (IEEE, New York, 2018), pp. 1–10
3. A. Ahmed, F. Farahmandi, P. Mishra, Directed test generation using concolic testing on RTL models, in *2018 Design, Automation & Test in Europe Conference & Exhibition (DATE)* (2018), pp. 1538–1543
4. A. Ahmed, Y. Huang, P. Mishra, Cache reconfiguration using machine learning for vulnerability-aware energy optimization. ACM Trans. Embedd. Comput. Syst. **18**(2), 1–24 (2019)
5. N. Binkert, B. Beckmann, G. Black, S.K. Reinhardt, A. Saidi, A. Basu, J. Hestness, D.R. Hower, T. Krishna, S. Sardashti, R. Sen, K. Sewell, M. Shoaib, N. Vaish, M.D. Hill, D.A. Wood, The gem5 simulator. SIGARCH Comput. Archit. News **39**(2), 1–7 (2011)
6. T. Boraten, A.K. Kodi, Mitigation of denial of service attack with hardware Trojans in NoC architectures, in *2016 IEEE International Parallel and Distributed Processing Symposium (IPDPS)* (2016), pp. 1091–1100
7. T. Boraten, D. DiTomaso, A.K. Kodi, Secure model checkers for network-on-chip (NoC) architectures, in *2016 International Great Lakes Symposium on VLSI (GLSVLSI)* (2016), pp. 45–50

8. J. Callas et al., OpenPGP message format. Technical report (2007)
9. S. Charles, P. Mishra, Reconfigurable network-on-chip security architecture. ACM Trans. Des. Autom. Electron. Syst. **25**(6), 1–25 (2020). Article 53. https://doi.org/10.1145/3406661
10. S. Charles, P. Mishra, Lightweight and trust-aware routing in NoC-based SoCs, in *2020 IEEE Computer Society Annual Symposium on VLSI (ISVLSI)* (2020), pp. 160–167
11. S. Charles, P. Mishra, Securing network-on-chip using incremental cryptography, in *2020 IEEE Computer Society Annual Symposium on VLSI (ISVLSI)* (2020), pp. 168–175
12. S. Charles, H. Hajimiri, P. Mishra, Proactive thermal management using memory-based computing in multicore architectures, in *2018 Ninth International Green and Sustainable Computing Conference (IGSC)* (2018), pp. 1–8
13. S. Charles, C.A. Patil, U.Y. Ogras, P. Mishra, Exploration of memory and cluster modes in directory-based many-core CMPs, in *2018 Twelfth IEEE/ACM International Symposium on Networks-on-Chip (NOCS)* (2018), pp. 1–8
14. S. Charles, Y. Lyu, P. Mishra, Real-time detection and localization of dos attacks in NoC based SoCs, in *2019 Design, Automation Test in Europe Conference Exhibition (DATE)* (2019), pp. 1160–1165
15. S. Charles, A. Ahmed, U.Y. Ogras, P. Mishra, Efficient cache reconfiguration using machine learning in NoC-based many-core CMPs. ACM Trans. Des. Autom. Electron. Syst. **24**(6), 1–23 (2019)
16. S. Charles, M. Logan, P. Mishra, Lightweight Anonymous Routing in NoC based SoCs, in *2020 Design, Automation & Test in Europe Conference & Exhibition (DATE)* (IEEE, New York, 2020)
17. S. Charles, Y. Lyu, P. Mishra, Real-time detection and localization of distributed dos attacks in NoC based SoCs, in *IEEE Transactions on Computer-Aided Design of Integrated Circuits and Systems* (2020)
18. M. Chen, X. Qin, H.-M. Koo, P. Mishra, *System-Level Validation: High-Level Modeling and Directed Test Generation Techniques* (Springer, New York, 2012)
19. T. Dierks, The Transport Layer Security (tls) protocol version 1.2. (2008)
20. D. Engels, X. Fan, G. Gong, H. Hu, E. Smith, Ultra-lightweight cryptography for low-cost RFID tags: hummingbird algorithm and protocol. Centre for Applied Cryptographic Research (CACR) Technical Reports, 29, 01 (2009)
21. F. Farahmandi, P. Mishra, Automated debugging of arithmetic circuits using incremental gröbner basis reduction, in *2017 IEEE International Conference on Computer Design (ICCD)* (IEEE, New York, 2017), pp. 193–200
22. F. Farahmandi, P. Mishra, FSM anomaly detection using formal analysis, in *2017 IEEE International Conference on Computer Design (ICCD)* (IEEE, New York, 2017), pp. 313–320
23. F. Farahmandi, P. Mishra, Automated test generation for debugging multiple bugs in arithmetic circuits. IEEE Trans. Comput. **68**(2), 182–197 (2018)
24. F. Farahmandi, Y. Huang, P. Mishra, Trojan localization using symbolic algebra. in *2017 22nd Asia and South Pacific Design Automation Conference (ASPDAC)* (IEEE, New York, 2017), pp. 591–597
25. F. Farahmandi, Y. Huang, P. Mishra, *System-on-Chip Security: Validation and Verification* (Springer Nature, New York, 2019)
26. U. Gupta, C.A. Patil, G. Bhat, P. Mishra, U.Y. Ogras, DyPo: Dynamic pareto-optimal configuration selection for heterogeneous MpSoCs. ACM Trans. Embedd. Comput. Syst. **16**(5s), 1–20 (2017)
27. Y. Huang, P. Mishra, Vulnerability-aware energy optimization for reconfigurable caches in multitasking systems. IEEE Trans. Comput.-Aided Des. Integr. Circ. Syst. **38**(5), 809–821 (2019)
28. Y. Huang, S. Bhunia, P. Mishra, MERS: statistical test generation for side-channel analysis based Trojan detection, in *ACM SIGSAC Conference on Computer and Communications Security (CCS)* (2016), pp. 130–141
29. Y. Huang, S. Bhunia, P. Mishra, Scalable test generation for Trojan detection using side channel analysis. IEEE Trans. Inf. Forensics Secur. **13**(11), 2746–2760 (2018)

30. M. Hussain, A. Malekpour, H. Guo, S. Parameswaran, EETD: an energy efficient design for runtime hardware Trojan detection in untrusted network-on-chip, in *2018 IEEE Computer Society Annual Symposium on VLSI (ISVLSI)* (2018), pp. 345–350
31. L.S. Indrusiak, J. Harbin, M.J. Sepulveda, Side-channel attack resilience through route randomisation in secure real-time networks-on-chip, in *2017 12th International Symposium on Reconfigurable Communication-centric Systems-on-Chip (ReCoSoC)* (2017), pp. 1–8
32. H.K. Kapoor, G. Bhoopal Rao, S. Arshi, G. Trivedi, A security framework for NoC using authenticated encryption and session keys. Circ. Syst. Signal Process. **32**(6), 2605–2622 (2013)
33. S. Li, J.H. Ahn, R.D. Strong, J.B. Brockman, D.M. Tullsen, N.P. Jouppi, McPAT: an integrated power, area, and timing modeling framework for multicore and many-core architectures. in *2009 42nd Annual IEEE/ACM International Symposium on Microarchitecture (MICRO)* (2009), pp. 469–480
34. Y. Lyu, P. Mishra, A survey of side-channel attacks on caches and countermeasures. J. Hardw. Syst. Secur. **2**(1), 33–50 (2018)
35. Y. Lyu, P. Mishra, Efficient test generation for Trojan detection using side channel analysis, in *Design Automation & Test in Europe Conference (DATE)* (2019), pp. 408–413
36. Y. Lyu, P. Mishra, Automated test generation for activation of assertions in RTL models, in *2020 25th Asia and South Pacific Design Automation Conference (ASPDAC)* (IEEE, New York, 2020), pp. 223–228
37. Y. Lyu, P. Mishra, Automated test generation for Trojan detection using delay-based side channel analysis, in *2020 Design, Automation & Test in Europe Conference & Exhibition (DATE)* (2020), pp. 1031–1036
38. Y. Lyu, P. Mishra, Automated trigger activation by repeated maximal clique sampling. in *Asia and South Pacific Design Automation Conference (ASPDAC)*, pp. 482–487 (2020)
39. Y. Lyu, P. Mishra, Scalable activation of rare triggers in hardware Trojans by repeated maximal clique sampling, in
textitIEEE Transactions on Computer-Aided Design of Integrated Circuits and Systems (2020)
40. Y. Lyu, P. Mishra, Scalable concolic testing of RTL models. IEEE Trans. Comput. (2020). https://doi.org/10.1109/TC.2020.2997644
41. Y. Lyu, X. Qin, M. Chen, P. Mishra, Directed test generation for validation of cache coherence protocols. IEEE Trans. Comput.-Aided Des. Integr. Circ. Syst. **38**(1), 163–176 (2018)
42. Y. Lyu, A. Ahmed, P. Mishra, Automated activation of multiple targets in RTL models using concolic testing, in *2019 Design, Automation & Test in Europe Conference & Exhibition (DATE)* (IEEE, New York, 2019), pp. 354–359
43. M. Li, Q.-A. Zeng, W.-B. Jone, DyXy - a proximity congestion-aware deadlock-free dynamic routing method for network on chip, in *2006 43rd ACM/IEEE Design Automation Conference* (2006), pp. 849–852
44. J.Y.V. Manoj Kumar, A.K. Swain, S. Kumar, S.R. Sahoo, K. Mahapatra, Run time mitigation of performance degradation hardware Trojan attacks in network on chip, in *2018 IEEE Computer Society Annual Symposium on VLSI (ISVLSI)* (2018), pp. 738–743
45. P. Mishra, F. Farahmandi, *Post-Silicon Validation and Debug*. (Springer, New York, 2019)
46. P. Mishra, S. Bhunia, M. Tehranipoor, *Hardware IP security and Trust* (Springer, New York, 2017)
47. A. Monemi, J.W. Tang, M. Palesi, M.N. Marsono, ProNoC: a low latency network-on-chip based many-core system-on-chip prototyping platform. Microprocess. Microsyst. **54**, 60–74 (2017)
48. S. Murali, T. Theocharides, N. Vijaykrishnan, M.J. Irwin, L. Benini, G. De Micheli, Analysis of error recovery schemes for networks on chips. IEEE Des. Test Comput. **22**(5), 434–442 (2005)
49. Z. Pan, P. Mishra, Automated test generation for hardware Trojan detection using reinforcement learning, in *Asia and South Pacific Design Automation Conference (ASPDAC)* (2021)
50. Z. Pan, J. Sheldon, P. Mishra, Test generation using reinforcement learning for delay-based side channel analysis, in *IEEE/ACM International Conference on Computer-Aided Design (ICCAD)* (2020)

51. J.S. Rajesh, D.M. Ancajas, K. Chakraborty, S. Roy, Runtime detection of a bandwidth denial attack from a rogue network-on-chip, in *Proceedings of the 9th International Symposium on Networks-on-Chip, NOCS '15*, New York, NY, 2015 (Association for Computing Machinery, New York, 2015)
52. Security on Arm TrustZone. https://www.arm.com/products/security-on-arm/trustzone (2008). [Online]
53. J. Sepúlveda, A. Zankl, D. Flórez, G. Sigl, Towards protected MPSoC communication for information protection against a malicious NoC. Proc. Comput. Sci. **108**, 1103–1112 (2017). *International Conference on Computational Science, ICCS 2017*, 12–14 June 2017, Zurich
54. Y.L. Sun, Z. Han, W. Yu, K.J.R. Liu, A trust evaluation framework in distributed networks: vulnerability analysis and defense against attacks. in *Proceedings IEEE INFOCOM 2006. 25TH IEEE International Conference on Computer Communications* (2006), pp. 1–13
55. W. Wang, P. Mishra, S. Ranka, *Dynamic Reconfiguration in Real-Time Systems*. (Springer, New York, 2012)
56. P. Zimmerman, Pretty good privacy (1995)

Chapter 6
Lightweight Anonymous Routing for On-chip Interconnects

Subodha Charles and Prabhat Mishra

6.1 Introduction

The growth of general purpose as well as embedded computing devices has been remarkable over the past decade. This was mainly enabled by the advances in manufacturing technologies that allowed the integration of many heterogeneous components on a single System-on-Chip (SoC). The drastic increase in SoC complexity has led to a significant increase in SoC design and validation complexity [3, 17, 21, 23, 39, 43–46]. The tight time-to-market deadlines and increasing complexity of modern SoCs have led manufacturers to outsource intellectual property (IP) cores from potentially untrusted third-party vendors [25, 30]. Therefore, the trusted computing base of the SoC should exclude the third-party IPs. In fact, measures should be taken since malicious third-party IPs (M3PIP) can launch passive as well as active attacks on the SoC [2, 22, 24, 25, 41, 42, 47, 49]. Such attacks are possible primarily because the on-chip interconnection network that connects SoC components together, popularly known as Network-on-Chip (NoC), has visibility of the entire SoC and the communications between IP cores. Previous efforts have developed countermeasures against stealing information [9], snooping attacks [53], and even causing performance degradation by launching denial-of-service (DoS) attacks [13]. In this chapter, we discuss countermeasures for M3PIPs operating under the following architecture and threat models.

S. Charles (✉)
University of Moratuwa, Colombo, Sri Lanka
e-mail: s.charles@ieee.org

P. Mishra
University of Florida, Gainesville, FL, USA
e-mail: prabhat@ufl.edu

P. Mishra, S. Charles (eds.), *Network-on-Chip Security and Privacy*,
https://doi.org/10.1007/978-3-030-69131-8_6

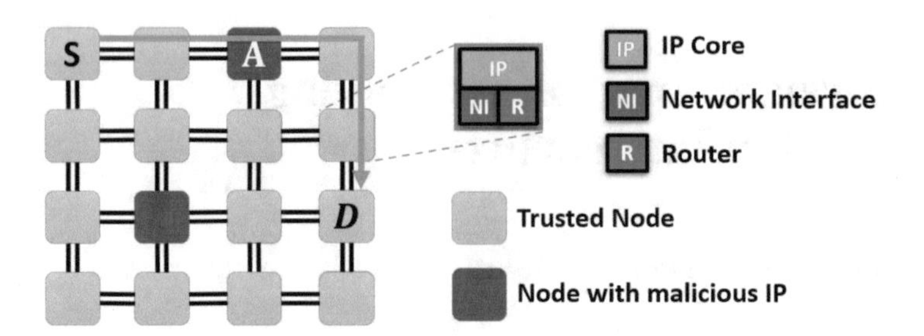

Fig. 6.1 Overview of a typical SoC architecture with IPs integrated on a Mesh NoC

Threat Model Figure 6.1 shows an SoC with heterogeneous IPs integrated on a Mesh NoC. The two nodes marked as *S* (source) and *D* (destination) are trusted IPs communicating with each other. M3PIPs integrated on the SoC (nodes shown in red) have the following capabilities when packets pass through their routers:

- They can steal information if data is sent as plaintext.
- If data is encrypted and header information is kept as plaintext, they can gather packets generated from the same source and intended to the same destination and launch complex attacks such as linear/differential cryptanalysis since they belong to the same communication session.
- When multiple M3PIPs are present on the same NoC, they can share information and trace messages.
- An M3PIP can compromise the router attached to it and gather information stored in the router. This can leak routing information. Assuming only some of the IPs are acquired from untrusted third-party vendors, all routers will never be compromised at the same time.

It is not feasible to utilize traditional security methods (encryption, authentication, etc.) in resource-constrained embedded devices. Previous studies explored lightweight security architectures to mitigate threats. Previous work on lightweight encryption proposed smaller block and key sizes, less rounds of encryption, and other hardware optimizations [31]. This chapter proposes a *Lightweight Encryption and Anonymous Routing protocol for NoCs* (LEARN) that requires only few addition and multiplication operations for encryption. The proposed approach is capable of eliminating the traditional encryption methods consisting of ciphers and keys entirely by using the *secret sharing* approach proposed by Shamir [57] without compromising the security guarantees. Furthermore, the framework supports anonymous routing such that an intermediate node can neither detect the origin nor the destination of a packet. Major contributions of this chapter can be summarized as follows:

- An anonymous routing scheme is proposed that hides both source and destination information making the packets untraceable. Launching attacks on encrypted data passing through a given router becomes more difficult when the packets are untraceable and origins are unknown.
- A lightweight encryption scheme is developed that is based on secret sharing.
- The efficiency of the solution is discussed in detail compared to other existing solutions.

The remainder of this chapter is organized as follows. Section 6.2 introduces some key concepts used in this chapter. Section 6.3 discusses other related research efforts in lightweight encryption and anonymous routing. Section 6.4 motivates the need for this work. Section 6.5 describes the lightweight encryption and anonymous routing protocol. Section 6.6 presents the experimental results. Section 6.7 discusses possible further enhancements to the approach. Finally, Sect. 6.8 summarizes the chapter.

6.2 Background

This section introduces some of the key concepts used in the proposed framework.

6.2.1 Symmetric and Asymmetric Encryption

Symmetric Encryption A symmetric encryption scheme takes the same key K for both decryption and encryption. The encryption algorithm E produces the *ciphertext* C by taking the key K and a *plaintext* M as inputs. This is denoted by $C \leftarrow E_K(M)$. Similarly, the *decryption* algorithm D denoted by $M \leftarrow D_K(C)$ takes a key K and a ciphertext C and returns the corresponding M. The correctness of the scheme is confirmed when any sequence of messages M_1, ..., M_u encrypted under a given key K produces $C_1 \leftarrow E_K(M_1)$, $C_2 \leftarrow E_K(M_2)$,..., $C_u \leftarrow E_K(M_u)$, and is related as $D_K(C_i) = M_i$ for each C_i.

Asymmetric Encryption In asymmetric encryption, also known as public key encryption, different keys are used for encryption and decryption. Encryption is done using the *public key* that is publicly known by all the entities in the environment. An entity B that wants to send a message M to another entity A will encrypt the message using A's public key (with public key PK_A) to produce ciphertext C denoted by $C \leftarrow E_{PK_A}(M)$. The ciphertext can only be decrypted by A's *secret key (private key)* SK_A corresponding to PK_A. SK_A is known by only A, and therefore, only A can decrypt C to produce M denoted by $M \leftarrow D_{SK_A}(C)$.

6.2.2 Authenticated Encryption with Associated Data

To encrypt packets in real-time embedded systems, the encryption scheme should support high-speed encryption with low performance and power overhead. To achieve this, the operation mode of the encryption scheme must support pipelined and parallelized implementations. Furthermore, due to the nature of packets transferred and routing protocols used in the NoC, some of the packet fields such as addresses, sequence numbers, and ports need to be transferred in plaintext. These fields are mainly the header fields of the packet. This has motivated the development of AEAD *(Authenticated Encryption with Associated Data)* schemes where the packet payload is encrypted and header information is sent as plaintext. The "Counter Mode" is a commonly used operation mode for encryption which supports AEAD. Figure 6.2 shows an overview of counter mode including both encryption and authentication components.

AEAD schemes can protect sensitive data from eavesdroppers while ensuring the integrity of packets. However, since the header fields are sent as plaintext, attackers can use this header information to identify packets from the same information flow and launch more complex attacks as discussed under the threat model. If the header field is encrypted as a solution to this vulnerability, the intermediate routers have to decrypt headers to learn the next hop of the packet, which can lead to unacceptable performance and energy overhead.

6.2.3 Secret Sharing with Polynomial Interpolation

Shamir's secret sharing [57] is based on a property of *Lagrange polynomials* known as the (k, n) threshold. It specifies that a certain secret M can be broken into n parts and M can only be recovered if at least k $(k \leq n)$ parts are retrieved. The knowledge

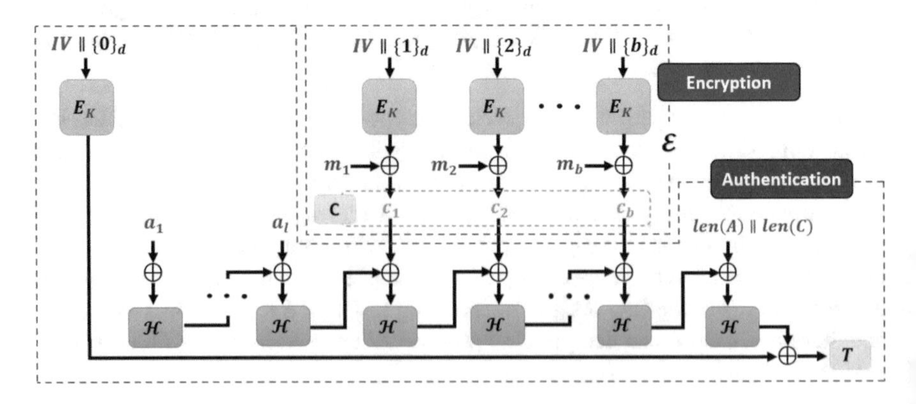

Fig. 6.2 Encryption and authentication in counter mode

of less than k parts leave M completely unknown. Lagrange polynomials meet this property with $k = n$. A Lagrange polynomial is comprised of some k points (x_0, y_0), ..., (x_{k-1}, y_{k-1}) where $x_i \neq x_j$ $(0 \leq i, j \leq k-1)$. A unique polynomial of degree $k-1$ can be calculated from these points:

$$L(x) = \sum_{j=0}^{k-1} l_j(x) \cdot y_j, \tag{6.1}$$

where

$$l_j(x) = \prod_{i=0, i \neq j}^{k-1} \frac{x - x_i}{x_j - x_i} \tag{6.2}$$

Any attempt to reconstruct the polynomial with less than k or incorrect points will give the incorrect polynomial with the wrong coefficients and/or wrong degree.

$L(x)$ forms the interpolated Lagrange polynomial, and $l_j(x)$ is the Lagrange basis polynomial. In order to share a secret using this method, a random polynomial of degree $k-1$ is chosen. It takes the form of $L(x) = a_0 + a_1 x + a_2 x^2 + \ldots + a_{k-1} x^{k-1}$. The shared secret M should be set as $a_0 = M$, and all the other coefficients are chosen randomly. Then a simple calculation at $x = 0$ would yield the secret $(M = L(0))$. In this case, k points on the curve are chosen at random and distributed together with their respective $l_j(0)$ values—the Lagrangian coefficients. To retrieve M, all the parties should share their portions of the secrets. Once all of the k points and $l_j(0)$ coefficients are combined, then the secret can be computed as:

$$M = \sum_{j=0}^{k-1} l_j(0) \cdot y_j \tag{6.3}$$

This method makes it easier to compute M without having to recalculate each $l_j(x)$.

6.2.4 Router and Routing Protocol

The routers in an NoC comprise input buffers that accept packets from the local IP via the NI or from other routers connected to it (in the Mesh topology, except for the routers in the border, each router is connected to the local IP and four other routers). Figure 6.3 shows an overview of the NoC packet traversal process. Based on the addresses in the packet header and the routing protocol, the crossbar switch routes data from the input buffers to the appropriate output port. Buffers are allocated for virtual channels which help avoid deadlock. The switch allocator handles input port arbitration for output ports [18].

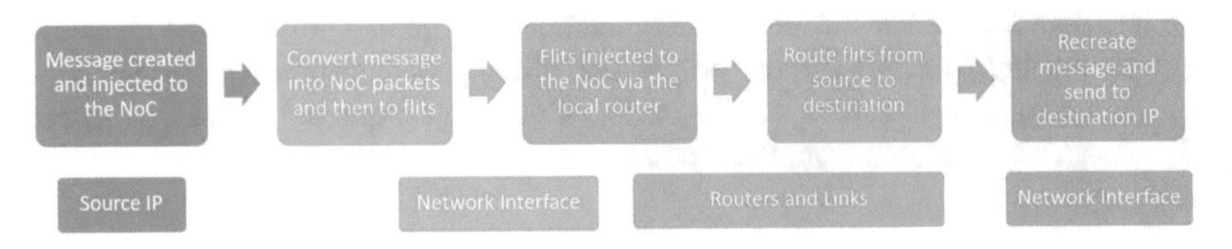

Fig. 6.3 Overview of an NoC traversal

The routing protocol defines the path a flit should take in a given topology. Routing protocols are broadly classified as (1) deterministic routing protocols and (2) adaptive routing protocols. In deterministic routing, each packet traversing from S to D follows the same path. X-Y routing is one common example for deterministic routing. In X-Y routing, packets use X-directional links first, before using Y-directional links [19]. Adaptive routing takes network states such as congestion, security, reliability into account and takes the flits through different paths based on the current state of the network [65].

6.2.5 Anonymous Communication using Onion Routing

Onion routing is widely used in the domain of computer networks when routing has to be done while keeping the sender anonymous. Each message is encrypted several times (layers of encryption) analogous to layers of an onion. Each intermediate router from source to destination (called onion routers) "peels" a single layer of encryption revealing the next hop. The final layer is decrypted and message is read at the destination. The identity of the sender is preserved since each intermediate router only knows the preceding and the following routers. The overhead of onion routing comes from the fact that the sender has to do several rounds of encryption before sending the packet to the network and each intermediate router has to do a decryption before forwarding it to the next hop. While this can be done in computer networks, adopting this in resource-constrained NoCs leads to unacceptable performance overhead as illustrated in Sect. 6.4.

6.3 Related Work

The current state of the art in NoC security revolves around protecting information traveling in the network against side channel [54], physical [52], and software attacks [5]. Other attacks such as denial-of-service [8, 13, 16] and buffer overflow [26, 36] have also been explored. Recent efforts try to combine the advantages of logic testing and side-channel analysis for effective Trojan detection in integrated circuits [29, 30, 37, 38, 40, 50]. However, developing efficient and flexible solutions

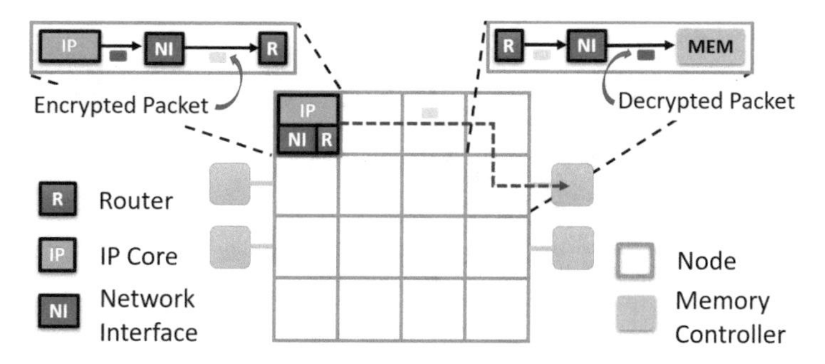

Fig. 6.4 NoC-based many-core architecture connecting IPs on a single SoC using a 4 × 4 Mesh topology. Communication between two IPs (in this case, a processor IP and a memory controller) is encrypted so that an eavesdropper cannot extract the packet content

at lower costs and minimal impact on performance as well as how to certify these solutions remain as challenges to the industry. It is not feasible to adopt the security mechanisms used in the computer networks domain in NoC-based SoCs due to the resource-constrained nature of embedded devices [10, 12, 14]. Security has to be considered in the context of other non-functional requirements such as performance, power, and area. Designers explore complementary directions. They try to employ various techniques to improve energy efficiency in NoC-based SoCs [4, 11, 14, 27, 28, 60]. They also try to optimize these security mechanisms to fit the performance and power budgets of embedded systems. This thought process has led to prior efforts on securing NoC-based SoC [5, 8, 15, 55, 56], which tried to eliminate complex encryption schemes such as AES and replace them with lightweight encryption schemes. Figure 6.4 shows a typical NoC-based many-core architecture which encrypts packets transferred between IP cores. The packets are encrypted at the network interface before being injected into the network and at the destination, decryption is done before passing to the destination IP.

Intel's TinyCrypt, a cryptographic library with a small footprint, is built for resource-constrained devices [59]. It provides basic functionality to build a secure system with minor overhead. It provides SHA-256 hash functions, message authentication, a pseudo-random number generator which can run using minimal memory, encryption, and the ability to create nonces[1] and challenges. Apart from Intel TinyCrypt, several researches have proposed other lightweight encryption solutions in the Internet-of-Things (IoT) domain [6, 48]. All of these solutions follow the traditional encryption method which takes a key and a plaintext as inputs to produce the ciphertext. In contrast, this chapter outlines a method where each router along the routing path contributes a portion of the message such that the message changes at each router and only the destination receives the entire message. This can be

[1] A **nonce** is a random number that is used only once during the lifetime of a cryptographic operation.

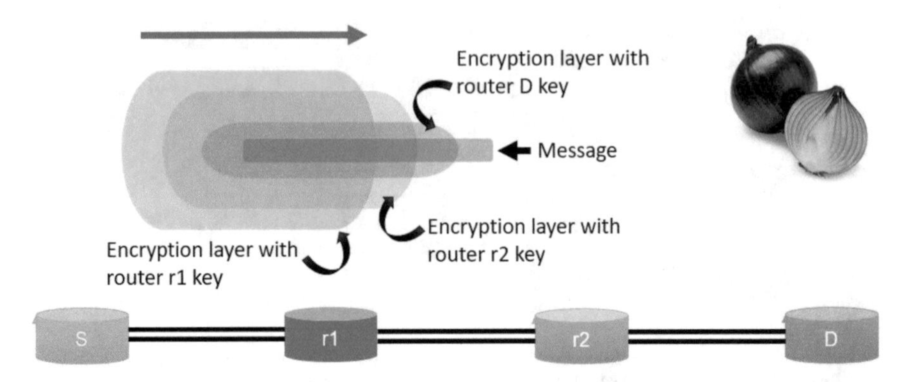

Fig. 6.5 Illustrative example for Onion routing

implemented using very few addition and multiplication operations leading to a lightweight solution for secure communication.

Even if the data in a packet are encrypted, typically, source and destination addresses are sent as plaintext for faster routing (e.g., AEAD schemes). Therefore, eavesdroppers can easily extract this information. To prevent such attacks, the route has to be anonymized. Hiding the source and destination information of NoC packets can ensure that the malicious agents in the NoC are unable to select the target application to eavesdrop. Existing work on anonymous routing (e.g., onion routing, mix-nets, dining cryptographers, etc.) considers mobile ad-hoc networks (MANETS) [33, 35, 51] as well as computer and vehicular networks [64]. The idea behind the widely used *onion routing* is explained in Sect. 6.2.5. An illustrative example is shown in Fig. 6.5. The main challenge in using these anonymous routing protocols in resource-constrained SoCs is that the protocol uses decryption ("peeling the onion") at each hop leading to unacceptable performance overhead. Optimized anonymous routing protocols in MANETS (e.g., [51]) use an on-demand lightweight anonymous routing protocol that eliminates per-hop decryption. However, the MANETS environment is fundamentally different from an NoC. Their work cannot address the unique communication requirements of an NoC as well as not designed for task migration and context switching.

Ancajas et al. [5] presented a lightweight solution that is compatible with the NoC context. Their work proposed to migrate applications periodically to another node in the SoC. The SoC firmware maintains the relevant information and initiates a seamless migration periodically. As a result, the source and destination addresses are decoupled making it harder for a malicious agent to launch the attack. In contrast, this chapter outlines a method which defines the routing path using a three-way handshake that only exposes the preceding and following routers at each hop. The entire routing path as well as the source and destination details are hidden.

6.4 Motivation

Security and performance is always a trade-off in resource-constrained systems. While computer networks with potentially unlimited resources can accommodate very strong security techniques such as AES encryption and onion routing, utilizing them in resource-constrained NoCs can lead to unacceptable overhead. To evaluate this impact, FFT, RADIX (RDX), FMM, and LU benchmarks from the SPLASH-2 benchmark suite [62] were run on an 8×8 Mesh NoC-based SoC with 64 IPs using the gem5 simulator [7] considering three scenarios:

- **No-Security:** NoC does not implement encryption or anonymous routing.
- **Enc-only:** NoC secures data by encrypting before sending into the network. However, it does not support anonymous routing.
- **Enc-and-AR:** Data encryption as well as anonymous routing achieved by onion routing.

A 12-cycle delay was assumed for encryption/decryption when simulating Enc-only and Enc-and-AR according to the evaluations in [55]. More details about the experimental setup is given in Sect. 6.6.1. Results are shown in Fig. 6.6. The values are normalized to the scenario that consumes the most time. Enc-only shows 42% (40% on average) increase in NoC delay (total NoC traversal delay for all packets) and 9% (7% on average) increase in execution time compared to the No-Security implementation. Enc-and-AR gives worse results with 83% (81% on average) increase in NoC delay leading to a 41% (33% on average) increase in execution time when compared with No-Security. In other words, Enc-and-AR leads to approximately 1.5X performance degradation. When security is considered, No-Security leaves the data totally vulnerable to attackers, Enc-only secures the data by encryption and Enc-and-AR provides an additional layer of security

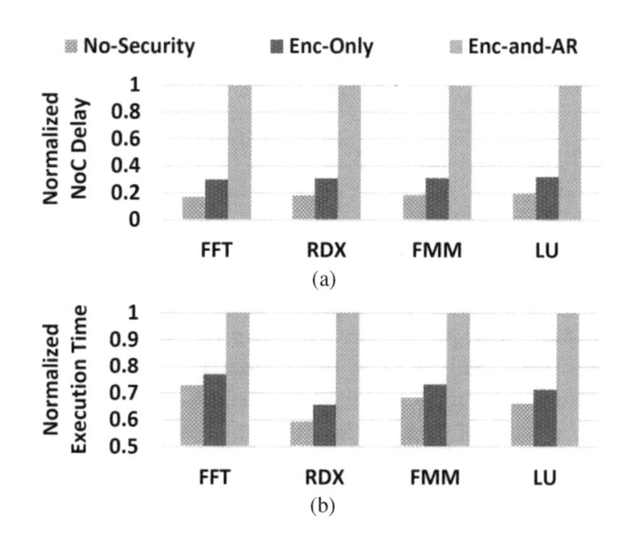

Fig. 6.6 NoC delay and execution time comparison across different levels of security. (**a**) NoC delay. (**b**) Execution time

with anonymous routing. The overhead of Enc-only is caused by the complex mathematical operations, and the number of cycles required to encrypt each packet. Onion routing used in Enc-and-AR aggravates this by requiring several rounds of encryption before injecting the packet into the network as well as decryption at each hop (router). Added security has less impact on execution time compared to NoC delay since execution time also includes the time for instruction execution and memory operations in addition to NoC delay. In many embedded systems, it would be unacceptable to have security at the cost of 1.5X performance degradation. It would be ideal if the security provided by Enc-and-AR can be achieved while maintaining performance comparable to No-security. The approach outlined in this chapter tries to achieve this goal by introducing a lightweight encryption and anonymous routing protocol as described in the next section.

6.5 Lightweight Encryption and Anonymous Routing Protocol

This section describes the proposed approach—Lightweight Encryption and Anonymous Routing protocol for NoCs (LEARN). By utilizing secret sharing based on polynomial interpolation [57], LEARN negates the need for complex cryptographic operations to encrypt messages. A forwarding node would only have to compute the low overhead addition and multiplication operations to hide the contents of the message. As the message passes through the forwarding path, its appearance is changed at each node, which makes the message's content and route safe from eavesdropping attackers as well as internal ones. The following sections describe the approach in detail. First, we provide an overview of the framework in Sect. 6.5.1. Next, Sects. 6.5.2 and 6.5.3 describe the two major components of the proposed routing protocol (route discovery and data transfer). Finally, Sect. 6.5.4 outlines how to efficiently manage relevant parameters during anonymous routing.

6.5.1 Overview

LEARN has two main phases as shown in Fig. 6.7. When an IP wants to communicate with another IP, it first completes the "Route Discovery" phase. The route discovery phase sends a packet and discovers the route, distributes the parameters among participants. Then the "Data Transfer" phase transfers the message securely and anonymously. The route discovery phase includes a three-way handshake between the sender and the destination nodes. The handshake uses 3 out of the 4 main types of packets sent over the network with the fourth type being used in the second phase. The four main packet types are:

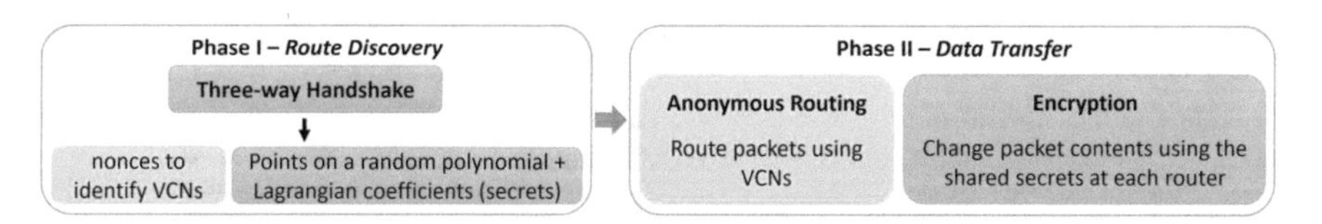

Fig. 6.7 Overview of the proposed framework (LEARN)

1. RI (Route Initiate)—flooded packet from sender S to destination D to initialize the conversation.
2. RA (Route Accept)—packet sent from D to accept new connection with S.
3. RC (Route Confirmation)—sent from S to distribute configuration parameters with intermediate nodes.
4. DT (Data)—the data packet from S to D that is routed anonymously through the NoC.

Algorithm 1 outlines the major steps of LEARN. During the three-way handshake, a route between S and D is discovered. Each router along the routing path is assigned with few parameters that are used when transferring data—(i) random nonces to represent preceding and following routers (line 2), and (ii) a point in a random polynomial together with its Lagrangian coefficient (line 3). This marks the end of the first phase which enables the second phase—"Data Transfer." The second phase uses the parameters assigned to each router to forward the original message through the route anonymously while hiding its contents. Anonymous routing is achieved by using the random nonces which act as virtual circuit numbers (VCN). When transferring data packets, the intermediate routers will only see the VCNs corresponding to the preceding router and the following router which reveals no information about the source or the destination (line 7). Encryption is achieved using the points in the random polynomial and their corresponding Lagrangian coefficients. Each router along the path changes the contents of the message in such a way that only the final destination will be able to retrieve the entire message (line 6).

Algorithm 1: Major steps of LEARN

```
   Phase I - Route Discovery
1: for all r ∈ routers do
2:     r ← υ_i, υ_j                         ▷ nonces to identify VCNs
3:     r ← (x_k, y_k, b_k)                  ▷ a point in a random polynomial
4: end for
   Phase II - Data Transfer
5: while r ≠ destination do
6:     m ← F(m, (x_k, y_k, b_k))            ▷ modify message
7:     r ← getNextHop(υ_i, υ_j)             ▷ get next hop
8: end while
```

Table 6.1 Notations used to illustrate LEARN

Notation	Description
$OPK_S^{(i)}$	One-time public key (OPK) used by the source to uniquely identify an RA packet
$OSK_S^{(i)}$	Private key corresponding to $OPK_S^{(i)}$
ρ	Random number generated by the source
PK_D	The global public key of the destination
SK_D	The private key corresponding to PK_D
$TPK_A^{(i)}$	Temporary public key of node A
$TSK_A^{(i)}$	The private key corresponding to $TPK_A^{(i)}$
K_{S-A}	Symmetric key shared between S and A
υ_A	Randomly generated nonce by node A
b_i	Lagrangian coefficient of a given point (x_i, y_i)
$E_K(M)$	A message M encrypted using the key K

LEARN improves performance by replacing complex cryptographic operations with addition/multiplication operations that consume significantly less time during the data transfer phase. The overhead occurs during the first phase (route discovery) that requires cryptographic operations. However, this is performed only a constant number of times (once per communication session). Since the route discovery phase happens only once in the beginning of a communication session, the cost for route discovery gets amortized over time. This leads to significant performance improvement.

Note that the route discovered at the route discovery stage will remain the same for the lifetime of the task. In case of context switching and/or task migration, the first phase will be repeated before transferring data. Each IP in the SoC that uses the NoC to communicate with other IPs follows the same procedure. The next two sections describe these two phases in detail. A list of notations used to illustrate the idea is listed in Table 6.1. The superscript "i" is used to indicate that the parameter is changed for each packet of a given packet type.

6.5.2 Route Discovery

The route discovery phase performs a three-way handshake between the sender S and destination D. This includes broadcasting the first packet—RI from S with the destination D, getting a response (RA) from D acknowledging the reception of RI, and finally, sending RC with the parameters required to implement polynomial interpolation based secret sharing. Figure 6.8 shows an illustrative example of parameters (using only four nodes) shared and stored during the handshake.

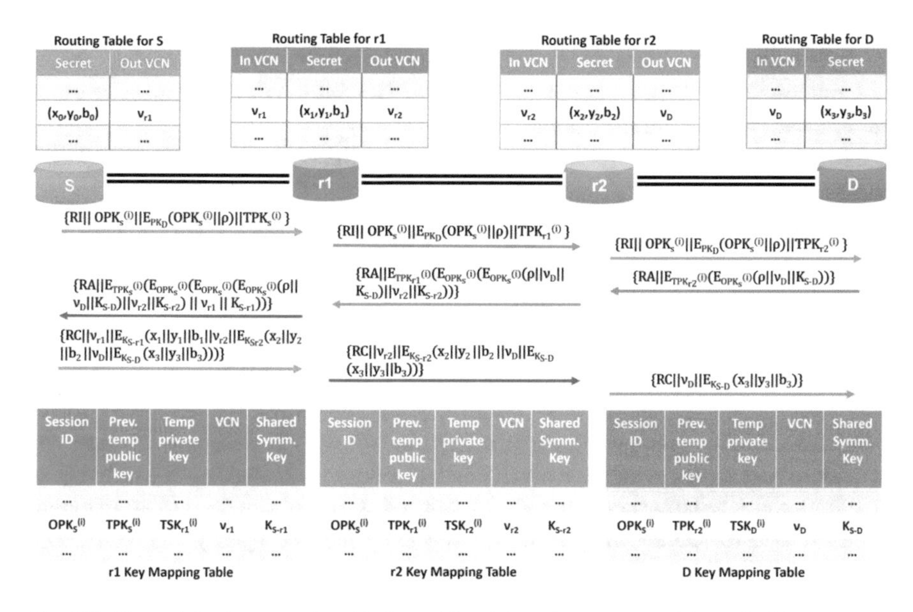

Fig. 6.8 Steps of the three-way handshake and the status of parameters at the end of the process

The initial route initiate packet (RI) takes the form:

$$\left\{RI \parallel OPK_S^{(i)} \parallel E_{PK_D}(OPK_S^{(i)} \parallel \rho) \parallel TPK_S^{(i)}\right\}$$

The first part of the message indicates the type of packet being sent, RI in this case. $OPK_S^{(i)}$ refers to the one-time public key associated with the sender node. This public key together with its corresponding private key $OSK_S^{(i)}$ change with each new conversation or RI. This change allows for a particular conversation to be uniquely identified by these keys, which are saved in its route request table. ρ is a randomly generated number by the sender that is concatenated with the $OPK_S^{(i)}$ and then encrypted with the destination node's public key PK_D as a global trapdoor [32]. Since PK_D is used to encrypt, only the destination is able to open the trapdoor using SK_D. Then the $TPK_S^{(i)}$ is attached to show the temporary key of the forwarding node, which is initially the sender. The temporary keys are also implemented as one-time trapdoors to ensure security.

The next node, $r1$, to receive the RI messages goes through a few basic steps. Firstly, it checks for the $OPK_S^{(i)}$ in its key mapping table, which would indicate a duplicated message. Any duplicates are discarded at this step. Next, $r1$ will attempt to decrypt the message and retrieve ρ. Success would indicate that $r1$ was the intended recipient D. If not, $r1$ replaces $TPK_S^{(i)}$ with its own temporary public key $TPK_{r1}^{(i)}$ and broadcasts:

$$\left\{RI \parallel OPK_S^{(i)} \parallel E_{PK_D}(OPK_S^{(i)} \parallel \rho) \parallel TPK_{r1}^{(i)}\right\}$$

$r1$ also logs $OPK_S^{(i)}$ and $TPK_S^{(i)}$ from the received message and $TSK_{r1}^{(i)}$ corresponding to $TPK_{r1}^{(i)}$ in its key mapping table. This information is used later when an RA message is received from D.

D will eventually receive the RI message and will decrypt using SK_D. This will allow D to retrieve $OPK_S^{(i)}$ and ρ from $E_{PK_D}(OPK_S^{(i)} \parallel \rho)$. Then to verify that the RI has not been tampered with, D will compare the plaintext $OPK_S^{(i)}$ and the now decrypted $OPK_S^{(i)}$. If they are different, the RI is simply discarded. Otherwise, D sends a RA (route accept) message:

$$\left\{RA \parallel E_{TPK_{r2}^{(i)}}(E_{OPK_S^{(i)}}(\rho \parallel \upsilon_D \parallel K_{S-D}))\right\} \tag{6.4}$$

RA, like RI in the previous message, is there to indicate message type. D generates a random nonce, υ_D, to serve as a VCN and a randomly selected key K_{S-D} to act as a symmetric key between S and D. D stores υ_D and K_{S-D} in its key mapping table. It also makes an entry in its routing table indexed by υ_D, the VCN. The concatenation of ρ, υ_D, and K_{S-D} is then encrypted with the $OPK_S^{(i)}$, so that only S can access that information. Then the message is encrypted again by $TPK_{r2}^{(i)}$, $r2$'s temporary public key, with $r2$ being the node that delivered RI to D.

Once $r2$ receives the RA, it decrypts it using its temporary private key, $TSK_{r2}^{(i)}$, and follows the same steps as D. It generates its own nonce, υ_{r2}, and shared symmetric key, K_{S-r2}, to be shared with S. Both the nonce and symmetric key are then concatenated to the RA message and encrypted by S's public key, $OPK_S^{(i)}$, so that only S can retrieve that data. This adds another layer of encrypted content to the message for S to decrypt using $OSK_S^{(i)}$. Similar to D, $r2$ also stores υ_{r2} and K_{S-r2} in its key mapping table and routing table. It then finds the temporary public key for the previous node in the path from its key mapping table—$TPK_{r1}^{(i)}$ and encrypts the message. The message sent out by $r2$ looks like:

$$\left\{RA \parallel E_{TPK_{r1}^{(i)}}\left(E_{OPK_S^{(i)}}\left(E_{OPK_S^{(i)}}(\rho \parallel \upsilon_D \parallel K_{S-D}) \parallel \upsilon_{r2} \parallel K_{S-r2}\right)\right)\right\} \tag{6.5}$$

This process is repeated at each node along the path until the RA packet makes its way back to S. The entire message at that point is encrypted with $TPK_S^{(i)}$, which is stripped away using $TSK_S^{(i)}$. Then S can "peel" each layer of the encrypted message by $OSK_S^{(i)}$ to retrieve all the VCNs, shared symmetric keys, and also, ρ. ρ is used to authenticate that the entire message came from the correct destination and was not changed during the journey.

Once S completes authentication of the received RA packet, it randomly generates $k+1$ points $(x_0, y_0), (x_1, y_1), ..., (x_k, y_k)$ on a k degree polynomial $L(x)$ as shown in Fig. 6.9. $k+1$ is the number of nodes in the path from S to D. S then uses these points to calculate the Lagrangian coefficients, $b_0, b_1, ..., b_k$, using:

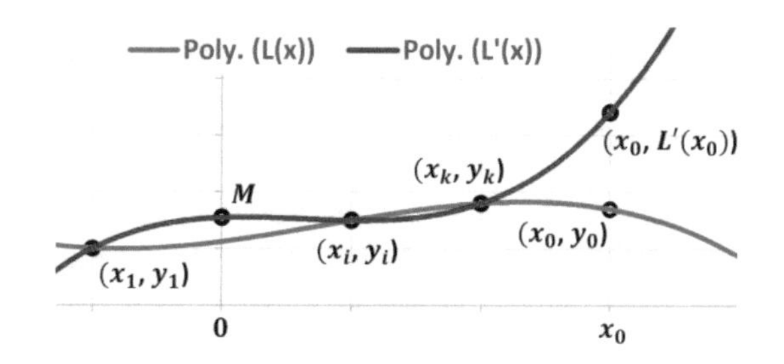

Fig. 6.9 Lagrangian polynomials $L(x)$ and $L'(x)$ together with the selected points

$$b_j = \prod_{i=0, i \neq j}^{k} \frac{x_i}{x_i - x_j} \tag{6.6}$$

Using the generated data, S constructs a route confirmation (RC) packet:

$$\begin{aligned} \{RC \parallel \upsilon_{r1} \parallel E_{K_{S-r1}}(x_1 \parallel y_1 \parallel b_1 \\ \parallel \upsilon_{r2} \parallel E_{K_{S-r2}}(x_2 \parallel y_2 \parallel b_2 \parallel \upsilon_D \parallel E_{K_{S-D}}(x_3 \parallel y_3 \parallel b_3)))\} \end{aligned} \tag{6.7}$$

Similar to the case in RA and RI, RC in the packet refers to the packet type. The rest of the message is layered much like the previous RA packet. Each layer contains the υ_* for each node concatenated with secret information that is encrypted with the shared key K_{S-*}, where * corresponds to $r1$, $r2$, or D in the example (Fig. 6.8). The (υ_*, K_{S-*}) pair was generated by each node during the RA packet transfer phase and the values were stored in the key mapping tables as well as entries indexed by the VCNs created in the routing table. Therefore, each node can decrypt one layer, store incoming and outgoing VCNs together with the secret, and pass it on to the next node to do the same. For example, $r1$ receiving the packet can observe that the incoming VCN is υ_{r1}. It then decrypts the first layer using the symmetric key K_{S-r1}, that is already stored in the key mapping table, and recovers the secret (x_1, y_1, b_1) as well as the outgoing VCN υ_{r2}. It then updates the entry indexed by υ_{r1} in its routing table with the secret tuple and the outgoing VCN. Similarly, each router from S to D can build its routing table.

6.5.3 Data Transfer

The path set up can now be used to transfer messages from S to D anonymously. For each conversation, $k + 1$ points were generated on a random curve $L(x)$ chosen by S. During the last step of the route discovery phase (RC packet), S kept (x_0, y_0, b_0) for itself and distributed each node on the discovered path a different point, (x_i, y_i) (where $1 \leq i \leq k$), with the corresponding Lagrangian coefficient b_i. If S wants to send the message M to D, S has to generate a new k degree polynomial $L'(x)$

which is defined by the k points distributed to nodes except for (x_0, y_0), i.e., points (x_i, y_i) where $(1 \leq i \leq k)$ and a new point $(0, M)$. This makes $L'(0) = M$ with M as the secret message, according to the explanation in Sect. 6.2.3. S then changes its own point (x_0, y_0) to (x_0, y_0') where $y_0' = L'(x_0)$, making sure the point retained by S is also on the curve L'(x) as shown in Fig. 6.9. It is important to note that every coefficient b_i and every point distributed to nodes along the route remain unchanged. For this scenario, considering Eq. (6.3), we can derive:

$$M = y_0' b_0 + \sum_{i=1}^{k} b_i \cdot y_i \tag{6.8}$$

To transfer a secret message, M, from S to D anonymously, S constructs data transfer (DT) packet with the form:

$$\{DT \parallel \upsilon_{r1} \parallel y_0' b_0\} \tag{6.9}$$

DT, like every other packet, has an indicator of packet type at the front of the packet—DT. υ_{r1} is the VCN of the next node. $y_0' b_0$ is the portion of the message M that is constructed by S. Once $r1$ receives the DT packet, it adds its own portion of the message, $y_1 b_1$, to $y_0' b_0$. It also uses its routing table to find the VCN of the next node and replaces the incoming VCN by the outgoing VCN in the DT packet. Therefore, the message received by $r2$ has the form:

$$\{DT \parallel \upsilon_{r2} \parallel y_0' b_0 + y_1 b_1\} \tag{6.10}$$

Next, $r2$ repeats the same process and forwards the packet:

$$\{DT \parallel \upsilon_D \parallel y_0' b_0 + y_1 b_1 + y_2 b_2\} \tag{6.11}$$

to D. Eventually, D will be able to retrieve the secret message, $M = y_0' b_0 + y_1 b_1 + y_2 b_2 + y_3 b_3$ by adding the last portion $y_3 b_3$ constructed using the part of the secret D shared. Using this method, neither an intermediate node nor an eavesdropper in the middle will be able to see the full message since the message M is incomplete at every intermediate node and is fully constructed only at the destination D.

6.5.4 Parameter Management

To ensure the efficient implementation of LEARN, an important aspect needs to be addressed—the generation and management of keys and nonces. Many previous studies have addressed this problem in several ways. One such example is the work done by Lebiednik et al. [34]. In their work, a separate IP called the key distribution center (KDC) handles the distribution of keys. Each node in the network negotiates a new key with the KDC using a pre-shared portion of memory that is known by only the KDC and the corresponding node. The node then communicates with the KDC using this unique key whenever it wants to obtain a new key. The KDC can then allocate keys depending on whether it is symmetric/asymmetric encryption, and inform other nodes as required. The key request can delay the communication. But once keys are established, it can be used for many times depending on the length of the encrypted packet before refreshing to prevent linear distinguishing attacks. In this approach, the keys are only used during the route discovery phase, and the discovered route will remain the same for the lifetime of the task unless context switching or task migration happens. Therefore, key refreshing will rarely happen and the cost for the initial key agreement as well as the route discovery phase will be amortized.

6.6 Experiments

This section presents results to evaluate the efficiency of the approach (LEARN). We first describe the experimental setup. Next, we compare the performance of LEARN with traditional encryption and anonymous routing protocols introduced in Sect. 6.4. Finally, we discuss the area overhead and security aspects of LEARN.

6.6.1 Experimental Setup

Extending the results presented in Fig. 6.6, LEARN was tested on an 8×8 Mesh NoC-based SoC with 64 IPs using the gem5 cycle-accurate full-system simulator [7]. The NoC was built using the "GARNET2.0" model that is integrated with gem5 [1, 12]. The route discovery phase of the approach relies on the RI, RA, and RC packets traversing along the same path to distribute the keys and nonces. Therefore, the topology requires bidirectional links connecting the routers. While the experiment were conducted on a Mesh NoC, there are many other NoC topologies that can adopt LEARN where all links are bidirectional as evidenced by academic research [1] as well as commercial SoCs [58].

Each encryption/decryption is modeled with a 12-cycle delay [55]. Computations related to generating the random polynomial and deciding the k points are assumed to consume 200 cycles. To accurately capture congestion, the NoC was modeled with 3-stage (buffer write, route compute + virtual channel allocation + switch allocation, and link traversal) pipelined routers with wormhole switching and 4 virtual channel buffers at each input port. Each link was assumed to consume one cycle to transmit packets between neighboring routers. The delays were chosen to be consistent with the delays of components in the gem5 simulator.

The default gem5 and Garnet2.0 configurations were used for packet sizes, virtual channels, and flow control. In addition to the four main types of packets described in Sect. 6.5.1, the DT packets can be further divided into two categories as control and data packets. For example, in case of a cache miss, a memory request packet (control packet) is injected into the NoC and the memory response packet (data packet) consists of the data block from the memory. The address portion of a control DT packet consists of 64 bits. In the data DT packet, in addition to the 64-bit address, 512 bits are reserved for the data block. A credit-based, virtual channel flow control was used in the architecture. Each data VC and control VC was allocated buffer depths of 4 and 1, respectively.

LEARN was tested using 6 real benchmarks (FFT, RADIX, FMM, LU, OCEAN, CHOLESKY) from the SPLASH-2 benchmark suite and 6 synthetic traffic patterns: uniform random (URD), tornado (TRD), bit complement (BCT), bit reverse (BRS), bit rotation (BRT), transpose (TPS). Out of the 64 cores, 16 IPs were chosen at random and each one of them instantiated an instance of the task. The packets injected into the NoC when running the real benchmarks were the memory requests/responses. 8 memory controllers were used that provided the interface to off-chip memory which were placed on the boundary of the SoC. This memory controller placement adheres to commercial SoC architectures such as Intel's Knights Landing (KNL) [58]. An example to illustrate the IP placement is shown in Fig. 6.10.

When running real benchmarks, the packets get injected to the NoC when there are private cache misses and the frequency of that happening depends on the characteristics of the benchmark. When running synthetic traffic patterns, packets were injected into the NoC at the rate of 0.01 packets/node/cycle. For synthetic traffic patterns, the destinations of injected packets were selected based on the traffic pattern. For example, uniform random selected the destination from the remaining IPs with equal probability whereas bit complement, complemented the

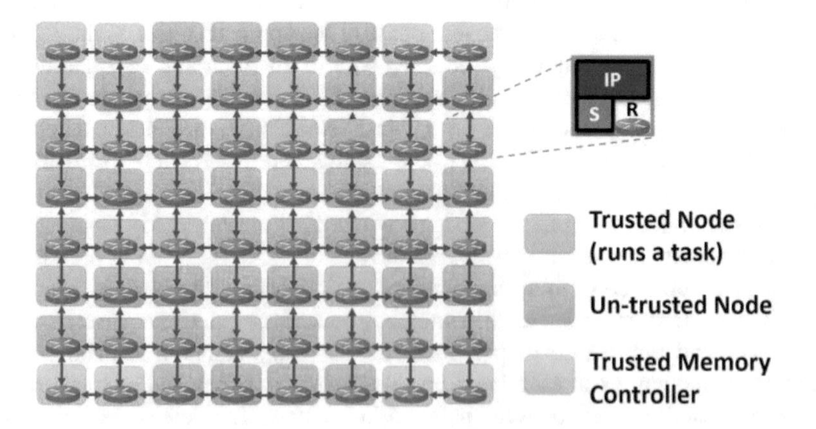

Fig. 6.10 8×8 Mesh NoC architecture used to generate results including trusted nodes running the tasks and communicating with memory controllers while untrusted nodes can potentially have malicious IPs

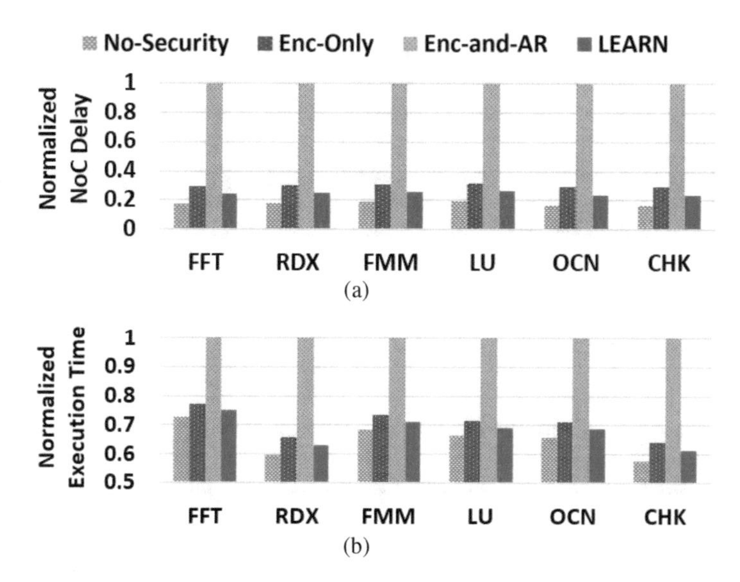

Fig. 6.11 NoC delay and execution time comparison across different security levels using real benchmarks. (**a**) NoC delay. (**b**) Execution time

bits of the source address to get the destination address, etc. The choices made in the experimental setup were motivated by the architecture/threat model and the behavior of the gem5 simulator. However, LEARN can be used with any other NoC topology and task/memory controller placement.

6.6.2 Performance Evaluation

Figure 6.11 shows performance improvement LEARN can gain when running real benchmarks. The results from LEARN were compared against the three scenarios considered in Fig. 6.6. Compared to the No-Security scenario, LEARN consumes 30% more time (28% on average) for NoC traversals (NoC delay) and that results in only 5% (4% on average) increase in total execution time. Compared to Enc-and-AR which also implements encryption and anonymous routing, LEARN improves NoC delay by 76% (74% on average) and total execution time by 37% (30% on average). We can observe from the results that the performance of LEARN is even better than Enc-Only, which provides encryption without anonymous routing. Overall, LEARN can provide encryption and anonymous routing consuming only 4% performance overhead compared to the NoC that does not implement any security features.

The same experiments were carried out using synthetic traffic traces, and results are shown in Fig. 6.12. Since synthetic traffic patterns only simulate NoC traffic and do not include instruction execution and memory operations, only NoC delay is

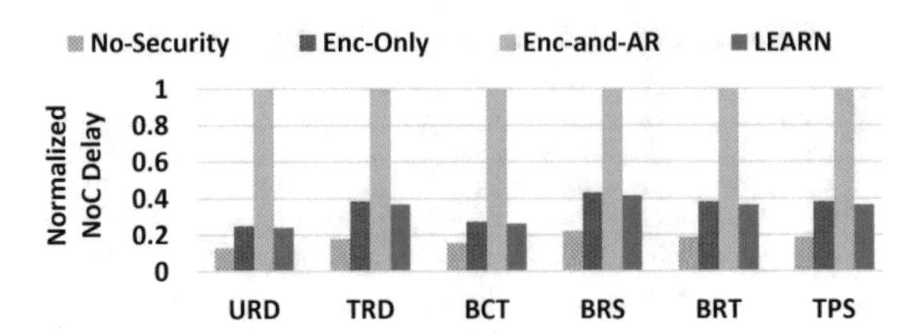

Fig. 6.12 NoC delay comparison across different levels of security when running synthetic traffic patterns

shown in the figure. Compared to Enc-and-AR, LEARN improves performance by 76% (72% on average).

The performance improvement of LEARN comes from the fact that once the path has been set up for the communication between any two IPs, the overhead caused to securely communicate between the two IPs (data transfer phase) while preserving route anonymity is much less. The notable overhead occurs at the route discovery phase due to complex cryptographic operations. The intermediate nodes encrypt/decrypt packets to exchange parameters securely. Yet, these complex cryptographic operations are performed only a constant number of times. Majority of the work is done at the source which selects points to be distributed among intermediate nodes after constructing a curve, calculates the Lagrangian coefficients of the selected points, and performs several rounds of encryption/decryption during the three-way handshake. Once the routing path is set up, packets can be forwarded from one router to the other by a simple table look-up. No per-hop encryption and decryption is required to preserve anonymity. The security of a message is ensured by changing the original message at each node using a few addition and multiplication operations which incur significantly fewer extra delays. Since the route discovery phase happens only once during the lifetime of a task unless context switching and/or task migration happens, and there is only a limited number of communications going on between IPs in an SoC, the cost during the route discovery phase gets amortized over time. When running real benchmarks, the packet ratio was observed to be 1:1:1:6325 on average for $RI : RA : RC : DT$, respectively. For synthetic traffic patterns, the same ratio was 1:1:1:1964. This leads to a significant performance improvement compared to the traditional methods of encryption and anonymous routing.

6.6.3 Area Overhead of the Key Mapping Table

The key mapping table is an extra table compared to No-Security approach used to implement the anonymous routing protocol. The key mapping table adds a row for each session. Therefore, the size of the key mapping table is linearly proportional to the number of sessions. If at design time, it is decided to have a fixed size for the key mapping table, it is possible for the key mapping table at a router to be full after adding sessions, and in that case, new sessions cannot be added through that router. Therefore, the size has to be decided according to the communication requirements.

The maximum number of communication pairs in an 8×8 Mesh is $\binom{64}{2} \times 2 = 4032$ (assuming two-way communication between any pair out of the 64 nodes). Depending on the address mapping, only some node pairs (out of all the possible node pairs) communicate. The simulations consisted of 256 unique node pairs. In the worst case, if each communication session is assumed to have one common router, the key mapping table should be $256 \times rowsize$ big. If each entry in the key mapping table is 128 bits, the total size becomes 20 kB. However, in reality, not all communication sessions overlap. It is also important to note that except for the Session ID in the key mapping table, the other entries can be overwritten once route discovery phase is complete. Therefore, it is possible to allocate a fixed size key mapping table during design time and yet keep the area overhead low.

6.6.4 Security Analysis

In this section, we discuss the security and privacy of messages transferred on the NoC using LEARN.

Security of Messages The security of messages is preserved by the (k, n) threshold property of Lagrangian polynomials discussed in Sect. 6.2.3. Therefore, unless an intermediate node can gather all points distributed among the routers in the routing path together with their Lagrangian coefficients, the original message cannot be recovered. The threat model states that the source and destination are trusted IPs, and also, only some of the IPs are untrusted. Therefore, all routers along the routing path will never be compromised at the same time. The threat comes from malicious IPs sitting on the routing path and eavesdropping to extract security critical information. LEARN ensures that intermediate nodes that can be malicious, cannot recover the original message during the data transfer phase by changing the message at each hop. The complete message can only be constructed at the destination. During route discovery phase, each packet is encrypted such that only the intended recipient can decrypt it. The key and nonce exchange is also secured according to the mechanism proposed in Sect. 6.5.4. Therefore, LEARN ensures that no intermediate M3PIP can gather enough data to recover the plaintext from messages.

Anonymity of Nodes in the Network LEARN preserves the anonymity of nodes in the network during all of its operational phases. When the source sends the initial RI packet to initiate the three-way handshake, it does not use the identity of the destination. Instead, the source uses the global public key of the destination (PK_D) and sends a broadcast message on the network. When the RI packet propagates through the network, each intermediate node saves a temporary public key of its predecessor. This temporary public key is then used to encrypt data when propagating the RA packet so that unicast messages can be sent to preceding nodes without using their identities. Random nonces and symmetric keys are assigned to each node during the RA packet propagation which in turn is used by the RC packet to distribute points and Lagrangian coefficients to each node. Data transfer is done by looking up the routing table that consists of the nonces representing incoming and outgoing VCNs. Therefore, the identities of the nodes are not revealed at any point during communication.

Anonymity of Routes Taken by Packets In addition to preserving the anonymity of nodes, LEARN also ensures that the path taken by each packet is anonymous. Anonymity of the routing path is ensured by two main characteristics. (1) The message is changed at each hop. Therefore, even if there are two M3PIPs on the same routing path, information exchange among the two M3PIPs will not help in identifying whether the same message was passed through both of them. The same message appears as two completely different messages when passing through two different nodes. (2) The routing table contains only the preceding and following nodes along the routing path. An M3PIP compromising a router will only reveal information about the next hop and the preceding hop. Therefore, the routing paths of all packets remain anonymous.

6.7 Discussion

In this section, we discuss possible alternatives to the design choices from both design overhead and security perspectives. Most importantly, we discuss security solutions to defend against attacks when an attacker is aware that LEARN is implemented a security mechanism.

6.7.1 Feasibility of a Separate Service NoC

Modern SoCs use multiple physical NoCs to carry different types of packets [58, 61]. The KNL architecture used in Intel Xeon-Phi processor family uses four parallel NoCs [58]. The Tilera TILE64 architecture uses five Mesh NoCs, each used to transfer packets belonging to a certain packet type such as main memory, communication with I/O devices, and user-level scalar operand and stream communication between tiles [61]. The decision to implement separate physical NoCs is dependent on the performance versus area trade-off. If only one physical NoC is used to carry all types of packets, the packets must contain header fields such as RI, RA, RC, DT to distinguish between different types. The buffer space is shared between different packet types. The SoC performance can deteriorate significantly due to these factors coupled with the increasing number of IPs in an

SoC. On the other hand, contrary to intuition, due to the advancements in chip fabrication processes, additional wiring between nodes incur minimal overhead as long as the wires stay on-chip. Furthermore, when wiring bandwidth and on-chip buffer capacity is compared, the more expensive and scarce commodity is the on-chip buffer area. If different packet types are carried on NoC using virtual channels and buffer space is shared [20], the increased buffer spaces and logic complexity to implement virtual channels become comparable to another physical NoC. A comprehensive analysis of having virtual channels versus several physical NoCs is given in [63].

It is possible to use two physical NoCs—one for data (DT) packet transfers and the other to carry packets related to the handshake (RI, RA, RC). However, in LEARN, the potential performance improvement from a separate service NoC is not enough to justify the area and power overhead. LEARN is envisioned to be a part of a suite of NoC security countermeasures that can address other threat models such as denial-of-service, buffer overflow, etc. The service NoC will be effective in such a scenario where more service type packets (e.g., DoS attack detection related packets [13]) are transferred through the NoC.

6.7.2 *Obfuscating the Added Secret*

An attacker who is aware of the security mechanism can try to infer a communication path by observing the incoming and outgoing packets at a router.

Since each intermediate node adds a constant value ($y_i b_i$) to the received DT packet, the difference between incoming and outgoing DT packets at each node will be the same for a given virtual circuit. For this attack to take place, two consecutive routers have to be infected by attackers and they have to collaborate. Alternatively, a Trojan in a router has to have the ability to observe both incoming and outgoing packets at the router. While these are strong security assumptions, it is important to address this loophole. In this Section, a countermeasure is presented against such an attack. Even in the presence of such an attack, the secret message cannot be inferred since the complete message is only constructed at the destination and according to the threat model, it is assumed that the source and destination IPs are trustworthy.

This can be solved by changing the shared secret at each node for each message. However, generating and distributing secrets for each node per message can incur significant performance overhead. Therefore, a solution is proposed based on each node updating its own secret. According to Eq. (6.6), to derive a new Lagrangian coefficient b_i, the x coordinates should be changed. The source can easily do it for each message by changing both x_0 and y_0 when a new message needs to be sent. In other words, rather than changing the point (x_0,y_0) to (x_0,y'_0), it should be changed to (x'_0,y'_0). However, the new x'_0 now has to be sent to each intermediate node for them to be able to calculate the new secrets using:

$$b'_j = b_j \cdot \frac{x'_0}{x'_0 - x_j} \cdot \frac{x_0 - x_j}{x_0} \tag{6.12}$$

Such communications should be avoided for performance as well as security concerns. An alternative is to use a function $\mathcal{F}(x_0, \delta)$ that can derive the next x-coordinate starting from the initial x_0.

$$x'_0 = \mathcal{F}(x_0, \delta) \tag{6.13}$$

where $\mathcal{F}(x_0, \delta)$ can be a simple incremental function such as $\mathcal{F}(x_0, \delta) = x + \delta$. δ can be a constant. To increase security, δ can be picked using a pseudo-random number generator (PRNG) seeded with the same value at each iteration. Using such a method will change the shared secrets at each iteration and that will remove correlation between incoming and outgoing packets at a node.

6.7.3 Hiding the Number of Layers

Another potential vulnerability introduced by LEARN is that attackers who are aware of the security mechanism, can infer how far they are from the source and destination based on the size of the RA and RC packets. However, except for the corner case where the source/destination is at the edge of a certain topology, there can be more than one choice for potential source/destination candidates. Experiments presented in this chapter used the Mesh topology in which from the perspective of any node, there can be more than one node that is at distance d away. However, the attacker can reduce the set of possible source/destination candidates for a given communication stream. Therefore, depending on the security requirements, this vulnerability can be addressed using the mechanism proposed in this section.

After receiving the RI Packet, when the RA packet is initiated at D, D generates m $\langle$nonce, key$\rangle$ pairs $(\langle \upsilon_D^1, K_{S-D}^1\rangle, \langle \upsilon_D^2, K_{S-D}^2\rangle, ..., \langle \upsilon_D^m, K_{S-D}^m\rangle)$ and adds m layers to the packet. As a result, the RA packet sent from D to $r2$ takes the form:

$$\{RA \parallel E_{TPK_{r2}^{(i)}}(E_{OPK_S^{(i)}}(...E_{OPK_S^{(i)}}(E_{OPK_S^{(i)}}(\rho \parallel \upsilon_D^1 \parallel K_{S-D}^1) \parallel \upsilon_D^2 \parallel K_{S-D}^2)... \parallel \upsilon_D^m \parallel K_{S-D}^m))\}$$

D stores the $\langle$nonce, key$\rangle$ pairs in its key mapping table. When S receives the RA packet, S cannot distinguish whether the m pairs were generated from multiple nodes or one node. Therefore, when the RC packet is generated at S, instead of generating $k + 1$ points (corresponding to the number of nodes in the path), the number of generated points depends on the number of $\langle$nonce, key$\rangle$ pairs received. During RC packet transfer, each intermediate node along the routing path stores

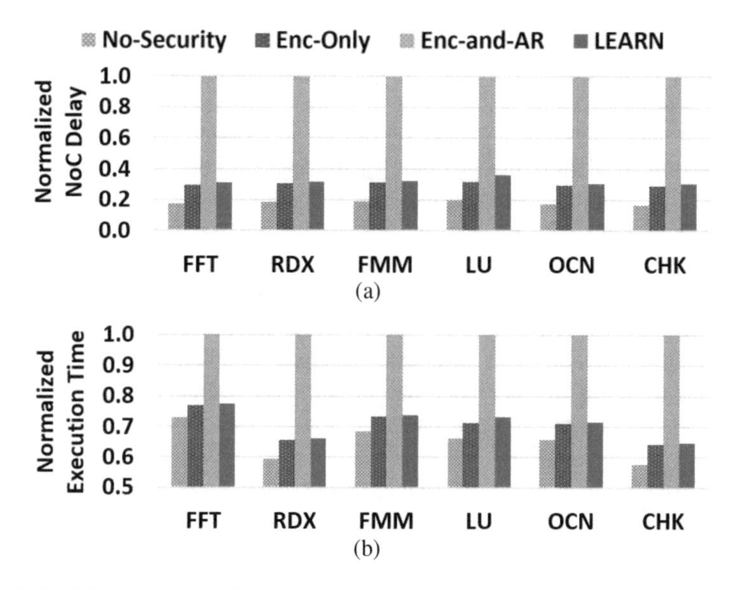

Fig. 6.13 NoC delay and execution time comparison across different security levels using real benchmarks considering the enhanced security features outlined in Sects. 6.7.2 and 6.7.3. (**a**) NoC delay. (**b**) Execution time

points (VCNs and secrets) corresponding to the nonces stored in the key mapping table. As a result, nodes can receive multiple secrets which can then be used during the data transfer phase. Depending on the required level of security, m can vary and also, each intermediate node can add multiple layers to the RA packet.

This method hides the correlation between the number of nodes and the length of the routing path, and therefore, eliminates the said vulnerability. However, this increases the performance penalty. Figure 6.13 shows an extension of Fig. 6.11 which considers the modification proposed in Sects. 6.7.2 and 6.7.3. LEARN improves NoC delay by 69% (67% on average) and total execution time by 34% (27% on average). Comparing with the results in Sect. 6.6.2, the average total execution time improvement has been reduced by 3% (from 30% on average to 27% on average) to accommodate the added security. Even then, LEARN enables significant performance improvement compared to traditional approaches.

6.8 Summary

Security and privacy are paramount considerations during electronic communication. Unfortunately, we cannot implement well-known security solutions from computer networks on resource-constrained SoCs in embedded systems and IoT devices. Specifically, these security solutions can lead to unacceptable performance overhead. In this chapter, we outline a lightweight encryption and anonymous routing protocol that addresses the classical trade-off between security and performance. The approach uses a secret sharing based mechanism to securely transfer data in an NoC-based SoC. Packets are changed at each hop and the complete packet is constructed only at the destination. Therefore, an eavesdropper along the routing path is unable to recover the plaintext of the intended message. Data is secured using only a few addition and multiplication operations which allows us to eliminate complex cryptographic operations that cause significant performance overhead. This anonymous routing protocol achieves superior performance compared to traditional anonymous routing methods such as onion routing by eliminating the need for per-hop decryption. Experimental results demonstrated that implementation of existing security solutions on NoC can introduce significant (1.5X) performance degradation, whereas this approach can provide the desired security requirements with minor (4%) impact on performance.

Acknowledgments This work was partially supported by the National Science Foundation (NSF) grant SaTC-1936040.

References

1. N. Agarwal, T. Krishna, L. Peh, N.K. Jha. Garnet: a detailed on-chip network model inside a full-system simulator, in *Proceedings of the 2009 IEEE International Symposium on Performance Analysis of Systems and Software* (2009), pp. 33–42
2. A. Ahmed, F. Farahmandi, Y. Iskander, P. Mishra, Scalable hardware trojan activation by interleaving concrete simulation and symbolic execution, in *Proceedings of the 2018 IEEE International Test Conference (ITC)* (IEEE, New York, 2018), pp. 1–10
3. A. Ahmed, F. Farahmandi, P. Mishra, Directed test generation using concolic testing on RTL models, in *Proceedings of the 2018 Design, Automation and Test in Europe Conference and Exhibition (DATE)* (2018), pp. 1538–1543
4. A. Ahmed, Y. Huang, P. Mishra, Cache reconfiguration using machine learning for vulnerability-aware energy optimization. ACM Trans. Embed. Comput. Syst. **18**(2), 1–24 (2019)
5. D.M. Ancajas, K. Chakraborty, S. Roy, Fort-NoCs: mitigating the threat of a compromised NoC, in *Proceedings of the 2014 51st ACM/EDAC/IEEE Design Automation Conference (DAC)* (2014), pp. 1–6
6. S. Babar, A. Stango, N. Prasad, J. Sen, R. Prasad, Proposed embedded security framework for internet of things (IoT), in *Proceedings of the 2011 2nd International Conference on Wireless Communication, Vehicular Technology, Information Theory and Aerospace Electronic Systems Technology (Wireless VITAE)* (2011), pp. 1–5
7. N. Binkert, B. Beckmann, G. Black, S.K. Reinhardt, A. Saidi, A. Basu, J. Hestness, D.R. Hower, T. Krishna, S. Sardashti, R. Sen, K. Sewell, M. Shoaib, N. Vaish, M.D. Hill, D.A. Wood, The gem5 simulator. SIGARCH Comput. Archit. News **39**(2), 1–7 (2011)
8. S. Charles, P. Mishra, Lightweight and trust-aware routing in NoC-based SoCs, in *Proceedings of the 2020 IEEE Computer Society Annual Symposium on VLSI (ISVLSI)* (2020), pp. 160–167
9. S. Charles, P. Mishra, Securing network-on-chip using incremental cryptography, in *Proceedings of the 2020 IEEE Computer Society Annual Symposium on VLSI (ISVLSI)* (2020), pp. 168–175
10. S. Charles, P. Mishra, Reconfigurable network-on-chip security architecture. ACM Trans. Des. Autom. Electron. Syst. **25**(6), 1–25 (2020)

11. S. Charles, H. Hajimiri, P. Mishra, Proactive thermal management using memory-based computing in multicore architectures, in *Proceedings of the 2018 Ninth International Green and Sustainable Computing Conference (IGSC)* (2018), pp. 1–8
12. S. Charles, C.A. Patil, U.Y. Ogras, P. Mishra, Exploration of memory and cluster modes in directory-based many-core CMPs, in *Proceedings of the 2018 Twelfth IEEE/ACM International Symposium on Networks-on-Chip (NOCS)* (2018), pp. 1–8
13. S. Charles, Y. Lyu, P. Mishra, Real-time detection and localization of dos attacks in NoC based SoCs, in *Proceedings of the 2019 Design, Automation Test in Europe Conference Exhibition (DATE)* (2019), pp. 1160–1165
14. S. Charles, A. Ahmed, U.Y. Ogras, P. Mishra, Efficient cache reconfiguration using machine learning in NoC-based many-core CMPs. ACM Trans. Des. Autom. Electron. Syst. (TODAES) **24**(6), 1–23 (2019)
15. S. Charles, M. Logan, P. Mishra, Lightweight Anonymous Routing in NoC based SoCs, in *Proceedings of the 2020 Design, Automation and Test in Europe Conference and Exhibition (DATE)* (IEEE, New York, 2020)
16. S. Charles, Y. Lyu, P. Mishra, Real-time detection and localization of distributed dos attacks in NoC based SoCs. IEEE Trans. Comput. Aided Des. Integr. Circuits Syst. **39**(12), 4510–4523 (2020)
17. M. Chen, X. Qin, H.-M. Koo, P. Mishra, *System-level Validation: High-level Modeling and Directed Test Generation Techniques* (Springer, Berlin, 2012)
18. É. Cota, A. de Morais Amory, M.S. Lubaszewski, NoC basics, in *Reliability, Availability and Serviceability of Networks-on-Chip* (Springer, Berlin, 2012), pp. 11–24
19. A.V. de Mello, L.C. Ost, F.G. Moraes, N.L.V. Calazans, Evaluation of routing algorithms on mesh based NoCs. PUCRS, Av. Ipiranga, 22 (2004)
20. J. Diguet, S. Evain, R. Vaslin, G. Gogniat, E. Juin, NoC-centric security of reconfigurable SoC, in *Proceedings of the First International Symposium on Networks-on-Chip (NOCS'07)* (2007), pp. 223–232
21. F. Farahmandi, P. Mishra, Automated debugging of arithmetic circuits using incremental gröbner basis reduction, in *Proceedings of the 2017 IEEE International Conference on Computer Design (ICCD)* (IEEE, New York, 2017), pp. 193–200
22. F. Farahmandi, P. Mishra, FSM anomaly detection using formal analysis, in *Proceedings of the 2017 IEEE International Conference on Computer Design (ICCD)* (IEEE, New York, 2017), pp. 313–320
23. F. Farahmandi, P. Mishra, Automated test generation for debugging multiple bugs in arithmetic circuits. IEEE Trans. Comput. **68**(2), 182–197 (2018)
24. F. Farahmandi, Y. Huang, P. Mishra, Trojan localization using symbolic algebra, in *Proceedings of the 2017 22nd Asia and South Pacific Design Automation Conference (ASP-DAC)* (2017), pp. 591–597
25. F. Farahmandi, Y. Huang, P. Mishra, *System-on-Chip Security: Validation and Verification* (Springer, Berlin, 2019)
26. L. Fiorin, G. Palermo, C. Silvano, *A Security Monitoring Service for NoCs* (Association for Computing Machinery, New York, 2008), pp. 197–202
27. U. Gupta et al. Dypo: Dynamic pareto-optimal configuration selection for heterogeneous MpSoCs. TECS **16**(5s), 1–20 (2017)
28. Y. Huang, P. Mishra, Vulnerability-aware energy optimization for reconfigurable caches in multitasking systems. IEEE Trans. Comput. Aided Des. Integr. Circuits Syst. **38**(5), 809–821 (2019)
29. Y. Huang, S. Bhunia, P. Mishra, Mers: statistical test generation for side-channel analysis based trojan detection, in *Proceedings of the 2016 ACM SIGSAC Conference on Computer and Communications Security* (2016), pp. 130–141
30. Y. Huang, S. Bhunia, P. Mishra, Scalable test generation for trojan detection using side channel analysis. IEEE Trans. Inf. Forensics Secur. **13**(11), 2746–2760 (2018)
31. H.K. Kapoor, G.B. Rao, S. Arshi, G. Trivedi, A security framework for NoC using authenticated encryption and session keys. Circuits Syst. Signal Process. **32**(6), 2605–2622 (2013)

32. J. Katz, A.J. Menezes, P.C. Van Oorschot, S.A. Vanstone, *Handbook of applied cryptography* (CRC press, New York, 1996)
33. J. Kong, X. Hong, Anodr: Anonymous on demand routing with untraceable routes for mobile ad-hoc networks, in *ACM International Symposium on Mobile Ad Hoc Networking and Computing*, MobiHoc '03 (Association for Computing Machinery, New York, 2003), pp. 291–302
34. B. Lebiednik, S. Abadal, H. Kwon, T. Krishna, Architecting a secure wireless network-on-chip, in *Proceedings of the 2018 Twelfth IEEE/ACM International Symposium on Networks-on-Chip (NOCS)* (2018), pp. 1–8
35. W. Liu, M. Yu, AASR: Authenticated anonymous secure routing for MANETs in adversarial environments. IEEE Trans. Veh. Technol. **63**(9), 4585–4593 (2014)
36. S. Lukovic, N. Christianos, Enhancing network-on-chip components to support security of processing elements, in *Proceedings of the 5th Workshop on Embedded Systems Security*, WESS '10 (Association for Computing Machinery, New York, 2010)
37. Y. Lyu, P. Mishra, A survey of side-channel attacks on caches and countermeasures. J. Hardware Syst. Secur. **2**(1), 33–50 (2018)
38. Y. Lyu, P. Mishra, Efficient test generation for trojan detection using side channel analysis, in *Proceedings of the 2019 Design, Automation and Test in Europe Conference and Exhibition (DATE)* (IEEE, New York, 2019), pp. 408–413
39. Y. Lyu, P. Mishra, Automated test generation for activation of assertions in RTL models, in *Proceedings of the 2020 25th Asia and South Pacific Design Automation Conference (ASPDAC)* (IEEE, New York, 2020), pp. 223–228
40. Y. Lyu, P. Mishra, Automated test generation for trojan detection using delay-based side channel analysis, in *Proceedings of the 2020 Design, Automation and Test in Europe Conference and Exhibition (DATE)* (2020), pp. 1031–1036
41. Y. Lyu, P. Mishra, Automated trigger activation by repeated maximal clique sampling, in *Proceedings of the Asia and South Pacific Design Automation Conference (ASPDAC)* (2020), pp. 482–487
42. Y. Lyu, P. Mishra, Scalable activation of rare triggers in hardware trojans by repeated maximal clique sampling. IEEE Trans. Comput. Aided Des. Integr. Circuits Syst. (2020), p. 1
43. Y. Lyu, P. Mishra, Scalable concolic testing of RTL models. IEEE Trans. Comput. (2020), p. 1
44. Y. Lyu, X. Qin, M. Chen, P. Mishra, Directed test generation for validation of cache coherence protocols. IEEE Trans. Comput. Aided Des. Integr. Circuits Syst. **38**(1), 163–176 (2018)
45. Y. Lyu, A. Ahmed, P. Mishra, Automated activation of multiple targets in RTL models using concolic testing, in *Proceedings of the 2019 Design, Automation and Test in Europe Conference and Exhibition (DATE)* (IEEE, New York, 2019), pp. 354–359
46. P. Mishra, F. Farahmandi, *Post-Silicon Validation and Debug* (Springer, Berlin, 2019)
47. P. Mishra, S. Bhunia, M. Tehranipoor, *Hardware IP security and Trust* (Springer, Berlin, 2017)
48. E.R. Naru, H. Saini, M. Sharma, A recent review on lightweight cryptography in IoT, in *Proceedings of the 2017 International Conference on I-SMAC (IoT in Social, Mobile, Analytics and Cloud) (I-SMAC)* (2017), pp. 887–890
49. Z. Pan, P. Mishra, Automated test generation for hardware trojan detection using reinforcement learning, in *Proceedings of the Asia and South Pacific Design Automation Conference (ASPDAC)* (2021)
50. Z. Pan, J. Sheldon, P. Mishra, Test generation using reinforcement learning for delay-based side channel analysis, in *Proceedings of the IEEE/ACM International Conference on Computer-Aided Design (ICCAD)* (2020)
51. Y. Qin, D. Huang, V. Kandiah, OLAR: on-demand lightweight anonymous routing in MANETs, in *Proceedings of the Fourth International Conference Mobile Computing and Ubiquitous Networking (ICMU'08)* (Citeseer, New York, 2008), pp. 72–79

52. V.Y. Raparti, S. Pasricha, Lightweight mitigation of hardware trojan attacks in NoC-based manycore computing, in *Proceedings of the 2019 56th ACM/IEEE Design Automation Conference (DAC)* (2019), pp. 1–6
53. S.V. Reddy Chittamuru, I.G. Thakkar, V. Bhat, S. Pasricha, SOTERIA: exploiting process variations to enhance hardware security with photonic NoC architectures, in *Proceedings of the 2018 55th ACM/ESDA/IEEE Design Automation Conference (DAC)* (2018), pp. 1–6
54. C. Reinbrecht, A. Susin, L. Bossuet, G. Sigl, J. Sepúlveda, Side channel attack on NoC-based MPSoCs are practical: NoC prime+probe attack, in *Proceedings of the 2016 29th Symposium on Integrated Circuits and Systems Design (SBCCI)* (2016), pp. 1–6
55. K. Sajeesh, H.K. Kapoor, An authenticated encryption based security framework for NoC architectures, in *Proceedings of the 2011 International Symposium on Electronic System Design* (2011), pp. 134–139
56. J. Sepúlveda, A. Zankl, D. Flórez, G. Sigl, Towards protected MPSoC communication for information protection against a malicious NoC. Procedia Comput. Sci. **108**, 1103–1112 (2017). International Conference on Computational Science, ICCS 2017, 12–14 June 2017, Zurich, Switzerland
57. A. Shamir, How to share a secret. Commun. ACM **22**(11), 612–613 (1979)
58. A. Sodani, R. Gramunt, J. Corbal, H. Kim, K. Vinod, S. Chinthamani, S. Hutsell, R. Agarwal, Y. Liu, Knights landing: second-generation Intel Xeon Phi product. IEEE Micro **36**(2), 34–46 (2016)
59. Using TinyCrypt Library, Intel Developer Zone, Intel, 2016. https://software.intel.com/en-us/node/734330 [Online]
60. W. Wang, P. Mishra, A. Gordon-Ross, Dynamic cache reconfiguration for soft real-time systems. ACM Trans. Embed. Comput. Syst. **11**(2), 1–31 (2012)
61. D. Wentzlaff, P. Griffin, H. Hoffmann, L. Bao, B. Edwards, C. Ramey, M. Mattina, C. Miao, J.F. Brown III, A. Agarwal, On-chip interconnection architecture of the tile processor. IEEE Micro **27**(5), 15–31 (2007)
62. S.C. Woo, M. Ohara, E. Torrie, J.P. Singh, A. Gupta, The splash-2 programs: characterization and methodological considerations, in *Proceedings of the 22nd Annual International Symposium on Computer Architecture* (1995), pp. 24–36
63. Y.J. Yoon, N. Concer, M. Petracca, L.P. Carloni, Virtual channels and multiple physical networks: two alternatives to improve NoC performance. IEEE Trans. Comput. Aided Des. Integr. Circuits Syst. **32**(12), 1906–1919 (2013)
64. W. Yuan, An anonymous routing protocol with authenticated key establishment in wireless ad hoc networks. Int. J. Distrib. Sens. Netw. **10**(1), 212350 (2014)
65. H. Zhu, P.P. Pande, C. Grecu, Performance evaluation of adaptive routing algorithms for achieving fault tolerance in NoC fabrics, in *Proceedings of the 2007 IEEE International Conf. on Application-specific Systems, Architectures and Processors (ASAP)* (2007), pp. 42–47

Chapter 7
Secure Cryptography Integration: NoC-Based Microarchitectural Attacks and Countermeasures

Johanna Sepúlveda

7.1 Introduction

High VLSI integration levels and demanding requirements of the current embedded applications promote the development and widespread adoption of Multi-Processor System-on-Chips (MPSoCs) in the Internet-of-Things (IoT) environments. Such computation paradigm is transforming many domains, including the automation industry, automotive, avionics, and healthcare [33]. It is expected that by 2021, 25 billion IoT devices are deployed [33]. In such a scenario, the MPSoCs, also known as systems of systems, are pervading our lives by blending seamlessly with the environment while processing and storing all kinds of applications and data. Such a dynamic and multi-tenant platform provides all the means for high performance and reliable operation. However, it brings security as a major requirement. Due to the intrinsic complexity and heterogeneity of the MPSoC, it is very challenging to guarantee the MPSoC security.

MPSoCs are complete computational systems that integrate hardware and software into a single die. The hardware MPSoC components are comprised of multiple Intellectual Property (IP) hardware cores for computation, storage, and communication. For modern MPSoCs that integrate a large amount of hardware IP cores, the Network-on-Chip (NoC) has become an attractive alternative for on-chip interconnection.

In order to mitigate possible MPSoC attacks and meet the system security requirements, many works have addressed the MPSoC security through a wide number of software and hardware techniques. The goal is to deploy the MPSoC security policy, a set of rules that establish the security characteristics of the system.

J. Sepúlveda (✉)
Airbus Defence and Space GmbH and Technical University of Munich, Taufkirchen, Germany
e-mail: johanna.sepulveda@airbus.com

P. Mishra, S. Charles (eds.), *Network-on-Chip Security and Privacy*,
https://doi.org/10.1007/978-3-030-69131-8_7

These rules protect the MPSoC assets (e.g., passwords, bitstreams, applications) against possible attacks. In order to determine the exposure to attacks, a threat model must be used. It identifies the capabilities of the attacker and possible vulnerabilities of the system. The threat model and the security policy will define the type of security services and mechanisms that are required in order to protect the MPSoC assets. A security mechanism is defined as a technique that provides one or multiple of the following security services: (1) *confidentiality*, which ensures the data secrecy; (2) *integrity*, which assures that data is kept unchanged without authorization; (3) *authentication*, which validates the identity of the initiator of the operation/communication; (4) *access control*, which allows or denies the use of a particular resource; (5) *availability*, to ensure that system resources are ready to be used; and (6) *non-repudiation*, which maintains evidence of system events.

One of the most preferred ways to support security services is through cryptographic primitives. Symmetric Key Cryptography (e.g., Advanced Encryption Standard—AES [21]) and Public Key Cryptography (e.g., Rivest–Shamir–Adleman—RSA, or Elliptic Curve Cryptography—ECC) are the foundation for many security transformations. Such cryptographic primitives can be integrated into the MPSoC to: (1) protect the internal MPSoC information, such as bitstreams [24]; (2) secure the MPSoC operation, including the secure booting [90], dynamic reconfiguration [87] and secure resources management [17, 62]; and (3) implement the different secure communication protocols, required to integrate the MPSoC to the different ubiquitous environment [95]. In many networked environments (e.g., industrial, IoT, aerial, automotive-V2X), the communication follows the guidelines established by the standardization bodies. Some of them are the Internet Engineering Task Force (IETF) [40], European Telecommunications Standards Institute (ETSI) [25], National Institute of Standards and Technology (NIST) [59], International Organization for Standardization (ISO) [43], and IEEE Standards Association [37].

A wide research and engineering work has been performed to provide the required level of security and performance. Many standards advise the use of the AES cipher, for example, to provide confidentiality in the protocols such as Transport Layer Security (TLS) [39] and Internet Protocol Security (IPsec) [38]. On the other hand, the authentication process is usually supported by RSA or ECC public key cryptography. Finally, the authentication process usually relies on a hash operation, such as the Secure Hash Algorithm (SHA) [56].

These symmetric and public key encryption algorithms can be implemented into the MPSoC through software and/or hardware techniques. However, empowering MPSoCs to support Symmetric Key Cryptography (SkC) and Public Key Cryptography (PKC) is a challenging task. MPSoCs design is usually pressed by the limited on-chip resources, strict performance and flexibility requirements, and short time-to-market. The integration of such cryptographic functionalities requires mathematical elements and operations which are usually not easy to implement on standard CPU architectures. To cope with the application performance and costs (e.g., area and power) requirements, while providing crypto-agility, different techniques are used to deploy on-chip cryptography. These techniques include the use of generic cryptography libraries, the use of pre-computed functions and values stored in tables/memory regions, and the use of hardware/software co-design strategies. In such a case, the goal is to find an optimal assignment of tasks between software running on processors and custom hardware accelerators that act as co-processors for performing computation intensive operations [8, 30, 52, 88].

However, the implementation of cryptographic primitives is prone to attacks. During the execution of a cryptographic operation, the secret key may passively be revealed through the so-called side-channels. One of the major threats appears from the high MPSoC resource sharing, where many applications are using shared computation, memory and communication structures. The behavior exhibited by such components varies according to their microarchitectural organization and the data that is being processed, stored, or exchanged. When malicious and sensitive processes are executed together on the MPSoC, characteristics such as the time, power, memory access pattern, and electromagnetic emanation of the MPSoC during the sensitive process operation can leak information to the attacker regarding the data used within the sensitive process.

While side-channel attacks through the exploitation of the MPSoC computation and the memory structures have been widely discussed, communication side-channels have been sidelined. Communication characteristics can be exploited to retrieve sensitive information. By exploiting the communication pattern, defined by the connectivity (i.e., the distribution of the source–destination communication parties), packet sizes, injection rates, and throughput, an attacker can perform two actions [83]: (1) reveal the sensitive information when the communication is data dependent. In addition the NoC microarchitecture can be used to perform illegal monitoring; and/or (2) identify vulnerable operation points to trigger powerful attacks, such as cache-based attacks. The latter is known as communication-enhanced attacks [80] and are very attractive for their capability of providing a very refined information regarding the sensitive traffic to/from the shared caches, improving for example the time measurement resolution capability of an attacker, which is critical for performing successful attacks.

To this end, in this chapter the NoC-based communication-enhanced attacks for retrieving the secret information of cryptographic operations executed on the MPSoC are discussed. This study further explores this vulnerability and highlights the security implications of a shared communication channel. It includes discussions and examples regarding the NoC-based exploitation for attacking symmetric and public key cryptography that are already in standards as well as the future cryptography that are in the process of standardization at NIST: lightweight cryptography and post-quantum cryptography. In addition, countermeasures able to be integrated into the NoC are presented. The present study provides a helpful input for the design and use of modern MPSoCs. In addition, it calls the attention to the need of effective hardware-based countermeasures against NoC-enhanced attack. This is critical for the adoption of such NoC-based MPSoC in current and future critical applications. Particularly, developers and users must be aware of the critical role of the interconnectivity of NoCs and their security implications.

This chapter is divided in seven sections. Section 7.2 presents the basic concepts regarding the MPSoC organization. Section 7.3 describes the present and future cryptographic approaches. The considered threat model is described in Sect. 7.4. Sections 7.5 and 7.6 describe the current microarchitectural attacks on MPSoCs and the NoC-enhanced attacks, respectively. Finally, the conclusions are presented in Sect. 7.7.

7.2 MPSoC Organization

7.2.1 General Description

Multi-Processors Systems-on-Chip (MPSoCs) are complete computational systems which integrate software and multiple hardware Intellectual Property (IP) elements in a single die. It includes, for example, the operating system, firmware, and drivers (software components), as well as multiples processors, memories, co-processors, and I/O controllers (hardware components). MPSoCs appeared as a natural evolution of the SoCs, shaped to meet the ever-increasing highly demanding applications characterized by very high computation rates (e.g., video applications) [3], low-power operation (e.g., IoT applications) [44], huge data rates handling, strict real-time constraints (e.g., automotive and aerospace applications) [45, 101], strict time-to-market, and a high need for adaptability [47]. MPSoCs are multi-tenant systems, that is, multiple applications with different criticality and performance requirements are spread through the MPSoC resources. Despite the different applications sharing the MPSoC resources, they are not aware of each other. Multi-tenancy is a crucial characteristic of the MPSoC, offering flexibility, cost, and area efficiency. MPSoCs are comprised of three main structures:

- ***Computation:*** Includes all the components that perform the data transformation and processing. Specifically processors, co-processors, and specific hardware accelerators (e.g., cryptographic accelerators [29, 53], Convolution Neural Network accelerators [85]) that are related to the domain of the application.
- ***Storage:*** Determines the data storing organization on the MPSoC. Usually it comprises several levels of small cache memories (L1 to L4) and DRAM distributed on the chip. This storage organization is widely adopted in current state-of-the-art market MPSoCs such as Tile-Mx100 from Tilera [1], MPPA from Kalray [48], or SCC from Intel[96]. Memory organization has a strong impact on the overall MPSoC performance, cost (area), and power.
- ***Communication:*** Comprises all the components required for the data exchange among the different IP cores. There are different alternatives (e.g., point-to-point connection, buses, crossbars, and Network-on-Chip). The selection of one of these alternatives is driven by the number of interconnected hardware IP cores and communication requirements (e.g., volume, traffic pattern).

MPSoC design is a process that is based on three premises [100]. First, the *adoption of different levels of abstraction*, which allows to develop models that describe the functional and non-functional system specifications with different levels of detail. Where each model is a refinement of the previous one and demands different design decisions. Second, the *reuse of components*, promoting the integration of pre-designed and pre-tested IP components. Finally, the *orthogonality*, which allows the splitting of the SoC design in the computation and communication structures, as well as in hardware and software components.

MPSoC structures are defined by a discrete set of components and parameters which may be configured in order to meet the application requirements and constraints. These components are usually supplied as IP core components (hardware or software) from third-party vendors or developed in-house. MPSoCs are organized in a tile-based construction, each constituted by a single or multiple computation and storage IP cores. While the intra-tile communication (i.e., the communication among the IP hardware cores encapsulated into a single tile) is usually performed through a bus structure, inter-tile communication (i.e., the communication among IP hardware cores mapped on different tiles) usually is implemented through crossbars or Networks-on-Chip (NoCs). NoCs are structures that use routers and links to communicate data between a pair of IPs, the IP_{source} (which injects the packet) and $IP_{destination}$ (which receives the packet).

7.2.2 Computation Structure

MPSoCs integrate a wide variety of IP hardware cores that are able to transform data using different types of logic units. According to the usability, such elements can be classified into two categories: (1) General purpose processors, which are able to perform the operations (instructions) that are specified in the instruction set architecture (ISA). It includes different registers, data types, and memory-related information (control, consistency rules, addressing modes, virtual memory). The realization of such operations defines the architecture of the processor. Central Processing Unit (CPU) and Graphics Processing Unit (GPU) are ruled by their own set of instructions; and (2) Application-Specific processors, which integrate a set of hardware accelerators (co-processor) that implement the entire/subset functionality of an application. Two main models of hardware accelerators can be identified. First, the *Loosely coupled accelerators*, which are implemented as a stand-alone hardware component linked to a processor through a point-to-point connection or through a bus. Second, the *Tightly coupled accelerators*, which are deeply embedded within the processor architecture as application-specific functional units.

Processing Elements (PEs) present different characteristics. Besides from their different ISAs, PEs are differentiated by their data width (e.g., from 8-bit to 64-bit), number of cores (e.g., from 8-bit to 64-bit) allowing multiple execution contexts simultaneously, type of cores (e.g., symmetrical or asymmetrical), and computation capabilities (e.g., real-time, microcontroller profiles) [5]. Most processing elements have two modes of operation: kernel mode and user mode. The Operating System (OS) is the most fundamental piece of software. It runs in kernel mode (also called supervisor mode) and provides the user programs with a better and simpler model of the computer resources [91]. This abstraction is the key to managing all the computer complexity. Supervisor mode allows the complete access to the hardware resources and can execute any instruction. The rest of the software is executed in *user mode* and can only execute a subset of the machine instructions.

Some modern processors include a set of security-related instruction codes (e.g., Intel Software Guard Extensions—SGX) that allow the definition of private regions of memory (i.e., enclaves) by the software executed in supervisor and user modes. Such isolation promises to avoid any memory accesses (read or write) by processes outside the enclave, even when running at higher privilege levels. Such isolation favors the multi-tenant environments, where multiple applications are being executed in the same processor.

7.2.3 Memory Structure

Multi-level memory hierarchy, including different caches (e.g., L1–L4) and DRAMs, is widely used to improve the performance of the MPSoCs. It exploits the two attributes of the application software: (1) temporal locality, where the same set of memory locations are likely to be repeatedly accessed; and (2) spatial locality, in which adjacent memory blocks are likely to be accessed. Since accessing the main memory requires a large number of clock cycles, to store the frequently accessed data on smaller and faster cache memories closer to the processors reduces the memory access bottleneck. Each time a process requests a memory access, every memory hierarchy is queried in order, from the closest and smallest memory in the processor, through the different cache memories, till the DRAM. If the requested data is found in a memory, a *cache hit* takes place and the data can be delivered to the initiator process. Otherwise, a *cache miss* occurs and the cache coherency mechanism initiates an access to the next memory level and eventually till the DRAM until the requested data is found.

For the tile-based MPSoC architecture, the queries to the memory hierarchy require the data exchange through the different communication structures: through bus transactions, when the memory is located inside the tile, or through NoC transactions, otherwise.

7.2.4 Communication Structure: Network-on-Chip (NoC)

NoCs are an attractive communication alternative due to their scalability, predictability, interoperability, and electrical regularity [36]. Figure 7.1 shows a MPSoC that integrates nine tiles T_1 to $T9$ and other IP cores (DDR controller, L3,

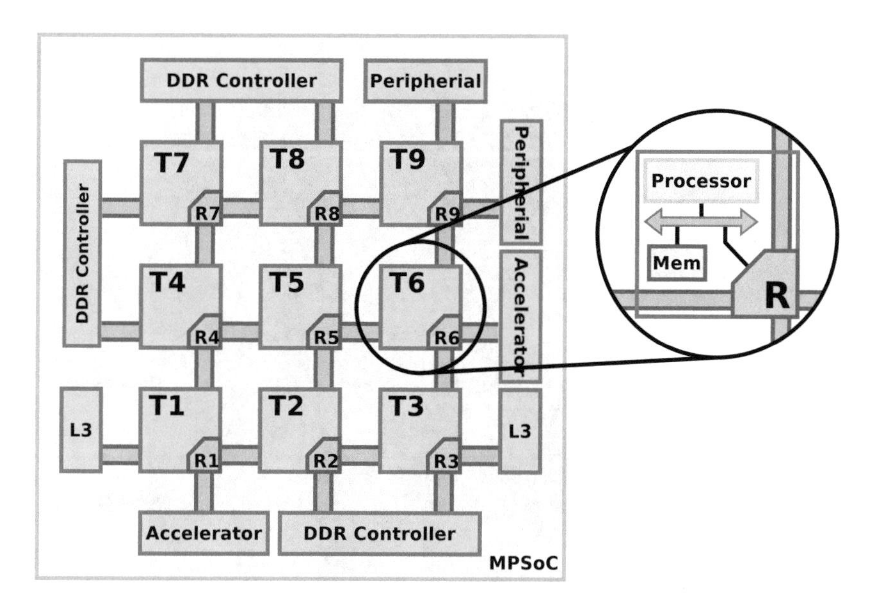

Fig. 7.1 Multi-Processor System-on-Chip. The detailed architecture of the Tile 6 (T6) is presented

peripheral, accelerator) which are interconnected through a 3 × 3 mesh-based NoC composed by 9 routers R_1 to R_9. Each tile is either composed of a single IP core or a set of IP cores that communicate through a bus. The tile can include heterogeneous processing units, storage components, peripherals, hardware accelerators, and other IP hardware cores.

NoC architectures are comprised of routers and links that exchange information wrapped as packets. Routers are switches that link input ports to any output port according to some commutation or routing algorithm. The routers are linked to the IP cores and tiles of the MPSoCs through network interfaces. While the interfaces implement the communication protocol for injecting/ejecting packets to/from the NoC, the NoC routers commute the packets from the communication source (IP_{source} or $Tile_{source}$) to the destination (IP_{sink} or $Tile_{sink}$). The general architecture of the router is shown in Fig. 7.2. Internally, routers integrate five main components:

1. **Input buffers**, which store the complete or a part of a packet incoming to the router through one of the input ports. Usually organized as a FIFO (First-In-First-Out policy);
2. **Arbitration logic**, implementing the policy employed for granting the utilization of the crossbar switch to one of the input buffers;
3. **Routing algorithm**, which selects the router output port to be employed for redirecting the incoming data. Such selection is implemented using logic-based distributed routing mechanism (removing the need for any routing tables);

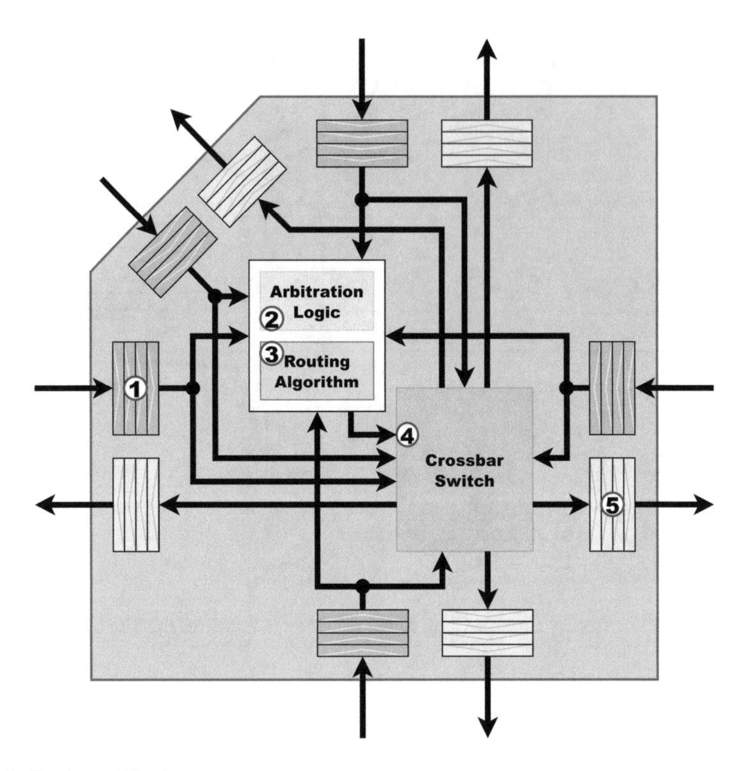

Fig. 7.2 Router architecture

4. **Crossbar switch**, which links input to output ports of the router and performs the commutation of packets; and
5. **Output buffers**, to store the information before being commuted to the neighbor router or network interface.

Messages injected from a $IP_{source}/Tile_{source}$ are translated by the network interface into packets compliant to the protocol used within the NoC. Packets are usually composed by three main chunks: (1) *header*, which includes the information required for driving the data from the source to the destination (e.g., source, destination) and for controlling the data exchange (e.g., size of the packet); (2) *payload*, including the actual information to be exchanged (e.g., operation, memory address, data type, data, role); and (3) *terminator*, to implement different communications services (e.g., error-correction [32, 73, 92], security [4, 23, 26, 34, 75, 77, 81]).

7.3 Cryptographic Implementation

Cryptography is the foundation for establishing secure communication between multiple parties. At its simplest level, cryptography is defined as any method that transforms information so that it can be kept secret and secure [21]. That is, such information should be unintelligible, difficult to forge, and therefore useless to those who are not meant to have access to it.[1] A cryptographic mechanism relies on two aspects. The first is the algorithm, which specifies the transformation steps that must be performed upon the original message (also called *plaintext*). Second, the variable cryptographic *key*, determined by a random data string. Both aspects are used together to secure data by means of encryption or digital signatures [31]. Cryptographic algorithms can be used for block cipher encryption, digital signatures, asymmetric key-establishment algorithms, message authentication codes, hash functions, or random data generators, for example.

7.3.1 Basic Concepts

According to the nature of the cryptographic key, cryptographic algorithms can be classified into two categories: symmetric and public key cryptography.

- ***Symmetric Key Cryptography (SKC)***, also known as secret key algorithm. It uses a single secret key which should be shared (i.e., identical) for all the communication parties (sender and receivers). The value of the secret key may change during the operation time. Thus, all the communication parties must have the updated secret key. Symmetric algorithms are usually faster that the public key cryptography, being able to handle thousands of keys with very little computing overhead. However, a key must be kept secret, and yet has to be transmitted to the receivers through a secure channel.
- ***Public Key Cryptography (PKC)***, also known as asymmetric key cryptography. It uses a pair of keys: (1) a public key, known by everybody; and (2) a private key, which should remain secret and only known by its owner. Both keys are mathematically related. However, to know the public key does not allow to retrieve the private key.

While symmetric key cryptography is most often used to protect the data confidentiality or to authenticate the data integrity, public key cryptography is used to protect the integrity and authenticity of information and to securely establish symmetric keys.

[1]This section provides a wide overview of the cryptographic techniques that are currently integrated on common MPSoCs and that will be integrated in the near future (less than 10 years). For a rigorous presentation and further details regarding the algorithm description and implementations, the reader can consult the suggested references.

Some of the main operations used in cryptography:

- ***Key Generation***: It creates a random string called *key* (single key for SKC or a pair public/private for PKC). According to the lifetime, cryptographic keys may be divide into two groups: (1) *static*, which is used for long periods of time. According to the application it varies from days to years; or (2) *ephemeral*, which is used only for a single session or transaction. In general, long term keys increase the vulnerability to attacks. It is critical to perform a re-keying process when required.
- ***Encryption***: It is a special form of computation that transforms a *plaintext* into a *ciphertext* using a cryptographic algorithm and key.
- ***Decryption***: It retrieves the original *plaintext* from a *ciphertext* by using a decryption algorithm and key.
- ***Signature***: It is a string of bits which are computed based on a set of rules and parameters for allowing: (1) source/entity authentication; (2) data integrity authentication; and/or (3) Support for signer non-repudiation. Stored and transmitted data can be signed.
- ***Verification***: It checks the validity and authenticity of the signatures (i.e., includes identity of the signatory and the integrity of the data) and other integrity keying materials, such as certification authority and controls of completeness and correctness.

Modern cryptography transforms data based on mathematical problems that are very hard to solve with our current and most powerful computational capabilities.

7.3.2 Current and Future Cryptography

7.3.2.1 Symmetric Cryptography

In general, the SKC is more efficient than PKC. Therefore, it is the preferred solution to ensure the data confidentiality and integrity for communicated and stored information which is characterized by large volumes of information. The currently used cryptographic algorithms and their integration for different applications are standardized by different organizations, such as NIST, ETSI, IETF, ISO, IEEE. The algorithms that have been standardized includes: Triple DES—TDES (NIST/FIPS 880-67), Advanced Encryption Standard—AES (NIST/FIPS 197), Secure Hash Algorithms—SHA-1/SHA-2 (NIST FIPS 180), SHA-3 (NIST FIPS 202), Simon (ISO/29167-21), Speck (ISO/29167-22), Trivium and Encoro (ISO/IEC 29192-3), and PRESENT (ISO/IEC 29192-2).

The two most popular algorithms are TDES, which is still widely used in banking applications, and its successor AES, which is massively distributed in a wide variety of applications. AES encrypts 128 bits of data with key lengths of 128/192/256 bits using 10/12/14 rounds, respectively. In AES-128, the intermediate states are represented as a 4×4 state matrix. This matrix is processed iteratively for 10 rounds, each composed of four round operations: (1) *AddRoundKey*, XORing the state matrix with the current round key; (2) *SubByte*, performing a byte substitution based on a non-linear function called Substitution box (S-Box). It is a multiplicative inverse in the finite field $GF(2^8)$ followed by an affine transformation over $GF(2)$; (3) *ShiftRow*, a byte transposition that moves the bytes to the neighboring positions; and (4) *MixColumn*, which performs a column-wise matrix multiplication between the current state and a fixed matrix. Each column is thereby treated as a polynomial with coefficients being elements of $GF(2^8)$.

In order to improve the performance of symmetric cryptography, look-up tables are widely used [10]. The goal is that instead of computing everything just in time, for some of the steps of the cryptographic transformations larger pre-computed values are stored in memory. The access to such tables usually is driven by the value of the secret key.

Performance optimized implementations of AES use **transformation tables** (T-tables), where the SubByte, ShiftRow, and MixColumn operations are reduced to four look-up tables (T0, T1, T2, and T3) whose entries are simply XOR'ed. In this implementation, one round consists of 16 table look-ups. Rounds 1–9 can be calculated as in Eq. (7.1). The last round of AES does not contain the MixColumn operation and is sometimes implemented with a fifth table T4. The fundamental mechanism exploited by the cache attack is that T-tables are accessed depending on the secret key.

$$
\begin{aligned}
(x_0^{r+1}, x_1^{r+1}, x_2^{r+1}, x_3^{r+1}) &\leftarrow T_0[x_0^r] \oplus T_1[x_5^r] \oplus T_2[x_{10}^r] \oplus T_3[x_{15}^r] \oplus k_0^{r+1} \\
(x_4^{r+1}, x_5^{r+1}, x_6^{r+1}, x_7^{r+1}) &\leftarrow T_0[x_4^r] \oplus T_1[x_9^r] \oplus T_2[x_{14}^r] \oplus T_3[x_3^r] \oplus k_1^{r+1} \\
(x_8^{r+1}, x_9^{r+1}, x_{10}^{r+1}, x_{11}^{r+1}) &\leftarrow T_0[x_8^r] \oplus T_1[x_{13}^r] \oplus T_2[x_2^r] \oplus T_3[x_7^r] \oplus k_2^{r+1} \\
(x_{12}^{r+1}, x_{13}^{r+1}, x_{14}^{r+1}, x_{15}^{r+1}) &\leftarrow T_0[x_{12}^r] \oplus T_1[x_1^r] \oplus T_2[x_6^r] \oplus T_3[x_{11}^r] \oplus k_3^{r+1}
\end{aligned}
\tag{7.1}
$$

In the AES T-Table implementation, the indexes used during the first round are defined as in Eq. (7.2). Each byte i of the intermediate value x involves a plaintext byte P_i, a table look-up T_i (which returns a 4-byte value), and the key K_i. The simple relation of the plaintext to the secret key in this round is exploited in the attack which is described later on in Sect. 7.5.

$$x_i^0 = P_i \oplus K_i \tag{7.2}$$

In order to perform the AES decryption, which retrieves the plaintext from a ciphertext, these round transformations are inverted and applied in reverse order.

The cryptographic functionalities also must adapt to the needs of the modern use cases. NIST has recently started a competition with the objective to standardize ***lightweight cryptography (LWC)*** [57, 58]. This process will select one or more authenticated encryption and hashing schemes suitable for constrained environments. The winning schemes will be deployed in Internet-of-Things (IoT) devices. To this end, new block ciphers will emerge as a standard for lightweight authenticated encryption in the upcoming years. Among the 32 candidates of the second competition round, 4 are based on AES as their underlying cipher[71] and 7 are based on the GIFT cipher [58]. GIFT is based on a substitution-permutation network (SPN) and was proposed in [6] as an improvement to the well-known PRESENT cipher [11]. LWC high performance implementations also use pre-computed tables. Despite some of the LWC relying on S-Boxes smaller than 64 bytes [20, 22, 94], which fits entirely on one cache line, sub-cache-line timing effects [104] are still possible. The authors in [68] showed for the first time the feasibility of such an attack. As a result, it was possible to retrieve the full key in a short time. Candidates with larger S-Box [19, 93] (256-byte S-Box) might still be vulnerable to traditional cache attacks. Further research is required [80].

7.3.2.2 Public Key Cryptography

PKC is more commonly used for the establishment of an initial symmetric key through key-agreement or key-encapsulation. Traditional PKC algorithms such as the *Rivest–Shamir–Adleman* (RSA) cryptosystem [72], which is based on the factorization of larger numbers, or Elliptic Curve Cryptography (ECC), which is based on the discrete logarithm problem, are considered secure today.

RSA is simply based on the modular exponentiation, where the modulus n is the product of two large prime's p and q. These primes are random integer numbers and have similar magnitude and length. In addition, d is an integer number, selected to be random and relatively prime to $(p - 1) \times (q - 1)$. The public key and private key are obtained as in Eq. (7.3).

$$e = d^{-1} \mod \Phi(n) \tag{7.3}$$

where the ciphertext C is obtained by transforming the *plaintext* M together with the public key (n and e) as in Eq. (7.4).

$$C = M \times e \mod n \tag{7.4}$$

where $0 < M < nand\ C$ can be decrypted to retrieve the *plaintext* by using the private key d and n as in Eq. (7.5).

$$M = C \times d \mod n \tag{7.5}$$

The implementation of PKC, e.g., RSA, requires the integration of different arithmetic blocks, including modular multiplication and exponentiation. In order to increase the performance of the RSA implementation, pre-computed tables are used. The authors of [54] proposed algorithms for performing modular reduction and modular multiplication to improve the speed of RSA software implementation. Both algorithms deal with performing modular arithmetic operations on very large numbers. They use look-up tables to perform the arithmetic computations on a byte basis. In addition, OpenSSL 0.9.7c, as example, uses a sliding-window exponentiation algorithm proposed in [12], which pre-computes values used during the exponentiation (known as multipliers). The access pattern to the multipliers depends on the value of the exponent, which, in the case of decryption and digital signature operations, should be kept secret [14, 103]. This vulnerability has been exploited by [63].

However, the future PKC may look different since powerful quantum computers will be able to solve the hard mathematical problems that PKC is based on. A new set of algorithms, based on hard mathematical problems able to resist traditional and quantum attacks, are required. Such algorithms are known as ***post-quantum cryptography (PQC)*** and are described upon hard problems derived from five fields (families): codes, hashes, multivariate, isogenies, and lattices. As soon as a capable quantum computer becomes available, current cryptographic algorithms will be threatened. All PKC algorithms can be attacked in polynomial time by executing the Shor's algorithm [86]. In addition, SKC is also affected by quantum computers, in particular by executing the Grover's quantum algorithm [51] the SKC key space is halved. However, SKC can easily be adapted by choosing larger key sizes [35]. Current pre-quantum 256-bit symmetric ciphers are enough to maintain the symmetric cryptography security in the post-quantum era.

Therefore, the scenario of powerful quantum computers poses a significant problem for applications with long life-cycles (cars, airplanes, satellites, industries, smart cities, health care) where devices are hard to update. As a reaction, the *National Security Agency* (NSA) announced in 2016 the intention to transition towards post-quantum cryptography for governmental usage in the foreseeable future [55]. NIST started in 2017 the process of post-quantum standardization. The goal is to select a set of appropriate post-quantum solutions able to meet the security, performance, cost, and adaptability that current and future applications demand. The candidates are going to be under scrutiny from the academia, industrial, and governmental communities in a round-like process that is expected to last up to 5 years. From the 82 PQC candidates submitted in the first round in 2017, in 2020, the number of candidates was reduced to 7 (finalists) and 8 (alternates) [60]. From the finalists a set of algorithms will be selected for the PQC standard, expected in 2022. The most severe drawback of post-quantum secure systems, compared to classical cryptography, is the significantly larger key size. Some schemes have been proposed that achieve a comparable key size but their security levels are not yet analyzed in full detail. In general, look-up tables are a valuable option for PQC software implementations for the different cryptographic algorithms. The 32- and 64-bit wide architectures tolerate the additional code size and gain run-time improvements, not least due to the speed-up from cache hierarchies. However, most of the memory accesses are driven by the value of the secret key, which can expose the system vulnerable to attacks when cache-based or communication-enhanced cache-based attacks are performed. This is especially effective in embedded devices. During the operation of the different cryptographic algorithms, the data stored in the tables is spread along the memory hierarchies. When caches are shared, an attacker can gain information regarding the secret keys [82].

7.4 Threat Model

The system considered in this work is the NoC-based MPSoC shown in Fig. 7.1. A closer look of this system (Tile 1 to Tile 3) is provided in Fig. 7.3. It shows three tiles (T_1, T_2, and T_3) which are communicating through a NoC. Only three routers (R_1, R_2, and R_3) connected in a mesh-based topology are shown. Each one of the tiles T_1 and T_2 is comprised of a processor IP (IP_{Victim} for T_1 and $IP_{Attacker}$ for T_2) which is connected to a private *L1* cache (*Mem*) through a bus. In contrast tile T_3 is composed only of a shared *L2* data cache (*Shared Memory*). Router R_3 also communicates with a *DDR* memory.

In this scenario, both processors IPs (IP_{Victim} and $IP_{Attacker}$) are able to read from and write to *L2*. Five (5) events can be identified as shown in Fig. 7.3. The process target of the attack **crypto/victim** (i.e., cryptographic functionality) is executed on IP_{Victim} at T_1, as in (1). The cryptographic functionality **crypto/victim** (e.g., AES transformation table implementation [66] lightweight GIFT [68], NTRU-Encrypt post-quantum lattice-based [82]) is implemented using tables, a common implementation technique to speed-up encryption [80].

During the execution of the **crypto/victim**, the tables are stored in the memory hierarchy of the system (*L1*, *L2*, *DRAM*). That is, each time **crypto/victim** requests a memory access, the memory hierarchy is queried (i.e., from *L1*, to *L2* and finally to *DDR*), as in (2) and (4b). Moreover, a ***spy*** process, a malware piece of code, is executed in the $IP_{Attacker}$ at T_2. **Crypto/victim** and **spy** processes are therefore physically isolated on different IP cores (IP_{Victim} and $IP_{Attacker}$) and different tiles (T_1 and T_2), respectively.

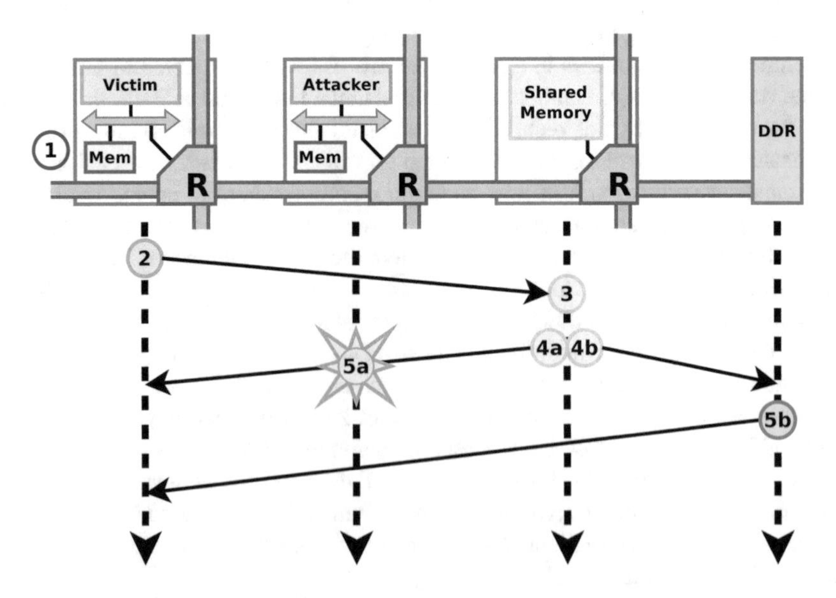

Fig. 7.3 Communication vulnerability

The attack assumes that **crypto/victim** and **spy** processes are executed in parallel. The goal of the **spy** is to detect whether the table entries stored on cache lines in *L2* have been evicted or not as a consequence of the **crypto/victim** requests. The **spy** only uses legitimate memory read and write operations that affect the *L2* data cache. Moreover, the **spy** does not have elevated privileges (i.e., supervisor). The **spy** is able to infer accesses to the *L2* data cache by the **crypto/victim**, because both processes use the same NoC (R_2 and R_3) to communicate with the cache. That is, the **spy** and the **crypto/victim** are simultaneously generating NoC transactions and communication collisions between both communication flows take place, as in (5a). The router can only perform a single commutation at time. When a single transaction is injected to the NoC, the router can immediately serve the communication request and the time required to complete the transaction is low. However, when multiple transactions are generated, a competition for the NoC communication resources is created. Thus, an arbitration process must take place. The winner transaction is allowed to use the router while the loser transaction must remain waiting and, thus, the time required to complete this transaction is high. Therefore, any reduction in the **spy's** communication throughput (i.e., the spy transaction must wait for the communication service) reveals a communication collision and thus a possible access to *L2* by the **crypto/victim**. In order to implement the NoC-based communication attack, the following requirements must be fulfilled:

- There are trusted and non-trusted nodes in the system. The attacker can infect an IP core within the SoC;
- Sensitive applications are only executed on trusted nodes. Any other application/process is executed on the non-trusted nodes;
- The attacker can generate random plaintexts and trigger encryptions;
- The trusted and non-trusted nodes share a cache (i.e., *L2* in the example);
- The *L2* data cache is inclusive with respect to the *L1* data cache.

The infection of a processor IP core is realistic, given the complexity of software running on modern embedded application processors. They typically host full scale operating systems and execute multiple tasks. Generating random plaintexts is straightforward and triggering encryptions on a different IP core can be a common use-case, e.g., if IP_{victim} provides security-related services for the rest of the system. Inclusive cache hierarchies are also common, as they can reduce maintenance and coherency efforts.

7.5 Microarchitectural Attacks

7.5.1 Computation Attacks

Common side-channel attacks are derived from the measurement of the execution time to complete an operation, power consumption, and electromagnetic radiation of the cryptographic function on the MPSoC. However, the advances of the understanding of the on-chip processing and storage capabilities have paved the way to new and more sophisticated side-channel attacks, also known as *microarchitectural side-channels*. Such threats arise from the wide amount of sharing resources and the multi-tenant characteristic of the MPSoC. By exploiting such characteristics of the system organization, the attacks are able to break the in-use software-isolation based security and to retrieve sensitive data [13, 46].

As shown in Sect. 7.2, MPSoCs shared a wide amount of resources, which can be the target of attacks when the victim and spy processes are executed (see Sect. 7.4). The first attacks that raise the awareness regarding the microarchitectural vulnerabilities of current embedded **processing elements** were the powerful Spectre [49] and Meltdown [50] attacks. They show that by exploiting the exception or branch misprediction it is possible to extract secret-dependent traces in the microarchitectural state of the processing units. This observation led to a proliferation of new microarchitectural attack variants [15, 16, 97].

7.5.2 Cache Attacks

Another common target of microarchitectural attacks is the **cache hierarchies**. While caches allow to speed-up the execution of applications, including the execution of cryptographic operations, caches also allow the mutual interference of processes executed on the MPSoC. When malicious and sensitive (victim) processes are executed together, this becomes a major threat. Microarchitectural attacks aim to retrieve sensitive data or gain the control of the system through either the exploitation of a wide variety of **side-channels** or by **corrupting data** using hardware vulnerabilities.

Cache attacks that exploit *side-channel* information can be commonly categorized based on the available information to the adversary: execution time (time-driven attacks [7, 10]), sequences of cache operations like hits and misses (trace-driven attacks [2]), and cache access patterns (access-driven attacks [61]). On the other side, microarchitectural attacks also can perform the *data corruption* on caches through the exploitation of their physical implementation. A classical example is the Rowhammer attack, which is able to modify data stored in victim memory row by repeatedly and frequently accessing physically adjacent memory rows. This controlled data modification can break the memory isolation between processes, exposing sensitive data to unauthorized and imperceptible corruption.

The access-driven cache attack is one of the most powerful and efficient techniques to retrieve the secret key of a cryptographic function, especially for those high performance software implementations (see Sect. 7.3) using tables with precomputed values that are accessed during the encryption/decryption operations. As discussed, for many of the SKC and PKC such accesses are secret-dependent. To identify the accessed memory location allows an attacker to retrieve the secret key. Two of the most successful cache attacks are the *Prime+Probe (P+P)* [61] and *Flush+Reload (P+R)* [102].

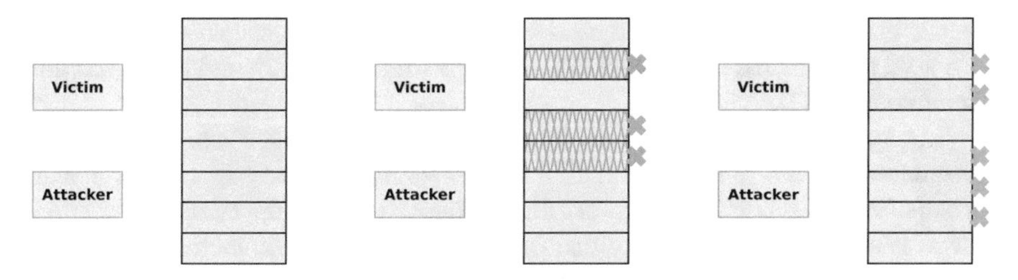

Fig. 7.4 Three steps of the Prime+Probe access-based cache attack. It includes: (i) Prime, to prepare the cache; (ii) Trigger, to execute the victim algorithm; and (iii) Probe, to measure the cache behavior

Note that the *Attacker* (which executes the **spy** process) and the *Victim* (which executes the cryptographic operation) share a cache (e.g., *L2*). In addition (as specified in the threat model), the *Attacker* is able to request to the *Victim* to encrypt data. Figure 7.4 shows the action of the victim and the attacker over the shared cache *L2* for each one of the three steps required to perform the Prime+Probe attack. The different colors represent the owner of the cache lines (victim is green and attacker is orange). The Prime+Probe attack is executed in three steps:

- ***Prime:*** In order to ensure that the cryptographic tables are not in victim's *L1*, the attacker *Primes* the *L2* cache by performing different requests that will result in the eviction of the data. As shown in Fig. 7.4a, the *L2* lines are all orange now. Due to the cache inclusiveness property, priming *L2* sets also causes the eviction of cache lines from the victim's L1 sets (i.e., the cryptographic tables are now stored in a higher level in the memory hierarchy).
- ***Trigger Cryptographic Operation:*** After the memory priming, the attacker requests the victim to encrypt a plaintext. The encryption process leads to the retrieval of the cryptographic tables and thus the eviction of some cache lines owned by the attacker. As shown in Fig. 7.4b, now, some of the lines are green.
- ***Probe:*** The attacker re-accesses (probes) the data previously brought into the cache at the prime step. The goal is to detect the accessed cache sets during encryption. If the victim accessed the cache line, the attacker's data (or parts of it) is removed from *L2* cache, producing a cache miss. Figure 7.4c shows the Probe step. As an effect, the attacker will experience a longer time to retrieve the requested data. Otherwise, (i.e., the victim does accessed the line), a cache hit will take place and the attacker's access time is faster.

The key recovery analysis process is described in [61, 80]. In general, by performing the *Prime+Probe*, the attacker becomes aware of the location of the cache sets and the accessed sets. Based on this information, the spy is able to discover which indexes have been used during the encryption by the victim. The method of the exploitation of the knowledge of the used indexes depends on each cryptographic algorithm. For the AES, an attacker exploits Eqs. (7.1) and (7.2) to obtain a set of key candidates. This search is further refined when the attack is further executed using different random or crafted plaintexts. *Prime+Probe* also was used to attack the *lightweight cryptography* [68], RSA [63], and *Post-Quantum Cryptography*[82].

Among all the types of access-driven attacks, the *Prime+Probe* technique is considered to be a generic variant. The *Flush+Reload (F+R)* is very similar to *Prime+Probe*. The difference is that *Flush+Reload* exploits the *clflush* instruction, which is able to evict a memory line from all the cache levels, including from the Last-Level-Cache (LLC). Attacks to SKC and PKC based on *Flush+Reload* are demonstrated in [42, 105].

7.5.3 NoC Attacks

MPSoC communication structure is based on NoCs. The goal of the NoC router is to switch packets from an input to an output port. An incorrect NoC router implementation might be exploited to perform different attacks. Figure 7.5 shows a general classification of the NoC Attacks. The circles represent the attacks and the blue labels represent the microarchitectural NoC features that can be exploited in order to provoke the attack. Three types of attacks can be identified (Data alteration, Denial-of-service, Timing attack). The formalization of such attacks is presented in [74].

- ***Data Alteration attacks***, which change the data embodied into the packets that are traversing the router. As a consequence, the data path and control path are modified. It includes: (1) *Modification*, which may result in deviation from their original intended destination (destination field modified), memory corruption at the destination (memory address modified), and violation of the access-control security rules (e.g., by accessing forbidden memory areas); (2) *Duplication*, which allows an attacker to capture the information and to perform replay attacks; and (3) *Drop*, which results in loss of information that may cause the disruption of sensitive traffic and data or control flow of an application. The data alteration attacks can be performed through data overwrite, injection of false data, and by exploiting buffer vulnerabilities (e.g., overflows, wrong control);
- ***Denial-of-Service (DoS) attack***, in which the communication resources are flooded by the intense router utilization. This prevents that authorized communications take place. By performing router DoS, the attacker is able to cause deadlocks, blocking all the NoC communications [18].
- ***Timing Side-Channel Attack***, which exploits the communication pattern of the sensitive data. By observing the timing (throughput or latency) required to inject traffic into the router, an attacker may gain such information (traffic pattern and communication volume) regarding the sensitive traffic. The timing attack can be effectively executed on a single NoC router. By detecting the optimal point of attack (e.g., communication event between a cryptographic co-processor and a shared memory), the NoC timing attack is able to optimize cache attacks as in [69].

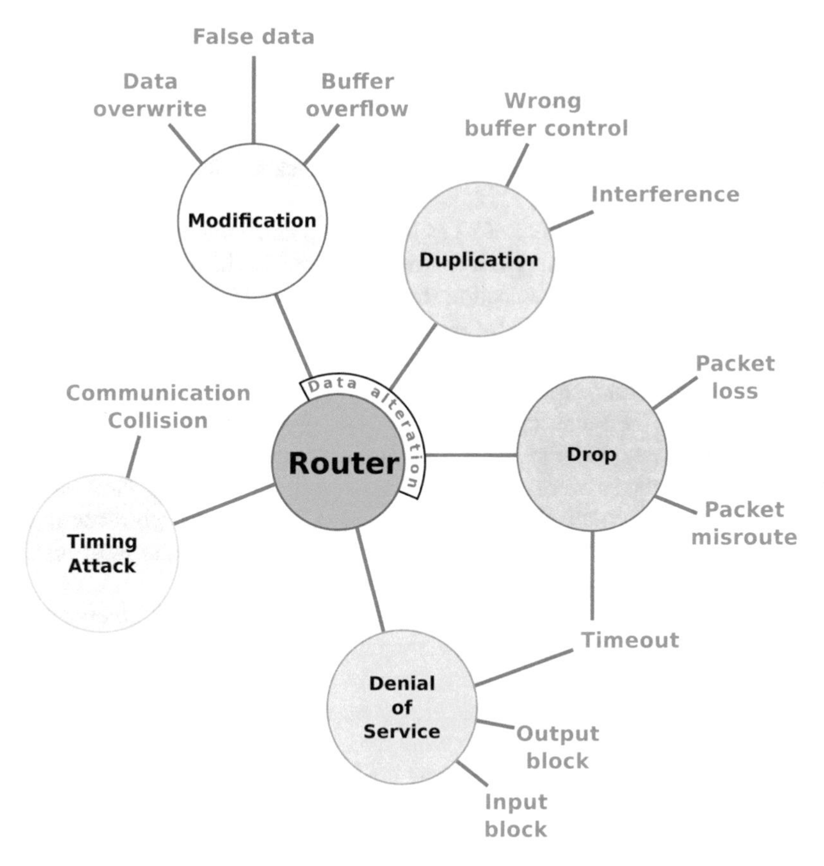

Fig. 7.5 Communication attacks

7.6 NoC-Enhanced Cache Attacks

7.6.1 Description

The efficiency of the *Prime+Probe* and other cache attacks heavily depends on the amount of noise in the attacker's measurements. For each cryptographic primitive there is a so-called *sweet spot* in which the cryptographic operation is on its most vulnerable point, since it offers to the attacker almost *noise free* measurements. Noisy measurements will demand a higher effort from the attacker (e.g., larger amount of *Prime+Probe* attack rounds) to avoid the false-positives key candidates and to be able to retrieve the complete key. The perfect moment of attack varies according to the security primitive (Victim). For the AES the *sweet spot* is after the first round, where the first 16 tables accesses to the shared cache *L2* took place. A *Probe* in such a stage will contain a clear information regarding the used indexes during the cryptographic operation. The cache sets accessed in the subsequent rounds will just add additional noise to the attacker measurement.

The NoC, as the heart of the MPSoC and main shared MPSoC structure, is an attractive structure for attackers. Previous works have demonstrated several types of attacks, which have exploited the NoC communication and the implementation weaknesses on the NoC routers [18, 83].

Communication leakage in SoCs has first been pointed out in the seminal works of [83, 98, 99]. Years later, in [70] it was demonstrated for the first time the potential of exploiting the NoC communication structure for magnifying cache-based attacks (access-based). By using the NoC-enhanced cache attack, the efficiency of the attack has been improved by up to 500%, when compared to a simple cache-based attack. Therefore, NoC-enhanced attacks offer the attackers a practical mean to easily retrieve the secret keys of the MPSoC. Later in [78, 80], this approach was also used to exploit the bus-based on-chip communication structure. Other types of cache attacks were also explored. In [65, 66] and [79], NoC-enhanced cache timing and trace attacks were demonstrated, respectively. However, from all the three alternatives, the access-based cache attack is one of the most preferred techniques by attackers due to their high efficiency.

The NoC-enhanced attacks exploit collisions in the network communication to detect the *sweet spot* of the cryptographic operation. The collisions appear from the shared nature of the routers. Usually two facts are exploited in a NoC-enhanced attack: (1) sensitive information (i.e., sensitive/victim flow) has a higher communication priority; and (2) the table retrieval of the cryptographic operation from the shared cache is characterized by the injection of long packets to the NoC. As explained previously, the goal of the attack is to identify the *sweet spot* of the sensitive/victim communication flow. The target characteristic of the sensitive flow (e.g., source/destination pattern, communication volume) will vary according to the cryptographic primitive. For instance, for AES it is the end of the first round of the encryption process, which for the NoC point of view translates into the detection of a large packet being injected from the shared cache *L2* to the victim tile [70, 83].

The attacker may be physically located either **directly** on the sensitive path (e.g., as shown on the Fig. 7.5), where the router that links the attacker to the MPSoC is used by the sensitive flow, or **outside** of the path. In the later case, higher NoC detection noise is expected. During all the attack time, the attacker is **injecting small packets** to the NoC and **measuring the achieved throughput**. In order to detect the massive sensitive packets, the attacker should search for collisions with the sensitive flow (intersection of flows). When the attacker is directly located on the sensitive path, the attacker and the victim have a direct collision on the router that links the attacker to the NoC. Otherwise, the effect can still be achieved but other reinforcement techniques should be used to further clear the noise of the NoC measurements.

As routers are shared, the communication collisions between attacker's and victim's traffic may reveal the sensitive behavior (e.g., mapping, topology, routing, transmission pattern, and volume of communication). The **collisions are detected by the reduction of throughput of the attacker**. After its detection, the *Probe* step is performed immediately. The key analysis and retrieval is performed as discussed in Sect. 7.5.2.

The seminal work in [28] has shown that by integrating communication services and techniques to improve the safety and real-time characteristics of the applications executed on the MPSoC, security backdoors may be opened. Preemptive NoCs, which allow that high priority communication flows preempt low priority packets on the NoC router, can be exploited. Further research on this area is required so as to protect the MPSoC while guaranteeing the real-time capabilities.

7.6.2 Countermeasures

However, many MPSoC attacks can be avoided by modifying only the NoC router architecture. Therefore, ensuring the correctness and security of the NoC routers is critical and needs to be addressed already in the early stages of the MPSoC design. Despite functional correctness has been widely explored, security has often been neglected. NoC verification, testing, and simulation techniques may contribute to ensure the router correctness, expanding the test patterns and coverage for possible threat scenarios. While simulation based analysis cannot be considered complete, some efforts have been conducted towards the formal verification of NoC security [67, 74].

Current countermeasures against the NoC-enhanced cache attacks are based on increasing the noise of the measurements performed by the NoC-based attack technique. These countermeasures modify the configuration parameters of the router. Figure 7.6 shows the different proposed countermeasures separated by the modified router component. *At the buffers*, the dynamic buffer allocation [76] and large buffer dimension [28] are used to avoid the direct correlation between the

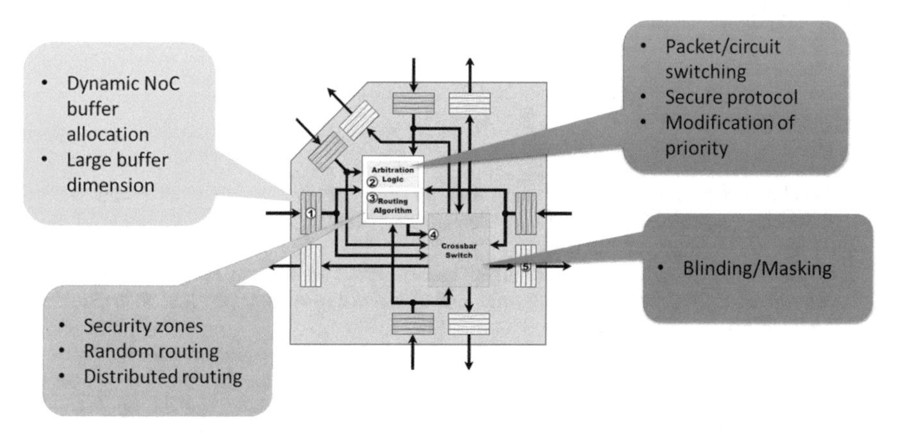

Fig. 7.6 State of the art of countermeasures against NoC-enhanced cache attacks

degradation of throughput and the communication collision between sensitive and attacker flow. *At the arbitration logic*, different techniques are proposed in order to isolate the traffic. It includes the mixed packet/switching data commutation [9, 64], the integration of protocols that include traffic monitoring [70], and the integration of hard Quality-of-Service (QoS) mechanisms to isolate the sensitive information through arbitration management (high priority [98], bounded priority [99], random priority [83]). *At the routing*, it includes the security-aware traffic management guided by security zones [27, 84], the use of random/dynamic and distributed routing [41, 70, 89]. *At the crossbar*, by blinding and masking the commutations inside the router [28].

7.7 Summary and Conclusions

MPSoCs are a key technology for satisfying the needs of current and future applications which have turned to be targets of attacks. In order to be able to use such a technology, security must be guaranteed. While cryptography, as a mean to ensure the security of the system is usually employed, their secure implementation is critical for achieving the security goals of the system. Current MPSoCs are target of microarchitectural attacks which exploit the shared computation, storage, and communication structures. While cache attacks are already very powerful, the NoCs can be further exploited to magnify those attacks, thus, giving to the attacker a very attractive means for retrieving the secret information. In order to avoid such a powerful attack, NoC security must be considered. A secure NoC by design will be possible only when NoC designers understand the critical role of this structure in the overall system security. This chapter calls the attention to such an issue and points to the need to protect current and future cryptographic implementations on the MPSoC. The NoC must be carefully crafted to not only meet the performance requirements, but also the security needs of the system.

References

1. Accelerating the Data Plane With the TILE-Mx Manycore Processor. http://www.tilera.com/files/drim__EZchip_LinleyDataCenterConference_Feb2015_7671.pdf. Accessed 14 Nov 2020
2. O. Acıiçmez, O., Ç.K. Koç, Trace-driven cache attacks on AES, in *Proceedings of the 8th International Conference on Information and Communications Security, ICICS'06* (Springer, berlin, 2006), pp. 112–121
3. ALDEC: Aldec 4k ultrahd imaging solutions (2020). https://www.aldec.com/en/solutions
4. D.M. Ancajas, K. Chakraborty, S. Roy, Fort-NoCs: mitigating the threat of a compromised NoC, in *Proceedings of the 51st Annual Design Automation Conference*, 2014, pp. 1–6
5. ARM: Arm Cortex-R Cortex-M Series Processors (2020). https://www.arm.com/
6. S. Banik, S.K. Pandey, T. Peyrin, Y. Sasaki, S.M. Sim, Y. Todo, Gift: a small present. Cryptology ePrint Archive, Report 2017/622 (2017). https://eprint.iacr.org/2017/622

7. D.J. Bernstein, Cache timing attacks on AES (2005)
8. G. Bertoni, L. Breveglieri, I. Koren, P. Maistri, An efficient hardware-based fault diagnosis scheme for AES: performances and cost, in *Proceedings of the 19th IEEE International Symposium on Defect and Fault Tolerance in VLSI Systems, 2004. DFT 2004* (IEEE, Piscataway, 2004), pp. 130–138
9. A.K. Biswas, Efficient timing channel protection for hybrid (packet/circuit-switched) network-on-chip. IEEE Trans. Parallel Distrib. Syst. **29**(5), 1044–1057 (2017)
10. A. Bogdanov, T. Eisenbarth, C. Paar, M. Wienecke, Differential cache-collision timing attacks on AES with applications to embedded CPUs, in *Proceedings of the Cryptographers' Track at the RSA Conference 2010*, San Francisco, CA, USA, March 1–5, 2010 (Springer, Berlin, 2010), pp. 235–251
11. A. Bogdanov, L.R. Knudsen, G. Leander, C. Paar, A. Poschmann, M.J.B. Robshaw, Y. Seurin, C. Vikkelsoe, Present: an ultra-lightweight block cipher, in *Cryptographic Hardware and Embedded Systems—CHES 2007*, ed. by P. Paillier, I. Verbauwhede (Springer, Berlin, 2007), pp. 450–466
12. J. Bos, M. Coster, Addition chain heuristics, in *Proceedings on Advances in Cryptology, CRYPTO '89* (Springer, Berlin, 1989), p. 400–407
13. F. Brasser, U. Müller, A. Dmitrienko, K. Kostiainen, S. Capkun, A.R. Sadeghi, Software grand exposure: SGX cache attacks are practical, in *11th USENIX Workshop on Offensive Technologies (WOOT 17)* (USENIX Association, Vancouver, 2017). https://www.usenix.org/conference/woot17/workshop-program/presentation/brasser
14. A. Cabrera Aldaya, C. Pereida Garcia, L.M. Alvarez Tapia, B.B. Brumley, Cache-timing attacks on RSA key generation. IACR Trans. Cryptogr. Hardw. Embed. Syst. **2019**(4), 213–242 (2019). https://doi.org/10.13154/tches.v2019.i4.213-242. https://tches.iacr.org/index.php/TCHES/article/view/8350
15. C. Canella, J.V. Bulck, M. Schwarz, M. Lipp, B. von Berg, P. Ortner, F. Piessens, D. Evtyushkin, D. Gruss, A systematic evaluation of transient execution attacks and defenses (2018). abs/1811.05441. http://arxiv.org/abs/1811.05441
16. C. Canella, D. Genkin, L. Giner, D. Gruss, M. Lipp, M. Minkin, D. Moghimi, F. Piessens, M. Schwarz, B. Sunar, J. Van Bulck, Y. Yarom, Fallout: leaking data on meltdown-resistant CPUs, in *Proceedings of the ACM SIGSAC Conference on Computer and Communications Security (CCS)* (ACM, New York, 2019)
17. S. Charles, P. Mishra, Securing network-on-chip using incremental cryptography, in 2020 IEEE *Computer Society Annual Symposium on VLSI (ISVLSI)* (2020), pp. 168–175. https://doi.org/10.1109/ISVLSI49217.2020.00039
18. C.G. Chaves, S.P. Azad, T. Hollstein, J. Sepulveda, A distributed DoS detection scheme for NoC-based MPSoCs, in *2018 IEEE Nordic Circuits and Systems Conference (NORCAS): NORCHIP and International Symposium of System-on-Chip (SoC)* (2018), pp. 1–6. https://doi.org/10.1109/NORCHIP.2018.8573524
19. B. Christof, J. Jéfemy, K. Stefan, L. Gregor, M. Amir, P. Thomas, S. Yu, S. Pascal, M.S. Siang, Skinny family of block ciphers (2019). https://sites.google.com/site/skinnycipher/home
20. D. Christoph, F.M. Maria Eichlseder, M. Schläffer, Ascon lightweight authenticated encryption & hashing (2019). https://ascon.iaik.tugraz.at/index.html
21. J. Daemen, V. Rijmen, *The Design of Rijndael* (Springer, New York, 2002)
22. D. Goudarzi, J. Jean, S. Kölbl, T. Peyrin, M. Rivain, Y. Sasaki, S.M. Sim, Pyajamask cipher (2019). https://pyjamask-cipher.github.io/
23. J.P. Diguet, S. Evain, R. Vaslin, G. Gogniat, E. Juin, NoC-centric security of reconfigurable SoC, in *First International Symposium on Networks-on-Chip (NOCS'07)* (IEEE, Piscataway, 2007), pp. 223–232
24. A. Duncan, F. Rahman, A. Lukefahr, F. Farahmandi, M. Tehranipoor, M., FPGA bitstream security: a day in the life, in *2019 IEEE International Test Conference (ITC)* (2019), pp. 1–10. https://doi.org/10.1109/ITC44170.2019.9000145
25. ETSI: European telecommunications standards institute (2020). https://www.etsi.org/

26. S. Evain, J.P. Diguet, From NoC security analysis to design solutions, in *IEEE Workshop on Signal Processing Systems Design and Implementation, 2005* (IEEE, Piscataway, 2005), pp. 166–171
27. R. Fernandes, C. Marcon, R. Cataldo, J. Sepúlveda, Using smart routing for secure and dependable NoC-based MPSoCs. IEEE/ACM Trans. Netw. **28**, 1158–1171 (2020)
28. B. Forlin, C. Reinbrecht, J. Sepúlveda, Security aspects of real-time MPSoCs: the flaws and opportunities of preemptive NoCs, in *VLSI-SoC: New Technology Enabler*, ed. by C. Metzler, P.E. Gaillardon, G. De Micheli, C. Silva-Cardenas, R. Reis (Springer, Cham, 2020), pp. 209–233
29. T. Fritzmann, J. Sepúlveda, Efficient and flexible low-power NTT for lattice-based cryptography, in *2019 IEEE International Symposium on Hardware Oriented Security and Trust (HOST)* (2019), pp. 141–150. https://doi.org/10.1109/HST.2019.8741027
30. T. Fritzmann, G. Sigl, J. Sepúlveda, RISQ-V: tightly coupled RISC-V accelerators for post-quantum cryptography. IACR Cryptol. ePrint Arch. **2020**, 446 (2020)
31. S.D. Galbraith, *Mathematics of Public Key Cryptography* (Cambridge University Press, Cambridge, 2012)
32. A. Ganguly, P.P. Pande, B. Belzer, Crosstalk-aware channel coding schemes for energy efficient and reliable NoC interconnects. IEEE Trans. Very Large Scale Integr. VLSI Syst. **17**(11), 1626–1639 (2009)
33. Gartner Inc., IoT penetration statistics (2018). https://www.gartner.com/
34. C.H. Gebotys, R.J. Gebotys, A framework for security on NoC technologies, in *Proceedings of the IEEE Computer Society Annual Symposium on VLSI, 2003* (IEEE, Piscataway, 2003), pp. 113–117
35. M. Grassl, B. Langenberg, M. Roetteler, R. Steinwandt, Applying Grover's algorithm to AES: quantum resource estimates, in *International Workshop on Post-Quantum Cryptography* (Springer, Berlin, 2016), pp. 29–43
36. J. Hu, R. Marculescu, Energy-and performance-aware mapping for regular NoC architectures. IEEE Trans. Comput. Aided Des. Integr. Circuits Syst. **24**(4), 551–562 (2005)
37. IEEE, IEEE standards association (2020). https://standards.ieee.org/
38. IETF, Rfc6071: IP security (IPSEC) and internet key exchange (IKE) document roadmap (2011). https://tools.ietf.org/pdf/rfc6071.pdf
39. IETF, Rfc8446: the transport layer security (TLS) protocol version 1.3 (2018). https://tools.ietf.org/pdf/rfc8446.pdf
40. IETF, Internet engineering task force (IETF) (2020). https://www.ietf.org/
41. L.S. Indrusiak, J. Harbin, C. Reinbrecht, J. Sepúlveda, Side-channel protected MPSoC through secure real-time networks-on-chip. Microprocess. Microsyst. **68**, 34–46 (2019). https://doi.org/10.1016/j.micpro.2019.04.004. http://www.sciencedirect.com/science/article/pii/S0141933118302965
42. G. Irazoqui, T. Eisenbarth, B. Sunar, *s$a*: a shared cache attack that works across cores and defies VM sandboxing—and its application to AES, in *2015 IEEE Symposium on Security and Privacy* (IEEE, Piscataway, 2015), pp. 591–604
43. ISO, ISO—international Organization for Standardization (2020). https://www.iso.org/
44. IWAVE, iWave Systems ultra-high-performance FPGA platforms for AI/ML accelerated edge computing in IoT applications (2020). https://www.iwavesystems.com/news/ai-ml-accelerated-edge-computing-in-iot/
45. Jamie Whitney, New frontiers in real-time software (2020). https://www.militaryaerospace.com/computers/article/14180254
46. Y. Jang, J. Lee, S. Lee, T. Kim, *SGX-Bomb: Locking Down the Processor via Rowhammer Attack* (2017), pp. 1–6. https://doi.org/10.1145/3152701.3152709
47. A.A. Jerraya, W. Wolf, Chapter 1—the what, why, and how of MPSoCs, in *Multiprocessor Systems-on-Chips, Systems on Silicon*, ed. by A.A. Jerraya, W. Wolf (Morgan Kaufmann, San Francisco, 2005), pp. 1–18. https://doi.org/10.1016/B978-012385251-9/50014-1. http://www.sciencedirect.com/science/article/pii/B9780123852519500141

48. KALRAY MPPA: a new era of processing. https://de.slideshare.net/infokalray/kalray-sc13-external3. Accessed 14 Nov 2020
49. P. Kocher, J. Horn, A. Fogh, D. Genkin, D. Gruss, W. Haas, M. Hamburg, M. Lipp, S. Mangard, T. Prescher, M. Schwarz, Y. Yarom, Spectre attacks: exploiting speculative execution, in *40th IEEE Symposium on Security and Privacy (S&P'19)* (2019)
50. M. Lipp, M. Schwarz, D. Gruss, T. Prescher, W. Haas, A. Fogh, J. Horn, S. Mangard, P. Kocher, D. Genkin, Y. Yarom, M. Hamburg, Meltdown: reading kernel memory from user space, in *27th USENIX Security Symposium (USENIX Security 18)* (2018)
51. K.G. Lov, A fast quantum mechanical algorithm for database search, in *28th Annual ACM Symposium on the Theory of Computing*, p. 212 (1996)
52. S.A. Manavski, Cuda compatible GPU as an efficient hardware accelerator for AES cryptography, in *2007 IEEE International Conference on Signal Processing and Communications* (IEEE, Piscataway, 2007), pp. 65–68
53. S.K. Mathew, F. Sheikh, M. Kounavis, S. Gueron, A. Agarwal, S.K. Hsu, H. Kaul, M.A. Anders, R.K. Krishnamurthy, 53 Gbps native gf(2^4) composite-field AES-encrypt/decrypt accelerator for content-protection in 45 nm high-performance microprocessors. IEEE J. Solid State Circuits **46**(4), 767–776 (2011)
54. C. Mitchell, A. Selby, Algorithms for software implementations of RSA. IEE Proc. Comput. Digital Tech. **136**, 166–170 (1989)
55. NSA says it "must act now" against the quantum computing threat (2016). https://www.technologyreview.com/s/600715/nsa-says-it-must-act-now-against-the-quantum-computing-threat/
56. NIST, Federal information processing standard (FIPS) 202 (2015). https://tools.ietf.org/pdf/rfc6071.pdf
57. NIST, Lightweight cryptography (2020). https://csrc.nist.gov/projects/lightweight-cryptography
58. NIST, Lightweight cryptography round 2 candidates (2020). https://csrc.nist.gov/Projects/lightweight-cryptography/round-2-candidates
59. NIST, NIST: National institute of standards and technology (2020). https://www.nist.gov/
60. NIST: Status report on the second round of the NIST post-quantum cryptography standardization process (2020). https://csrc.nist.gov/publications/detail/nistir/8309/final
61. D.A. Osvik, A. Shamir, E. Tromer, *Cache Attacks and Countermeasures: The Case of AES* (Springer, Berlin, 2006)
62. S. Payandeh Azad, M. Tempelmeier, G. Jervan, J. Sepúlveda, CAESAR-MPSoC: Dynamic and efficient MPSoC security zones, in *2019 IEEE Computer Society Annual Symposium on VLSI (ISVLSI)* (2019), pp. 477–482. https://doi.org/10.1109/ISVLSI.2019.00092
63. C. Percival, Cache missing for fun and profit, in *Proceedings of the BSDCan 2005* (2005)
64. C. Reinbrecht, A. Aljuffri, S. Hamdioui, M. Taouil, B. Forlin, J. Sepulveda, Guard-NoC: a protection against side-channel attacks for MPSoCs, in *2020 IEEE Computer Society Annual Symposium on VLSI (ISVLSI)* (2020), pp. 536–541. https://doi.org/10.1109/ISVLSI49217.2020.000-1
65. C. Reinbrecht, B. Forlin, J. Sepulveda, Cache timing attacks on NoC-based MPSoCs. Microprocess. Microsyst. **66**, 9 (2019). https://doi.org/10.1016/j.micpro.2019.01.007. http://www.sciencedirect.com/science/article/pii/S0141933118302898
66. C. Reinbrecht, B. Forlin, A. Zankl, J. Sepúlveda, Earthquake—a NoC-based optimized differential cache-collision attack for MPSoCs, in *2018 Design, Automation & Test in Europe Conference & Exhibition (DATE)* (IEEE, Piscataway, 2018), pp. 648–653
67. C. Reinbrecht, S. Hamdioui, M. Taouil, B. Niazmand, T. Ghasempouri, J. Raik, J. Sepúlveda, LiD-CAT: a lightweight detector for cache attacks, in *2020 IEEE European Test Symposium (ETS)* (2020), pp. 1–6. https://doi.org/10.1109/ETS48528.2020.9131603
68. C. Reinbrecht, S. Hamdioui, M. TaouilForlin, J. Sepúlveda, Grinch: a cache attack against gift lightweight cipher, in *2021 Design, Automation & Test in Europe Conference & Exhibition (DATE)* (IEEE, Piscataway, 2021)

69. C. Reinbrecht, A. Susin, L. Bossuet, S. Georg, J. Sepúlveda, Timing attack on NoC-based systems: Prime+Probe attack and NoC-based protection. Microprocess. Microsyst. Embed. Hardw. Design **52**, 556–565 (2017)
70. C. Reinbrecht, A. Susin, L. Bossuet, G. Sigl, J. Sepúlveda, Side-channel attack on NoC-based MPSoCs are practical: NoC Prime+Probe attack, in *2016 29th Symposium on Integrated Circuits and Systems Design (SBCCI)* (2016), pp. 1–6. https://doi.org/10.1109/SBCCI.2016.7724051
71. B. Rezvani, F. Coleman, S. Sachin, W. Diehl, Hardware implementations of NIST lightweight cryptographic candidates: a first look. Cryptology ePrint Archive, Report 2019/824 (2019). https://eprint.iacr.org/2019/824
72. R. Rivest, A. Shamir, L. Adleman, A method for obtaining digital signatures and public-key cryptosystems. Commun. ACM **21**(2), 120–126 (1978)
73. D. Rossi, P. Angelini, C. Metra, Configurable error control scheme for NoC signal integrity, in *13th IEEE International On-Line Testing Symposium (IOLTS 2007)* (IEEE, Piscataway, 2007), pp. 43–48
74. J. Sepulveda, D. Aboul-Hassan, G. Sigl, B. Becker, M. Sauer, Towards the formal verification of security properties of a network-on-chip router, in *2018 IEEE 23rd European Test Symposium (ETS)* (2018), pp. 1–6. https://doi.org/10.1109/ETS.2018.8400692
75. J. Sepúlveda, D. Flórez, G. Gogniat, Reconfigurable security architecture for disrupted protection zones in NoC-based MPSoCs, in *2015 10th International Symposium on Reconfigurable Communication-centric Systems-on-Chip (ReCoSoC)*, pp. 1–8. IEEE (2015)
76. J. Sepúlveda, D. Florez, M. Soeken, J. Diguet, G. Gogniat, Dynamic NoC buffer allocation for MPSoC timing side channel attack protection, in *IEEE 7th Latin American Symposium on Circuits & Systems, LASCAS 2016*, Florianopolis, Brazil, February 28–March 2, 2016 (IEEE, 2016), pp. 91–94. DOI 10.1109/LASCAS.2016.7451017
77. J. Sepúlveda, G. Gogniat, D. Florez, J.P. Diguet, C. Zeferino, M. Strum, Elastic security zones for NoC-based 3d-MPSoCs, in *2014 21st IEEE International Conference on Electronics, Circuits and Systems (ICECS)* (IEEE, Piscataway, 2014), pp. 506–509
78. J. Sepúlveda, M. Gross, A. Zankl, G. Sigl, Exploiting bus communication to improve cache attacks on systems-on-chips, in *2017 IEEE Computer Society Annual Symposium on VLSI, ISVLSI 2017*, Bochum, Germany, July 3–5, 2017 (IEEE Computer Society, 2017), pp. 284–289. https://doi.org/10.1109/ISVLSI.2017.57
79. J. Sepúlveda, M. Gross, A. Zankl, G. Sigl, Towards trace-driven cache attacks on systems-on-chips—exploiting bus communication, in *12th International Symposium on econfigurable Communication-centric Systems-on-Chip, ReCoSoC 2017*, Madrid, Spain, July 12–14, 2017 (IEEE, Piscataway, 2017), pp. 1–7. https://doi.org/10.1109/ReCoSoC.2017.8016150
80. J. Sepúlveda, M. Gross, A. Zankl, G. Sigl, Beyond cache attacks: exploiting the bus-based communication structure for powerful on-chip microarchitectural attacks. ACM Trans. Embed. Comput. Syst. **20**(2), 23. 17 Feb (2021)
81. J. Sepulveda, R. Pires, G. Gogniat, W.J. Chau, M. Strum, QoSS hierarchical NoC-based architecture for MPSoC dynamic protection. Int. J. Reconfigurable Comput. **2012**, 578363 (2012)
82. J. Sepulveda, A. Zankl, O. Mischke, Cache attacks and countermeasures for NTRUEncrypt on MPSoCs: post-quantum resistance for the IoT. in *2017 30th IEEE International System-on-Chip Conference (SOCC)* (IEEE, Piscataway, 2017), pp. 120–125
83. M.J. Sepúlveda, J. Diguet, M. Strum, G. Gogniat, NoC-based protection for SoC time-driven attacks. IEEE Embed. Syst. Lett. **7**(1), 7–10 (2015). https://doi.org/10.1109/LES.2014.2384744
84. M.J. Sepúlveda, D. Flórez, V. Immler, G. Gogniat, G. Sigl, Efficient security zones implementation through hierarchical group key management at NoC-based MPSoCs. Microprocess. Microsyst. **50**, 164–174 (2017). https://doi.org/10.1016/j.micpro.2017.03.002
85. Y. Shen, M. Ferdman, P. Milder, Maximizing CNN accelerator efficiency through resource partitioning, in *2017 ACM/IEEE 44th Annual International Symposium on Computer Architecture (ISCA)* (IEEE, IEEE, 2017), pp. 535–547

86. P.W. Shor, Polynomial-time algorithms for prime factorization and discrete logarithms on a quantum computer. SIAM J. Comput. **26**(5), 1484–1509 (1997)
87. P. Socha, J. Brejník, J. Balasch, M. Novotný, N. Mentens, Side-channel countermeasures utilizing dynamic logic reconfiguration: protecting AES/Rijndael and serpent encryption in hardware. Microprocess. Microsyst. **78**, 103208 (2020). https://doi.org/10.1016/j.micpro.2020.103208
88. F.X. Standaert, G. Rouvroy, J.J. Quisquater, J.D. Legat, A methodology to implement block ciphers in reconfigurable hardware and its application to fast and compact AES RIJNDAEL, in *Proceedings of the 2003 ACM/SIGDA Eleventh International Symposium on Field Programmable Gate Arrays* (2003), pp. 216–224
89. R. Stefan, K. Goossens, Enhancing the security of time-division-multiplexing networks-on-chip through the use of multipath routing, in *Proceedings of the 4th International Workshop on Network on Chip Architectures, NoCArc 11* (2011)
90. F.J. Streit, F. Fritz, A. Becher, S. Wildermann, S. Werner, M. Schmidt-Korth, M. Pschyklenk, J. Teich, Secure boot from non-volatile memory for programmable SOC architectures (2020)
91. A.S. Tanenbaum, H. Bos, *Modern Operating Systems* (Pearson, London, 2015)
92. T.S. Teja, T.S. Kiran, T.S. Narayana, M. Vinodhini, N. Murty, Joint crosstalk avoidance with multiple bit error correction coding technique for NoC interconnect, in *2018 International Conference on Advances in Computing, Communications and Informatics (ICACCI)* (IEEE, 2018), pp. 726–731
93. I. Tetsu, K. Khairallah, M. Minematsu, P. Thomas, Romulus authenticated encryption (2019). https://romulusae.github.io/romulus/
94. B. Tim, Y.L. Chen, C. Dobraunig, B. Mennink, Elephant lightweight authenticated encryption scheme (2019). https://www.esat.kuleuven.be/cosic/elephant/
95. Towards holistic secure networking in connected vehicles through securing can-bus communication and firmware-over-the-air updating. J. Syst. Archit. **109**, 101761 (2020). https://doi.org/10.1016/j.sysarc.2020.101761. http://www.sciencedirect.com/science/article/pii/S1383762120300552
96. Using Intel's Single-Chip Cloud Computer (SCC). https://communities.intel.com/docs/DOC-19269. Accessed 14 Nov 2020
97. S. van Schaik, A. Milburn, S. Österlund, P. Frigo, G. Maisuradze, K. Razavi, H. Bos, C. Giuffrida, RIDL: rogue in-flight data load, in *S&P* (2019)
98. Wang, Y., Suh, G.E.: Efficient timing channel protection for on-chip networks, in *2012 IEEE/ACM Sixth International Symposium on Networks-on-Chip* (2012), pp. 142–151. https://doi.org/10.1109/NOCS.2012.24
99. H.M.G. Wassel, Y. Gao, J.K. Oberg, T. Huffmire, R. Kastner, F.T. Chong, T. Sherwood, Networks on chip with provable security properties. IEEE Micro **34**(3), 57–68 (2014). https://doi.org/10.1109/MM.2014.46
100. W. Wolf, A.A. Jerraya, G. Martin, Multiprocessor system-on-chip (MPSoC) technology. IEEE IEEE Trans. Comput. Aided Des. Integr. Circuits Syst. **27**(10), 1701–1713 (2008). https://doi.org/10.1109/TCAD.2008.923415
101. XILINX, Automotive grade Zynq ultrascale+ MPSoCs (2020). https://www.xilinx.com/products/silicon-devices/soc/xa-zynq-ultrascale-mpsoc.html
102. Y. Yarom, K. Falkner, Flush+reload: a high resolution, low noise, l3 cache side-channel attack, in *23rd USENIX Security Symposium (USENIX Security 14)* (USENIX Association, San Diego, 2014), pp. 719–732. https://www.usenix.org/conference/usenixsecurity14/technical-sessions/presentation/yarom
103. Y. Yarom, D. Genkin, N. Heninger, Cachebleed: a timing attack on OpenSSL constant time RSA. Cryptology ePrint Archive, Report 2016/224 (2016). https://eprint.iacr.org/2016/224
104. Y. Yarom, D. Genkin, N. Heninger, Cachebleed: a timing attack on OpenSSL constant-time RSA. J. Cryptogr. Eng. **7**(2), 99–112 (2017)
105. X. Zhang, Y. Xiao, Y. Zhang, Return-oriented flush-reload side channels on arm and their implications for android devices, in *Proceedings of the 2016 ACM SIGSAC Conference on Computer and Communications Security* (2016), pp. 858–870

Part III
Runtime Monitoring Techniques

Chapter 8
Real-Time Detection and Localization of DoS Attacks

Subodha Charles and Prabhat Mishra

8.1 Introduction

The drastic increase in System-on-Chip (SoC) complexity has led to a significant increase in SoC design and validation complexity [3, 20, 24, 26, 44, 48–51]. SoC design using third-party intellectual property (IP) blocks is a common practice today due to both design cost and time-to-market constraints. These third-party IPs, gathered from different companies around the globe, may not be trustworthy. Integrating these untrusted IPs can lead to security threats. A full-system diagnosis for potential security breaches may not be possible due to lack of design details shared by the vendors. Even if they do, any malicious modifications (e.g., hardware Trojans) can still go undetected since it may be infeasible to detect stealthy triggers [2, 25, 27, 28, 46, 47, 52, 55]. Recent efforts try to combine the advantages of logic testing and side-channel analysis for effective Trojan detection in integrated circuits [34, 35, 42, 43, 45, 56]. The problem gets aggravated due to the presence of Network-on-Chip (NoC) in today's complex and heterogeneous SoCs. Figure 8.1 shows a typical NoC-based many-core architecture with heterogeneous IPs. As NoC has direct access to all the components in an SoC, malicious third-party IPs can leverage the resources provided by the NoC to attack other legitimate components. It can slow down traffic causing performance degradation, steal information, corrupt data, or inject power viruses to physically damage the chip. The problem of NoC security has been explored in two directions: (1) trusted NoC is used to secure the SoC from other untrusted IPs [5, 13, 61], and (2) NoC is untrustworthy and security

S. Charles (✉)
University of Moratuwa, Colombo, Sri Lanka
e-mail: s.charles@ieee.org

P. Mishra
University of Florida, Gainesville, FL, USA
e-mail: prabhat@ufl.edu

P. Mishra, S. Charles (eds.), *Network-on-Chip Security and Privacy*,
https://doi.org/10.1007/978-3-030-69131-8_8

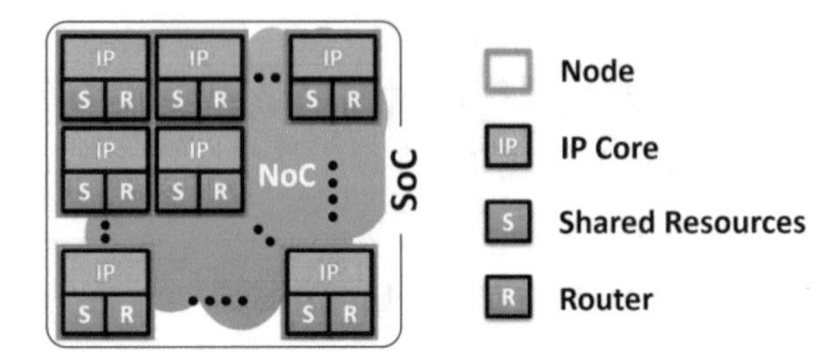

Fig. 8.1 NoC-based many-core architecture connecting heterogeneous IPs on a single SoC. Each IP connects to a router via a network interface. Depending on the selected topology, routers will be arranged across the NoC

countermeasures are required to secure the SoC [11, 12, 18]. This chapter is mainly focused on the first scenario where the NoC is trustworthy.

Denial-of-Service (DoS) in a network is an attack preventing legitimate users from accessing services and information. In an NoC setup, DoS attacks can happen from malicious third-party IPs (MIP) manipulating the availability of on-chip resources by flooding the NoC with packets. The performance of an SoC can heavily depend on few components. For example, a memory intensive application will send many requests to memory controllers, and as a result, routers connected to them will experience heavy traffic [16, 19]. If an MIP targets the same node, the SoC performance will suffer significant degradation [58]. Distributed DoS (DDoS) is a type of DoS attack where multiple compromised IPs are used to target one or more components in the SoC causing a DoS attack. We use “DoS attacks” to indicate both DoS and DDoS attacks in the rest of this chapter. The main solution presented in this chapter mitigates both DoS (single attacker) and DDoS (multiple attackers) attacks.

Unlike microcontroller based designs in the past, even resource constrained embedded and IoT (Internet-of-Things) devices nowadays incorporate one or more NoC-based SoCs. Many embedded and IoT systems have to deal with real-time requirements with soft or hard deadlines, where variations in applications as well as usage scenarios (inputs) are either well defined or predictable. In other words, if the applications are not predictable, it is impossible to provide any real-time guarantees. As expected, the communication patterns are known at design time for such systems. In fact, these assumptions are observed in a wide variety of prior research efforts involving soft [15, 66] as well as hard real-time systems [67]. These embedded and IoT devices can be one of the main targets of DoS attacks due to their real-time requirements with task deadlines. Early detection of DoS attacks in such systems is crucial as increased latencies in packet transmission can lead to deadline violations.

Importance of NoC security has led to many prior efforts to mitigate DoS attacks in an NoC such as traffic monitoring [30, 58] and formal verification-based methods [8]. Other real-time traffic monitoring mechanisms have also been discussed in non-NoC domains [67]. As outlined in Sect. 8.2.1, it is a major challenge to detect and localize a malicious IP in real-time. The problem is more challenging in the presence of multiple malicious IPs, and it gets further aggravated when multiple attackers help each other to mount the DoS attack. In this chapter, we present an efficient method that focuses on detecting changes in the communication behavior in real-time to identify DoS attacks. It is a common practice to encrypt critical data in an NoC packet and leave only few fields as plain text [64].[1]Therefore, this approach monitors communication patterns without analyzing the encrypted contents of the packets.

Major contributions of this chapter can be summarized as follows:

1. We present a detailed discussion on "flooding-type" DoS attacks in NoC-based SoCs and outline a real-time and lightweight DoS attack detection technique. The routers store statically profiled traffic behavior and monitor packets in the NoC to detect any violations in real-time.
2. A lightweight approach is developed to localize the MIP(s) in real-time once an attack is detected.
3. The effectiveness of this approach is evaluated against different NoC topologies using both real benchmarks and synthetic traffic patterns considering DoS attacks originating from a single malicious IP as well as from multiple malicious IPs.
4. The applicability of this approach is further evaluated by using an architecture model similar to one of the commercially available SoCs-Intel's KNL architecture [62].

The remainder of the chapter is organized as follows. Section 8.2 discusses the threat model and communication model used in this framework. Section 8.3 discusses other related research efforts in flooding-type DoS attack mitigation. Section 8.4 describes the real-time attack detection and localization methodology. Section 8.5 presents the experimental results. Section 8.6 presents the case study using KNL. Section 8.7 discusses the applicability and limitations of the proposed approach. Finally, Sect. 8.8 summarizes the chapter.

8.2 System and Threat Models

8.2.1 Threat Model

Previous works have explored two main types of DoS attacks on NoCs [29]—(1) MIPs flooding the network with useless packets frequently to waste bandwidth and cause a higher communication latency causing saturation, and (2) draining attack which makes the system execute high-power tasks and causes fast draining of battery. An illustrative example is shown in Fig. 8.2 to demonstrate the first type of DoS attack. As a result of the injected traffic from the malicious IPs to the victim IP (this can be a critical NoC component such as a memory controller), routers in that area of the NoC get congested and responses experience severe delays.

[1]On-chip encryption schemes introduce the notion of *authenticated encryption with associated data* in which the data is encrypted and associated data (initialization vectors, routing details, etc.) are sent as plain text [64].

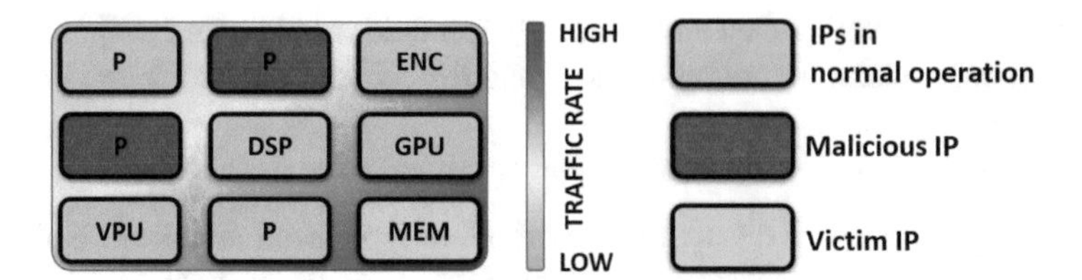

Fig. 8.2 Example DDoS attack from malicious IPs to a victim IP in a Mesh NoC setup. The thermal map shows high traffic near the victim IP (MEM). P-processor, DSP-digital signal processor, VPU-vector processing unit, GPU-graphics processing unit, ENC-encoder, MEM-memory controller

A practical example of a draining attack was shown in [54]. A malware known as a worm spread through Bluetooth and multimedia messaging services (MMS) and infected the recipient's mobile phone. The code is crafted in such a way that it sends continuous requests to the Bluetooth module for paging and to scan for devices. Power consumption in the infected phone was increased up to 500% compared to the idle state causing significant degradation of battery lifetime. There are instances of draining attacks where even though the computation overhead increases, the communication traffic does not increase. Such attacks cannot be detected using a security mechanism implemented at the NoC. Moreover, these attacks can be successful even if there are energy-efficient design solutions [4, 15, 17, 31, 32, 66].

The threat model is generic, it does not make any assumption about the placement or the number of malicious IPs or victim IPs. Figure 8.3 shows four illustrative examples of malicious/victim IP placements that can lead to different communication patterns. Figure 8.3a shows a scenario involving one malicious IP and one victim IP. The other three examples represent scenarios where the packets injected from the malicious IPs to victim IPs are routed through paths that (b) partially overlap, (c) completely overlap, and (d) form a loop. The approach proposed in this chapter is capable of both detecting and localizing all the malicious IPs in all these scenarios.

8.2.2 Communication Model

Since each packet injected in the NoC goes through at least one router, the router is identified to be an ideal NoC component for traffic monitoring. The router also has visibility to the packet header information related to routing. Packet arrivals at a router can be viewed as "events" and captured using arrival curves [9]. The set of all packets passing through router r during a program execution is denoted as a "packet stream" P_r. Figure 8.4 shows two packet streams within a specific time interval $[1, 17]$. The stream P_r (blue) shows packet arrivals in normal operation and $\widetilde{P_r}$ (red) depicts a compromised stream with more arrivals within the same time interval. The packet count $N_{p_r}[t_a, t_b)$ gives the number of packets arriving at router r within the

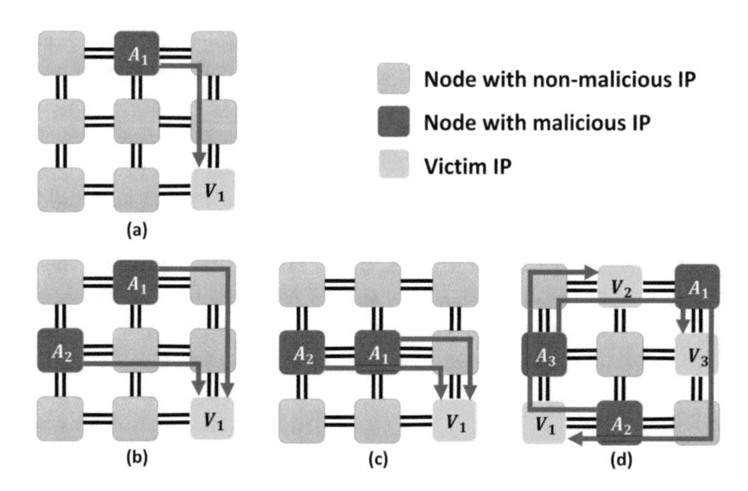

Fig. 8.3 Different scenarios of malicious and victim IP placement. Packet routing paths from malicious IPs to victim IPs shown in blue. (**a**) only one attacker is present, (**b**) paths partially overlap, (**c**) paths completely overlap, (**d**) paths form a loop

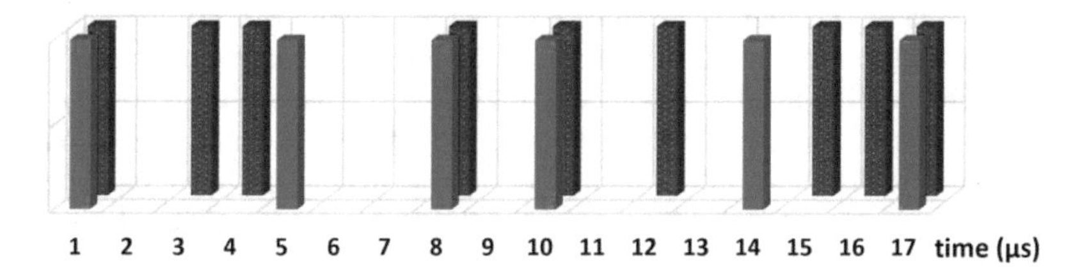

Fig. 8.4 Example of two event traces. Six blue event arrivals represent an excerpt of a regular packet stream P_r and nine red event arrivals represent a compromised packet stream $\widetilde{P_r}$

half-closed interval $[t_a, t_b)$. Equation (8.1) formally defines this using $N_{p_r}(t_a)$ and $N_{p_r}(t_b)$—maximum number of packet arrivals up to time t_a and t_b, respectively. $\forall t_a, t_b \in \mathbb{R}^+, t_a < t_b, n \in \mathbb{N}:$

$$N_{p_r}[t_a, t_b) = N_{p_r}(t_b) - N_{p_r}(t_a) \tag{8.1}$$

8.3 Related Work

Several other research efforts discussed performance degradation by flooding the network with additional packets [30, 57]. In [57], Trojans embedded in the router inject additional packets to the network to cause congestion. Additional packets are generated by sending illegal packet requests to the switch allocator when the cores are idling (core idle times during application execution). Fang et al. explored the effects of a DoS attack in an NoC architecture with mesh topology. They showed that with changes to the attack traffic rate (i.e., severity of attack), different routing protocols will get affected differently [23].

Countermeasures for DoS attacks both in terms of bandwidth and connectivity have been studied in an NoC context. One such method tries to stop the hardware Trojan which causes the DoS attack from triggering by obfuscating flits through shuffling, inverting, and scrambling [8]. If the Trojan gets triggered, there should be a threat detection mechanism. Other studies explored latency monitoring [58], centralized traffic analysis [30], security verification techniques [8], and design guidelines to reduce performance impacts caused by DoS attacks [23]. In [30], probes attached to the network interface gather NoC traffic data and send it to a central unit for analysis. In contrast, the method in [58] relies on injecting additional packets to the network and observing their latencies.

DoS attacks have been extensively studied in computer networks as well as mobile ad-hoc networks. In the computer network field, DoS attacks can be categorized as brute force attacks and semantic attacks. Brute force attacks overwhelm the system or the targeted resource with a flood of requests similar to the threat model used in this chapter. This can be achieved by techniques such as the attacker sending a large number of ICMP packets to the broadcast address of a network or by launching a DNS amplification attack [37]. It is common to use *botnets* rather than few sources to maximize the impact of the attacks. Semantic attacks on the other hand exploit some artificial limit of the system to deny services. Two popular examples are Ping-of-Death [22] and TCP SYN flooding [21]. Figure 8.5 shows an overview of a DoS attack in computer networks domain. Techniques such as botnet fluxing [69], back propagation neural networks [40], and TCP blocking [10] have been used to mitigate these attacks. However, using these techniques in SoC domain is not feasible due to the resource constrained nature and the architectural differences.

Waszecki et al. [67] discussed network traffic monitoring in an automotive architecture by monitoring message streams between electronic control units (ECU) via the controller area network (CAN) bus. Since multiple ECUs are connected on the same bus, it is difficult to localize the origin of attack, and therefore, the solution is presented only as a detection mechanism. Moreover, this architecture is bus-based and fundamentally different from an NoC. In this chapter, we outline a lightweight and real-time mechanism to detect and localize DoS attacks in an NoC-based SoC.

8.4 Real-Time Attack Detection and Localization

Figure 8.6 shows the overview of the proposed security framework to detect and localize DoS attacks originating from one or more MIPs. The first stage (upper part of the figure) illustrates the DoS attack detection phase while the second stage (lower part of the figure) represents the localization of MIPs. During the detection

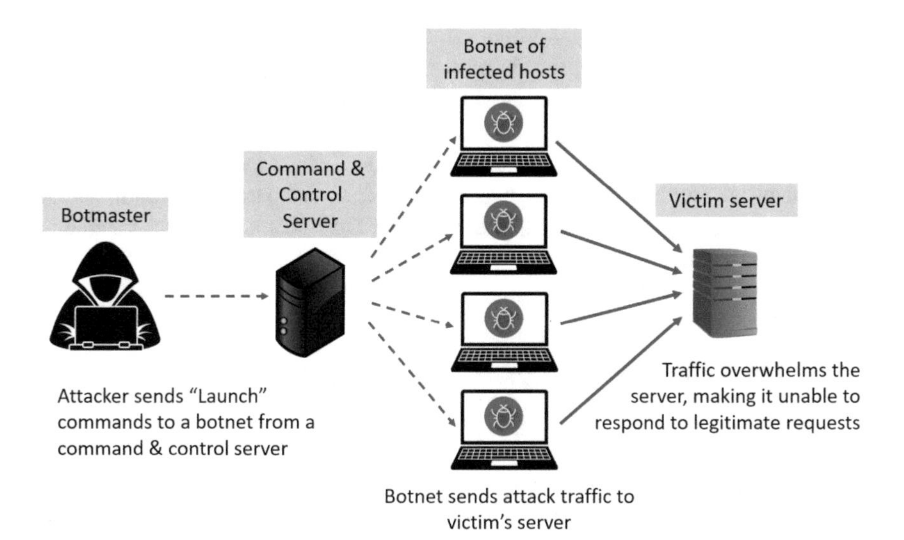

Fig. 8.5 Overview of a DoS attack in the computer networks domain

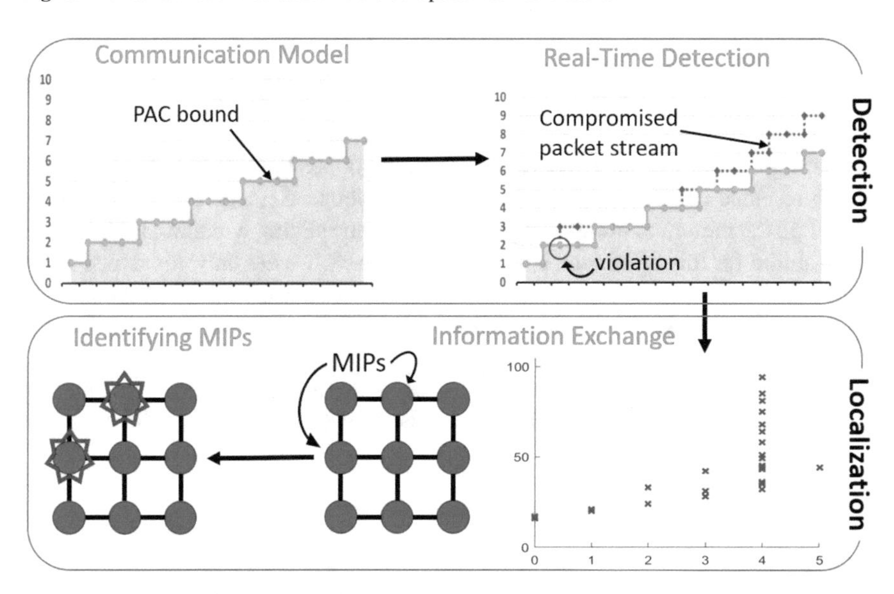

Fig. 8.6 Overview of the proposed framework: the system specification is analyzed to obtain the necessary packet arrival curves and detection parameters. These are used to design the real-time attack detection and localization framework

phase, the network traffic is statically analyzed and communication patterns are parameterized during design time to obtain the upper bound of "packet arrival curves" (PAC) at each router and "destination packet latency curves" (DLC) at each IP. The PACs are then used to detect violations of communication bounds in real-time. Once a router flags a violation, the IP attached to that router (local IP) takes responsibility of diagnosis. It looks at its corresponding DLC and identifies packets with abnormal latencies. Using the source addresses of those delayed packets, the local IP communicates with routers along that routing path to get their congestion information to localize the MIPs. The remainder of this section is organized as follows. The first two sections describe parameterization of PAC and DLC. Section 8.4.3 elaborates the real-time DoS attack detection mechanism implemented at each router. Section 8.4.4 describes the localization of MIPs.

8.4.1 Determination of Arrival Curve Bounds

To determine the PAC bounds, the packet arrivals are statically profiled and the upper PAC bound ($\lambda^u_{p_r}(\Delta)$) is constructed at each router. The maximum number of packets arriving at a router within an arbitrary time interval $\Delta (= t_b - t_a)$ is captured for this purpose. This is done by sliding a window of length Δ across the packet stream P_r and recording the maximum number of packets as formally defined in Eq. (8.2).

$$\lambda^u_{p_r}(\Delta) = \max_{t \geq 0}\{N_{P_r}(t + \Delta) - N_{P_r}(t)\} \quad (8.2)$$

Repeating this for several fixed Δ constructs the upper PAC bound. These bounds are represented as step functions. A lower PAC bound can also be constructed by recording the minimum number of packets within the sliding window. However, it is not required for this discussion since in a DoS attack, we are only concerned about violating the upper bound. An example PAC bound and two PACs corresponding to the packet streams in Fig. 8.4 are shown in Fig. 8.7. During normal execution, the PACs should fall within the shaded area.

While NoCs in general-purpose SoCs may exhibit dynamic and unpredictable packet transmissions, for vast majority of embedded and IoT systems, the variations

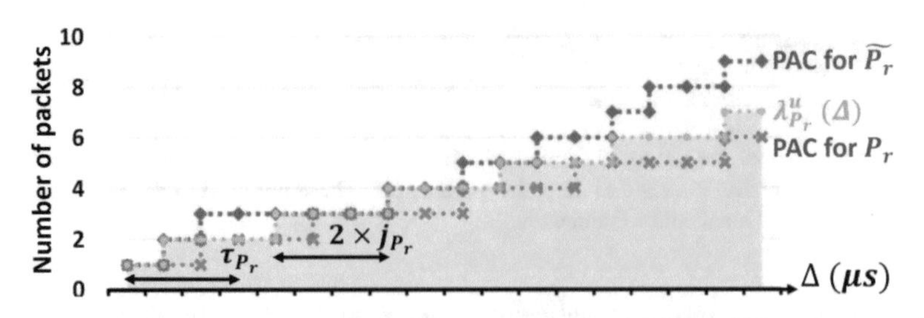

Fig. 8.7 Graph showing upper bound ($\lambda^u_{p_r}(\Delta)$) of PACs (green line with green markers) and the normal operational area shaded in green. The blue and red step functions show PACs corresponding to P_r and $\widetilde{P_r}$, respectively

in applications as well as usage scenarios (inputs) are either well defined or predictable. Therefore, the network traffic is expected to follow a specific trend for a given SoC. SoCs in such systems allow the reliable construction of PAC bounds during design time. To get a more accurate model, it is necessary to consider delays that can occur due to NoC congestion, task preemption, changes of execution times, and other delays. To capture this, the packet streams are considered to be periodic with jitter. The jitter corresponds to the variations of delays. Equation (8.3) represents the upper PAC bound for a packet stream P_r with maximum possible jitter j_{P_r} and period τ_{P_r} [63].

$$\forall \tau_{P_r}, j_{P_r} \in \mathbb{R}^+, \Delta > 0 : \lambda^u_{P_r}(\Delta) = \left\lceil \frac{\Delta + j_{P_r}}{\tau_{P_r}} \right\rceil \tag{8.3}$$

The equation captures the shift of the upper PAC bound because of the maximum possible jitter j_{P_r} relative to a nominal period τ_{P_r}. This method of modeling upper PAC bounds is validated by the studies in modular performance analysis (MPA) that uses real-time calculus (RTC) as the mathematical basis. MPA is widely used to analyze the best and worst case behavior of real-time systems. Capturing packet arrivals as event streams allows the packet arrivals to be abstracted from the time domain and represented in the interval domain (Fig. 8.7) with almost negligible loss in accuracy [63]. The same model is used in the MATLAB RTC toolbox [65].

8.4.2 *Determination of Destination Latency Curves*

Similar to the PACs recorded at each router, each destination IP records a DLC. An example DLC in normal operation is shown in Fig. 8.8a. The graph shows the latency against hop count for each packet arriving at a destination IP D_i. The distribution of latencies for each hop count is stored as a normal distribution, which can be represented by its mean and variance. Mean and variance of latency distribution at destination D_i for hop count k are denoted by $\mu_{i,k}$ and $\sigma_{i,k}$, respectively. In the example (Fig. 8.8a), $\mu_{i,4}$ is 31 cycles and $\sigma_{i,4}$ is 2. During the static profiling stage, upon reception of a packet, the recipient IP extracts the timestamp and hop count from the packet header, and plots the travel time (from the source to the recipient IP) against the number of hops. The mean and variance are derived after all the packets have been received. The illustrative example considered one malicious IP four hops away from the victim IP launching the DoS attack. No other IP is communicating with the victim IP in a path that overlaps with the congested path. Therefore, the increased delay is observed only at hop count 4 (Fig. 8.8b). In general, when multiple IPs send packets with destination D_i, and the paths overlap with the congested path, the increased delay will be reflected in several hop counts in the DLC. This scenario is not shown for the ease of illustration. However, such overlapping paths are considered in the experiments.

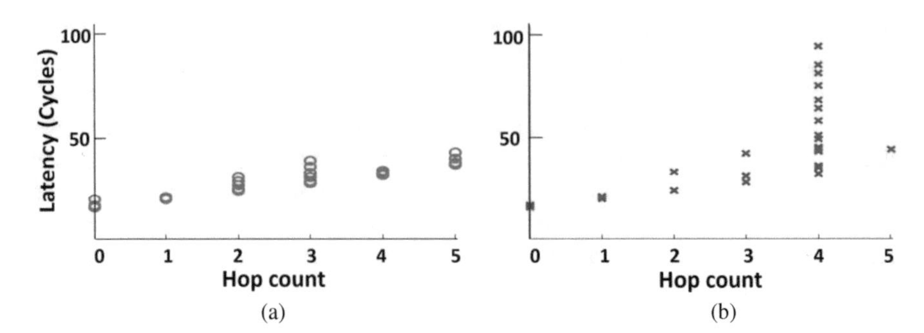

Fig. 8.8 Destination packet latency curves at an IP. The large variation in latency at hop count 4 in (**b**) compared to (**a**), contributes to identifying the malicious IP. (**a**) Normal operation. (**b**) Attack scenario

8.4.3 Real-Time Detection of DoS Attacks

Detecting an attack in a real-time system requires monitoring of each message stream continuously in order to react to malicious activity as soon as possible. For example, each router should observe the packet arrivals and check whether the pre-defined PAC bound is violated. The attack scenario can be formalized as follows:

$$\exists t \in \mathbb{R}^+ : \lambda^u_{p_r}(\Delta) < \max_{t \geq 0}\{N_{\widetilde{P}_r}(t+\Delta) - N_{\widetilde{P}_r}(t)\} \tag{8.4}$$

An obvious way to detect violations with the upper bound would be to construct the PAC and check if it violates the bound as shown in Fig. 8.7. However, to construct the PAC, the entire packet stream should be observed. In other words, all packet arrivals at a router during the application execution should be recorded to construct the PAC. While it is feasible during upper PAC bound construction at design time, it does not lead to a real-time solution. Therefore, an efficient method is needed to detect PAC bound violations during runtime.

To facilitate runtime detection of PAC bound violations, the "leaky bucket" algorithm is used. The algorithm considers packet arrivals and the history of packet streams and gives a real-time solution [39]. Once $\lambda^u_{p_r}(\Delta)$ is parameterized, the algorithm checks the number of packet arrivals within all time intervals for violations. Algorithm 8.1 outlines the leaky bucket approach where $\theta_{r,s}$ denotes the minimum time interval between consecutive packets in a staircase function s at router r, and $\omega_{r,s}$ represents the burst capacity or maximum number of packets within interval length zero. $\lambda^u_{p_r}(\Delta)$, which is modeled as a staircase function can be represented by n tuples—$(\theta_{r,s}, \omega_{r,s})$, $s \in \{1, n\}$ sorted in ascending order with respect to $\omega_{r,s}$. This assumes that each PAC can be approximated by a minimum on a set of periodic staircase functions [38].

Lines 2–5 initialize the timers ($\text{TIMER}_{r,s}$) to $\theta_{r,s}$ and packet counters at time zero ($\text{COUNTER}_{r,s}$) to corresponding initial packet numbers $\omega_{r,s}$, for each

Algorithm 8.1: Detecting compromised packet streams

```
1: Input: (θ_{r,s},ω_{r,s}) tuples containing parameterized PAC bound at router r.
2: for s ∈ {1, n} do
3:     TIMER_{r,s} = θ_{r,s}
4:     COUNTER_{r,s} = ω_{r,s}
5: end for
6: if packetReceived = TRUE then
7:     for s ∈ {1, n} do
8:         if COUNTER_{r,s} = ω_{r,s} then
9:             TIMER_{r,s} = θ_{r,s}
10:        end if
11:        COUNTER_{r,s} = COUNTER_{r,s} − 1
12:        if COUNTER_{r,s} < 0 then
13:            attacked(r) = TRUE
14:        end if
15:    end for
16: end if
17: for s ∈ {1, n} do
18:    if timeoutOccured(TIMER_{r,s}) = TRUE then
19:        COUNTER_{r,s} = min(COUNTER_{r,s} + 1, ω_{r,s})
20:        TIMER_{r,s} = θ_{r,s}
21:    end if
22: end for
```

staircase function and packet stream P_r. The DoS attack detection process (lines 6–16) basically checks whether the initial packet limits (COUNTER$_{r,s}$) have been violated. Upon reception of a packet (line 6), the counters are decremented (line 11), and if it falls below zero, a potential attack is flagged (line 13). If the received packet is the first within that time interval (line 8), the corresponding timer is restarted (line 9). This is done to ensure that the violation of PAC upper bound can be captured and visualized by aligning the first packet arrival to the beginning of the PAC bound. When the timer expires, values are changed to match the next time interval (lines 18–21). As demonstrated in Sect. 8.5, the algorithm allows real-time detection of DoS attacks under the threat model. Another important observation described in Sect. 8.5.4.1 drastically reduces the complexity of the algorithm allowing a lightweight implementation. The leaky bucket algorithm is originally proposed to check the runtime conformity of event arrivals in the context of network calculus. Its correctness is proven by Huang et al. [33].

8.4.4 Real-Time Localization of Malicious IPs

Figure 8.8b shows an example DLC during an attack scenario, where all IPs are injecting packets exactly the same way as shown in Fig. 8.8a except for one MIP, which injects a lot of packets to a node attached to a memory controller. Those

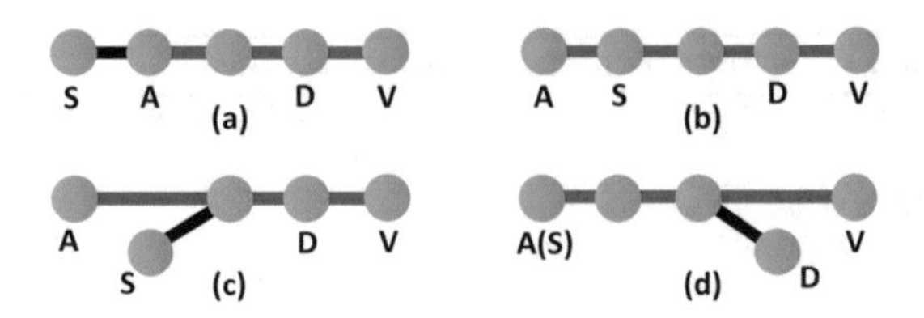

Fig. 8.9 Four scenarios of the relative positions of local IP (D), attacker IP (A), victim IP (V), and the candidate MIP (S) as found by D. The red line represents the congested path

two nodes are 4-hops apart in the Mesh topology. This makes the latency for 4-hop packets drastically higher than usual. For every hop count, the traffic distribution is maintained as a normal distribution using $\mu_{i,k}$ and $\sigma_{i,k}$. Once a potential threat is detected at a router, it sends a signal to the local IP. The local IP then looks at its DLC and checks if any of the curves have packets that took more than $\mu_{i,k}+1.96\sigma_{i,k}$ time (95% confidence level). One simple solution is to examine source addresses of those packets and conclude that the source with most number of packets violating the threshold is the MIP. However, this simple solution may lead to many false positives. As each IP is distributed and examines the latency curve independently, the IP found using this method may or may not be a real MIP (attacker). Therefore, it is called a "candidate MIP."

To illustrate the difference between an attacker and a candidate MIP, we first examine four scenarios with only one attacker as shown in Fig. 8.9. In these scenarios, the attacker A is sending heavy traffic to a victim IP V, and as a result, local IP D is experiencing large latency for packets from source S. The first three examples in Fig. 8.9 show examples where candidate MIP S is not the real attacker A. Since a large anomalous latency is triggered by the congestion in the network, the only conclusion obtained by the local IP from its DLC is that at least part of the path from candidate MIP to local IP is congested. The path from attacker A to victim V is called the "congested path."

In Fig. 8.9a, c, the false positives of the candidate MIP S can be removed with global information of congested paths, by checking the congestion status of path from S to its first hop. It is certain that S is not the attacker when this path is not congested. However, we cannot tell whether S is the attacker when the path of S is congested. For example, the routers of Fig. 8.9b, d are both congested, but S is not the attacker in Fig. 8.9b.

Things get much worse when multiple attackers are present. If we look at the example in Fig. 8.10, the path from candidate MIP S to local IP D is part of all paths along which three different attackers are sending packets to different victims. The "congested graph" is defined as the set of all congested paths and all the routers in the paths. Since each hop connecting two routers consists of two separate uni-directional links, a congested graph is a bi-directional graph as shown in Fig. 8.10. In order to detect attackers and avoid false positives, one simple solution would be building the entire congested graph by exchanging information from all the other

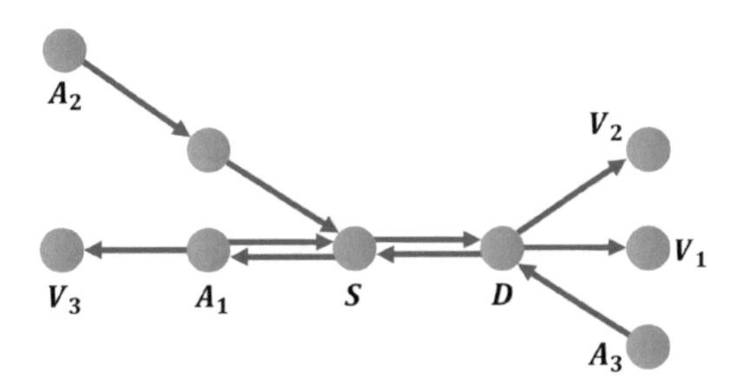

Fig. 8.10 Congested graph of three attackers

Algorithm 8.2: Event handler for router R

```
 1: upon event RESET:
 2: R.flag[p_i] = 0 for all ports p_i
 3: upon event attacked == TRUE:
 4: send a signal to local IP
 5: upon receiving a diagnostic message ⟨S, D⟩ from port p_i:
 6: start TIMEOUT if all R.flag == 0
 7: if S is local IP then
 8:     if flag[p_i] == 0 then
 9:         flag[p_i] = 1                                   ▷ local IP is the MIP
10:     end if
11:     if flag[p_i] == 2 then                     ▷ false positive, do nothing
12:     end if
13: else                                                  ▷ S is not local IP
14:     Let N be the neighbor of R that sits in the path from S to R
15:     if path from N to R is congested then
16:         sends a diagnostic message ⟨S, D⟩ to N indicating that S is a candidate attacker
17:         flag[p_i] = 2                                   ▷ other IP is the MIP
18:     else                                       ▷ false positive, do nothing
19:     end if
20: end if
21: upon event TIMEOUT:
22: if any flag in R.flag is 1 then
23:     broadcasting that its local IP is the attacker
24:     RESET
25: end if
```

routers and analyzing the graph to detect the actual MIPs. However, it would add a lot of burden on the already congested paths.

To overcome the bottlenecks, a distributed and lightweight protocol is implemented on the routers to detect the attackers. The event handler for each router for MIP localization is shown in Algorithm 8.2. The description of the steps of the complete protocol is shown below:

1. The router R detects an ongoing attack and sends a signal to the local IP (line 4). In Fig. 8.9, both D and V will send a signal to their local IPs.
2. The local IP D looks at its DLC and responds to its router with a diagnostic message $< S, D >$ indicating the address of the candidate MIP S and destination D. The local router then forwards the packet towards S.

3. Each port in each router maintains a three-state flag to identify the attacker. The flag is 0, 1, and 2 to denote the attacker is undefined, local IP, or others, respectively. When a diagnostic message $< S, D >$ comes in, R checks if the candidate MIP S is the local IP. If yes and its flag is not set yet, it will set the flag to be 1 (line 9). If S is not the local IP, it first finds out its neighbor N which sits in the path from S to R. If the one-hop path from N to R is congested, it sends the message to N (line 16) and sets the flag to 2, to indicate other IP as a potential attacker (line 17). Except for these two scenarios, the received message is a false positive and no action is taken (line 11 and 18), which will be explained in the examples. Note that the flag cannot decrease except for the reset signal which sets it to undefined (line 2). Therefore, if a diagnostic message already mentioned that other IPs may be the potential attackers, a new diagnostic message from the same port claiming that the local IP is the attacker will be ignored.
4. Each router maintains a timer. The timer starts as soon as any one of the router ports receives a diagnostic message. A pre-defined timeout period is used by each router. If the flag of any port is 1 after timeout, it broadcasts a message alerting that its local IP (line 23) is the attacker. Finally, a reset signal is triggered (line 24).

First, we will show that this approach works when a DoS attack is originating from only one MIP in the NoC. Later, we will describe how the proposed approach works in the presence of multiple MIPs mounting a DoS attack.

8.4.4.1 DoS Attack by a Single MIP

We use Fig. 8.9b to illustrate how this approach will localize the attacker when a DoS is caused by a single MIP. The router of S will receive two messages, one from the router of D saying that its local IP is a candidate MIP, and the other from the router of V saying that A is a candidate MIP, i.e., $< S, D >$ and $< A, V >$. Depending on the arrival time of these two messages, there are two scenarios. (a) $< S, D >$ comes first. It will change the flag of the corresponding port to 1 to denote that the local IP is the potential attacker. Then, S will receive $< A, V >$ through the same port. In this example, A is also the neighbor N. As the one-hop path from A to S is congested, the flag will be set to 2, denoting that the attacker is some other IP. (b) $< A, V >$ comes first. It will change the flag of the corresponding port to 2 to denote that the other IP is the potential attacker. Then, S will receive $< S, D >$ through the same port. As the flag is already set to 2, the received message is a false positive (line 11). When timeout occurs, nothing happens at the router of S. However, the router of A receives only the message from V indicating that its local IP is the potential attacker and its flag remains 1 when timeout occurs. A broadcast is sent indicating that A is the attacker.

For the case in Fig. 8.9a, A will receive a message from D indicating that S is a candidate MIP. However, when A checks the congestion status of the one-hop path from S to A, it will find out that the path is not congested. Therefore, the message is a false positive (line 18), and A will not change its flag. In other words, the flag of A will be set to 1 after receiving the message from V, and will not be changed by the message from D to S. After timeout, A will be identified as the attacker.

8.4.4.2 DoS Attack by Multiple MIPs

Before giving an illustrative example of how this approach will localize attacks by multiple malicious IPs, we outline the proof of the correctness of this approach.

Theorem 8.1 *If the congested graph contains no loops, Algorithm 8.2 can localize at least one attacker.*

Proof Merge multiple diagnostic messages with the same destination as one message and ignore all false positive messages detected in line 11 and line 18 of Algorithm 8.2. Define message φ_i as a diagnostic message which points out that A_i is a candidate MIP. Consider the port of any attacker A_i that receives message φ_i. Such a port always exists in a DoS attack scenario due to the fact that victim V_i will send a message φ_i to A_i saying that A_i is a candidate MIP. If φ_i is the only message received from this port, the algorithm can declare A_i as an attacker.

The algorithm fails only when all routers connected to the attackers have flags set to either 0 or 2 in each of their ports as illustrated in Algorithm 8.2. This can only happen when each port that receives a diagnostic message receives another diagnostic message which causes the flag to be set to 2. Assume that a port in router of A_i receives messages $MS_i = \{\varphi_i, \varphi_j, \ldots\}$. It will digest the message φ_i and send out the remaining ones. Construct a diagnostic message path in the following way. First, add A_i to the path. Then, select any message from MS_i other than φ_i, e.g., φ_j. Next, follow the diagnostic message path from A_i to A_j, and add all routers to the path. By the same process, select one message other than φ_j from MS_j, e.g., φ_k. Next, follow the path from A_j to A_k. This can be done one by one since for every message set MS_u at attacker A_u, there is at least one message other than φ_u to select from. Therefore, the constructed diagnostic message path contains an infinite number of attackers, as shown in Fig. 8.11. The infinite number of attackers implies that this path contains repeated attackers. Without loss of generality, it can be assumed that $A_k = A_i$. Since A_i cannot be sending out diagnostic messages MS_i through the same port that receives MS_i, the diagnostic path must form a loop. It is easy to see that diagnostic paths are the reverse of congested paths. As a result, there exists a loop in the congested graph, which contradicts the assumption made. Hence, Theorem 8.1 is proven. □

Thus, there always exists a port of the router connected to attacker A_i which receives only one diagnostic message φ_i given that there are no loops. This is a sufficient condition to detect A_i using Algorithm 8.2. Using this approach for localizing multiple malicious IPs gives rise to three cases that behave differently depending on how the MIPs are placed.

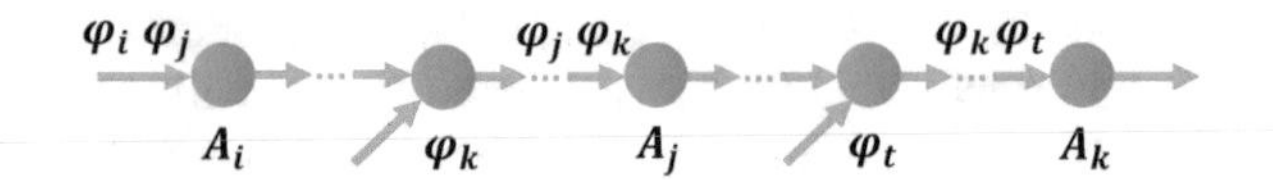

Fig. 8.11 An example of a diagnostic message path constructed by following the flow of a diagnostic message in each attacker

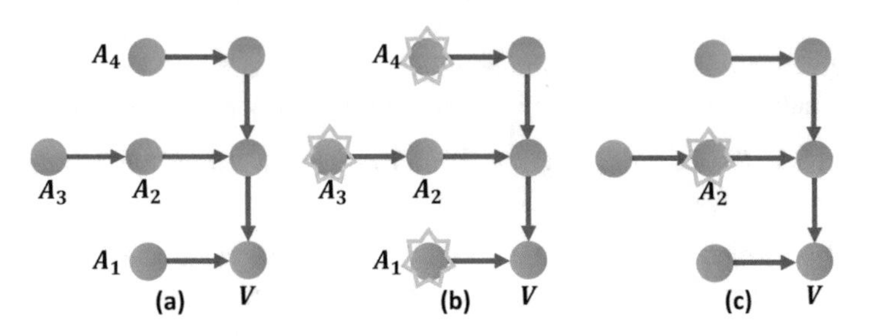

Fig. 8.12 Illustrative example to show how the detection and localization framework works. (**a**) Placement of attackers and victim that causes an overlap of congested paths of attackers A_2 and A_3. (**b**) Attacker(s) detected from first iteration. (**c**) Attacker(s) detected from second iteration

1. **Case 1:** If the congested paths do not overlap, all MIPs will be localized in one iteration using the process outlined above. This is the best case scenario for the approach and localizes MIPs in minimum time.
2. **Case 2:** If at least two paths overlap, it will need more than one iteration to localize all MIPs. To explain this scenario, an illustrative example is shown in Fig. 8.12. Figure 8.12a shows the placement of the four MIPs (A_1, A_2, A_3, A_4) attacking the victim IP (V). Once the attack is detected, in the first iteration, A_1, A_3, and A_4 are detected as shown in Fig. 8.12b. Due to the nature of this approach, A_2 is not marked as an attacker. This is caused by two diagnostic messages going in the paths $V \rightarrow A_2$ and $V \rightarrow A_3$. The router of A_2 will receive a message from the router of V saying that its local IP is a candidate MIP. It will change the flag of the corresponding port to 1 to denote that A_2 is the potential attacker. A_2 will receive another message from the router of V through the same port saying that A_3 is a candidate MIP. In this example, A_3 is also the neighbor of A_2. As the one-hop path from A_3 to A_2 is congested, the flag will be set to 2, denoting that the attacker is some other IP. When timeout occurs, nothing happens at the router of A_2. However, the router of A_3 receives only the message from V indicating that its local IP is the potential attacker and its flag remains 1 when timeout occurs. Therefore, A_3 is detected as an attacker whereas A_2 is not. In the case of A_1 and A_4, there is no overlap of congested paths and the two attackers are detected without any false negatives. Once the system resumes with only A_2 being malicious, the attacker will be detected and localized in the second iteration (Fig. 8.12c). This case consumes more time since an additional detection phase is required to localize all MIPs. The number of iterations will depend on how many overlapped paths can be resolved at each iteration. In the

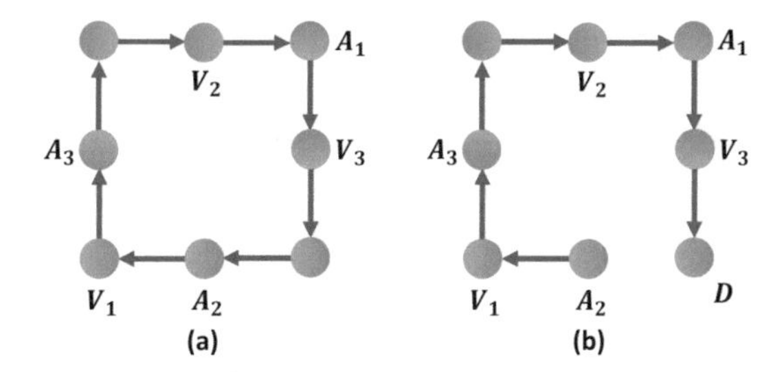

Fig. 8.13 (**a**) Three attackers cooperate and construct a loop in the congested graph. Algorithm 8.2 will fail to detect any attacker in the loop. (**b**) When a router randomly "stops working," an attacker A_2 is revealed after breaking the loop

worst case (where all congested paths can overlap and each iteration will resolve one path), the number of iterations will be equal to the number of MIPs. However, this approach is guaranteed to localize all MIPs.

3. **Case 3:** The proof of Theorem 8.1 had the assumption that the congested graph contains no loops. Therefore, using this approach as it is will not lead to localizing all MIPs if the congested graph forms a loop as shown in Fig. 8.13. One solution is that any router in the congested loop can randomly "stop working" and resume after a short while. By breaking the loop, this approach will detect attackers with the new congested graph. The router "stopping work" can be triggered by the system observing that a DoS attack is going on (during the detection phase), but no MIPs being localized.

In summary, this approach will detect one or more MIPs at each iteration depending on whether congested paths overlap. After detecting attackers(s) in the congested graph, their local router(s) can remove the attacker by dropping all its packets. Then, the process will be repeated with a new congested graph if more attackers exist. This approach continues to find more attackers until either all attackers have been found, or the congested graph forms a loop, which can be handled using the method outlined above (Case 3).

It is easy to see that the extra work for the router is minimal in this protocol because all computations are localized. It only needs to check the congestion status of connected paths (one hop away), and compute the flag which has two bits for each port. This protocol relies on the victim to pinpoint the correct attackers and the other routers to remove false positives. The timeout should be large enough for the victim to send messages to all the routers in the path of the attack. In practice, it can be the maximum communication latency between any two routers. The total time from detection to localization is the latency for packet traversal from the victim to attackers plus the timeout. Therefore, the time complexity for localization is linear in the worst case with respect to the number of IPs. It is important to note that most of the time, the diagnostic message path is the reverse of the congested path, and therefore, it is not congested.

8.5 Experiments

This chapter explores DoS attacks caused by a single MIP as well as multiple MIPs using the architecture shown in Fig. 8.14. In Sect. 8.6, the efficiency of this approach is evaluated using an architecture model similar to one of the commercially available SoCs [62].

8.5.1 *Experimental Setup*

The DoS attack detection and localization approach was evaluated by modeling an NoC-based SoC using the cycle-accurate full-system simulator—gem5 [7]. The interconnection network (NoC) was built on top of the "GARNET2.0" model that is integrated with gem5 [1]. The default gem5 source was modified to include the detection and localization algorithms. Experiments were conducted using several synthetic traffic patterns (uniform_random, tornado, bit_complement, bit_reverse, bit_rotation, neighbor, shuffle, transpose), topologies (Point2Point (16 IPs), Ring (8 IPs), Mesh4×4, Mesh8×8), and XY routing protocol to illustrate the efficiency of the approach across different NoC parameters. A total of 40 traffic traces were collected using the simulator by varying the traffic pattern and topology. Synthetic traffic patterns were only tested using one MIP in the SoC launching the DoS attack and an application instance running in 50% of the available IPs. These traffic traces act as test cases for the algorithms. The placement of the MIP, victim IP, and IP(s) running the traffic pattern was chosen at random for the 40 test cases.

The approach was also evaluated using real traffic patterns based on 5 benchmarks (FFT, RADIX, OCEAN, LU, FMM) from the SPLASH-2 benchmark suite [68] in Mesh 4×4 topology. Traffic traces from real traffic patterns were used to test both single-source DoS attacks as well as multiple-source DoS attacks.

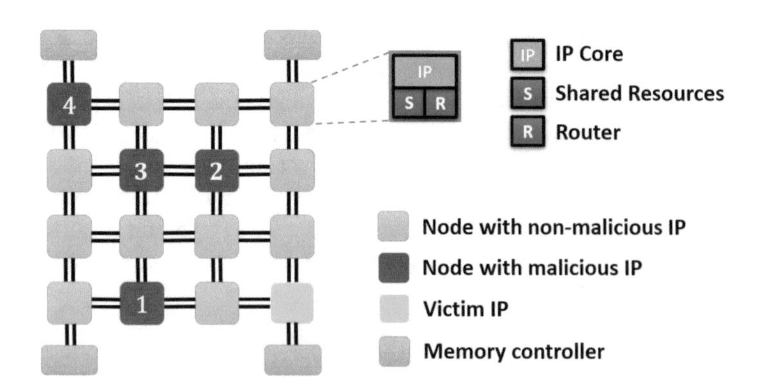

Fig. 8.14 MIP and victim IP placement when running tests with real benchmarks on a 4×4 Mesh NoC

The attack was launched at a node connected to a memory controller. Relative placements of the MIP and victim IP used to test the single-source DoS attack were the same as for the synthetic traces running on Mesh 4×4 topology (test case IDs 1 through 5 in Fig. 8.16). For the DoS attack involving multiple MIPs, experiments were done using the same set of benchmarks and topology with the victim and MIP placements as shown in Fig. 8.14. The placement captures both Case 1 and Case 2 discussed in Sect. 8.4.4.2. Each node with a non-malicious IP ran an instance of the benchmark while the four nodes in the four corners were connected to memory controllers. The jitter for all applications was calculated using the method proposed in [41].

8.5.2 Efficiency of Real-Time DoS Attack Detection

Before showing the results of experimental evaluation, we will first give an illustrative example to show how the parameters associated with the leaky bucket algorithm (Algorithm 8.1) are calculated and used in attack detection.

An important observation allows the reduction of the number of parameters required to model the PACs, and as a result, implement a lightweight scheme with much less overhead. The model in Eq. (8.3) is derived using the fact that the packet streams are periodic with jitter. As proposed in [67] and [6], for message streams with such arrival characteristics, the PACs can be parameterized by using only worst case jitter j_{P_r}, period τ_{P_r}, and an additional parameter ϵ_r which denotes the packet counter decrement amount. The relationship between these parameters is derived in [38] as shown in Eq. (8.5).

$$\theta_r = greatest\_common\_divisor(\tau_{P_r}, \tau_{P_r} - j_{P_r}) \tag{8.5a}$$

$$\omega_r = 2 \times \epsilon_r - \frac{\tau_{P_r} - j_{P_r}}{\theta_r} \tag{8.5b}$$

$$\epsilon_r = \frac{\tau_{P_r}}{\theta_r} \tag{8.5c}$$

To use these parameters, the only changes to Algorithm 8.1 are at line 11 ($\text{COUNTER}_{r,s}$ = $\text{COUNTER}_{r,s} - \epsilon_r$) and one tuple per packet stream instead of n tuples ($s \in \{1\}$). The illustrative example is based on this observation.

Illustrative Example Consider the example packet streams shown in Fig. 8.4. Assume that the packet steam P_r has a period $\tau_{P_r} = 3\,\mu\text{s}$ and jitter $j_{P_r} = 1.5\,\mu\text{s}$. During an attack scenario, this stream is changed to stream $\widetilde{P_r}$ with $\tau_{\widetilde{P_r}} = 2\,\mu\text{s}$ and no jitter. Using these values in Eq. (8.5) will give $\theta_r = 1.5\,\mu\text{s}$, $\omega_r = 3$, and $\epsilon_r = 2$, which are the parameters used in the leaky bucket algorithm. Therefore, $COUNTER_{r,s}$ is initialized with 3 (line 4, line 19) and decremented by 2 at each message arrival (line 11). $TIMER_{r,s}$ is initialized to 1.5 μs (line 3, line 20). Using

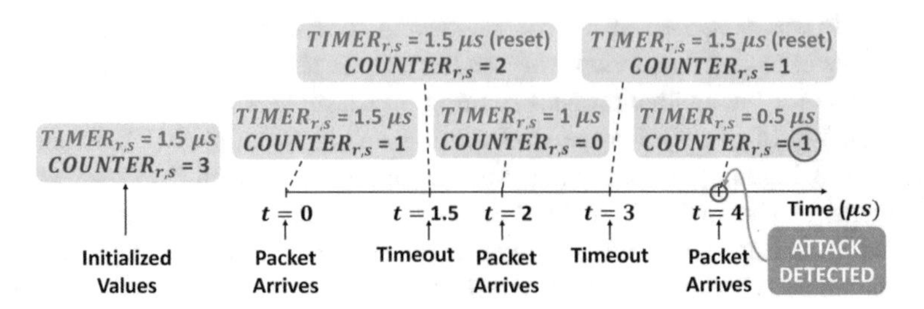

Fig. 8.15 Illustrative example of parameter changes in the leaky bucket algorithm with packet arrivals and timeouts

these values and running the detection algorithm during the attack scenario will lead to a detection time of 4 μs. Figure 8.15 shows the values of the parameters changing with each packet arrival and timeout leading to the detection of the attack at $t = 4$ μs.

The experimental evaluation follows the same process as the illustrative example using the experimental setup described in Sect. 8.5.1. Figure 8.16 shows the detection time across different topologies for synthetic traffic traces in the presence of one MIP. The 40 test cases are divided into different topologies, 10 each. The packet stream periods are selected at random to be between 2 and 6 μs. Attack periods are set to a random value between 10 and 80% of the packet stream period. The detection time is approximately twice the attack period in all topologies. This is expected according to Algorithm 8.1 and consistent with the observations in [67].

In addition to the time taken by the leaky bucket approach, the detection time also depends on the topology. For example, attack detection in Point2Point topology (Fig. 8.16a), where every node is one hop away, requires less time to detect compared to Mesh8×8 (Fig. 8.16d) where some nodes can be multiple hops away. The topology mainly affects attack localization time due to the number of hops from detector to attacker. But for detection, topology plays a relatively minor role since the routers are connected to each IP and detection mechanism neither takes into account the source nor the destination of packets. The routers only look at how many packets arrived in a given time interval. It is also important to note that any router in the congested path can detect the attack, not only the router connected to the victim IP. A combination of these reasons have led to the topology playing a relatively minor role in attack detection time. These results confirm that this approach can detect DoS attacks in real-time.

Results for DoS attack detection in the presence of multiple attacking MIPs are shown in Figs. 8.17 and 8.18. For all of these experiments, packet stream period is fixed at 2.5 μs and attack period is set to 1.5 μs. Figure 8.17 shows detection time variation in the presence of different number of IPs across benchmarks. The time to detect an ongoing attack in the multiple MIP scenario is typically less than the single MIP scenario. When more IPs are malicious, the detection time shows a decreasing trend. This is expected since multiple attackers flood the NoC faster and cause PAC

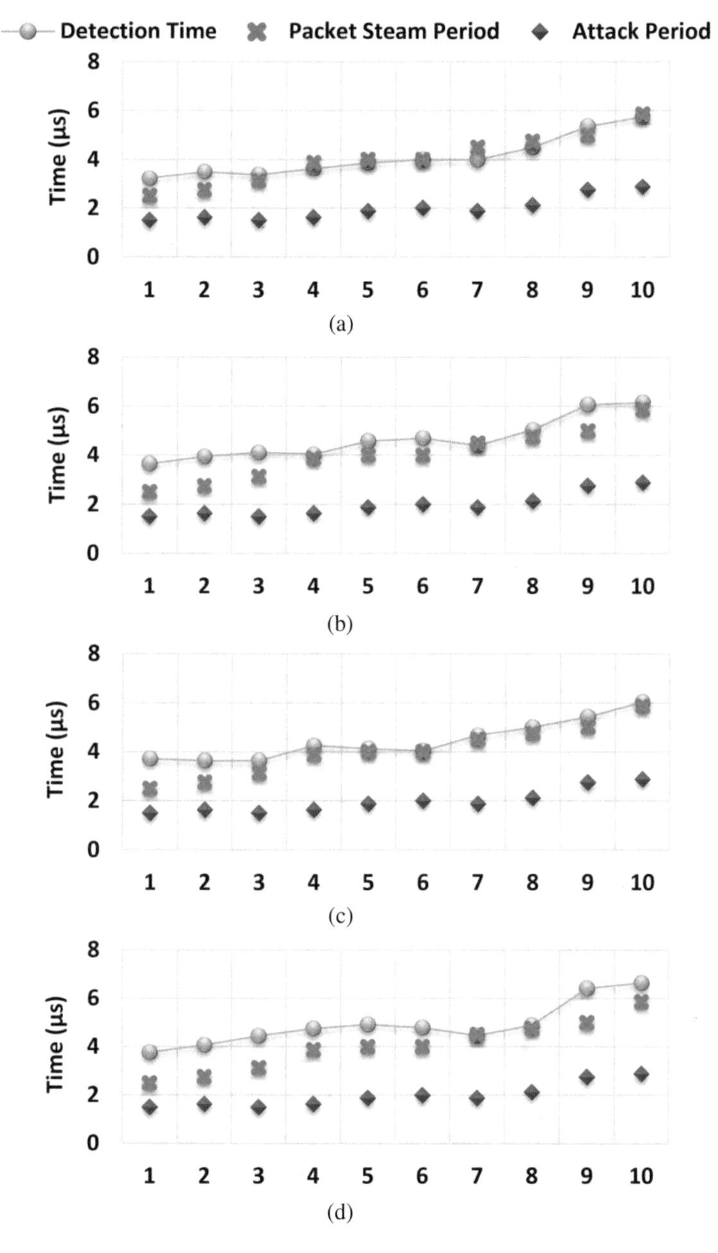

Fig. 8.16 Attack detection time for different topologies when running synthetic traffic patterns with the presence of one MIP. Each graph shows time in microseconds (y-axis) against test case ID (x-axis). (**a**) Point2Point. (**b**) Ring. (**c**) Mesh 4×4. (**d**) Mesh 8×8

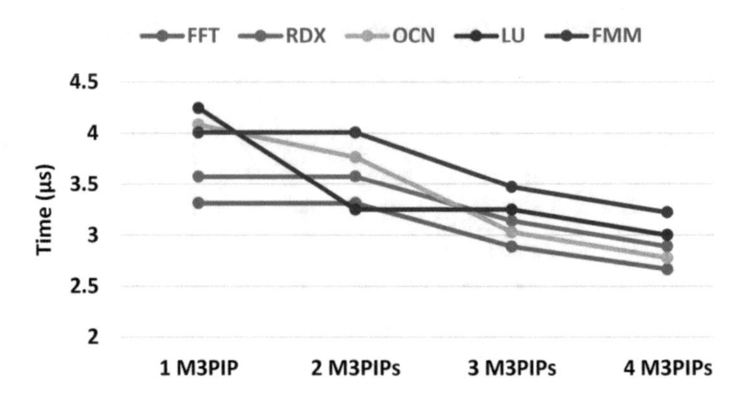

Fig. 8.17 Attack detection time when running real benchmarks with the presence of different number of MIPs

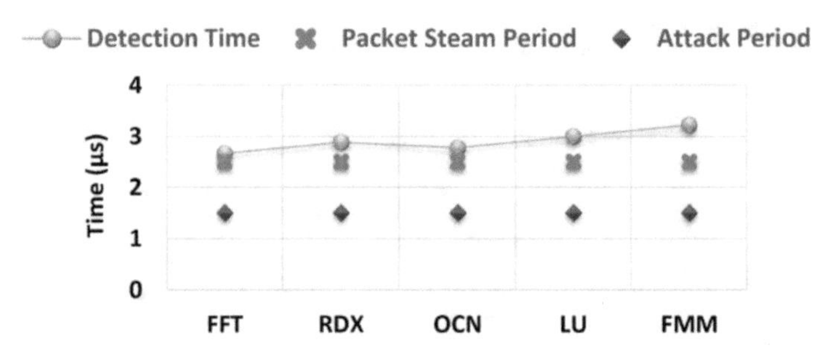

Fig. 8.18 Attack detection time when running real benchmarks with the presence of four MIPs

bound violations quicker. To compare detection time with packet stream period and attack period, the detection time variation is shown in the presence of four MIPs across benchmarks in Fig. 8.18.

8.5.3 Efficiency of Real-Time DoS Attack Localization

The efficiency of attack localization is evaluated by measuring the time it takes from detecting the attack to localizing the malicious IPs. According to this protocol, the localization time is mainly dominated by the latency for packet traversal from victim to attacker (V2AL) as well as the timeout (TOUT) described in Sect. 8.4.4. Figure 8.19 shows these statistics using the same set of synthetic traffic patterns for the single MIP scenario. The experimental setup for the localization results corresponds to the experimental results for the detection results in Fig. 8.16. Unlike the detection phase, since the localization time depends heavily on the time it takes for the diagnostic packets to traverse from the IPs connected to the routers that

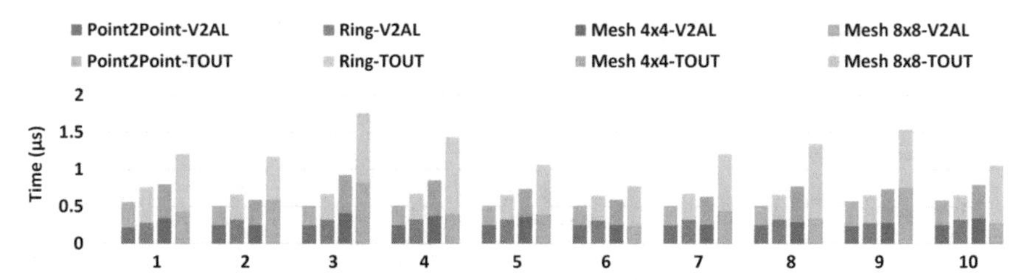

Fig. 8.19 Attack localization time for synthetic traffic patterns in the presence of one MIP. Figure shows time in microseconds (y-axis) against test case ID (x-axis) across different topologies. Test cases correspond to the test cases in Fig. 8.16

flagged the attack to the potentially malicious IPs, the localization time varies for each topology. For example, in a Point2Point topology, localization needs diagnostic message to travel only one hop, whereas a Mesh8×8 topology may require multiple hops. Therefore, localization is faster in Point2Point compared to Mesh8×8 as shown in Fig. 8.19. The localization time is less compared to detection time because the localization process completes once the small number of diagnostic packets reach all the potentially malicious IPs, whereas detection requires many packets before violating a PAC bound during runtime.

Results for DoS attack localization in the presence of multiple MIPs when running real benchmarks is shown in Fig. 8.20. Similar to the experiments done for DoS attack detection efficiency, localization results are shown for one, two, three, and four MIPs attacking the victim IP at the same time. The time is measured as the time it takes since launching the attack, until the localization of all MIPs. Once the first iteration of localization and detection is complete, the attack has to be detected again before starting the localization procedure. Therefore, the y-axis shows detection as well as localization time. For clarity of the graph, unlike in Fig. 8.19, total localization time is shown for each iteration rather than dividing the localization time as V2AL and TOUT. For both one and two MIP scenarios, only one iteration of detection and localization is required. When the third MIP is added, the two congested paths from victim to second MIP and from victim third MIP overlap. Therefore, only the first and third MIPs are localized during the first iteration leaving the second MIP to be detected during the second iteration. Similarly, in the four MIP scenarios, first, third, and fourth MIPs are localized during the first iteration and the second MIP, during the second iteration. This is consistent with the discussion presented in Sect. 8.4.4.2. The results show that both detection and localization can be achieved in real-time. If a system requires only detection, the architecture of this framework allows easy decoupling of the two steps.

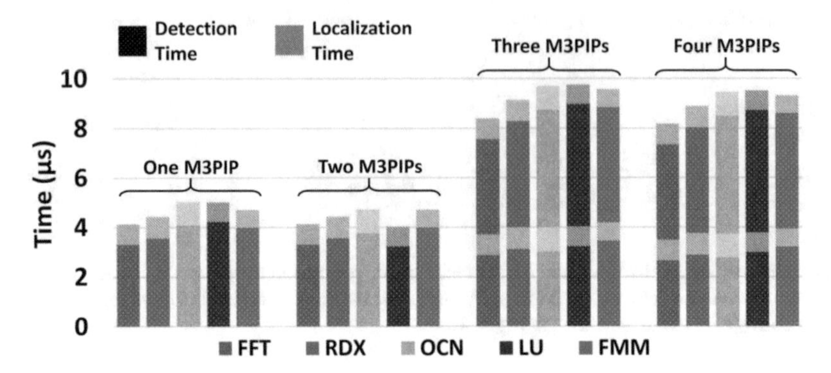

Fig. 8.20 Attack localization time when running real benchmarks with the presence of different number of MIPs

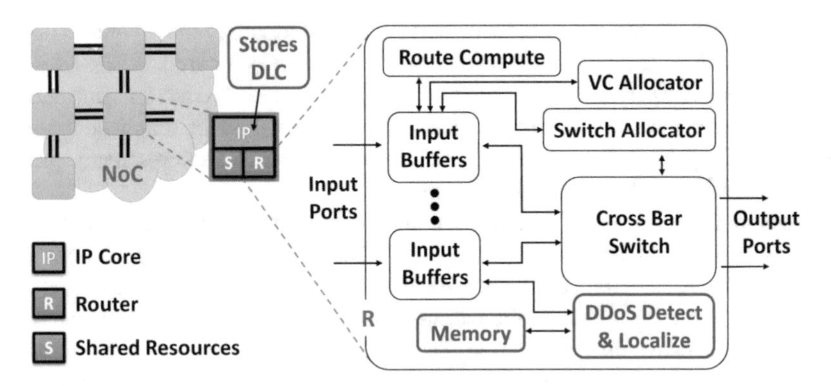

Fig. 8.21 Block diagram of NoC architecture showing additional hardware required to implement the security protocol in red

8.5.4 *Overhead Analysis*

The overhead is caused by the additional hardware that is required to implement the DoS attack detection and localization processes. The detection process requires additional hardware components and memory implemented at each router to monitor packet arrivals as well as store the parameterized curves. The localization process uses DLCs stored at IPs and the communication protocol implemented at the routers. Figure 8.21 shows an overview of how the security components are integrated into the NoC components. The observation made in Sect. 8.5.1 allows the reduction of the number of parameters required to model the PACs, and as a result, reduces the additional memory requirement and improves performance. The following sections evaluate the power, performance, and area overhead of the optimized algorithms.

8.5.4.1 Performance Overhead

In this work, a 5-stage router pipeline (buffer write, virtual channel allocation, switch allocation, switch traversal, and link traversal) was implemented in gem5. The computations related to the leaky bucket algorithm can be carried out in parallel to these pipeline stages once separate hardware is implemented. Therefore, no additional performance penalty for DoS attack detection.

During the localization phase, the diagnostic messages do not lead to additional congestion for two reasons. (1) As shown in Algorithm 8.2, the diagnostic message is transmitted along the reverse direction of the congested path. Since routers utilize two separate uni-directional links, the diagnostic messages are not sent along the congested path. (2) While it is unlikely, it is possible for multiple MIPs to carefully select multiple victims to construct a congested path in both directions. Even in this scenario, the number of diagnostic messages is negligible. This is because when an attack is flagged by the detection mechanism, diagnostic messages are sent to the source IPs which have violated the DLC threshold. Since the number of such source IPs can be at most the number of IPs communicating with the node that detected the attack, the performance impact by diagnostic messages is negligible.

8.5.4.2 Hardware Overhead

The overhead due to modifications in the router, packet header, and local IPs is as follows.

Router The proposed leaky bucket algorithm is lightweight and can be efficiently implemented with just three parameters per PAC bound as discussed above. The localization protocol requires two-bit flags at each port resulting in 10 bits of memory per router in Mesh topology. To evaluate the area and power overhead of adding the distributed DoS attack detection and localization mechanism at each router, the RTL of an open-source NoC Router [53] was modified. The design was synthesized with the 180 nm GSCLib Library from Cadence using the Synopsys Design Compiler. It gave area and power overhead of 6% and 4%, respectively, compared to the default router.

Packet Header In a typical packet header, the header flit contains basic fields such as source, destination addresses, and the physical address of the (memory) request. Some cache coherence protocols include special fields such as flags and timestamps in the header. If the header carries only the basic fields, the space required by these fields is much less compared to the wide bit widths of a typical NoC link. Therefore, most of the available flit header space goes unused [59]. This approach uses some of these bits to carry the timestamp to calculate latency. This eliminates the overhead of additional flits, making better utilization of bits that were being wasted. If the available header bit space is not sufficient, adding an extra “monitor tail flit” is an easily implementable alternative [59]. In most NoC protocols, the packet header has a hop count or time-to-live field. Otherwise, it can be derived from the source, destination addresses, and routing protocol details.

Local IP The DLPs are stored and processed by IPs connected to each node of an NoC. Since the IPs have much more resources than any other NoC component, the proposed lightweight approach has negligible power and performance overhead. $\mu_{i,k} + 1.96\sigma_{i,k}$ is stored as a 4-byte integer for each hop count. Therefore, the entire DLP at each IP can be stored using $1 \times m$ parameters where m is the maximum number of hops between any two IPs in the NoC. It gives a total memory space of just $1 \times m \times 4$ bytes.

These evaluations demonstrate that the area, power, and performance overhead introduced by this approach are negligible.

8.6 Case Study with Intel KNL Architecture

In the previous section, the DoS attack detection and localization method was evaluated using a regular 4×4 Mesh architecture (Fig. 8.14). In order to demonstrate the applicability of this approach across NoC architectures, in this section, we evaluate the efficiency of the approach in an architecture model similar to one of the commercially available SoCs—Intel's KNL architecture. Knights Landing (KNL) is the codename for the second generation Xeon-Phi processor introduced by Intel [62]. The architecture was modeled on gem5 according to a validated simulator model [14].

The KNL architecture, which is designed for highly parallel workloads, provide 36 tiles interconnected on a Mesh NoC. An overview of the KNL architecture is shown in Fig. 8.22. It implements a directory-based cache coherence protocol and supports two types of memory (1) multi-channel DRAM (MCDRAM) and (2) double data rate (DDR) memory. The architecture gives the option of configuring these two memories in several configurations which are called *memory modes*. Furthermore, the affinity between cores, directories, and memory controllers can be configured in three modes which are known as *cluster modes*. The memory and cluster modes allow configuration of the architecture depending on the application characteristics to achieve optimum performance and energy efficiency. Each combination of memory and cluster modes cause different traffic patterns in the NoC [17]. The goal is to simulate the NoC traffic behavior in a realistic architecture and evaluate how the security framework performs in it.

The gem5 model is adopted from previous work in [14] which validated the gem5 simulator statistics with the actual hardware behavior of a Xeon Phi 7210 platform [36]. In this model, 32 tiles connect on a Mesh NoC. Each tile is composed of a core that runs at 1.4 GHz, private L1 cache, tag directory, and a router. Each cache is split into data and instruction caches with 16kB capacity each. The complete set of simulation parameters are summarized in Table 8.1. The memory controllers are placed to match the architecture shown in Fig. 8.22. A few modeling choices were made that deviates from the actual KNL hardware due to the following reasons:

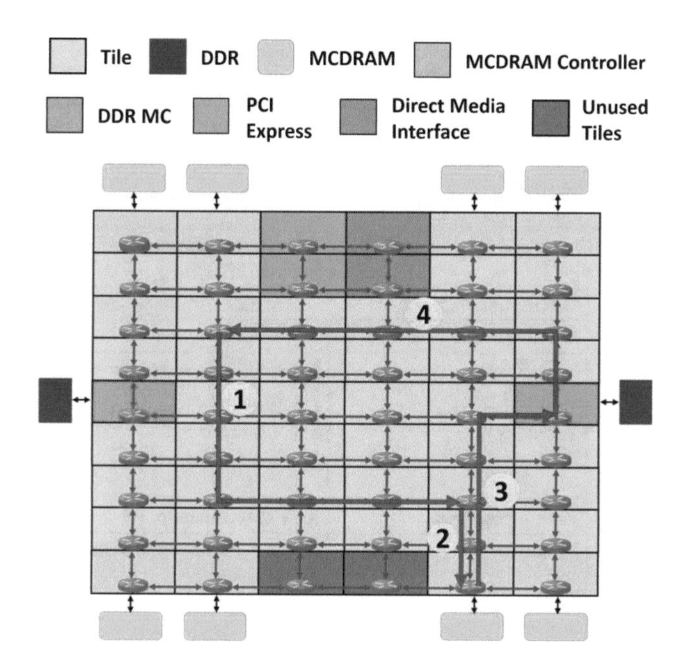

Fig. 8.22 Overview of the KNL architecture together with an example of MCDRAM miss in *cache* memory mode and *all-to-all* cluster mode: (1) L2 cache miss. Memory request sent to check the tag directory, (2) request forwarded to MCDRAM which acts as a cache after miss in tag directory, (3) request forwarded to memory after miss in MCDRAM, (4) data read from memory and sent to the requester [62]

- 32 tiles are used instead of the 36 in KNL since the number of cores in gem5 must be a power of 2. This can be considered as a use-case where the KNL hardware has switched off cores in four of its tiles.
- The cache sizes used in the model are less compared to the actual KNL hardware numbers. This was done to get 95% hit rate in L1 cache, which is usually the hit rate when running embedded applications for the benchmarks used. If a larger cache size was used, the L1 hit rate would be 100%, and NoC optimization will not affect cache performance.
- KNL runs AVX512 instructions whereas the gem5 model runs X86. gem5 is yet to support AVX512 instructions.
- Each tile in KNL consists of two cores. The detection mechanism is capable of detecting DoS attacks irrespective of whether one or both cores in a tile are active. However, the localization method can only pinpoint which tile is malicious. Since detection as well as localization happens at the router level, it is not possible to pinpoint the malicious core in a tile if both cores are active. Therefore, in the experimental setup, it was assumed that one core per tile is active simulating 50% utilization.

Table 8.1 System configuration parameters used when modeling KNL on gem5 simulator

Parameter class	Parameter	Value
Processor configuration	Number of cores	32
	Core frequency	1.4 GHz
	Instruction set architecture	×86
Memory system configuration	L1 cache	Private, separate instruction and data cache. Each 16 kB in size
	Cache coherence	Distributed directory-based protocol
	Memory size	4 GB DDR
	MCDRAM	Shared, direct mapped cache
	Access latency	300 cycles
Interconnection network configuration	Topology	8 × 4 Mesh
	Routing scheme	X–Y deterministic
	Router	4 port, 4 input buffer router with 5 cycle pipeline delay
	Link latency	1 cycle

Therefore, the gem5 model is a simplified version of the real KNL hardware. However, previous work has validated the model and related performance and energy results to show that it accurately captures relative advantages/disadvantages of using different memory and cluster modes [14]. To evaluate the security framework, out of the memory and cluster modes, the cache memory mode and all-to-all cluster mode were modeled.

- **Cache memory mode:** In the *cache* mode, MCDRAM acts as a last level cache which is placed in between the DDR memory and the private cache. All memory requests first go to the MCDRAM for a cache memory lookup, if there is a cache miss, they are sent to the DDR memory.
- **All-to-all cluster mode:** In this mode, there is no affinity between the core, memory controller, and directory. That is, a memory request can go from any directory to any memory controller.

The traffic flow when applications are running is defined by these modes. Figure 8.22 shows an example traffic flow.

The same real traffic patterns used in Sect. 8.5.1 were used to evaluate the KNL setup. To mimic the highly parallel workloads executable by the KNL architecture, 50% of the total available cores were utilized when running each application by running an instance of the benchmarks in each active core. The DDR address space was used uniformly for each benchmark. Attackers were modeled and placed randomly in 25% of the tiles that does not have an application instance. The DoS attack was launched at the memory controller that experienced highest traffic during normal operation. Given that the model has 32 cores, 16 of them ran instances of the

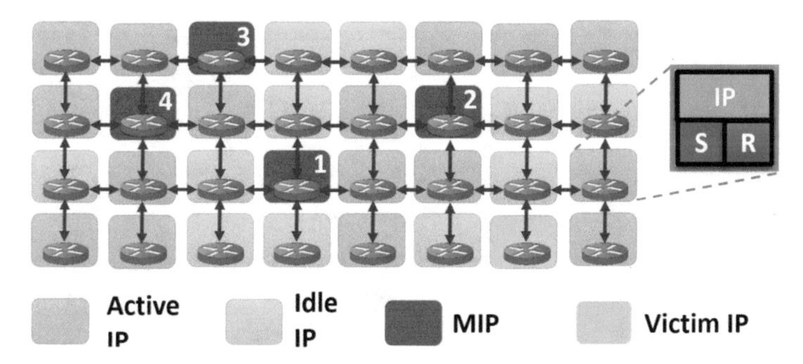

Fig. 8.23 4 × 8 Mesh NoC architecture used to simulate DoS attacks in an architecture similar to KNL

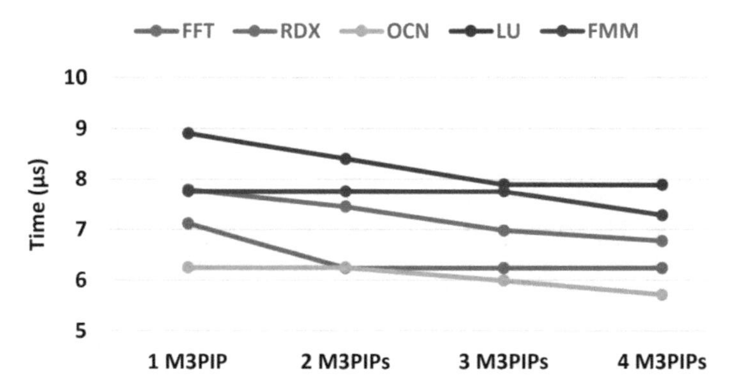

Fig. 8.24 Attack detection time when running real benchmarks on an architecture similar to KNL with the presence of different number of MIPs

benchmark and 4 of the non-active cores injected packets directed at the memory controller to simulate the behavior of malicious IPs launching a DoS attack. The packet stream period and attack period were selected as explained in Sect. 8.5.2. Figure 8.23 shows the placement of the four MIPs, cores running the benchmarks (active cores), and the victim IP when running the RADIX benchmark. The victim IP depends on the benchmark since it is the IP connected to the memory controller experiencing highest traffic during normal operation.

Similar to the experimental results presented in Sect. 8.5.1, the DoS attack detection results are shown in Figs. 8.24 and 8.25. Figure 8.24 shows detection time variation across benchmarks and number of MIPs. A zoomed-in version of the four MIP scenario is shown in Fig. 8.25. Attack localization results are shown in Fig. 8.26. Until the fourth MIP is added, there are no overlapping congested paths. Therefore, the MIPs are localized using only one iteration. Once the fourth MIP is added, the first, third, and fourth MIPs are localized during the first iteration and a second iteration is required to localize the second MIP. This is reflected in localization time in Fig. 8.26. From these as well as the previous results we notice

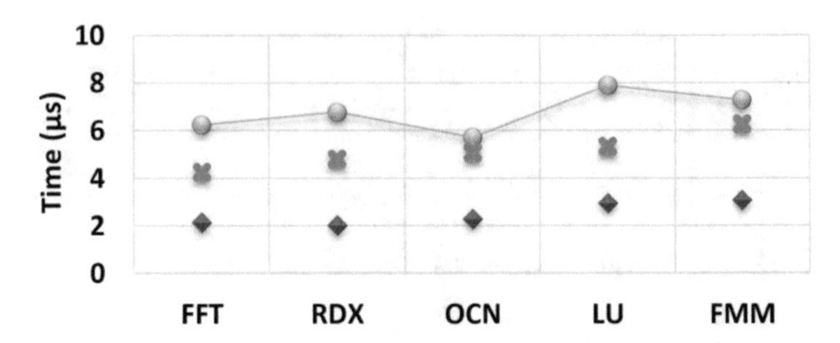

Fig. 8.25 Attack detection time when running real benchmarks on an architecture similar to KNL with the presence of four MIPs

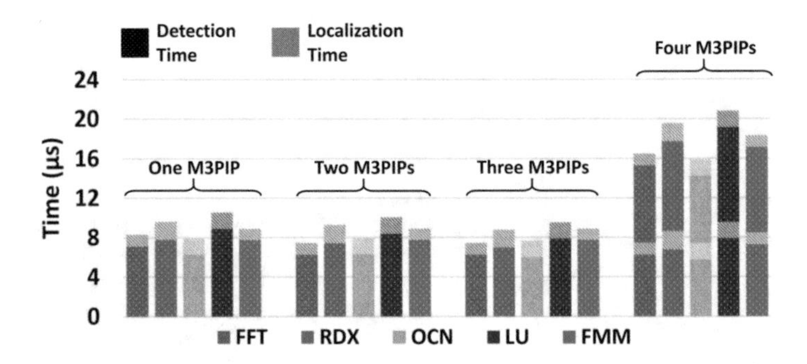

Fig. 8.26 Attack localization time when running real benchmarks on an architecture similar to KNL with the presence of different number of MIPs

that the DoS attack detection and localization framework gives real-time results across different topologies and architectures.

8.7 Discussion

This approach is designed for DoS attack detection and localization, and therefore, it is not suitable to capture other forms of security violations such as eavesdropping, snooping, and buffer overflow. Specific security attacks would require other security countermeasures which are not covered in this chapter. Due to the low implementation cost, this approach can be easily coupled with other security countermeasures. For example, [60] discussed a snooping attack in which the header of the packet is modified before injecting into the NoC. This will alter the source address of the packet. While this detection mechanism does not depend on any of the header information of the packet, since the localization method uses the source address to localize the MIPs, an address validation mechanism needs to be implemented at each router to accommodate header modification. The address validation can be implemented as follows. Before a router injects each packet that comes from the local IP into the NoC, the router can check the source address and if it is not the address of the local IP attached to that router, the router can drop it without injecting into the NoC.

This work is targeted for embedded systems with real-time constraints. Such systems allow only a specific set of scenarios in order to provide real-time guarantees. Features commonly observed in general-purpose computing such as task mapping, runtime task-migration, adaptive routing, and introduction of new applications during runtime cannot be addressed by this work. In order to apply this approach in general-purpose systems, we need to store PACs and DLCs corresponding to each scenario and select the respective curves during runtime. As discussed in Sect. 8.5.4, the hardware overhead to store the parameterized curves for each scenario is minimal, which consists of two major parts (1) overhead for storing the curves ($1 \times m \times 4$ bytes), and (2) overhead for runtime monitoring (6% of NoC area). For example, if we consider an 8×8 Mesh, the memory overhead to store the curves would be 56 bytes ($m = 14$). If N scenarios are considered, the overhead would be 6% + $N \times 56$. Therefore, it may be feasible to consider a small number of scenarios (e.g., $N < 10$) without violating area overhead constraints.

8.8 Summary

This chapter presented a real-time and lightweight DoS attack detection and localization mechanism for IoT and embedded systems. It relies on real-time network traffic monitoring to detect unusual traffic behavior. This chapter described a real-time and efficient technique for detection of DoS attacks originating from multiple malicious IPs in NoC-based SoCs. Once an attack is detected, this approach is also capable of real-time localization of the malicious IPs using the latency data in the NoC routers. The effectiveness of the approach is demonstrated using several NoC topologies and traffic patterns. Experimental results showed that all the attack scenarios can be detected and localized in a timely manner. Overhead calculations have revealed that the area overhead is less than 6% to implement the proposed framework on a realistic NoC model. This framework can be easily integrated with existing security mechanisms that address other types of attacks such as buffer overflow and information leakage.

Acknowledgement This work was partially supported by the National Science Foundation (NSF) grant SaTC-1936040.

References

1. N. Agarwal, T. Krishna, L. Peh, N.K. Jha, Garnet: a detailed on-chip network model inside a full-system simulator, in *2009 IEEE International Symposium on Performance Analysis of Systems and Software* (2009), pp. 33–42
2. A. Ahmed, F. Farahmandi, Y. Iskander, P. Mishra, Scalable hardware trojan activation by interleaving concrete simulation and symbolic execution, in *2018 IEEE International Test Conference (ITC)* (IEEE, Piscataway, 2018), pp. 1–10
3. A. Ahmed, F. Farahmandi, P. Mishra, Directed test generation using concolic testing on RTL models, in *2018 Design, Automation & Test in Europe Conference & Exhibition (DATE)* (2018), pp. 1538–1543
4. A. Ahmed, Y. Huang, P. Mishra, Cache reconfiguration using machine learning for vulnerability-aware energy optimization. ACM Trans. Embed. Comput. Syst. **18**(2), 1–24 (2019)
5. Alteris FlexNoC Resilience Package, www.arteris.com/flexnoc-resilience-package-functional-safety [Online]
6. S. Baruah, G. Buttazzo, S. Gorinsky, G. Lipari, Scheduling periodic task systems to minimize output jitter, in *Proceedings Sixth International Conference on Real-Time Computing Systems and Applications. RTCSA'99 (Cat. No.PR00306)* (1999), pp. 62–69
7. N. Binkert, B. Beckmann, G. Black, S.K. Reinhardt, A. Saidi, A. Basu, J. Hestness, D.R. Hower, T. Krishna, S. Sardashti, R. Sen, K. Sewell, M. Shoaib, N. Vaish, M.D. Hill, D.A. Wood, The gem5 simulator. SIGARCH Comput. Archit. News **39**(2), 1–7 (2011)
8. T. Boraten, D. DiTomaso, A.K. Kodi, Secure model checkers for network-on-chip (NOC) architectures, in *2016 International Great Lakes Symposium on VLSI (GLSVLSI)* (2016), pp. 45–50
9. S. Chakraborty, S. Kunzli, L. Thiele, A general framework for analysing system properties in platform-based embedded system designs, in *2003 Design, Automation and Test in Europe Conference and Exhibition* (2003), pp. 190–195
10. Z. Chao-Yang, DOS attack analysis and study of new measures to prevent, in *2011 International Conference on Intelligence Science and Information Engineering* (2011), pp. 426–429
11. S. Charles, P. Mishra, Lightweight and trust-aware routing in NoC-based SoCs, in *2020 IEEE Computer Society Annual Symposium on VLSI (ISVLSI)* (2020), pp. 160–167
12. S. Charles, P. Mishra, Securing network-on-chip using incremental cryptography, in *2020 IEEE Computer Society Annual Symposium on VLSI (ISVLSI)* (2020), pp. 168–175
13. S. Charles, P. Mishra, Reconfigurable network-on-chip security architecture. ACM Trans. Des. Autom. Electron. Syst. **25**(6), 1–25 (2020)
14. S. Charles, C.A. Patil, U.Y. Ogras, P. Mishra, Exploration of memory and cluster modes in directory-based many-core CMPs, in *2018 Twelfth IEEE/ACM International Symposium on Networks-on-Chip (NOCS)* (2018), pp. 1–8
15. S. Charles, H. Hajimiri, P. Mishra, Proactive thermal management using memory-based computing in multicore architectures, in *2018 Ninth International Green and Sustainable Computing Conference (IGSC)* (2018), pp. 1–8
16. S. Charles, Y. Lyu, P. Mishra, Real-time detection and localization of dos attacks in NoC based SoCs, in *2019 Design, Automation Test in Europe Conference Exhibition (DATE)* (2019), pp. 1160–1165
17. S. Charles, A. Ahmed, U.Y. Ogras, P. Mishra, Efficient cache reconfiguration using machine learning in NoC-based many-core CMPs. ACM Transact. Des. Autom. Electron. Syst. **24**(6), 1–23 (2019)
18. S. Charles, M. Logan, P. Mishra, Lightweight anonymous routing in NoC based SoCs, in *2020 Design, Automation & Test in Europe Conference & Exhibition (DATE)* (IEEE, Piscataway, 2020)

19. S. Charles, Y. Lyu, P. Mishra, Real-time detection and localization of distributed DoS attacks in NoC based SoCs. IEEE Trans. Comput. Aided Des. Integr. Circuits Syst. **39**(12), 4510–4523 (2020)
20. M. Chen, X. Qin, H.-M. Koo, P. Mishra, *System-Level Validation: High-Level Modeling and Directed Test Generation Techniques* (Springer, Berlin, 2012)
21. W. Eddy, TCP SYN flooding attacks and common mitigations. Technical report, 2007
22. K. Elleithy, D. Blagovic, W. Cheng, P. Sideleau, Denial of service attack techniques: analysis, implementation and comparison. J. Syst. Cybern. Inf. **3**(01), 66–71 (2006)
23. D. Fang, H. Li, J. Han, X. Zeng, Robustness analysis of mesh-based network-on-chip architecture under flooding-based denial of service attacks, in *2013 IEEE Eighth International Conference on Networking, Architecture and Storage* (2013), pp. 178–186
24. F. Farahmandi, P. Mishra, Automated debugging of arithmetic circuits using incremental gröbner basis reduction, in *2017 IEEE International Conference on Computer Design (ICCD)* (IEEE, Piscataway, 2017)
25. F. Farahmandi, P. Mishra, FSM anomaly detection using formal analysis, in *2017 IEEE International Conference on Computer Design (ICCD)* (IEEE, Piscataway, 2017), pp. 313–320
26. F. Farahmandi, P. Mishra, Automated test generation for debugging multiple bugs in arithmetic circuits. IEEE Trans. Comput. **68**(2), 182–197 (2018)
27. F. Farahmandi, Y. Huang, P. Mishra, Trojan localization using symbolic algebra, in *22nd Asia and South Pacific Design Automation Conference (ASP-DAC)* (2017), pp. 591–597
28. F. Farahmandi, Y. Huang, P. Mishra, *System-on-Chip Security: Validation and Verification* (Springer Nature, Cham, 2019)
29. L. Fiorin, C. Silvano, M. Sami, Security aspects in networks-on-chips: overview and proposals for secure implementations, in *10th Euromicro Conference on Digital System Design Architectures, Methods and Tools (DSD 2007)* (2007), pp. 539–542
30. L. Fiorin, G. Palermo, C. Silvano, *A Security Monitoring Service for NoCs* (Association for Computing Machinery, New York, 2008), pp. 197–202
31. U. Gupta et al., DyPO: dynamic pareto-optimal configuration selection for heterogeneous MpSoCs. ACM Trans. Embed. Comput. Syst. **16**(5s), 1–20 (2017)
32. Y. Huang, P. Mishra, Vulnerability-aware energy optimization for reconfigurable caches in multitasking systems. IEEE Trans. Comput. Aided Des. Integr. Circuits Syst. **38**(5), 809–821 (2019)
33. K. Huang, C. Buckl, G. Chen, A. Knoll, Conforming the runtime inputs for hard real-time embedded systems, in *DAC Design Automation Conference 2012* (2012), pp. 430–436
34. Y. Huang, S. Bhunia, P. Mishra, MERS: statistical test generation for side-channel analysis based trojan detection, in *Proceedings of the 2016 ACM SIGSAC Conference on Computer and Communications Security* (2016), pp. 130–141
35. Y. Huang, S. Bhunia, P. Mishra, Scalable test generation for trojan detection using side channel analysis. IEEE Trans. Inf. Forensics Secur. **13**(11), 2746–2760 (2018)
36. Intel Xeon Phi Processor 7210, http://ark.intel.com/products/94033/Intel-Xeon-Phi-Processor-7210-16GB-1_30-GHz-64-core [Online]
37. S. Kumar, Smurf-based distributed denial of service (DDoS) attack amplification in internet, in *Second International Conference on Internet Monitoring and Protection (ICIMP 2007)* (IEEE, Piscataway, 2007), p. 25
38. K. Lampka, S. Perathoner, L. Thiele, Analytic real-time analysis and timed automata: a hybrid method for analyzing embedded real-time systems, in *Proceedings of the Seventh ACM International Conference on Embedded Software* (ACM, New York, 2009), pp. 107–116
39. J.-Y. Le Boudec, P. Thiran, *Network Calculus: A Theory of Deterministic Queuing Systems for the Internet*, vol. 2050, 06 (Springer, Berlin, 2004)
40. J. Li, Y. Liu, L. Gu, DDoS attack detection based on neural network, in *2010 2nd International Symposium on Aware Computing* (2010), pp. 196–199
41. M. Lukasiewycz, S. Steinhorst, S. Chakraborty, Priority assignment for event-triggered systems using mathematical programming, in *2013 Design, Automation Test in Europe Conference Exhibition (DATE)* (2013), pp. 982–987

42. Y. Lyu, P. Mishra, A survey of side-channel attacks on caches and countermeasures. J. Hardw. Syst. Secur. **2**(1), 33–50 (2018)
43. Y. Lyu, P. Mishra, Efficient test generation for trojan detection using side channel analysis, in *2019 Design, Automation & Test in Europe Conference & Exhibition (DATE)* (IEEE, Piscataway, 2019), pp. 408–413
44. Y. Lyu, P. Mishra, Automated test generation for activation of assertions in RTL models, in *2020 25th Asia and South Pacific Design Automation Conference (ASPDAC)* (IEEE, Piscataway, 2020), pp. 223–228
45. Y. Lyu, P. Mishra, Automated test generation for trojan detection using delay-based side channel analysis, in *2020 Design, Automation & Test in Europe Conference & Exhibition (DATE)* (2020), pp. 1031–1036
46. Y. Lyu, P. Mishra, Automated trigger activation by repeated maximal clique sampling, in *Asia and South Pacific Design Automation Conference (ASPDAC)* (2020), pp. 482–487
47. Y. Lyu, P. Mishra, Scalable activation of rare triggers in hardware trojans by repeated maximal clique sampling. IEEE Trans. Comput. Aided Des. Integr. Circuits Syst. (2020). https://doi.org/10.1109/TCAD.2020.3019984
48. Y. Lyu, P. Mishra, Scalable concolic testing of RTL models. IEEE Trans. Comput. (2020). https://doi.org/10.1109/TC.2020.2997644
49. Y. Lyu, X. Qin, M. Chen, P. Mishra, Directed test generation for validation of cache coherence protocols. IEEE Trans. Comput. Aided Des. Integr. Circuits Syst. **38**(1), 163–176 (2018)
50. Y. Lyu, A. Ahmed, P. Mishra, Automated activation of multiple targets in RTL models using concolic testing, in *2019 Design, Automation & Test in Europe Conference & Exhibition (DATE)* (IEEE, Piscataway, 2019), pp. 354–359
51. P. Mishra, F. Farahmandi, *Post-Silicon Validation and Debug* (Springer, Berlin, 2019)
52. P. Mishra, S. Bhunia, M. Tehranipoor, *Hardware IP Security and Trust* (Springer, Berlin, 2017)
53. A. Monemi, J.W. Tang, M. Palesi, M.N. Marsono, ProNoC: a low latency network-on-chip based many-core system-on-chip prototyping platform. Microprocess. Microsyst. **54**, 60–74 (2017)
54. J. Niemela, F-secure virus descriptions, in *F-Secure*, December 2007
55. Z. Pan, P. Mishra, Automated test generation for hardware trojan detection using reinforcement learning, in *Asia and South Pacific Design Automation Conference (ASPDAC)* (2021)
56. Z. Pan, J. Sheldon, P. Mishra, Test generation using reinforcement learning for delay-based side channel analysis, in *IEEE/ACM International Conference on Computer-Aided Design (ICCAD)* (ACM, New York, 2020)
57. N. Prasad, R. Karmakar, S. Chattopadhyay, I. Chakrabarti, Runtime mitigation of illegal packet request attacks in networks-on-chip, in *2017 IEEE International Symposium on Circuits and Systems (ISCAS)* (IEEE, Piscataway, 2017), pp. 1–4
58. J.S. Rajesh, D.M. Ancajas, K. Chakraborty, S. Roy, Runtime detection of a bandwidth denial attack from a rogue network-on-chip, in *Proceedings of the 9th International Symposium on Networks-on-Chip, NOCS '15* (Association for Computing Machinery, New York, 2015)
59. M. Ramakrishna, V. Kodati, P.V. Gratz, A. Sprintson, GCA:global congestion awareness for load balance in networks-on-chip. IEEE Trans. Parallel Distrib. Syst. **27**(07), 2022–2035 (2016)
60. V.Y. Raparti, S. Pasricha, Lightweight mitigation of hardware trojan attacks in NoC-based manycore computing, in *2019 56th ACM/IEEE Design Automation Conference (DAC)* (2019), pp. 1–6
61. A. Saeed, A. Ahmadinia, M. Just, C. Bobda, An ID and address protection unit for NoC based communication architectures, in *Proceedings of the 7th International Conference on Security of Information and Networks* (ACM, New York, 2014), p. 288
62. A. Sodani, R. Gramunt, J. Corbal, H. Kim, K. Vinod, S. Chinthamani, S. Hutsell, R. Agarwal, Y. Liu, Knights landing: second-generation intel xeon phi product. IEEE Micro **36**(2), 34–46 (2016)
63. U. Suppiger, S. Perathoner, K. Lampka, L. Thiele, A simple approximation method for reducing the complexity of modular performance analysis. Tech. Rep. 329, 2010

64. Using TinyCrypt Library, Intel Developer Zone, Intel, 2016. https://software.intel.com/en-us/node/734330 [Online]
65. E. Wandeler, L. Thiele, Real-time calculus (RTC) toolbox (2006). http://www.mpa.ethz.ch/Rtctoolbox [Online]
66. W. Wang, P. Mishra, A. Gordon-Ross, Dynamic cache reconfiguration for soft real-time systems. ACM Trans. Embed. Comput. Syst. **11**(2), 1–31 (2012)
67. P. Waszecki, P. Mundhenk, S. Steinhorst, M. Lukasiewycz, R. Karri, S. Chakraborty, Automotive electrical and electronic architecture security via distributed in-vehicle traffic monitoring. IEEE Trans. Comput. Aided Des. Integr. Circuits Syst. **36**(11), 1790–1803 (2017)
68. S.C. Woo, M. Ohara, E. Torrie, J.P. Singh, A. Gupta, The splash-2 programs: characterization and methodological considerations, in *Proceedings 22nd Annual International Symposium on Computer Architecture* (1995), pp. 24–36
69. L. Zhang, S. Yu, D. Wu, P. Watters, A survey on latest botnet attack and defense, in *2011 IEEE 10th International Conference on Trust, Security and Privacy in Computing and Communications* (2011), pp. 53–60

Chapter 9
Securing on-Chip Communication Using Digital Watermarking

Subodha Charles and Prabhat Mishra

9.1 Introduction

Design considerations for roads in a city involve accessibility, traffic distribution, and handling of specific scenarios. For example, an important objective in the design of a network of roads is to ensure ease of access to popular and important places in the city such as offices, schools, parks, etc. If prominent places are all located in the same area, the roads in that area will be congested while roads in other areas will remain (relatively) empty. An architect should ensure that the traffic is as uniformly distributed as possible or the main roads have enough lanes to mitigate congestion. A System-on-chip (SoC) designer faces similar challenges when designing the communication infrastructure connecting all the SoC components, i.e., processor cores, memories, controllers, input/output, etc. As the complexity of SoCs increases, more and more Intellectual Property (IP) cores are integrated on the same SoC. State-of-the-art SoCs have hundreds of components. For example, a typical automotive SoC may include 100–200 diverse IP cores. The demand for scalable and high-throughput interconnects has made Network-on-chip (NoC) the standard interconnection solution for complex SoCs [67].

Due to time-to-market constraints, it is a common practice for manufacturers to outsource IPs to third-party vendors. Typically, manufacturers produce only a few important IPs in-house and integrate them with third-party IPs to obtain the final SoC. As a result of this distributed supply chain, it is feasible for an attacker to insert malicious implants, such as hardware Trojans, into the IPs [5, 18, 58].

S. Charles (✉)
University of Moratuwa, Colombo, Sri Lanka
e-mail: s.charles@ieee.org

P. Mishra
University of Florida, Gainesville, FL, USA
e-mail: prabhat@ufl.edu

P. Mishra, S. Charles (eds.), *Network-on-Chip Security and Privacy*,
https://doi.org/10.1007/978-3-030-69131-8_9

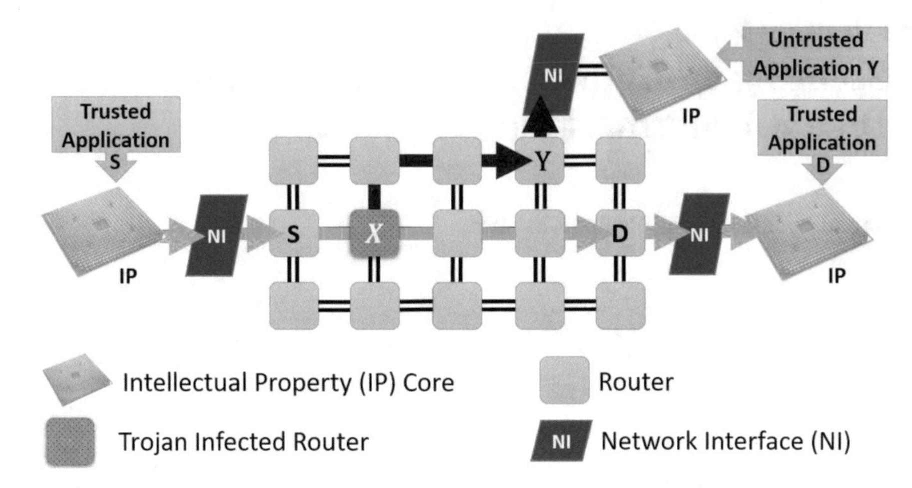

Fig. 9.1 Illustration of an eavesdropping attack through colluding hardware and software. A hardware Trojan integrated in a router (X) copies packets passing through it and sends them to a malicious application running on an IP (Y). An NI and an IP core are connected to each router (For clarity, only three such pairs are shown)

A recent occurrence of a hardware security breach due to third-party vendors aiming at industrial espionage raised concerns across top US authorities [10]. The attack was facilitated by a hardware Trojan that acted as a covert backdoor and spied on computer servers used by more than 30 companies in USA, including Amazon and Apple.

A similar attack scenario can be considered in the NoC context. A hardware Trojan integrated in the NoC IP launches an attack to eavesdrop on the NoC packets. The goal is to exfiltrate information while remaining hidden, and thus the Trojan will not perform any action that would reveal its presence, such as corrupting packets to cause SoC malfunction (data integrity attacks) or degrade performance causing denial-of-service (DoS) attacks. Existing literature has explored the most effective way of launching an eavesdropping attack in NoC, considering attack effectiveness and difficulty to detect the Trojan. It identified Trojan(s) inserted in NoC component(s) colluding with another malicious IP(s) as the strongest attack model. An illustrative example of this scenario is shown in Fig. 9.1, where a hardware Trojan-infected router and an accomplice application launch an eavesdropping attack where the infected router copies packets passing through it and sends them to the accomplice application running on another malicious IP. This hardware-software collusion attack is similar to the Illinois Malicious Processor (IMP) [41]. This setting and related threat models have been the focus of [5] as well as several prior studies [11, 19, 35, 49, 58, 64].

NoC security research has proposed authenticated encryption (AE) as a solution to eavesdropping attacks [11, 35, 64]. With AE, packets are encrypted to ensure confidentiality and an authentication tag is appended to each packet to ensure integrity (and detect re-routed packets). However, the use of AE as the defense to eavesdropping attacks is sub-optimal for two reasons. First, it incurs significant performance degradation on resource-constrained devices (as shown experimentally in Sect. 9.3). Second, authentication tags may be unnecessarily complex if used only for the purpose of detecting eavesdropping attackers who seek to remain undetected as long as possible—and thus are unlikely to interfere with data integrity.

In this chapter, we ask a fundamental question: *is it possible to replace authenticated encryption with a lightweight defense while maintaining security against eavesdropping attacks?* Specifically, the method replaces the costly computation of authentication tags with a lightweight eavesdropping attack detection mechanism based on *digital watermarking*. The attack detection capabilities achieved by digital watermarking is coupled with encryption to ensure data confidentiality.

The remainder of this chapter is organized as follows. Section 9.2 gives an overview of related research and describes the threat model in detail. Section 9.3 motivates the need for a lightweight alternative under the given threat model by comparing with other approaches. Section 9.4 introduces the watermarking-based attack detection method. Section 9.5 provides theoretical guarantees on performance and security of the approach followed by experimental results in Sect. 9.6. Section 9.7 discusses additional security considerations. Finally, Sect. 9.8 concludes the chapter.

9.2 Threat Model and Related Work

In this section, we discuss related research and describe the threat model.

9.2.1 Related Work

NoC Security State-of-the-art NoC security revolves around protecting information traveling in the network against physical, software, and side-channel attacks [12–14, 16, 18–20]. There are many side-channel analysis methods for effective Trojan detection in integrated circuits [33, 34, 44–46, 54]. While detecting hardware Trojans in NoC IPs during design time is still in its infancy, most solutions aim to detect/mitigate the threat of hardware Trojans during runtime. To identify most prominent threats in NoC-based SoCs, we surveyed 25 related papers published in the last 10 years and categorized them into five widely studied categories of NoC security attacks: (1) eavesdropping, (2) spoofing and data integrity, (3) denial-of-service, (4) buffer overflow and memory extraction, and (5) side-channel attacks. Results are shown in Table 9.1.

The survey makes it evident that eavesdropping attacks are indeed one of the most widely explored threat models related to security in NoC-based SoC. The threat model used in this work is well-established and has been considered in

Table 9.1 Summary of NoC security papers found in literature categorized by attack class and defense type. **Attack Class**: Eavesdropping (EAV), Spoofing/Data Integrity (SDI), Denial-of-service (DOS), Buffer Overflow and Memory Extraction (BOM), and Side-Channel Attacks (SCA). **Defense Type**: Obfuscation (OBF), Detection (DET), and Localization (LOC)

Paper	Attack class	Defense type
Sajeesh and Kapoor [62]	EAV	OBF, DET
Porquet et al. [55]	BOM	OBF
Wang and Suh [69]	SCA	OBF
Kapoor et al. [40]	EAV	OBF, DET
Yu and Frey [73]	SDI	OBF
Ancajas et al. [5]	EAV	OBF
Saeed et al. [61]	BOM	DET
Sepúlveda et al. [63]	BOM	OBF, DET
Rajesh et al. [57]	DOS	DET
Biswas et al. [9]	DOS	DET
Reinbrecht et al. [59]	SCA	OBF, DET
Boraten and Kodi [11]	EAV	OBF
Prasad et al. [56]	DOS	DET
Sepúlveda et al. [64]	EAV	OBF
Frey and Yu [27]	DOS	OBF, DET
Indrusiak et al. [37]	SCA	OBF
Sepúlveda et al. [65]	DOS	DET
Hussain et al. [35]	EAV	DET, LOC
Kumar et al. [49]	DOS	OBF
Chittamuru et al. [21]	EAV	OBF, DET
Lebiednik et al. [42]	EAV	OBF
Indrusiak et al. [38]	SCA	OBF
Charles et al. [18]	DOS	DET, LOC
Raparti and Pasricha [58]	EAV	DET, LOC
Charles et al. [19]	EAV	OBF

previous work that proposed solutions to protect the SoC from a compromised NoC IP eavesdropping on data [5, 11, 19, 21, 35, 49, 58, 64]. Ancajas et al. proposed a combination of data scrambling, packet authentication, and node obfuscation to prevent eavesdropping attacks [5]. In [58], a combination of threshold voltage degradation and an encoding based packet duplication detector was proposed. Charles et al. proposed to increase the difficulty of information extraction by introducing anonymous routing in the NoC [19]. Manor et al. attempted to reduce the effectiveness of hardware Trojans trying to manipulate data packets using bit shuffling and Hamming error correction codes [49]. When eavesdropping attacks are considered, packet authentication combined with encryption (authenticated encryption) is the most popular countermeasure [5, 11, 21, 35, 40, 62, 64].

Digital Watermarking The process of hiding information related to digital data in the data itself is called digital watermarking. An overview of a typical watermarking mechanism is shown in Fig. 9.2. It has been widely used in domains such as broadcast monitoring, copyright identification, transaction tracking, and copy control. For example, in the movie industry, a unique watermark can be embedded in

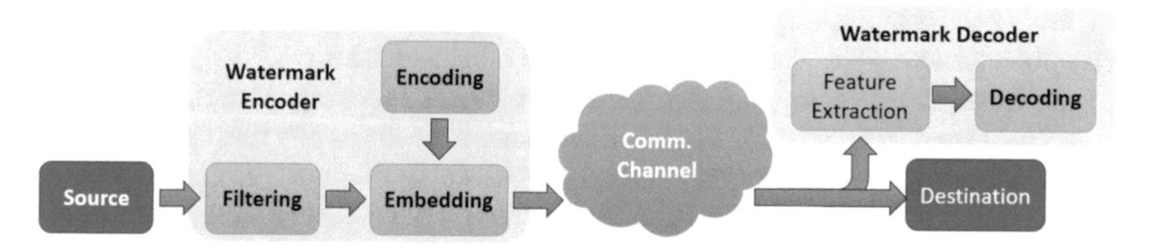

Fig. 9.2 Overview of a typical watermarking mechanism

every movie. If the movie later gets published on the internet illegally, the embedded watermark can be used to identify the person who leaked it. Biswas et al. [8] presented a technique called circular path-based fingerprinting using fingerprint embedding against NoC IP stealing attacks. However, the threat model used in this chapter—eavesdropping attacks, cannot be addressed using their approach. Network flow watermarking is one possible solution to prevent eavesdropping attacks [36]. In network flow watermarking, watermarks are embedded into the packet flow using packet content [22], timing information [70], or packet size [43]. This can be used for tracing botmasters in a botnet [30], tracing other network-based attacks [31], and service dependency detection [74].

9.2.2 Threat Model

The global trend of distributed design, validation, and fabrication has raised concerns about security vulnerabilities [2, 24–26, 47, 48, 52, 53]. Malicious implants, such as hardware Trojans, can be inserted into the RTL or into the netlist of an IP core with the intention of launching attacks without being detected at the post-silicon verification stage or during runtime [2, 24–26, 47, 48, 52, 53]. Insertion of Trojans can happen in many places of the long, distributed supply chain such as by an untrusted CAD tool or designer or at the foundry via reverse engineering [52]. As evidence of the globally distributed supply chain of NoC IPs, iSuppli, an independent market research firm, reports that the FlexNoC on-chip interconnection architecture [4] is used by four out of the top five Chinese fabless semiconductor OEM (original equipment manufacturer) companies [66]. In fact, Arteris, the company that developed FlexNoC, achieved a sales growth of 1002% over a 3-year time period through IP licensing [66]. Therefore, there is ample opportunity for attackers to integrate hardware Trojans in the NoC IP and compromise the SoC. NoC IPs are ideal candidates to insert hardware Trojans due to several reasons: (1) the complexity of NoC IPs makes it extremely difficult to detect hardware Trojans during functional verification as well as runtime [58], (2) extracting data from NoC packets allows attackers to obtain confidential information without relying on memory access or hacking into individual IPs, and (3) the distributed nature of NoC components across the SoC makes it easier to launch attacks.

We focus on eavesdropping attacks, also known as snooping attacks, which pose a serious threat to applications running on many-core SoCs. IPs that are integrated on the same SoC use the NoC IP when communicating through message passing as well as through shared memory. For example, the Intel Knights Landing architecture prompts memory requests/responses from cores to traverse the NoC for shared cache look-ups and for off-chip memory accesses [67]. Therefore, eavesdropping on data transferred through the NoC allows adversaries to extract confidential information.

Adversarial Model In this chapter, we consider an adversary consisting of a hardware Trojan-infected router and a colluding malicious application running on an IP. The goal of the adversary is to exfiltrate confidential information by observing NoC traffic *without being detected.* Remaining hidden is key for the adversary to exfiltrate as much information as possible. Because the adversary must remain hidden, we assume that the adversary does not interfere with the normal operation of the NoC. For example, this means that the adversary does not modify the content of packets (attack on integrity) or cause large delays in processing of packets (denial-of-service) as either would likely lead to detection.

Attack Scenario Eavesdropping attacks by malicious NoC IPs rely on the hardware Trojan creating duplicate packets with modified headers (specifically, destination address in the header) and sending them into the NoC for an accomplice application to receive them [5, 58]. Figure 9.1 shows an illustrative example. A commonly used 2D Mesh NoC topology is considered where IPs are connected to the NoC, more specifically to the router, via a network interface (NI). When the NI receives a message from the local IP, the message is packetized and injected into the network.[1] Packets injected into the NoC are routed using the hop-by-hop, turn-based XY routing algorithm and received by the destination router. The NI then combines the packets to form the message which is passed to the intended destination IP. In the example (Fig. 9.1), two trusted applications running in nodes S and D are communicating with each other, and an eavesdropping attack is launched to steal confidential information. The attack is carried out by two main components: (1) a Trojan-infected router, and (2) an IP running a malicious application. The malicious router (X) copies packets passing through it and sends them to the IP running the malicious program at node Y, which reads the confidential information. To facilitate this attack, several steps should be carried out by the attacker. First, the hardware Trojan is inserted by the third-party NoC IP provider during design time. The Trojan is designed such that it can act upon commands sent by the malicious application. Once the SoC is deployed, the malicious application sends commands at a desired time to launch the attack. The Trojan then starts copying and sending packets to the

[1] Most NoCs facilitate flits, which is a further breakdown of a packet used for flow control purposes. We stick to the level of packets for the ease of explanation as this method remains the same at the flit level as well.

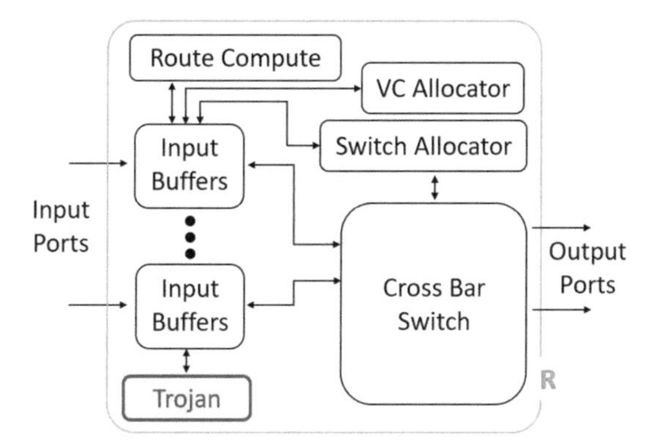

Fig. 9.3 Router infected with a hardware Trojan

malicious application. The malicious application can also send commands to pause the attack to avoid being detected.

Figure 9.3 shows a block diagram of a router design infected with the Trojan that launches the attack described in the threat model [5]. The Trojan copies packets arriving at the input buffer, changes the header information so that the new destination of the packet is where the malicious application is (node Y according to the illustrative example), and injects the new packet back to the input buffers so that it gets routed through the NoC to reach Y. The Trojan does not tamper with any other part of the packet, except for the header to re-route the packet, due to two reasons: (1) the goal is to extract information, so corrupting data defeats the purpose, and (2) corrupting data increases chances of the Trojan getting detected. Since the original packet is not tampered with and is routed to the intended destination D, the normal operation of the SoC is preserved. The Trojan also has a very small area and power footprint. Ancajas et al. [5] used a similar threat model in their work and reported 4.62 and 0.28% area and power overheads, respectively, when compared with the router design without the Trojan. The performance overhead when copying and routing packets to the malicious application is less than 1% [5]. Therefore, the likelihood of the Trojan being detected is very small unless additional security mechanisms (such as the one discussed in this chapter) are implemented.

9.3 Motivation

As explained in Sect. 9.2.1, AE is a widely accepted countermeasure against eavesdropping attacks. Figure 9.4 shows a block diagram of how authenticated encryption can be implemented in an NoC as proposed in existing literature. The block diagram closely resembles the Galois Counter Mode (GCM) based encryption and authentication [51]. Packets originating from each IP are encrypted and an authentication tag is appended to each packet at the NI before injecting it to the NoC. The entire packet, which consists of $H \parallel C \parallel T$ traverses the NoC and arrives

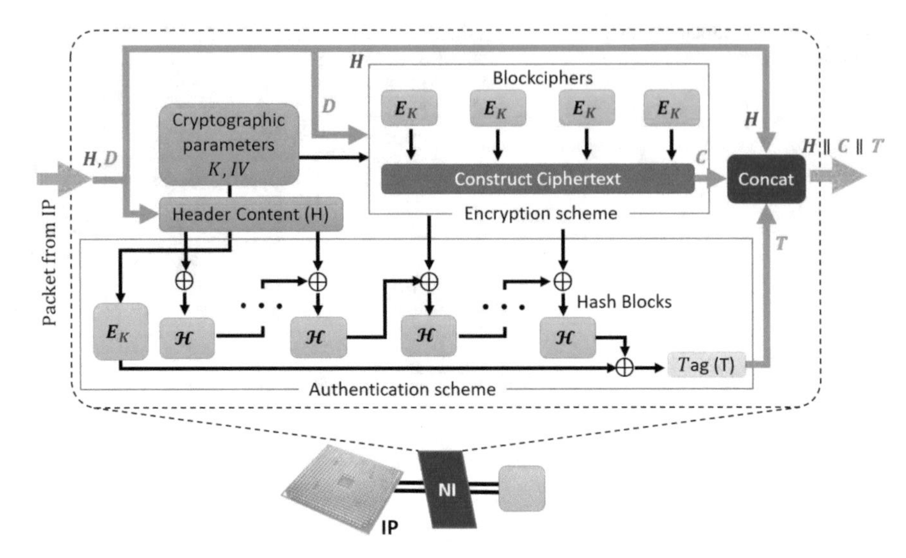

Fig. 9.4 Overview of an authenticated encryption scheme implemented to provide security to NoC

at the destination. The header H is sent as plaintext so that intermediate routers can use the header information for routing. At the destination, the tag T is validated and if valid, the ciphertext C is decrypted to send the plaintext to the desired IP. Encryption ensures that the plaintext of the secure information is not leaked and authentication detects any tampering with the packet including header information. Since the header is modified by the hardware Trojan in order to re-route the packet to the malicious application, the authentication tag validation fails and the attack is detected.

To analyze the performance overhead introduced by an AE scheme, FFT, RADIX (RDX), FMM, and LU benchmarks from the SPLASH-2 benchmark suite [72] were run on an 8×8 Mesh NoC-based SoC with 64 IPs using the gem5 simulator [7] considering two scenarios:

- **Default-NoC**: Bare NoC that does not implement encryption or authentication.
- **AE-NoC**: NoC that uses an authenticated encryption scheme.

More details about the experimental setup are given in Sect. 9.6.1. Results are shown in Fig. 9.5. A 12-cycle delay was assumed for encryption/decryption and authentication tag calculation when simulating AE-NoC according to the evaluations in [40]. The values are normalized to the scenario that consumes the most time. AE-NoC shows 59% (57% on average) increase in NoC delay (average NoC traversal delay for all packets) and 17% (13% on average) increase in execution time compared to the Default-NoC. The overhead for security has a relatively lower impact on execution time compared to the NoC delay since the execution time also includes the time for executing instructions and memory operations (in addition

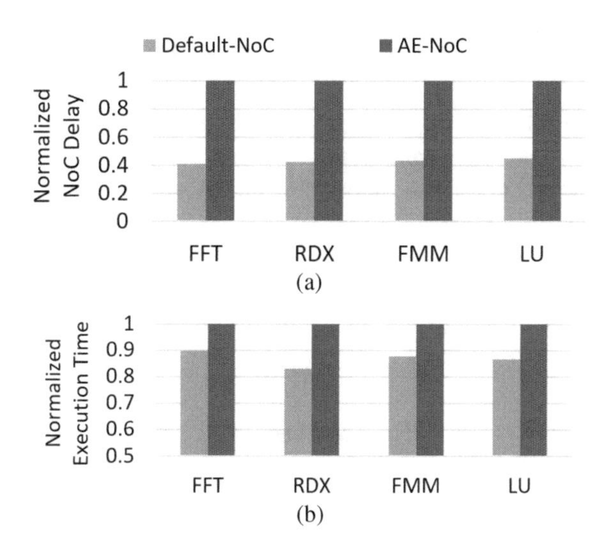

Fig. 9.5 (**a**) NoC delay and (**b**) execution time comparison across different levels of security for four SPLASH-2 benchmarks

to NoC delay). NoC delay in Default-NoC case is caused by delays at routers, links, and the NI. In AE-NoC, in addition to those delays, encryption/decryption delays and authentication tag calculation/validation delays are added to each packet. Additional delays are due to complex encryption/decryption operations and hash calculations for authentication.

When security is considered, Default-NoC leaves the data totally vulnerable to attacks, whereas AE-NoC ensures confidentiality and data integrity. For systems with real-time requirements, an execution time increase of 17% to accommodate a security mechanism is unacceptable. Furthermore, validating the authentication tag for each packet contributes to the SoC power consumption. While there are a wide variety of techniques to improve energy efficiency in NoC-based SoCs [3, 15, 17, 28, 32, 71], they are not suitable in this case. Since the Trojan is rarely activated and only the packet header is modified (packet data is not corrupted) to avoid detection, authenticating each packet becomes inefficient in terms of both performance and power consumption [35]. Clearly, authenticating to detect re-routed packets introduces unnecessary overhead. It would be ideal if the security provided by AE-NoC could be achieved while maintaining the performance of Default-NoC. However, in resource-constrained environments, there is always a trade-off between security and performance.

This motivates the need for the novel digital watermarking-based security mechanism that incurs minimal overhead while providing high security. The watermarking-based security mechanism replaces authentication by watermarking. Encryption is used to ensure data confidentiality. This method achieves a better trade-off than: (1) no authentication that is vulnerable to credible Trojan attacks, and (2) authenticated encryption, which incurs performance degradation prohibiting their use in applications with real-time constraints.

9.4 NoC Packet Watermarking

In this section, we first present a few key definitions and concepts used in the watermarking construction. We then describe the lightweight eavesdropping attack detection mechanism based on digital watermarking.

9.4.1 *Definitions*

9.4.1.1 Hoeffding's Inequality

Let $\{X_1, \ldots, X_n\}$ be a sequence of independent and bounded random variables with $X_i \in [a, b]$ for all i, where $-\infty < a \leq b < \infty$. Then

$$\Pr\left[\left|\frac{1}{n}\sum_{i=1}^{n}(X_i - \mathbb{E}[X_i])\right| \geq t\right] \leq e^{\left(-\frac{2nt^2}{(b-a)^2}\right)}$$

for all $t \geq 0$ [29]. By Hoeffding's Lemma, which says if $X_i \in [a, b]$ then $\mathbb{E}\left[e^{\lambda X}\right] \leq e^{\lambda^2(b-a)^2/8}$ for any $\lambda \geq 0$, a random variable bounded in $[a, b]$ is sub-Gaussian with variance proxy $\sigma^2 = \frac{(b-a)^2}{4}$. Therefore

$$\Pr\left[\left|\frac{1}{n}\sum_{i=1}^{n}(X_i - \mathbb{E}[X_i])\right| \geq t\right] \leq e^{\left(-\frac{nt^2}{2\sigma^2}\right)}$$

9.4.1.2 Bounds for Binary Codes

Let C be a binary code of length w, size M (i.e., having M codewords), and minimum Hamming distance δ between any two codewords denoted by (w, M, d). The distance distribution of C can be calculated as

$$B_i = \frac{1}{M}\sum_{c \in C}|c' \in C : \mathcal{D}(c, c') = i|, 0 \leq i \leq n$$

It is clear that $B_0 = 1$ and $B_i = 0$ for $0 < i < d$ [60].

Let $A(w, d)$ represent the maximum number of codewords M in any binary code of length w and minimum Hamming distance d between codewords. Finding optimum $A(w, d)$ for a given w and d is an NP-Hard problem [23]. However, exact solutions are known for few combinations of values and in the general case, upper and lower bounds of the maximum number of codewords are known [6].

9.4.2 Overview

The flow of packets sent from one IP (source) to another IP (destination) is called a *packet stream*. The detection mechanism relies on the following assumptions about the architecture and threat model.

- The Trojan does not tamper with the legitimate packet content as this may reveal its presence (Sect. 9.2.2). The Trojan only modifies the header of duplicated packets to change the destination (data fields of the duplicated packets are not tampered with) and it allows the legitimate packets to pass as usual.
- Packets are not dropped by intermediate routers and the order of packets in a packet stream is kept constant. This is reasonable as deadlock and livelock free XY routing is used together with FIFO buffers [58].
- When the attacker injects copied packets into the NoC, all the packets can get delayed due to congestion. While this delay is random, the maximum delay is bounded. We explore this assumption in detail in Sect. 9.5.2.

The proposed approach is to embed a unique watermark into every packet stream. Figure 9.6 shows an overview. The watermark encoder and decoder are included at the NI of each node. It is reasonable to assume that the NI can be trusted since it acts as the interface between all the IPs in the SoC and the NoC IP, and is typically designed in-house [19, 40]. The NI at source S encodes the watermark and the NI at destination D decodes it to identify that the packet stream is valid, or in other words, the packets in the packet stream are intended to be received by D. This process is followed by each source/destination pair in the NoC. In case of an attack, the watermark decoded by the NI of the receiving node (node Y according to the illustrative example) will be invalid and a potential attack is flagged. To ensure this behavior, the watermarking mechanism must have the following characteristics:

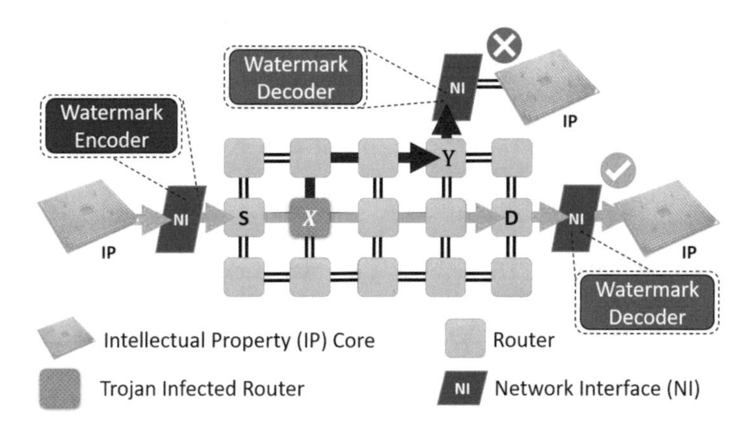

Fig. 9.6 Overview of the watermarking scheme where the watermark encoder and decoder are implemented at the NI

1. The watermark is unique to each packet stream.
2. There is a shared secret between S and D, which is "hard" for any other node to guess or deduce.

In addition to watermarking, this approach relies on encryption/decryption modules implemented at the NIs. The watermark is embedded in the encrypted packets and is decoded before the decryption process. Encrypting packets is required to provide data confidentiality during packet transfers and due to the nature of the watermarking scheme that allows the malicious application to receive some packets before detecting the attack. Proposing an encryption mechanism is beyond the scope of this chapter and several previous works have already proposed NoC-based SoC architectures with encryption/decryption modules implemented at the NI [19, 40, 62]. The proposed watermarking scheme can be implemented on top of those solutions. The performance improvement is achieved by replacing the authentication scheme with the lightweight digital watermarking scheme. The following sections describe the approach in detail. First, the concept behind probabilistic NoC packet watermarking is outlined (Sect. 9.4.3), and then the operation of the watermark encoder and decoder is discussed in detail (Sect. 9.4.4). Finally, we outline an effective method for managing secrets shared between nodes (Sect. 9.4.5).

9.4.3 Probabilistic Watermarking Concept

The watermark ω_{SD} is embedded by the NI of S before the packets are injected into the NoC. This approach uses a timing-based watermark (as opposed to size or content-based) for three reasons: (1) timing alterations are harder to detect by an attacker, (2) it allows a lightweight implementation as it is easy to manipulate, and (3) it does not alter the packet content allowing encryption schemes to be implemented together with watermarking. The watermark is embedded by slightly delaying certain packets in the stream. If ω_{SD} is unique, it should be correctly decoded at the NI of destination D with high probability. In contrast, the probability of decoding ω_{SD} as valid at any other NI should be very low.

Given n packets of a packet stream P_{SD} such that

$$P_{SD} = \{p_{SD,1}, p_{SD,1}, \ldots, p_{SD,i}, \ldots, p_{SD,n}\}$$

the inter-packet delay (IPD) between any two packets can be calculated as $\tau_{SD,i,i+1} = t_{SD,i+1} - t_{SD,i}$ where $t_{SD,i}$ is the timestamp of the packet $p_{SD,i}$. Without loss of generality, for the ease of illustration, we will remove "SD" from the notation and denote the packet stream P_{SD} as P and IPD $\tau_{SD,i,i+1}$ as τ_i.

The encoder selects $2m$ packets $\{p_{r_1}, p_{r_2}, \ldots, p_{r_{2m}}\}$ out of the n packets of packet stream P. The selected packets are paired with another $2m$ packets (outside of the initially selected $2m$ packets) to create $2m$ pairs such that each pair is constructed as $\{p_{r_z}, p_{r_z+x}\}$ where $x \geq 1$ and $z = 1, \ldots, 2m$. Therefore, it is assumed that the packet stream has at least $4m$ packets. The IPD between each pair of packets can be calculated as

$$\tau_{r_z} = t_{r_z+x} - t_{r_z} \tag{9.1}$$

Given that the $2m$ packets are selected independently and randomly, the IPDs are modeled as *independently and identically distributed (IID)* random variables with a common distribution. The IPD values are then divided into two groups. Since there are $2m$ pairs of packets, each group will have m IPD values. Let the IPD values of the two groups be denoted by τ_k^1 and τ_k^2 ($k = 1, \ldots, m$), respectively. It follows that both τ_k^1 and τ_k^2 are IID. Therefore, the expected values μ (and the variances) of the two distributions are equal. Let Δ be the average difference between the two IPD distributions:

$$\Delta = \frac{1}{m} \cdot \sum_{k=1}^{m} \frac{\tau_k^1 - \tau_k^2}{2} \tag{9.2}$$

Then, the expected value and variance of Δ can be calculated as

$$\mathbb{E}[\Delta] = \mathbb{E}\left[\tau_k^1\right] - \mathbb{E}\left[\tau_k^2\right] = 0\,, \qquad \mathrm{Var}(\Delta) = \frac{\sigma^2}{m}$$

Where σ^2 is the variance of the distribution $\frac{\tau_k^1 - \tau_k^2}{2}$. In other words, the distribution of Δ is symmetric and centered around zero. The parameter m is referred to as the *sample size*.

The core idea of the watermarking approach is to intentionally delay a selected set of packets to shift the Δ distribution left or right to encode the watermark bits in the timing information of the packets. Specifically, the distribution of Δ can be shifted along the x-axis to be centered on $-\alpha$ or α by decreasing or increasing Δ by α, where α is called the *shift amount*. As a result, the probability of Δ being negative or positive will increase. Concretely, to embed bit 0, decrease Δ by α. To embed bit 1, increase Δ by α. Decreasing Δ can be done by decreasing each $\frac{\tau_k^1 - \tau_k^2}{2}$ by α (Eq. (9.2)). Decreasing $\frac{\tau_k^1 - \tau_k^2}{2}$ can be achieved by decreasing each τ_k^1 by α and increasing each τ_k^2 by α. It is easy to see that increasing Δ can be done in a similar way. Decreasing or increasing one IPD (τ_k^1) is achieved by delaying the first packet or the second packet of the pair, respectively.

The encoded watermark can be detected by calculating Δ and checking if Δ is positive or negative. If $\Delta > 0$, bit 1 is decoded. Otherwise (if $\Delta \leq 0$) bit 0 is decoded. This scheme can be extended to a w-bit watermark (ω_{SD}) by repeating the above process w times. During the decoding process, a w-bit watermark (ω'_{SD}) is extracted from the packet stream and if the hamming distance between ω_{SD} and ω'_{SD} is lower than a pre-defined *error margin* δ, it can be concluded that the watermark embedded at the source S is detected at the receiver. If the watermark does not match, an attack is flagged.

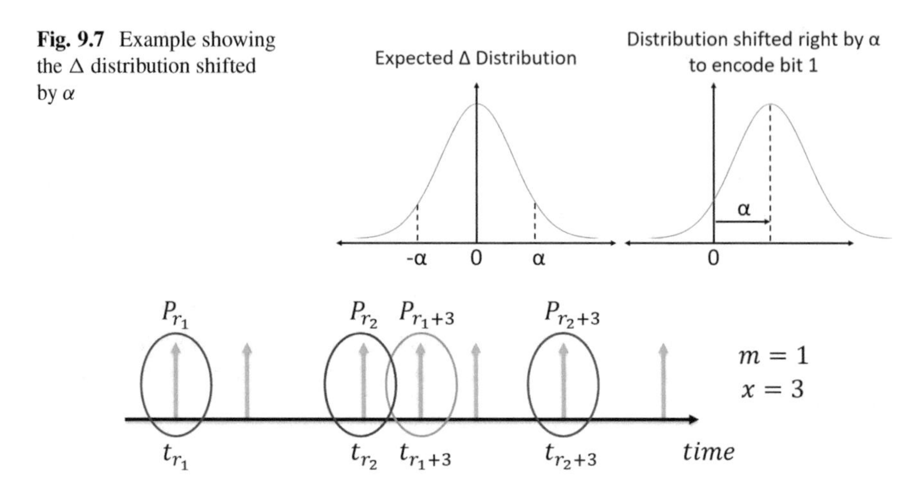

Fig. 9.7 Example showing the Δ distribution shifted by α

Fig. 9.8 Sample packet stream with $m = 1$ and $x = 3$

Figure 9.7 shows the distribution of Δ and the corresponding distribution after shifting it by $\alpha > 0$. Since this scheme is probabilistic, there is a probability that the embedded watermark bits will be incorrectly decoded, thus leading to false alarms (false positives) or missed detection (false negatives). This is because for any $\alpha > 0$, a small portion of the distribution of Δ falls outside the range $(-\infty, \alpha]$. Therefore, if bit 0 is embedded, there is a small probability that the bit will be incorrectly decoded as 1. It can be seen that this probability is the same as the probability that a sample from the unshifted distribution takes a value outside the range $(-\infty, \alpha]$. Similarly, a bit encoded to be 1 can be decoded incorrectly because samples from Δ have a small probability of falling outside the range $[-\alpha, \infty)$. However, it is possible to tune parameters m (sample size), α (shift amount), and δ (error margin) to achieve a very high (nearly 100%) decoding success rate as shown in Sect. 9.6.

To provide formal guarantees, the *bit decoding success rate* (BDSR) is defined as the probability of the embedded watermark bit being decoded correctly (for a shift amount of α). This quantity is denoted by $\Pr[\Delta < \alpha]$. Note that the BDSR also depends on m and σ^2, but this is not explicit in the notation $\Pr[\Delta < \alpha]$ because it is implicitly captured by Δ. We now give an illustrative example to further explain this concept.

Illustrative Example Figure 9.8 shows a sample packet stream in the time domain with packet injection times. For ease of explanation in this example, m is set to one and therefore, two packets ($2m$) are selected from the packet stream (P_{r_1} and P_{r_2}). Both packets are paired with two other packets that are x (=3) packets away in the packet stream (P_{r_1} with P_{r_1+3} and P_{r_2} with P_{r_2+3}). The IPD between each pair is calculated as $\tau_{r_1} = t_{r_1+3} - t_{r_1}$ and $\tau_{r_2} = t_{r_2+3} - t_{r_2}$. The two IPD values are then divided into two groups and Δ calculated according to Eq. (9.2) as $\frac{\tau_{r_1} - \tau_{r_2}}{2}$ (sum for all m and division by m not shown since $m = 1$). The process was

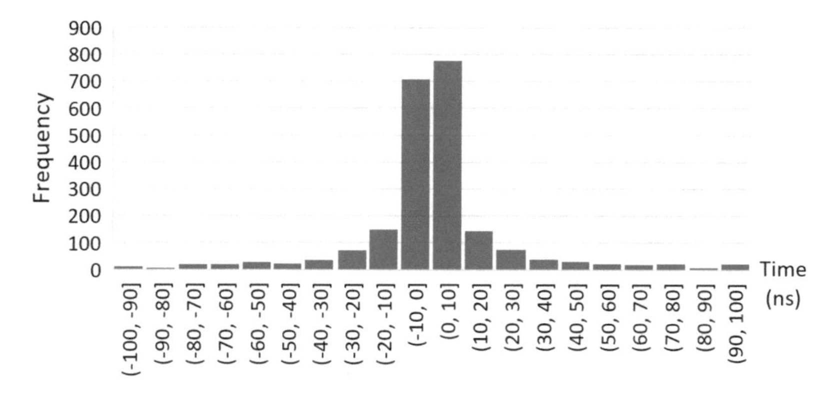

Fig. 9.9 Distribution of Δ with $m = 1$ and $x = 3$

repeated using a packet stream that had more than 3000 packets obtained by running a simulation using the gem5 architectural simulator [7] on a real benchmark. An 8×8 Mesh NoC was modeled using the Garnet2.0 [1] interconnection network model. The node in the top left corner (node S) ran the RADIX benchmark from the SPLASH-2 benchmark suite [72]. One memory controller was modeled and attached to the node in the bottom right corner (node D) so that the memory requests always traverse from S to D. Figure 9.9 shows the histogram collected at the NI of S for the distribution of Δ with $m = 1$ and $x = 3$. Packets were collected at random with the above parameter values to plot Δ. We can observe from Fig. 9.9 that the distribution closely approximates the distribution we expected. The calculated sample mean ($\mathbb{E}[\Delta]$) for this particular example was 0.0053, which is very close to zero. Increasing the number of selected packets ($2m$) further increases the likelihood of the sample mean being zero.

The next section describes the details of the watermark encoder and decoder operations.

9.4.4 *Watermark Encoder and Decoder*

As outlined in Sect. 9.4.2, the watermarking scheme includes a shared secret between S and D, which is "hard" for any other node to guess or deduce. In addition, several parameters are shared between S and D. Specifically, S and D share the tuple $\langle m, \alpha, w_{SD}, \mathbb{K} \rangle$. The first three parameters were introduced in Sect. 9.4.2 as the sample size (m), the shift amount (α), and the unique watermark that represents P_{SD} (w_{SD}). The length of w_{SD} (w) can be derived from w_{SD}. In addition, $\mathbb{K}$ is a secret which is used to derive a key for the encryption scheme and a seed $\mathbb{S}$ using a key derivation function. $\mathbb{S}$ is used to seed the pseudo-random number generator which selects the $2m$ IPDs. We assume the attacker does not know w_{SD} or $\mathbb{K}$, but may know m and α.

Algorithm 9.1: Selection function $\mathcal{F}$

Input: Seed $\mathbb{S}$
Output: Two IPD groups used to encode one watermark bit
Procedure: $\mathcal{F}$
1: $r_z, x \leftarrow PRNG(\mathbb{S})$
2: **for all** $k = 1, \ldots, 2m$ **do**
3: $\quad A \leftarrow$ selectNextWindow(P_{SD})
4: $\quad p_{r_z} \leftarrow A[r_z]$
5: $\quad p_{r_z+x} \leftarrow A[r_z + x]$
6: $\quad \tau_{r_z} \leftarrow t_{r_z+x} - t_{r_z}$
7: $\quad$ **if** $k$is odd **then**
8: $\quad\quad \tau_k^1 \leftarrow \tau_{r_z}$
9: $\quad$ **else**
10: $\quad\quad \tau_k^2 \leftarrow \tau_{r_z}$
11: $\quad$ **end if**
12: **end for**
13: **return** $[\{\tau_1^1, \tau_2^1, \ldots, \tau_m^1\}, \{\tau_1^2, \tau_2^2, \ldots, \tau_m^2\}]$

9.4.4.1 Watermark Encoding Process

When the watermark encoder, which is integrated in the NI of node S, receives packets from its local IP with the destination node D, it encodes the watermark according to the process outlined in Sect. 9.4.3 and the shared secret between S and D. The selection of the IPDs that construct the Δ distribution needs to be deterministic so that the process is identical for the watermark encoder and decoder, and it needs to ensure that an attacker cannot replicate the same behavior. To achieve this, we need a method to pair packets deterministically based on the shared secret, but that appears uniformly random to the attacker (who does not know the shared secret). This approach proposes to implement this using a pseudo-random number generator (PRNG) seeded (i.e., initialized) with $\mathbb{S}$ (or something derived from it). This ensures that the encoder and decoder produce the *same* sequence of random numbers. Further, an attacker (who does not know the seed) cannot predict the next PRNG output, even with the knowledge of the previous output [68].

Let $\mathcal{F}$ denote the selection function that given a packet stream, selects and divides $2m$ IPDs into two groups, each of size m. A window of packets is chosen and two random packets are paired together from each window. Therefore, to construct $2m$ IPDs, $2m$ such packet windows are required. The operation of $\mathcal{F}$ used in this method is outlined in Algorithm 9.1. The PRNG seeded with $\mathbb{S}$ is used to randomly generate two integers r_z and x (line 1) such that $0 \leq r_z \leq W - 1$ and $0 < x$ and $r_z + x \leq W - 1$, where W is the size of the window. This can be done using rejection sampling to ensure that $r_z \neq x$ and then calling the smaller integer r_z and the larger $r_z + x$. The packet at the index r_z (p_{r_z}) is paired with the packet that is x packets away giving the random pair $\{p_{r_z}, p_{r_z+x}\}$ (lines 4–5). The calculated IPD values are then evenly divided into two groups (lines 6–11).

Since $2m$ IPDs are required to encode a 1-bit watermark, w iterations of the procedure $\mathcal{F}$ are required to encode the w-bit watermark. When encoding one watermark bit, the distribution discussed in Sect. 9.4.3 holds only when each pair of packets is the same distance x apart from each other. Therefore, the same r_z and x values are used for each iteration of k. When encoding another watermark bit, another iteration of $\mathcal{F}$ is required in which another pair of r_z and x values will be generated by the PRNG. To ensure that the same r_z and x values are not generated for subsequent watermark bits, the PRNG must be seeded only once. An example to show how the selection function can be used to encode a w-bit watermark including how to select the window is given in Sect. 9.6.

9.4.4.2 Watermark Decoding Process

Node D upon examining the packet stream P_{SD} decodes the w-bit watermark w'_{SD} by following the process outlined in Sect. 9.4.3 and the shared secret tuple. The decoder concludes that the watermark is valid if the Hamming distance between w_{SD} (taken from the shared secret tuple) and w'_{SD} (decoded from the received packet stream P_{SD}) is less than or equal to the error margin δ. Formally, the watermark is valid if

$$\mathcal{D}(w_{SD}, w'_{SD}) \leq \delta \tag{9.3}$$

where $\mathcal{D}$ is the Hamming distance between two bit strings and $0 \leq \delta \leq w$. The reason for allowing an error margin δ and not looking for an exact match is that no matter how large the shift amount α is, there is a probability that the watermark is decoded incorrectly as discussed in Sect. 9.5.1. Tuning parameter δ allows us to minimize this probability. In addition, as shown in Sect. 9.5.2, it allows to minimize the impact of the attack.

9.4.5 Managing Shared Secrets

The watermark encoder and decoder operation introduced in Sect. 9.4.4 relies on shared secret tuples between nodes to make sure the watermarking scheme cannot be compromised. To facilitate this, an efficient way to generate and manage such secrets is required. Many previous studies have addressed the challenge of developing an efficient key management system in several ways. One such example is the key management system proposed by Lebiednik et al. [42]. In their work, a separate IP called the *key distribution center* (KDC) handles the distribution of keys. Each node in the network negotiates a new key with the KDC using a pre-shared portion of memory that is known by only the KDC and the corresponding node. The node then communicates with the KDC using this unique key whenever it wants to obtain a new key. The KDC can then allocate keys and inform other nodes as required. The digital watermarking scheme can be integrated with a similar key generation and management mechanism.

9.5 Theoretical Analysis

In this section, we discuss mathematical guarantees about the correctness and security of the watermarking scheme which are further validated with experimental results in Sect. 9.6. First, we discuss a bound on BDSR during normal operation (Sect. 9.5.1). Then the impact of an attacker on BDSR is evaluated (Sect. 9.5.2). Finally, a method to select the error margin δ is presented such that it maximizes the chance of successfully decoding the watermark while minimizing the chances of an attack if the attacker is aware of the detection method (Sect. 9.5.3).

9.5.1 Bit Decoding Success Rate During Normal Operation

Given this watermark encoding/decoding scheme, it is clear that larger the shift amount α is, the higher the bit decoding success rate (BDSR) will be. However, having arbitrarily large α is not feasible in systems with real-time constraints. In this section, we show that close to 100% BDSR can be achieved for arbitrarily small α by changing the sample size m.

As discussed in Sect. 9.4.3, a watermark bit can be decoded incorrectly if at the receiver's end, $|\Delta| > \alpha$. Therefore, we should analyze the behavior of $\Pr[|\Delta| > \alpha]$. There are several well-established statistical tools for this, but in particular we can use concentration results, also known as tail bounds. Since the IPDs are bounded and independent, we can use Hoeffding's inequality (introduced in Sect. 9.4.1.1) and equations from Sect. 9.4.3 related to the distribution of Δ:

$$\Pr[|\Delta| \geq \alpha] \leq e^{\left(-\frac{m\alpha^2}{2\sigma^2}\right)}$$

$$\text{Using symmetry;} \qquad \Pr[\Delta < \alpha] \geq 1 - \frac{1}{2}e^{\left(-\frac{m\alpha^2}{2\sigma^2}\right)} \tag{9.4}$$

Therefore, we can observe that the BDSR is lower bounded by a value that depends on α and m. The results show that irrespective of the distribution of the IPDs, for arbitrarily small α values, we can always take the BDSR close to 100% by increasing the sample size m. In other words, no matter how small the shift amount α needs to be to abide by the timing constraints of the system, we can still achieve high BDSR by selecting more packets in each IPD group.

9.5.2 *Impact of an Attack on the Bit Decoding Success Rate*

Having established mathematical guarantees about BDSR during normal operation, we shift our focus to explore how BDSR of legitimate packet streams can be affected by an attack. According to the threat model, the Trojan-infected router copies packets and sends them to a malicious application running on a different IP. As a result, more packets are introduced to the network which can cause congestion. All packets in the network can be delayed because of this. Therefore, the attack can introduce additional delays to the legitimate packet streams. It is safe to assume that these additional delays are finite. If the attacker delays packets indefinitely through congestion, the attack is no longer a snooping attack, but rather a flooding type of denial-of-service attack [18] that is beyond the scope of this threat model.

Therefore, assume that the attack introduces a delay ϵ_i $(i = 1, 2, \ldots, n)$ for each packet in packet stream P_{SD}. Since the delay is finite, we can denote $\epsilon_i \leq \xi$ where ξ is an upper-bound on the delay. Given that the Trojan-infected router does not know which packets were selected by the watermark encoder (as explained in Sect. 9.4.4), the delay introduced by the attacker (whatever it is) on the selected IPDs is IID from the perspective of S and D.

Let θ_k^1 and θ_k^2 be the impact on the IPDs of the groups τ_k^1 and τ_k^2 as a result of the delays, respectively. We observe that θ_k^1 and θ_k^2 are in the interval $(-\xi, \xi)$, and since they are IID, $\mathbb{E}\left[\theta_k^1\right] = \mathbb{E}\left[\theta_k^2\right]$ and $\mathrm{Var}(\theta_k^1) = \mathrm{Var}(\theta_k^2)$. Let $\mathrm{Var}(\theta_k^1) = \mathrm{Var}(\theta_k^2) = 2\sigma_d^2$ and Δ' denote the distribution after modifying Δ defined in Eq. (9.2) with the added delays.

$$\begin{aligned}
\Delta' &= \frac{1}{m} \cdot \sum_{k=1}^{m} \frac{(\tau_k^1 + \theta_k^1) - (\tau_k^2 + \theta_k^2)}{2} \\
&= \frac{1}{m} \cdot \sum_{k=1}^{m} \left[\frac{(\tau_k^1 - \tau_k^2)}{2} + \frac{(\theta_k^1 - \theta_k^2)}{2} \right] \\
&= \Delta + \frac{1}{m} \cdot \sum_{k=1}^{m} \left[\frac{(\theta_k^1 - \theta_k^2)}{2} \right]
\end{aligned}$$

It is easy to see that $\mathbb{E}\left[\Delta'\right] = 0$. From the properties of variance of IID variables, we can deduce

$$\mathrm{Var}(\Delta') = \frac{1}{m} \mathrm{Var}\left(\left[\frac{(\tau_k^1 - \tau_k^2)}{2} + \frac{(\theta_k^1 - \theta_k^2)}{2} \right] \right)$$

Let $A = \frac{(\tau_k^1 - \tau_k^2)}{2}$ and $B = \frac{(\theta_k^1 - \theta_k^2)}{2}$. According to the definitions above, $\mathrm{Var}(A) = \sigma^2$ and $\mathrm{Var}(B) = \sigma_d^2$. Therefore

$$\begin{aligned} m \cdot \mathrm{Var}(\Delta') &= \mathrm{Var}(A) + \mathrm{Var}(B) + 2Cov(A, B) \\ &\leq \sigma^2 + \sigma_d^2 + 2\sigma\sigma_d \\ &= (\sigma + \sigma_d)^2 \end{aligned}$$

Since the random delays and the packet timing of the legitimate packet flow are independent, we can assume those two variables are not correlated. To understand how BDSR has changed with Δ changing to Δ', we apply the Hoeffding's inequality to Δ' similar to what we did when deriving Eq. (9.4).

$$\Pr\left[|\Delta'| \geq \alpha\right] \leq e^{\left(-\frac{\alpha^2}{2\mathrm{Var}(\Delta')}\right)} = e^{\left(-\frac{m\alpha^2}{2(\sigma+\sigma_d)^2}\right)}$$

Using symmetry,

$$\Pr\left[\Delta < \alpha\right] \geq 1 - \frac{1}{2}e^{\left(-\frac{m\alpha^2}{2(\sigma+\sigma_d)^2}\right)} \tag{9.5}$$

Observe that the only change is the increase in variance caused by the attacker. We can choose σ_d depending on the amount of congestion the attacker is willing to cause without risking being detected. Similar to the argument we made when reasoning about the BDSR using Eq. (9.4), we can see that BDSR is lower bounded and by manipulating the sample size, we can make the BDSR be arbitrarily close to 100%. Therefore, the impact on the watermarking detection is a bounded increase of variance on an otherwise 100% successful watermarking scheme. As the illustrative example that calculates BDSR in Sect. 9.5.1 outlines, the success rate can be brought very close to 100% even with the selection of a modest value for m.

9.5.3 Optimal Error Margin Selection

As discussed in Sect. 9.4.4, the use of the error margin δ instead of an exact match between the decoded and the expected watermark enables tuning δ to maximize the *watermark detection success rate* (WDSR). Unlike BDSR, which refers to the success of decoding a single bit, WDSR considers the entire watermark with w bits. The probabilistic nature of this watermarking scheme leaves a small probability that the watermark will be incorrectly decoded irrespective of the values chosen for the parameters. While this probability is small, efficient selection of δ can push WDSR as close as possible to 100%. On the other hand, using a larger error margin also increases the success of potential attacks. Indeed, assuming that the attacker is aware of the detection strategy, the best strategy for an attacker to eavesdrop on data without being detected is to try to *forge* a watermark. If the attacker succeeds, then the duplicated packets will be accepted as valid by the node that runs the accomplice application and the watermarking-based defense will be defeated. We call the success probability of such a forging attack the *watermark forging success probability* (WFSP). The goal of the detection scheme is thus to set the parameters such that WDSR is maximized while minimizing WFSP. We explore how this can be achieved in this section.

9.5.3.1 Maximizing Watermark Detection Rate

The probability of incorrectly decoding a bit was formalized using the metric BDSR as $\Pr[\Delta < \alpha]$. Considering symmetry, let $\vartheta = \Pr[-\infty < \Delta < \alpha] = \Pr[-\alpha < \Delta < \infty]$. Then for a w-bit watermark, probability of accurately decoding all w bits will be ϑ^w. Therefore, the expected WDSR can be calculated as

$$\sum_{i=0}^{\delta} \binom{w}{i} \vartheta^{w-i}(1-\vartheta)^i \tag{9.6}$$

We can see that with a large δ, the expected WDSR increases. We observe from Eq. (9.6) that

$$\sum_{i=0}^{\delta} \binom{w}{i} \vartheta^{w-i}(1-\vartheta)^i \geq \vartheta^w$$

Therefore, it is possible to make the expected WDSR larger than the desired WDSR by increasing ϑ. Revisiting Eq. (9.5), we observe that ϑ can be made sufficiently close to 1 by increasing the sample size m irrespective of α, σ, and σ_d. Therefore, we can conclude that in theory, it is possible to make WDSR close to 100% even with a modest error margin.

9.5.3.2 Minimizing Risk of Watermark Forging Attacks

While increasing δ can increase WDSR, larger the δ, larger the expected WFSP will be. This is addressed in two steps. First, watermarks are selected such that under a given error margin δ, the probability that one watermark can be incorrectly decoded as another watermark (watermark collision) is minimized. Then, we discuss the case where an attacker, after knowing the detection mechanism, tries to inject duplicated packets such that the decoder at the receiver incorrectly validates the watermark (watermark forging) and accepts the duplicated packet steam as valid.

The problem of selecting distinct w-bit watermarks for each source–destination pair can be recast as the problem of selecting distinct codewords. This is a well-established problem that has been extensively studied in the information theory literature. Indeed, it is known that for any given set of distinct codewords, if the minimum Hamming distance between any two codewords is at least $2\delta + 1$, a nearest neighbor decoder will always decode correctly when there are δ or fewer errors [50]. Therefore, if the watermarks are chosen such that any two watermarks are at least $2\delta+1$ distance apart, the probability of a watermark collision is minimal. This approach selects the number of bits in the watermark w such that this property is satisfied using the method explained in Sect. 9.4.1.2. An example of how w is selected is given in Sect. 9.6.2.2.

Even if w is selected such that watermark collision probability is minimized, an attacker may still try to impersonate a legitimate sender. Assume that w_{SD} and w_{SY} are valid watermarks with distance $2\delta+1$ (minimum possible distance between two watermarks) between nodes S and D and S and Y, respectively. A Trojan-infected router in the path from S to D duplicates packets and sends to an accomplice application in node Y. For Y to accept the duplicated packet stream as a legitimate packet stream coming from S, the watermark of the duplicated packet stream should match w_{SY}. We refer to this attack as a *watermark forging* attack.

Sections 9.4.4 and 9.4.5 detailed how watermarks are kept unknown to any other parties, except for the sender and receiver in a packet stream, using shared secrets. Therefore, the attacker's method to forge a watermark can be reduced to a random bit flipping game with the goal of matching w_{SY}. Random bit flipping is achieved by randomly delaying the duplicated packets in P_{SD}. For the attacker to win the game, w_{SD} should change to w_{SY}. Since the minimum distance between any two watermarks is $2\delta + 1$, considering the error margin of δ, the minimum required number of bit flips is $\delta + 1$. Therefore, the attacker should flip at least $\delta + 1$ bits to win the game. However, flipping the wrong bits can take the target even further. Therefore, the best chance for the attacker to win the game is if it flips the correct $\delta+1$ bits of w_{SD} to match w_{SY} (to end up within the error margin of w_{SY}, i.e., within δ-Hamming distance of w_{SY}). The probability that the attacker flips the correct $\delta+1$ bits at any given round of the game is thus: $\binom{w}{\delta+1}^{-1}$. Assuming the attacker plays n times, the attacker's probability of winning, or in other words, the probability of successfully forging the watermark (WFSP) at least once (after n attempts) is

$$1 - \left[1 - \frac{1}{\binom{w}{\delta+1}}\right]^n \tag{9.7}$$

Observe that by manipulating w and δ, this probability can be made arbitrarily small. Furthermore, n cannot be arbitrarily large because if the probability of winning in the first few attempts is low, then the attacker will be detected before the attacker can successfully forge the watermark.

This allows the conclusion that it is possible to make WDSR close to 100% and WFSP close to 0%. Equations (9.5)–(9.7) combined give us the theoretical trade-off model between WDSR and WFSP. However, accommodating arbitrarily large m and w is not possible in practical scenarios. Therefore, the next section (Sect. 9.6) provides experimental evaluations and discusses realistic values that can be achieved under the threat model and architecture.

9.6 Experimental Results

In this section, an experimental evaluation of the theoretical models established in previous sections is presented. The experimentally selected parameters are then used to explore the performance gain achieved by using this method compared to traditional AE based schemes.

9.6.1 Experimental Setup

This approach is evaluated by modeling an NoC-based SoC using the cycle-accurate full-system simulator—gem5 [7]. "GARNET2.0" interconnection network model that is integrated with gem5 was used to model an 8×8 Mesh 2D NoC [1]. To ensure the accuracy of the simulator model when compared to real hardware, the simulator framework proposed in [16] was used, which has validated simulator results with results from the Intel Knights Landing (KNL) architecture (Xeon Phi 7210 hardware platform [39]). Figure 9.10 shows an overview of the NoC-based SoC model. Each IP was modeled as a processor core executing a given task at 1 GHz with a private L1 Cache. Eight memory controllers were modeled and attached to the IPs in the boundary providing the interface to off-chip memory. In case of a cache miss, the memory request/response messages were sent to/from memory controllers as NoC packets. The NoC was modeled with 3-stage (buffer write, route compute + virtual channel allocation + switch allocation, and link traversal) pipelined routers with wormhole switching and 4 virtual channel buffers at each input port. Packets are routed using the deadlock and livelock free, hop-by-hop, turn-based XY deterministic routing protocol.

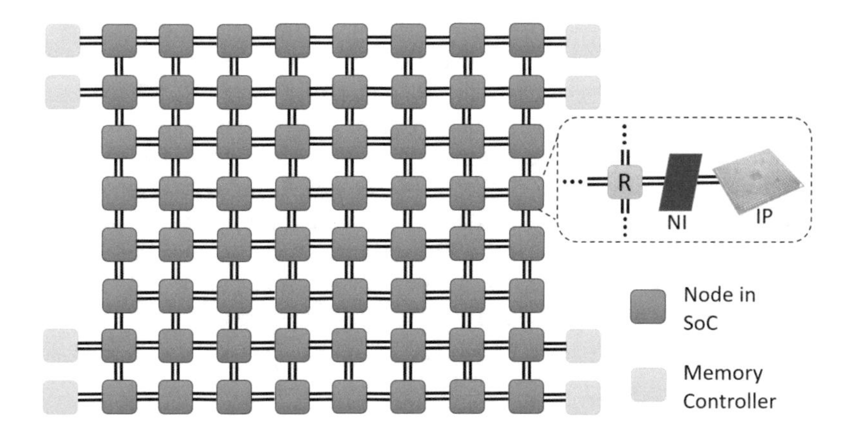

Fig. 9.10 8×8 Mesh NoC setup used to generate results

Each processor core in the SoC was assigned an instance out of FFT, RADIX (RDX), FFM, and LU benchmarks from the SPLASH-2 benchmark suite [72]. Each simulation round can in theory give $\binom{64}{2} \times 2 = 4032$ packet streams (assuming two-way communication between any pair out of the 64 nodes) and the number of iterations that depended on the number of benchmarks (four in this case) can give $4 \times \binom{64}{2} \times 2 = 16{,}128$ packet streams. However, depending on the address mapping, only some node pairs out of all the possible node pairs communicate. The experiments generated 3072 packet streams for all benchmarks between 1024 unique node pairs which were used to evaluate the method. However, to decide the number of bits in the watermark w, looking at only the number of unique node pairs is not sufficient because to avoid watermark collisions, the Hamming distance between any two watermarks should be at least $2\delta + 1$. According to Sect. 9.4.1.2, as δ increases, w increases as well. Therefore, more packets are required to encode the watermark and as a result, the time to detect an ongoing attack increases (more packets need to be observed before recognizing the watermark). Increasing m has a similar impact. Increasing α increases the application execution time and it takes longer to detect eavesdropping attacks. This motivates the exploration of optimum parameter (m, α and δ) values such that WDSR is maximized and attack detection time, execution time as well as WFSP are minimized.

9.6.2 *Parameter Tuning*

We first explore m and α when encoding a single watermark bit and then extend the discussion to consider WDSR, WFSP, execution time, and detection time.

9.6.2.1 Bit Decoding Success Rate Behavior with m and α

When embedding one watermark bit in a packet stream, Eq. (9.4) gives a theoretical estimate of the BDSR. To compare the theoretically expected BDSR with experimental results, a non-overlapping sliding window of λ packets is used and $2m$ IPDs are selected according to the method in Sect. 9.4.4.1. One bit is encoded in each of the 3072 selected packet steams following the same methodology and decoded at the receiver's side according to the method introduced in Sect. 9.4.3. $\lambda = 8$ is chosen to ensure adequate randomness in the IPD selection process. A detailed analysis of λ value selection is given in Sect. 9.7. $\alpha = 60\,\text{ns}$ is fixed and m is varied from 2 to 15. Results are shown in Fig. 9.11. We compare the outcome from the experiments with the theoretical model (Eq. (9.4)). For example, expected BDSR for $m = 4$, $\alpha = 60\,\text{ns}$, and $\sigma^2 = 2662$ is calculated as

$$\Pr[\Delta < 60] \geq 1 - \frac{1}{2}e^{\left(-\frac{4\times 60^2}{2\times 2662}\right)} \approx 0.967$$

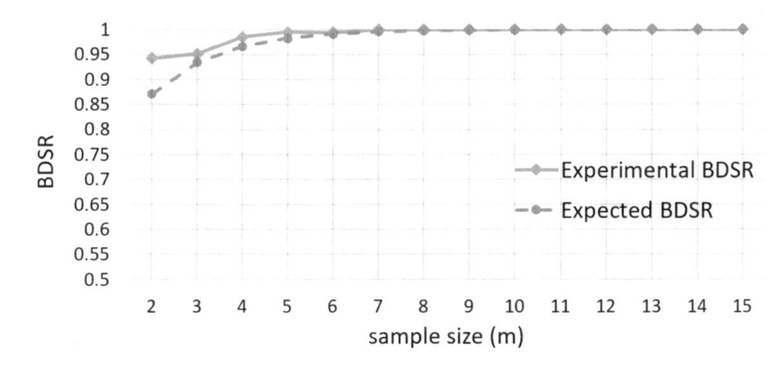

Fig. 9.11 BDSR variation with sample size m. $\alpha = 60\,\text{ns}$

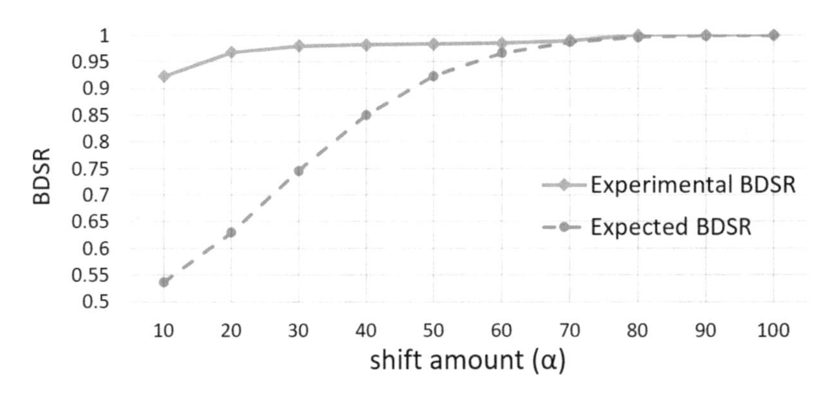

Fig. 9.12 BDSR variation with shift amount α. $m = 4$

m is now fixed at 4 and α is varied from 10 to 100 ns to explore BDSR variation with α. Figure 9.12 shows the comparison between the theoretical model (Eq. (9.4)) and results generated from the experiments. The experimental results in both Figs. 9.11 and 9.12 show that the theoretical model gives an accurate bound on BDSR. As α and m are increased, BDSR converges to 1. However, the goal is to detect any attack with high accuracy while incurring minimum performance overhead. Therefore, BDSR is not the only deciding factor. As α and m is increased, the execution time of the application/benchmark running with the attack detection mechanism increases as well. α and m should be chosen such that this trade-off is maintained.

While Figs. 9.11 and 9.12 show how BDSR varies with m and α, both figures had one parameter fixed while varying the other. To observe how both m and α effect the BDSR as well as the execution time, we did a grid search in the ranges $2 \leq m \leq 10$, $10 \leq \alpha \leq 80$, and $w = 20$ and eliminated cases where expected BDSR was less than 0.95 and execution time increase was more than 5%. These thresholds were chosen to achieve the optimum balance in the trade-off. Results are shown in Fig. 9.13. $w = 20$ is chosen because, to provide a unique watermark for

α \ m	2	3	4	5	6
50	X	X	X	0.952 0.989 4.27%	0.970 0.991 4.75%
60	X	X	0.967 0.985 4.17%	0.983 0.995 4.73%	X
70	X	0.968 0.971 3.89%	0.987 0.990 4.54%	X	X
80	0.955 0.960 3.42%	0.986 0.989 4.19%	X	X	X

Fig. 9.13 BDSR and execution time variation with m and α. w fixed at 20. The green cells show expected BDSR, purple show experimental BDSR, and yellow indicate execution time increase. Crosses indicate either expected BDSR or execution time increase falling beyond the selected thresholds

each communicating node pair (1024 in these experiments), 10 bits are required. 10 additional bits are kept to allow error margins as well as to avoid collisions. However, as discussed in Sect. 9.6.2.2, w can be further optimized leading to a better execution time. Execution time increase is measured as the average execution time increase as a percentage when benchmarks are run with this approach compared to Default-NoC introduced in Sect. 9.3. Out of the possible combinations in Fig. 9.13, we pick $m = 4$ and $\alpha = 60$ as it gives an adequate trade-off for the exploration.

9.6.2.2 Choosing δ and *w*

With the values selected for m and α, the impact of the error margin δ on WDSR is explored. To calculate expected WDSR according to Eq. (9.6), w should be decided. However, the value of w is dependent on the value selected for δ. Therefore, we explore the behavior of expected WDSR with respect to δ for several fixed w values ($w \in \{14, 16, 18, 20\}$). Results are shown in Fig. 9.14. $\delta = 0$ represents exact matches between the decoded watermark and the expected watermark without using an error margin. The importance of using δ is evident when the scenario of looking for exact matches ($\delta = 0$) is compared with any other δ value. For example, for the values $\vartheta = 0.967$ and $w = 20$, WDSR with exact matches is $\vartheta^w = 51.1\%$ whereas for the same ϑ and w values with an error margin of 2, WDSR is 97.3%.

As outlined in Sect. 9.5.3.2, the chosen δ value affects the chances of the attacker succeeding in a forging attack (WFSP). To evaluate the impact, we explored WDSR (Eq. (9.6)) and WFSP (Eq. (9.7)) values for different combinations of w and δ. However not all w and δ values can co-exist if watermark collisions are to be avoided. Assume that the chosen δ value is 2. As outlined in Sect. 9.5.3.2, for two watermarks not to collide, they should be at least $2\delta + 1 (= 5 \text{ if } \delta = 2)$ Hamming

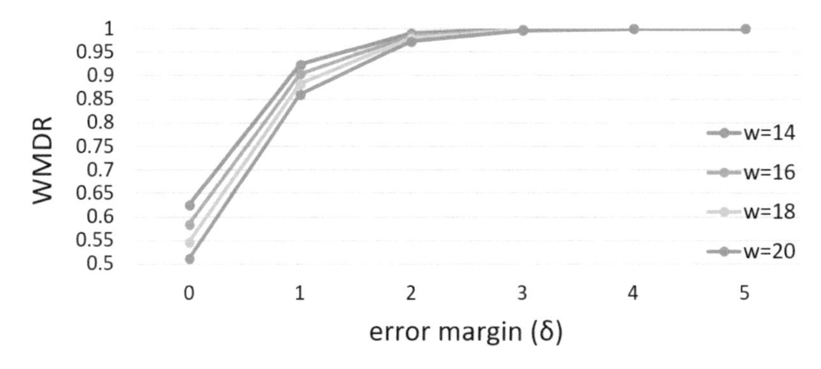

Fig. 9.14 Expected WDSR variation with error margin δ for several w values. m and α fixed at 4 and 60 ns, respectively

Table 9.2 WDSR, WFSP, and execution time increase for varying w and δ. $\vartheta = 0.967$, $n = 10$

δ	w	Expected WDSR	WFSP	Experimental WDSR	Execution time increase
1	14	0.9238	0.1046	0.9538	3.49%
1	15	0.9139	0.0912	0.9512	3.61%
2	18	0.9797	0.0121	0.9801	3.95%
2	19	0.9765	0.0102	0.97884	4.06%
3	21	0.9955	0.0075	0.9987	4.29%
3	22	0.9946	0.0064	0.9964	4.40%

distance apart. Since there are 1024 unique node pairs, w can be set as the minimum number of bits required to generate 1024 unique codewords such that the minimum Hamming distance between any two codewords is 5. In other words, we are looking for w such that $A(w, 5) \geq 1024$ according to Sect. 9.4.1.2. From [6], we can derive $w \geq 18$. Therefore, to ensure that there are no collisions between watermarks with an error margin of 2, at least 18 bits are required for the watermark. Similarly, we can derive $w \geq 21$, for $\delta = 3$, and $w \geq 14$ for $\delta = 1$. Since increasing w has an impact on execution time as well, for each δ value, we pick the two smallest possible w value such that there are no watermark collisions. Table 9.2 shows expected WDSR, WFSP values, experimental WDSR value, and execution time increase for the selected configurations.

These results strongly support the claim that WFSP can be made arbitrarily small by manipulating w and δ. We observe from Fig. 9.14 that WDSR converges to 1 starting $\delta = 2$. Furthermore, observing values in Table 9.2, we can pick $\delta = 2$ and $w = 18$ as a configuration that gives an adequate trade-off.

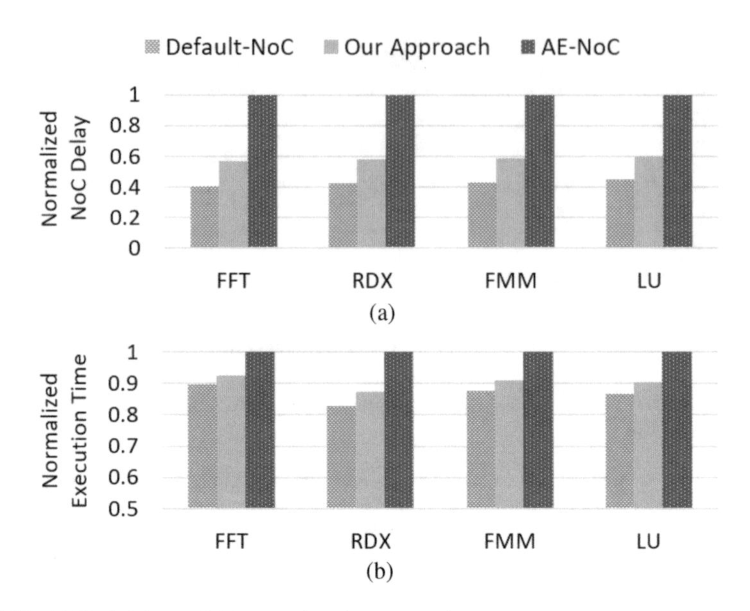

Fig. 9.15 (**a**) NoC delay and (**b**) execution time comparison

9.6.3 Performance Evaluation

With the selected parameters, $m = 4, \alpha = 60, \delta = 2, w = 18$, we explore the performance improvement achieved by this method compared to the traditional AE based defenses. Section 9.3 introduced two scenarios—*Default-NoC* and *AE-NoC* against which the performance of this approach (digital watermarking-based attack detection coupled with encryption) is evaluated. NoC delay and execution time comparison are shown in Fig. 9.15 considering Default-NoC, AE-NoC, and the watermarking-based attack detection method. This approach only increases the NoC delay by 27.9% (26.3% on average) and execution time by 5.2% (3.95% on average) compared to the default-NoC whereas AE-NoC increased NoC delay by 59% (57% on average) and execution time by 17% (13% on average). Therefore, this method has the ability to significantly improve performance compared to other state-of-the-art security mechanisms intended at preventing eavesdropping attacks.

In addition to execution time comparison, time taken to detect an ongoing attack (detection time) is also critical. Detection time is calculated as the time taken to decode the complete watermark from a packet stream. As soon as the w-bit watermark is decoded and validated, any eavesdropping attack can be detected. Table 9.3 shows detection time for each benchmark normalized to total execution time. This shows that the watermark detection scheme is capable of detecting any eavesdropping attacks in a timely manner.

Table 9.3 Attack detection time for different applications/benchmarks. Each value is normalized to the corresponding benchmark execution time

FFT	RDX	FMM	LU
6.56E−3	4.8E−5	1.9E−4	3.9E−4

In summary, these results validate the theoretical model and provide a framework to tune the parameters such that eavesdropping attacks can be detected quickly with high accuracy while providing a significant performance improvement compared to existing state-of-the-art solutions.

9.7 Discussion

The security of the watermarking scheme depends on the secrecy of some parameters (Sect. 9.4.4). Parameters include the watermark w_{SD} as well as the key $\mathbb{K}$ for each P_{SD}. A key distribution center (KDC) acts as a trusted dealer to distribute these parameters. In this section, we discuss security implications if some of these assumptions do not hold.

9.7.1 Eliminating the Trusted Dealer

In the absence of a trusted dealer, each communicating node pair will have to agree on a watermark and a key. While this can be facilitated by key-exchange protocols such as the Diffie–Hellman key exchange, the lack of a trusted dealer can cause duplicated watermarks (watermark collisions). If watermarks are selected uniformly at random to minimize the chances of collision, according to the birthday bound, the number of bits assigned to the watermark should be double of what is required. For example, if an 18-bit watermark is required in the presence of a trusted dealer, 36 bits are required in its absence because of the birthday bound. While this watermarking scheme can give better accuracy and less collisions for a 36-bit watermark, the execution time as well as the detection time will increase. Therefore, a designer needs to carefully select the size of the watermark to minimize the collision without violating the performance budget.

9.7.2 What Can Be Inferred from Packet Timing?

It is important to note that the watermark is encoded in the IPD values, not in the individual packet injection/received times. Furthermore, packet injection times can vary depending on the behavior of the application as well. There can be phases in the application execution where more packets are injected to the NoC whereas in some other phases, delay between packet injections is comparatively high. Therefore, "guessing" the watermark cannot be easily accomplished by merely observing packet arrival times. Moreover, the only way for an attacker to forge the watermark successfully is to know both the watermark and the PRNG seed.

Indeed, even if the watermark could be inferred from packet timing, the PRNG seed cannot be inferred from packet-timing information due to cryptographic guarantees of using a PRNG. In the next section, we assume that the watermark is known by the adversary but not the PRNG seed and analyze the probability that an attacker can forge the watermark. This probability can be reduced to a random bit flipping game (probability $= \frac{1}{2}$).

9.7.3 Watermark Is Not a Secret Anymore?

Assume that the attacker knows the watermark, but not the PRNG seed. To forge the watermark, the attacker must select the two correct packets (that forms the IPD) from each window. Observe that without the PRNG seed, the attacker's probability of correctly guessing the two packets from a given window is $1/\binom{\lambda}{2}$ (Case I). Similarly, we can derive that the probability of two packets chosen by the attacker partially overlapping with the correct two packets and the probability of the attacker not selecting either one of the two correct packets are $2(\lambda - 2)/\binom{\lambda}{2}$ (Case II) and $\binom{\lambda-2}{2}/\binom{\lambda}{2}$ (Case III), respectively. Therefore, the higher the value chosen for λ, the lower the chances of a successful attack. The probability of the attacker *not* selecting either one of the two packets correctly (Case III) goes above 0.5 at $\lambda = 8$. In the overlapping scenario, if the first packet selected by the attacker is the correct second packet (or vice versa), delaying it will give the incorrect watermark bit. However, to give a conservative estimate, we ignore that possibility and use $\lambda = 8$ so that the probability of selecting both packets incorrectly is at least $\frac{1}{2}$. This analysis shows that this watermarking scheme can be tuned to work even in scenarios with very strong security assumptions such as the watermark being leaked to the attacker. Additionally, for systems which require even stronger security, another layer of security can be added if the watermark assigned between each pair of nodes is rotated after some number of iterations.

9.8 Summary

In this chapter, we introduced a lightweight eavesdropping attack detection mechanism using digital watermarking in NoC-based SoCs. We consider a widely explored threat model in on-chip communication architectures where a hardware Trojan-infected router in the NoC IP copies packets passing through it, and re-routes the duplicated packets to an accompanying malicious application running on another IP in an attempt to leak information. Compared to existing authenticated encryption based methods, this approach offers significant performance improvement while providing the required security guarantees. Performance improvement is achieved by replacing authentication with packet watermarking that can detect duplicated packet streams at the network interface of the receiver. We discussed the accuracy and security of the approach using theoretical models and empirically validated them. Experimental results demonstrated that this approach can significantly outperform the state-of-the-art methods.

Acknowledgement This work was partially supported by the National Science Foundation (NSF) grant SaTC-1936040.

References

1. N. Agarwal, T. Krishna, L. Peh, N.K. Jha, Garnet: a detailed on-chip network model inside a full-system simulator, in *2009 IEEE International Symposium on Performance Analysis of Systems and Software* (2009), pp. 33–42
2. A. Ahmed, F. Farahmandi, Y. Iskander, P. Mishra, Scalable hardware trojan activation by interleaving concrete simulation and symbolic execution, in *2018 IEEE International Test Conference (ITC)* (IEEE, Piscataway, 2018), pp. 1–10
3. A. Ahmed, Y. Huang, P. Mishra, Cache reconfiguration using machine learning for vulnerability-aware energy optimization. ACM Trans. Embed. Comput. Syst. **18**(2), 1–24 (2019)
4. Alteris FlexNoC Resilience Package, www.arteris.com/flexnoc-resilience-package-functional-safety [Online]
5. D.M. Ancajas, K. Chakraborty, S. Roy, Fort-NoCs: mitigating the threat of a compromised NoC, in *2014 51st ACM/EDAC/IEEE Design Automation Conference (DAC)* (2014), pp. 1–6
6. M. Best, A. Brouwer, F. MacWilliams, A. Odlyzko, N.J.A.A. Sloane, Bounds for binary codes of length less than 25. IEEE Trans. Inf. Theory **24**(1), 81–93 (1978)
7. N. Binkert, B. Beckmann, G. Black, S.K. Reinhardt, A. Saidi, A. Basu, J. Hestness, D.R. Hower, T. Krishna, S. Sardashti, R. Sen, K. Sewell, M. Shoaib, N. Vaish, M.D. Hill, D.A. Wood, The gem5 simulator. SIGARCH Comput. Archit. News **39**(2), 1–7 (2011)
8. A.K. Biswas, Network-on-chip intellectual property protection using circular path–based fingerprinting. ACM J. Emerg. Technol. Comput. Syst. **17**(1), 1–22 (2020)
9. A.K. Biswas, S.K. Nandy, R. Narayan. Router attack toward NoC-enabled MPSoC and monitoring countermeasures against such threat. Circuits Syst. Signal Process. **34**(10), 3241–3290 (2015)
10. Bloomberg, *The Big Hack: How China Used a Tiny Chip to Infiltrate U.S. Companies*. https://www.bloomberg.com/news/features/2018-10-04/the-big-hack-how-china-used-a-tiny-chip-to-infiltrate-america-s-top-companies
11. T. Boraten, A.K. Kodi, Packet security with path sensitization for NoCs, in *2016 Design, Automation Test in Europe Conference Exhibition (DATE)* (2016), pp. 1136–1139
12. S. Charles, P. Mishra, Lightweight and trust-aware routing in NoC based SoCs, in *IEEE Computer Society Annual Symposium on VLSI (ISVLSI)* (2020)
13. S. Charles, P. Mishra, Reconfigurable network-on-chip security architecture. ACM Trans. Des. Autom. Electron. Syst. **25**(6), 1–25 (2020)
14. S. Charles, P. Mishra, Securing network-on-chip using incremental cryptography, in *IEEE Computer Society Annual Symposium on VLSI (ISVLSI)* (2020)

15. S. Charles, H. Hajimiri, P. Mishra, Proactive thermal management using memory-based computing in multicore architectures, in *International Green and Sustainable Computing Conference (IGSC)* (2018), pp. 1–8
16. S. Charles, C.A. Patil, U.Y. Ogras, P. Mishra, Exploration of memory and cluster modes in directory-based many-core CMPs, in *IEEE/ACM International Symposium on Networks-on-Chip (NOCS)* (2018), pp. 1–8
17. S. Charles, A. Ahmed, U.Y. Ogras, P. Mishra, Efficient cache reconfiguration using machine learning in NoC-based many-core CMPs. ACM Transact. Des. Autom. Electron. Syst. **24**(6), 1–23 (2019)
18. S. Charles, Y. Lyu, P. Mishra, Real-time detection and localization of DoS attacks in NoC based SoCs, in *Design Automation & Test in Europe (DATE)* (2019), pp. 1160–1165
19. S. Charles, M. Logan, P. Mishra, Lightweight anonymous routing in NoC based SoCs, in *Design Automation & Test in Europe (DATE)* (2020)
20. S. Charles, Y. Lyu, P. Mishra, Real-time detection and localization of distributed DoS attacks in NoC based SoCs. IEEE Trans. Comput. Aided Des. Integr. Circuits Syst. **39**(12), 4510– 4523 (2020)
21. S.V.R. Chittamuru, I.G. Thakkar, V. Bhat, S. Pasricha, Soteria: exploiting process variations to enhance hardware security with photonic NoC architectures, in *2018 55th ACM/ESDA/IEEE Design Automation Conference (DAC)* (2018), pp. 1–6
22. H. Deng, X. Sun, B. Wang, Y. Cao, Selective forwarding attack detection using watermark in WSNs, in *2009 ISECS International Colloquium on Computing, Communication, Control, and Management*, vol. 3 (IEEE, Piscataway, 2009), pp. 109–113
23. I. Dumer, D. Micciancio, M. Sudan, Hardness of approximating the minimum distance of a linear code. IEEE Trans. Inf. Theory **49**(1), 22–37 (2003)
24. F. Farahmandi, P. Mishra, FSM anomaly detection using formal analysis, in *2017 IEEE International Conference on Computer Design (ICCD)* (IEEE, Piscataway, 2017), pp. 313–320
25. F. Farahmandi, Y. Huang, P. Mishra, Trojan localization using symbolic algebra, in *2017 22nd Asia and South Pacific Design Automation Conference (ASPDAC)* (IEEE, Piscataway, 2017), pp. 591–597
26. F. Farahmandi, Y. Huang, P. Mishra, *System-on-Chip Security: Validation and Verification* (Springer Nature, Cham, 2019)
27. J. Frey, Q. Yu, A hardened network-on-chip design using runtime hardware trojan mitigation methods. Integr. VLSI J. **56**(C), 15–31 (2017)
28. U. Gupta, C.A. Patil, G. Bhat, P. Mishra, U.Y. Ogras, DyPo: dynamic pareto-optimal configuration selection for heterogeneous MpSoCs. ACM Trans. Embed. Comput. Syst. **16**(5s), 1–20 (2017)
29. W. Hoeffding, Probability inequalities for sums of bounded random variables, in *The Collected Works of Wassily Hoeffding* (Springer, Berlin, 1994), pp. 409–426
30. A. Houmansadr, N. Borisov, BotMosaic: collaborative network watermark for the detection of IRC-based botnets. J. Syst. Softw. **86**(3), 707–715 (2013)
31. A. Houmansadr, N. Kiyavash, N. Borisov, Rainbow: a robust and invisible non-blind watermark for network flows, in *Proceedings of the NDSS* (2009)
32. Y. Huang, P. Mishra, Vulnerability-aware energy optimization for reconfigurable caches in multitasking systems. IEEE Trans. Comput. Aided Des. Integr. Circuits Syst. **38**(5), 809–821 (2019)
33. Y. Huang, S. Bhunia, P. Mishra, MERS: statistical test generation for side-channel analysis based trojan detection, in *ACM SIGSAC Conference on Computer and Communications Security (CCS)* (2016), pp. 130–141
34. Y. Huang, S. Bhunia, P. Mishra, Scalable test generation for trojan detection using side channel analysis. IEEE Trans. Inf. Forensics Secur. **13**(11), 2746–2760 (2018)
35. M. Hussain, A. Malekpour, H. Guo, S. Parameswaran, EETD: an energy efficient design for runtime hardware trojan detection in untrusted network-on-chip, in *2018 IEEE Computer Society Annual Symposium on VLSI (ISVLSI)* (2018), pp. 345–350

36. A. Iacovazzi, Y. Elovici, Network flow watermarking: a survey. IEEE Commun. Surv. Tutor. **19**(1), 512–530 (2017)
37. L.S. Indrusiak, J. Harbin, M.J. Sepulveda, Side-channel attack resilience through route randomisation in secure real-time networks-on-chip, in *2017 12th International Symposium on Reconfigurable Communication-centric Systems-on-Chip (ReCoSoC)* (2017), pp. 1–8
38. L.S. Indrusiak, J. Harbin, C. Reinbrecht, J. Sepúlveda, Side-channel protected MPSoC through secure real-time networks-on-chip. Microprocess. Microsyst. **68**, 34–46 (2019)
39. Intel Xeon Phi Processor 7210, http://ark.intel.com/products/94033/Intel-Xeon-Phi-Processor-7210-16GB-1_30-GHz-64-core [Online]
40. H.K. Kapoor, G.B. Rao, S. Arshi, G. Trivedi, A security framework for NoC using authenticated encryption and session keys. Circuits Syst. Signal Process. **32**(6), 2605–2622 (2013)
41. S.T. King, J. Tucek, A. Cozzie, C. Grier, W. Jiang, Y. Zhou, Designing and implementing malicious hardware, in *Proceedings of the 1st Usenix Workshop on Large-Scale Exploits and Emergent Threats, LEET'08* (USENIX Association, Berkeley, 2008)
42. B. Lebiednik, S. Abadal, H. Kwon, T. Krishna, Architecting a secure wireless network-on-chip, in *2018 Twelfth IEEE/ACM International Symposium on Networks-on-Chip (NOCS)* (2018), pp. 1–8
43. Z. Ling, X. Fu, W. Jia, W. Yu, D. Xuan, J. Luo, Novel packet size-based covert channel attacks against anonymizer. IEEE Trans. Comput. **62**(12), 2411–2426 (2012)
44. Y. Lyu, P. Mishra, A survey of side-channel attacks on caches and countermeasures. J. Hardw. Syst. Secur. **2**(1), 33–50 (2018)
45. Y. Lyu, P. Mishra, Efficient test generation for trojan detection using side channel analysis, in *Design Automation & Test in Europe Conference (DATE)* (2019), pp. 408–413
46. Y. Lyu, P. Mishra, Automated test generation for trojan detection using delay-based side channel analysis, in *2020 Design, Automation & Test in Europe Conference & Exhibition (DATE)* (2020), pp. 1031–1036
47. Y. Lyu, P. Mishra, Automated trigger activation by repeated maximal clique sampling, in *Asia and South Pacific Design Automation Conference (ASPDAC)* (2020), pp. 482–487
48. Y. Lyu, P. Mishra, Scalable activation of rare triggers in hardware trojans by repeated maximal clique sampling. IEEE Trans. Comput. Aided Des. Integr. Circuits Syst. (2020). https://doi.org/10.1109/TCAD.2020.3019984
49. J.Y.V. Manoj Kumar, A.K. Swain, S. Kumar, S.R. Sahoo, K. Mahapatra, Run time mitigation of performance degradation hardware trojan attacks in network on chip, in *2018 IEEE Computer Society Annual Symposium on VLSI (ISVLSI)* (2018), pp. 738–743
50. A. May, I. Ozerov, On computing nearest neighbors with applications to decoding of binary linear codes, in *Annual International Conference on the Theory and Applications of Cryptographic Techniques* (Springer, Berlin, 2015), pp. 203–228
51. D. McGrew, J. Viega, The Galois/counter mode of operation (GCM), in *Submission to NIST Modes of Operation Process*, 20 (2004)
52. P. Mishra, S. Bhunia, M. Tehranipoor, *Hardware IP Security and Trust* (Springer, Berlin, 2017)
53. Z. Pan, P. Mishra, Automated test generation for hardware trojan detection using reinforcement learning, in *Asia and South Pacific Design Automation Conference (ASPDAC)* (2021)
54. Z. Pan, J. Sheldon, P. Mishra, Test generation using reinforcement learning for delay-based side channel analysis, in *IEEE/ACM International Conference on Computer-Aided Design (ICCAD)* (2020)
55. J. Porquet, A. Greiner, C. Schwarz, NoC-MPU: a secure architecture for flexible co-hosting on shared memory MPSoCs, in *2011 Design, Automation Test in Europe* (2011), pp. 1–4
56. N Prasad, R. Karmakar, S. Chattopadhyay, I. Chakrabarti, Runtime mitigation of illegal packet request attacks in networks-on-chip, in *2017 IEEE International Symposium on Circuits and Systems (ISCAS)* (IEEE, Piscataway, 2017), pp. 1–4
57. J.S. Rajesh, D.M. Ancajas, K.Chakraborty, S. Roy, Runtime detection of a bandwidth denial attack from a rogue network-on-chip, in *Proceedings of the 9th International Symposium on Networks-on-Chip, NOCS '15* (Association for Computing Machinery, New York, 2015)

58. V.Y. Raparti, S. Pasricha, Lightweight mitigation of hardware trojan attacks in NoC-based manycore computing, in *2019 56th ACM/IEEE Design Automation Conference (DAC)* (2019), pp. 1–6
59. C. Reinbrecht, A. Susin, L. Bossuet, J. Sepúlveda, Gossip NoC – avoiding timing side-channel attacks through traffic management, in *2016 IEEE Computer Society Annual Symposium on VLSI (ISVLSI)* (2016), pp. 601–606
60. R.M. Roth, G. Seroussi, Bounds for binary codes with narrow distance distributions. IEEE Trans. Inf. Theory **53**(8), 2760–2768 (2007)
61. A. Saeed, A. Ahmadinia, M. Just, C. Bobda, An ID and address protection unit for NoC based communication architectures, in *Proceedings of the 7th International Conference on Security of Information and Networks* (ACM, New York, 2014), p. 288
62. K. Sajeesh, H.K. Kapoor, An authenticated encryption based security framework for NoC architectures, in *2011 International Symposium on Electronic System Design* (2011), pp. 134–139
63. J. Sepúlveda, D. Flórez, G. Gogniat, Reconfigurable security architecture for disrupted protection zones in NoC-based MPSoCs, in *10th International Symposium on Reconfigurable Communication-centric Systems-on-Chip (ReCoSoC)* (IEEE, Piscataway, 2015), pp. 1–8
64. J. Sepúlveda, A. Zankl, D. Flórez, G. Sigl, Towards protected MPSoC communication for information protection against a malicious NoC. Procedia Comput. Sci. **108**, 1103–1112 (2017). International Conference on Computational Science, ICCS 2017, 12–14 June 2017, Zurich
65. J. Sepúlveda, D. Aboul-Hassan, G. Sigl, B. Becker, M. Sauer, Towards the formal verification of security properties of a network-on-chip router, in *2018 IEEE 23rd European Test Symposium (ETS)* (IEEE, Piscataway, 2018), pp. 1–6
66. K. Shuler, Majority of leading China semiconductor companies rely on arteris network-on-chip interconnect IP (2013)
67. A. Sodani, R. Gramunt, J. Corbal, H. Kim, K. Vinod, S. Chinthamani, S. Hutsell, R. Agarwal, Y. Liu, Knights landing: second-generation Intel Xeon Phi product. IEEE Micro **36**(2), 34–46 (2016)
68. A. Van Herrewege, I. Verbauwhede, Software only, extremely compact, Keccak-based secure PRNG on ARM Cortex-M, in *Proceedings of the 51st Annual Design Automation Conference (DAC)* (IEEE, Piscataway, 2014), pp. 1–6
69. Y. Wang, G.E. Suh, Efficient timing channel protection for on-chip networks, in *2012 IEEE/ACM Sixth International Symposium on Networks-on-Chip* (2012), pp. 142–151
70. X. Wang, D.S. Reeves, P. Ning, F. Feng, Robust network-based attack attribution through probabilistic watermarking of packet flows. Technical report, North Carolina State University, Dept. of Computer Science, 2005
71. W. Wang, P. Mishra, S. Ranka, *Dynamic Reconfiguration in Real-Time Systems* (Springer, Berlin, 2012)
72. S.C. Woo, M. Ohara, E. Torrie, J.P. Singh, A. Gupta, The splash-2 programs: characterization and methodological considerations, in *Proceedings 22nd Annual International Symposium on Computer Architecture* (1995), pp. 24–36
73. Q. Yu, J. Frey, Exploiting error control approaches for hardware trojans on network-on-chip links, in *2013 IEEE International Symposium on Defect and Fault Tolerance in VLSI and Nanotechnology Systems (DFTS)* (IEEE, Piscataway, 2013), pp. 266–271
74. A. Zand, G. Vigna, R. Kemmerer, C. Kruegel, Rippler: delay injection for service dependency detection, in *IEEE INFOCOM 2014-IEEE Conference on Computer Communications* (IEEE, Piscataway, 2014), pp. 2157–2165

Chapter 10
Network-on-Chip Attack Detection using Machine Learning

Chamika Sudusinghe, Subodha Charles, and Prabhat Mishra

10.1 Introduction

Network-on-chip (NoC) is widely used for on-chip communication in modern system-on-chips (SoC). NoC has allowed computer architects to fully utilize the computational power in an SoC by facilitating low-latency and high-throughput communication between intellectual property (IP) cores in a many-core SoC. As a result, NoC has become a critical component in state-of-the-art SoC designs [56, 59]. The drastic increase in SoC complexity has led to a significant increase in SoC design and validation complexity [3, 20, 22, 24, 37, 41–44]. With the increased complexity of SoCs, manufacturers have favored IP licensing and outsourcing where only a subset of IPs are manufactured in house and the rest is sourced from third-party vendors.

There are multiple avenues to introduce malicious implants (e.g., hardware Trojans) in designs during the long supply chain, such as by an untrusted CAD tool, a rouge designer or at the foundry via reverse engineering [45]. The Trojans are carefully crafted such that they can evade design-time detection by requiring a specific, usually very rare, trigger condition (such as time, input sequence, traffic pattern, thermal conditions, etc.) to be met before behaving maliciously [5]. A recent occurrence of a security breach caused by a hardware Trojan (HT) implanted by a third-party vendor impacted several blue-chip companies including Apple and Amazon [8]. Recent efforts try to combine the advantages of logic testing and side-channel analysis for effective Trojan detection in integrated circuits [33–36, 38, 48].

C. Sudusinghe (✉) · S. Charles
University of Moratuwa, Colombo, Sri Lanka
e-mail: chamika.sudusinghe@ieee.org; s.charles@ieee.org

P. Mishra
University of Florida, Gainesville, FL, USA
e-mail: prabhat@ufl.edu

P. Mishra, S. Charles (eds.), *Network-on-Chip Security and Privacy*,
https://doi.org/10.1007/978-3-030-69131-8_10

In spite of such extensive efforts, it is not feasible to capture all security vulnerabilities using security validation tools during design time [2, 23, 25, 26, 39, 40, 45, 46].

The NoC is at an elevated risk of being vulnerable to hardware attacks due to several reasons: (1) NoC interconnects IPs manufactured in house and/or sourced from trusted vendors (secure IPs) together with IPs from potentially untrusted vendors (non-secure IPs) allowing Trojan-infected malicious IPs (MIPs) to utilize NoC to launch attacks, (2) the distributed nature of NoC makes it easier to replicate an attack, and (3) the complexity of NoC design allows Trojans to hide without being detected. These vulnerabilities have motivated both industry and academic researchers to develop countermeasures to secure NoC-based SoCs [11–14, 17–19]. Existing research on NoC security has explored two orthogonal directions: (1) the NoC is assumed to be secure and security countermeasures are built utilizing secure NoC to prevent attacks on trusted IPs from other malicious IPs [17, 18], and (2) trusted IPs need to be protected from malicious IPs as well as Trojan-infected NoC [9, 19].

There are a wide variety of threats from MIPs such as eavesdropping attacks, data integrity attacks, denial-of-service (DoS) attacks, etc. In this chapter, the focus is on securing the SoC from DoS attacks using machine learning assuming a secure and trustworthy NoC architecture. The primary objective of a DoS attack is to prevent legitimate users from accessing services and information. In the context of NoC, MIPs sending unnecessary requests to IPs can delay legitimate requests leading to delay of service (e.g., deadline violations in real-time systems) or denial-of-service (e.g., temporary or permanent service failure). Such “flooding” type of DoS attacks can also cause congestion in the network, further degrading performance and energy efficiency [17, 18].

Some of the existing work on mitigating flooding type of DoS attacks explored traffic latency comparison [52] and packet arrival monitoring [17, 18]. These approaches made an unrealistic assumption, highly predictable NoC traffic patterns, which allowed the construction of linear statistical bounds to detect DoS attacks. Unfortunately, this assumption does not hold during many realistic scenarios that include task migration, task preemption, changing application characteristics due to major input variations, etc.

In this chapter, we discuss the feasibility of using machine learning (ML) for DoS attack detection as a potential solution to address runtime variations in NoC traffic. Major focus areas are as follows:

- We describe an ML-based DoS attack detection method that trains ML models during design time and uses the trained models to classify network traffic behavior as normal or attack during runtime to detect flooding type of DoS attacks.
- We outline features that can be extracted from NoC traffic as well as engineered features, and experimental evaluation of the most suitable features.
- We present a comprehensive exploration of 12 different ML models to select the best fit for the given architecture and threat models.

- We analyze the accuracy in DoS attack detection across different NoC traffic patterns caused by various applications and application mappings.
- This approach can detect DoS attacks in real-time with detection times comparable to previous work [17, 18] without requiring highly predictable traffic patterns.

The rest of the chapter is organized as follows. Section 10.2 outlines the threat model and related efforts. Section 10.3 motivates the need for ML-based detection of DoS attacks. Section 10.4 describes the ML-based DoS attack detection methodology. Section 10.5 presents the experimental results. Finally, Sect. 10.6 concludes the chapter.

10.2 Threat Model and Related Work

In this section, we first outline the threat model. Next, we provide a brief survey of related efforts.

10.2.1 Threat Model

Figure 10.1 shows the architecture model considered that includes a 4 × 4 mesh NoC connecting 16 IP cores. When a memory request (e.g., memory LOAD or STORE instruction) is initiated by a core during application execution, in case of a cache miss, a memory request is injected into the NoC in the form of NoC packets. Typically, the packets are further broken down into smaller units called *flits* to facilitate flow control mechanisms. The flits are routed in the appropriate virtual network (vnet) that matches the cache coherence request type, via routers and links. When the flits arrive at the memory controller, the memory fetch is initiated and once the operation is completed, the response is routed back to the original requestor. Intel's Knights Landing (KNL) architecture [56] and Tilera's Tile64 architecture both utilize their NoCs for similar memory transactions [59].

DoS attacks can happen from MIPs intentionally degrading SoC performance by flooding the NoC with packets. MIPs can target a component that is critical to SoC performance, such as a memory controller that provides the interface to off-chip memory, and inject unnecessary requests [18]. As a result, the legitimate requests can experience severe delays leading to deadline violations, performance degradation, and reduced energy efficiency. While there are a wide variety of techniques to improve energy efficiency in NoC-based SoCs [4, 15, 16, 31, 32, 57], they are not suitable in this scenario. Figure 10.1 shows a MIP at node 1 that targets its victim at node 7 and injects additional packets. The traffic rate in routers along the routing path is increased causing NoC congestion, which leads to performance degradation and reduced energy efficiency. Since the victim receives a lot more requests than it is designed to handle, responses are delayed and that can lead to

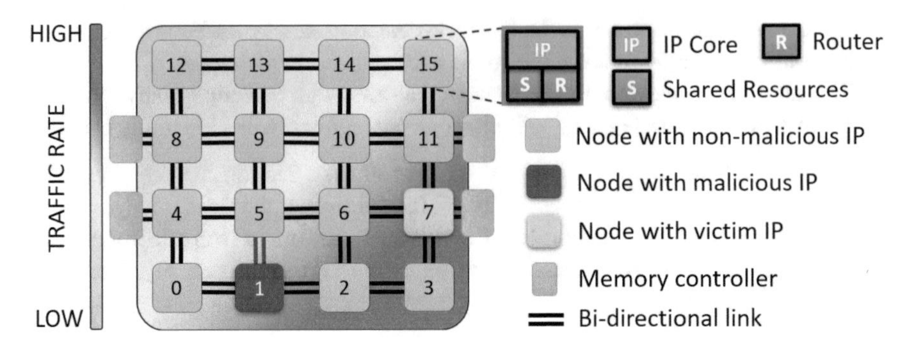

Fig. 10.1 Example DoS attack from a malicious IP to a victim IP in a mesh NoC setup. The thermal map shows high traffic near the victim IP

violation of task deadlines. Violation of real-time requirements can be catastrophic for safety-critical applications. A similar threat model was also used by previous work that explored DoS attacks in NoC-based SoCs [17, 18, 52].

10.2.2 Related Work

In this section, we describe the related efforts in three broad categories: DoS attacks in computer networks, DoS attacks in on-chip networks, and machine learning for network security.

10.2.2.1 DoS Attacks in Computer Networks

Before discussing DoS attacks in the NoC context, we outline two well-known DoS attack scenarios ("Bashlite" and "Mirai") from the computer networks domain [47, 49]. *Bashlite* (also known as Gafgyt) infects many devices to form a botnet. If a botnet is formed successfully, the attacker may remotely orchestrate DDoS (Distributed Denial-of-Service) attack and download other malware by sending commands to infected devices (also known as "bots"). This malware utilizes a client-server model in which the attacker's device functions as the command-and-control (CnC) server, and the infected devices function as clients. The client bots constantly poll for server commands. Large botnets can overwhelm target servers by simultaneously making requests when the attacker sends the command. *Mirai* is a more sophisticated version of Bashlite as shown in Fig. 10.2. Mirai includes a wider variety of commands, and can infect a wider variety of IoT devices. Because Mirai is compatible with more devices, it has the potential to build a larger botnet. The number of devices included in the botnet improves the botnet's ability to overwhelm

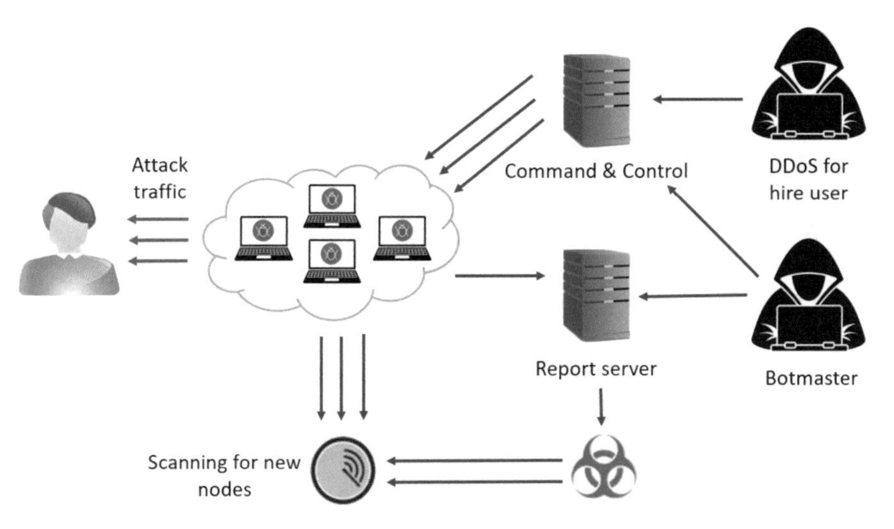

Fig. 10.2 Overview of the Mirai botnet operation [55]

target servers. The wider range of vulnerable devices also improves Mirai malware's ability to steal information from these devices.

10.2.2.2 DoS Attacks in NoC-based SoCs

DoS attack threat models in NoCs can be divided into three broad categories: (1) flooding [17, 18, 27, 50, 52], (2) packet corruption [10, 28], and (3) traffic flow manipulation [7, 53]. The focus of this chapter is on flooding type of DoS attacks. Existing efforts explored defenses against flooding type of DoS attacks by traffic flow monitoring [17, 18, 27, 52] and developing additional validation checks [50] as countermeasures. The traffic latency comparison method in [52] proposed to inject additional packets to the network to detect congested paths. However, it can further congest the network [52] due to the additional packets. Fiorin et al. introduced a countermeasure against DoS attacks that has an architecture similar to the method discussed in this chapter [27]. Their method is fundamentally different from this method since they monitor the bandwidth considering the data loaded/stored by an initiator from/to a specific memory block or range of addresses. Charles et al. proposed to statically profile the normal behavior of the SoC and detect DoS attacks during runtime [17, 18]. In their work, each router statically profiled NoC traffic behavior based on packet arrivals at routers and used that as an upper bound to detect attacks. While such methods are efficient when the applications and the application mappings are fixed, they are not suitable in many real-world scenarios where variations can alter the NoC traffic behavior.

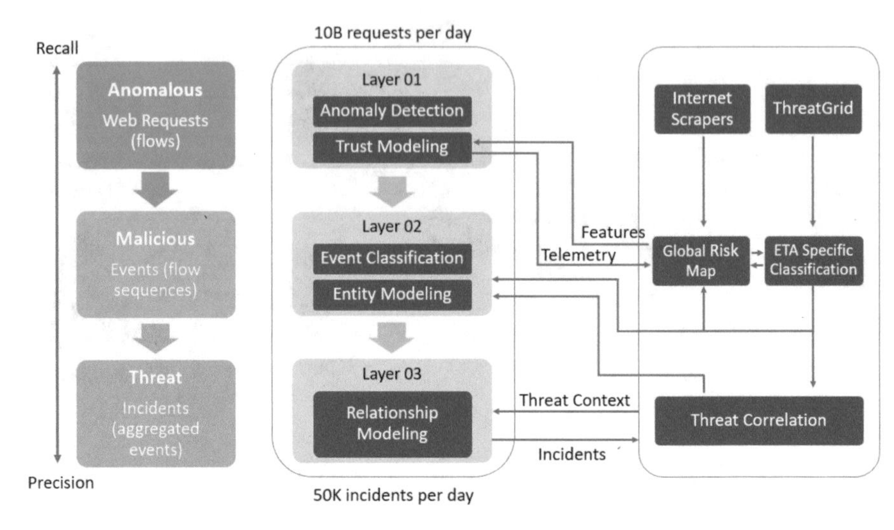

Fig. 10.3 Machine learning based network traffic analysis in Stealthwatch [29]

10.2.2.3 Securing Networks using Machine Learning

Machine learning (ML) has been widely adopted in various domains for efficient data processing and fast decision making. For example, ML can analyze encrypted HTTP traffic to differentiate between malicious and benign execution [29, 51, 54]. Cisco encrypted traffic analytics [29] and IBM QRadar security intelligence [30] are two state-of-the-art network security countermeasures developed by the industry. Figure 10.3 shows an overview of Cisco Stealthwatch that utilizes machine learning for encrypted traffic analysis to correlate traffic with global threat behaviors to automatically identify infected hosts, command-and-control communication, and suspicious traffic [29]. The usage of ML models in network security ranges from linear models such as logistic regression and naive Bayes to non-linear models such as decision trees, support vector machines with kernels, MLP neural networks, random forests and gradient boosting. Shekhawat et al. [54] showed that XGBoost, an algorithm based on gradient boosting, is able to classify encrypted traffic as malicious or benign with an accuracy of 99.15%. In the domain of NoC security, Wang et al. [58] proposed TSA-NoC, a framework based on artificial neural networks for runtime HT detection and an adaptive routing design based on deep reinforcement learning for HT mitigation. They have shown that TSA-NoC can detect HTs with an accuracy of 97% with an improvement of 70% in energy efficiency and 29% reduced network latency. While ML has shown promising results in computer network security in the past decade, there are limited efforts for utilizing ML to secure NoC-based SoCs.

10.3 Motivation

To evaluate the potential of using ML to detect DoS attacks in NoC-based SoCs, we simulated both malicious and benign programs using the gem5 cycle-accurate full-system simulator [6] and extracted features from NoC traffic. A 4 × 4 mesh NoC was modeled using the GARNET2.0 [1] interconnection network model. The mesh consisted of 16 IP cores and 4 memory controllers as shown in Fig. 10.4. Each router in the middle of the mesh is connected to four other routers in the four directions and to an IP core. Figure 10.4 shows the architecture model used for both normal and attack scenarios. During normal execution, two processor cores ran two instances of the FFT benchmark from the SPLASH-2 benchmark suite [60]. For the attack scenario, in addition to the two applications, a MIP was modeled to inject packets at the four memory controllers uniformly, increasing the overall network traffic by 50%. More details about the experimental setup are given in Sect. 10.5.1.

We extracted NoC traffic features and labeled them based on normal (target label = 0) and attack (target label = 1) scenarios. Figure 10.5 shows the correlation matrix of features extracted from NoC traffic. Each feature is denoted by a *feature ID* instead of the feature name. A detailed description of the features is given in Sect. 10.4. The highlighted column shows the correlation of each feature to the target label (feature ID—V). The values shown in Fig. 10.5 are the pairwise *Pearson Correlation Coefficients* (PCC) of all the features. PCC, calculated as:

$$\rho_{X,Y} = \frac{cov(X, Y)}{\sigma_X \cdot \sigma_Y}$$

gives a measure of the linear correlation between a pair of random variables (X, Y). PCC value ranges from -1 to 1. $\rho_{X,Y} = 1$ (light color shades) implies that X and Y

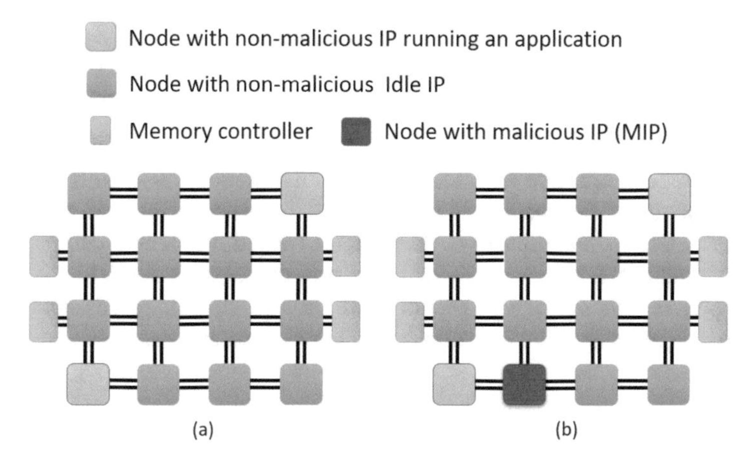

Fig. 10.4 Architecture models used to extract NoC traffic features. (**a**) Normal scenario. (**b**) Attack scenario

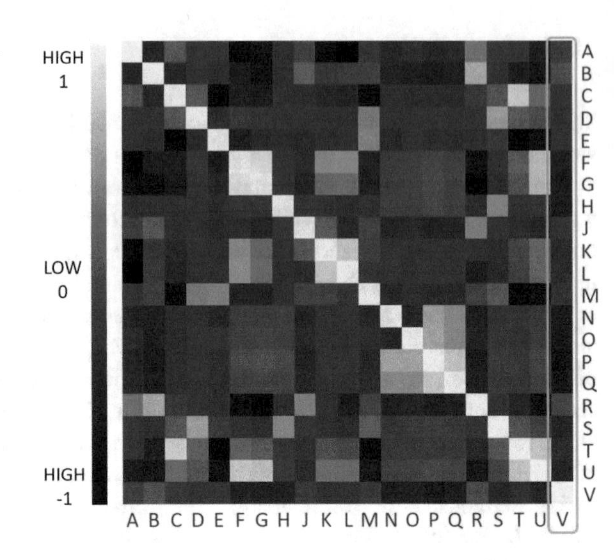

Fig. 10.5 Correlation matrix of extracted NoC traffic features

have a linear relationship where Y increases as X increases. $\rho_{X,Y} = -1$ (dark color shades) also implies a linear relationship, but in this case, Y decreases as X increases. $\rho_{X,Y} = 0$ implies that there is no linear correlation between the variables.

We can observe the following from Fig. 10.5:

- Most features are not perfectly correlated to each other and falls in the low to medium (0 ± 0.5) correlation range. In other words, the dataset does not exhibit "Multicollinearity". Multicollinearity, if existed, can severely affect performance of ML models outside of the original (training) dataset. Therefore, NoC features have the potential to train accurate classifiers.
- Since column V shows values in the range of (0 ± 0.3), we can conclude that the target label is not linearly correlated with the features. Therefore, a linear model, such as linear regression or naive bayes classifier, to differentiate normal and attack scenarios are unlikely to yield good results. Exploration of ML techniques that capture non-linear behavior, such as neural networks, decision trees or gradient boosting, is required.

These observations give us evidence that if an ML model is to be trained based on these features, it has the ability to distinguish between normal and attack traffic without causing multicollinearity. Therefore, a trained ML model can potentially detect DoS attacks during runtime irrespective of the location of the MIP(s). Based on this premise, in subsequent sections, we present the ML-based runtime DoS attack detection mechanism and empirically validate the approach.

10.4 NoC Attack Detection Using Machine Learning

In order to achieve high accuracy in detecting DoS attacks in the presence of runtime variations of NoC traffic, we have developed an efficient mechanism for ML-based DoS attack detection. An overview of the approach is shown in Fig. 10.6. During design time, NoC traffic is statically analyzed to gather the dataset that is used to train the ML models. Both normal and attack scenarios are emulated during this phase using a few known application mappings. The trained models are stored in a dedicated IP denoted as the *Security Engine* (SE). During runtime, NoC traffic data is gathered at each router using probes attached to routers and the collected data is sent to the SE using a separate physical *Service NoC*. The models at the SE use data collected within a predefined time window to make inferences about the condition of the NoC. Section 10.5 demonstrates that the ML-based approach is capable of classifying data as normal or attack, irrespective of the locations of cores running the applications and the locations of MIP(s).

The ML-based DoS attack detection mechanism relies on the following features of the architecture model.

- Probes attached to routers can gather data from NoC packets with minor performance and power overhead.
- The SoC architecture comprises two physical NoCs: (1) a *Data NoC* that is used to communicate between IPs for application execution, and (2) a *Service NoC* which transfers data collected from probes to the SE.

The remainder of this section is organized as follows. Section 10.4.1 presents the NoC traffic features and the ML models used to make inferences. Section 10.4.2 discusses the hardware implementation to have probes connected to routers that gather data and send to the SE via the Service NoC.

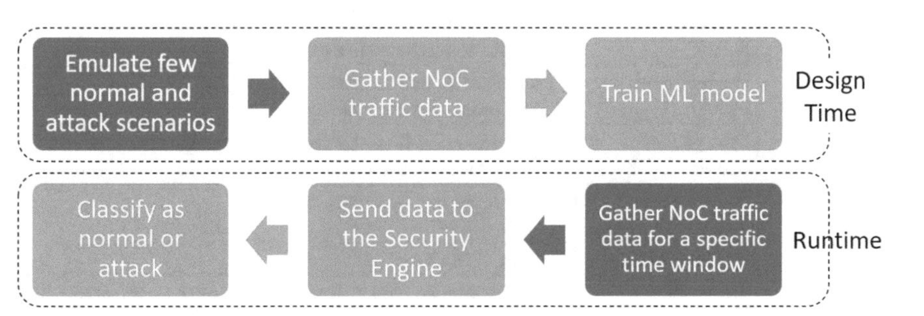

Fig. 10.6 Major steps of the ML-based DoS attack detection mechanism

10.4.1 Machine Learning Model

As outlined in Sect. 10.2.1, NoC packets/flits in the architecture model correspond to memory requests/responses between IPs running the applications and the memory controllers. Information is extracted when flits are transferred through routers. The features consist of data extracted from NoC packets as well as engineered features (marked with the symbol † in Table 10.1) created using the extracted data. A complete list of NoC traffic features used in the exploration is shown in Table 10.1. However, as elaborated in Sect. 10.5.3, some features are experimentally eliminated

Table 10.1 NoC traffic features used in the ML model

Feature ID	Feature name	Feature description
A	Outport*	Port used by the flit to exit the router (0-local,1-north,2-east,3-south,4-west)
B	Inport	Port used by the flit to enter the router (0-local,1-north,2-east,3-south,4-west)
C	cc type	Cache coherence type of the packet corresponding to the flit
D	Flit id	Identifier used to denote each flit of a packet
E	Flit type	Type of flit (head, tail, body)
F	vnet	Virtual network used by the flit
G	vc*	Virtual channel used by the flit
H	Traversal id*†	Identifier used to group all packet transfers related to one NoC traversal
J	Hop count*†	Number of hops from the source to the destination
K	Current hop†	Number of hops from the source to the current router
L	Hop percentage†	Ratio between the current hop and the hop count
M	Enqueue time*	Time spent inside the router by the flit
N	Packet count decr.*†	Cumulative no. of flit arrivals within time window τ (decremented as packets arrive)
O	Packet count incr.*†	Cumulative no. of flit arrivals within time window τ (incremented as packets arrive)
P	Max packet count*†	Maximum no. of flits transferred through the router within a given time window τ
Q	Packet count index*†	Packet count incr $\times$ packet count decr
R	Port index†	Outport $\times$ inport
S	Traversal index*†	Cache coherence type $\times$ flit id $\times$ flit type $\times$ traversal id
T	cc vnet index†	Cache coherence type $\times$ vnet
U	vnet vc cc index†	Cache coherence vnet index $\times$ vc

based on feature importance[1] in an attempt to find the optimum trade-off between the least number of features and the highest model accuracy. Feature IDs of the selected features, when running the final model, are marked with a star (*) in Table 10.1.

Gradient Boosting, a powerful technique to perform supervised ML classification, is used to classify normal and attack scenarios. It is an ensemble learner that creates the final model based on a collection of weak predictive models, decision trees in most instances, and that results in better overall prediction capabilities due to iterative learning from each model. The key concept of the algorithm is to create new base-learners having a maximum correlation with the negative gradient of the loss function of the entire ensemble. Weaker predictive models in the ensemble are trained gradually, additively, and sequentially, and their shortcomings are identified by the use of gradients in the loss function which indicates the acceptability of the model's coefficients at fitting the underlying data. The decision to use gradient boosting for the classification was made experimentally as outlined in Sect. 10.5.2.

10.4.1.1 Training the ML model

The ML model is trained statically, during design time. A few application mapping scenarios are chosen to train the model that includes both normal execution and attack scenarios. A list of all training and testing configurations is outlined in Sect. 10.5.1. Network traces are collected during application execution at each router. When flits pass through the routers, a feature vector is constructed including the selected features for each flit. Selected features are transformed using *MinMaxScaler* to fit into the range of 0–1, without distorting the shape of the original features. Transformed features are then used to tune the hyperparameters of the model using *Bayesian Optimization*, which outputs the best-optimized list of parameters while learning from previous iterations in each iteration. This process is repeated for all 16 routers separately to train 16 models, one per router.

10.4.1.2 Attack Detection

During runtime, probes attached to the routers gather data and send to the SE for evaluation. The SE aggregates data and constructs feature vectors corresponding to each router, following a process similar to that of during model training. Let $\mathcal{M}_i$ correspond to the model trained for router r_i using gradient boosting. Feature vectors that fall within a predefined time window τ_j is then used as input to each trained model, which gives a probability of an attack as the output. If $\mathcal{V}_{i,j}$ denotes the set of feature vectors constructed at r_i for τ_j, the probability of an attack is denoted by

[1] Feature importance gives a score to indicate how important a feature is in the decision making process of an ML model. In a trained model, the more a feature contributes to key decisions, the higher its relative importance.

$p_{i,j}$, where $p_{i,j} \leftarrow \mathcal{M}_i(\mathcal{V}_{i,j})$. The probability is calculated as the portion of feature vectors labeled as "attack" during τ_j. If all feature vectors are classified as "attack" by the model, the probability is 1. If all feature vectors are classified as "normal", the probability is 0. The overall probability of an attack for the time window τ_j is calculated after *pooling* all probabilities as;

$$\mathcal{P}_j = \frac{\sum_{\forall i}(p_{i,j} \cdot |\mathcal{V}_{i,j}|)}{\sum_{\forall i} |\mathcal{V}_{i,j}|} \tag{10.1}$$

The overall probability for the time window τ_j ($\mathcal{P}_j$) is a weighted average of probabilities from each model where the weights correspond to the number of flits transferred through each router within the given time window. If $\mathcal{P}_j$ is greater than a predefined threshold λ, an attack is flagged. This process is repeated for every τ_j during SoC operation to detect attacks that can be potentially initiated at any point in time.

Weights based on the number of flits indicate that when a model makes a decision based on a lot of data points, it can be trusted to give a more accurate result. The choice was motivated by the fact that there are no assumptions made about the placement of the secure and non-secure IPs. However, if more information is available, the weighted average can be adjusted so that some models contribute more to the final decision. For example, if the locations of the non-secure (potentially malicious) IPs are known, the probabilities of models corresponding to those routers can be given more weight and it would result in a better overall performance in distinguishing normal traffic from an attack scenario. How to combine different probabilities to arrive at a single conclusion under various assumptions is well studied in the area of *Opinion Pooling*, which is a part of probability theory, and can be used in the approach based on the assumptions made.

It is important to note that all the features that have been used in the method can be extracted from the packet header or by counting flits or as a combination of header and count information. Observing the packet payload (e.g., memory data block in case of a memory data fetch packet) is not required. Therefore, this approach can be used together with other NoC security mechanisms such as encryption and authentication.

10.4.2 Implementation of Hardware Components

This approach relies on collecting features at routers using probes and sending the data via a separate physical NoC (Service NoC) to the SE to make inferences. This section describes the implementation details for these hardware components.

10.4.2.1 Multiple Physical NoCs

Two main types of packets are identified to be transferred through the NoC to facilitate the ML-based DoS attack detection method: (1) packets related to application execution as introduced in Sect. 10.2.1, and (2) packets related to extracted NoC features transferred from probes at routers to the SE. Instead of using different virtual networks to carry the different packets types, it is proposed to use two separate physical NoCs (Data NoC and Service NoC) to carry the two main types of packets. The choice is motivated by state-of-the-art commercial NoC-based SoC architectures that follow the same practise of carrying different types of packets over multiple physical NoCs [56, 59].

There is a trade-off between area and performance when considering one versus multiple NoCs. When different packet types are facilitated through the same NoC, header fields must be added to distinguish between the packets' types. Furthermore, the buffer space must be shared between virtual networks. This can lead to performance degradation, especially when scaling to many-core processors. On the other hand, separate physical NoCs contribute to the area overhead. However, due to advances in manufacturing technologies, additional wiring to facilitate the NoCs incurs minimal overhead as long as the wires stay on-chip. On-chip buffer area has become the more scarce resource. If virtual networks are used, the increased buffer space due to sharing and the logic complexity to handle virtual networks can closely resemble to having a separate physical NoC. Intel and Tilera opted for separate physical NoCs for the same reasons. Yoon et al's work provides a comprehensive trade-off analysis [61]. When the analysis from [61] is applied to fit the parameters in this work, the power and area overhead of having two physical NoCs versus one NoC are 7% and 6%, respectively.

10.4.2.2 Probes at Routers and Security Engine

Hardware implementations for probes collecting data at routers and the SE have been explored in several prior work [21, 27]. Fiorin et al. [27] utilized probes attached to the network interfaces to collect data and send to a central processing element to detect DoS attacks. The runtime NoC monitoring and debugging framework proposed in [21] also used a similar setup where event related information is gathered at NoC routers and sent to a central unit for processing. This security mechanism is built using a similar architecture. In the framework, the probes are event triggered on flit arrival. The probes consist of a sniffer, an event generator and an interface to the Service NoC. The sniffer extracts the features from flits and sends to the event generator to create the timestamped messages. The network interface then packetizes the messages and sends to the SE via the Service NoC. The SE completes feature engineering and combines the engineered and extracted features to construct the final feature vectors. Previous work performed detailed overhead analysis and reported minimal area overhead, for example, the probes consumed 0.05 mm^2 compared to a 0.26 mm^2 router area when synthesized with 0.13 micron technology [21]. The overhead analysis is consistent with the analysis done in [21].

Table 10.2 Train and test configurations

	Train		Test
Iteration ID (IID)	Normal	Attack	Attack
1	N-0-15	N-0-15-A-1	N-0-15-A-7
			N-0-15-A-11
			N-0-15-A-12
2	N-0-15	N-0-15-A-1	N-0-15-A-7
	N-0-15	N-0-15-A-11	N-0-15-A-12
	N-0-9	N-0-9-A-1	N-0-9-A-7
	N-0-9	N-0-9-A-11	N-0-9-A-12
	N-0-6	N-0-6-A-1	N-0-6-A-7
	N-0-6	N-0-6-A-11	N-0-6-A-12
	N-0-4	N-0-4-A-1	N-0-4-A-7
	N-0-4	N-0-4-A-11	N-0-4-A-12
3	N-0-6-9-15	N-0-6-9-15-A-1-11	N-0-6-9-15-A-1-7
			N-0-6-9-15-A-7-11
			N-0-6-9-15-A-11-12
			N-0-6-9-15-A-7-12

10.5 Experiments

In this section, we evaluate the effectiveness of ML-based DoS attack detection. First, we describe the experimental setup (Sect. 10.5.1). Next, we explore several machine learning models to identify the best performing one and justify the choice of gradient boosting (Sect. 10.5.2). Then, we rank the feature importance according to the selected model and eliminate low priority features in an attempt to find the optimum trade-off between the number of features and model accuracy (Sect. 10.5.3). Finally, we show how the ML-based DoS attack detection mechanism performs across several training and testing configurations by exploring model accuracy for all the test cases in Table 10.2 (Sect. 10.5.4).

10.5.1 Experimental Setup

Following the realistic architecture model proposed in [14], the 4×4 mesh NoC was modeled using the "GARNET2.0" framework [1] that is integrated with the gem5 [6] cycle-accurate full-system simulator. The NoC model was implemented using X-Y routing with wormhole switching, 3-stage router pipeline (buffer write, route compute + virtual channel allocation + switch allocation, and link traversal) and 4 virtual channel buffers per input port. Each IP was modeled as a processor core executing a given task at 1 GHz with a private L1 cache. Processor cores used the NoC for memory operations as outlined in Sect. 10.2.1. The four memory controllers attached to four boundary nodes of the NoC provided the interface to off-chip memory. The address space was shared equally between the memory controllers. Four benchmarks from the SPLASH-2 benchmark suite [60] (FFT, LU, FMM, RADIX) were used as application instances. The same set of benchmarks has been used in [17, 18] that explored DoS attacks in NoC-based SoCs. First, the results using the FFT benchmark is presented and in Sect. 10.5.4, it is shown that similar trends can be observed from other benchmarks as well.

During normal operation, n IPs out of the 16 IPs in the 4×4 mesh, were chosen at random to run an instance of the benchmark (active IPs). To model the DoS attack scenario, an IP that did not run an instance of the benchmark injects memory request packets to the four memory controllers increasing the overall network traffic by 50%. Figure 10.1 shows one configuration of the random active, idle, and malicious IP placement where $n = 1$. A complete set of training and testing configurations are listed in Table 10.2. Iteration ID (IID) 1 indicates that the model has been trained with two datasets: (1) normal execution scenario with applications running on IPs 0 and 15 (N-0-15), and (2) attack scenario with an attacker at IP 1 launching a DoS attack while applications are running on IPs 0 and 15 (N-0-15-A-1). The trained model has been tested with three attack scenarios: (1) N-0-15-A-7, (2) N-0-15-A-11, and (3) N-0-15-A-12. The IP numbers correspond to the node numbers given in Fig. 10.1.

10.5.2 Machine Learning Model Comparison

To identify which ML model performs the best for the given architecture and threat models, a performance comparison was made between 12 ML models—Naive Bayes Classifier (NBC), Logistic Regression (LRN), 2-Layer Neural Network (2NN), 3-Layer Neural Network (3NN), 4-Layer Neural Network (4NN), 5-Layer Neural Network (5NN), 6-Layer Neural Network (6NN), K-Neighbors Classifier (KNN), LightGBM Classifier (LGB), Decision Tree Classifier (DCT), Random Forest Classifier (RFC), and XGBoost Classifier (XGB). Each model was trained using the training dataset of IID 2. Figure 10.7 shows training accuracy and validation accuracy measured using an 80:20 training:validation split from the dataset at router 0 (r_0). The model comparison results at other routers manifested a similar trend (omitted from Fig. 10.7 for clarity). It can be observed that non-linear ML models perform better than linear models with XGB showing the best results. XGBoost is an algorithm based on gradient boosting machines.

To evaluate the selected XGB model further, we use cross validation, which is a resampling process used to evaluate the performance of a trained ML model. We use StratifiedKFold cross validation method since it gives a better representation over

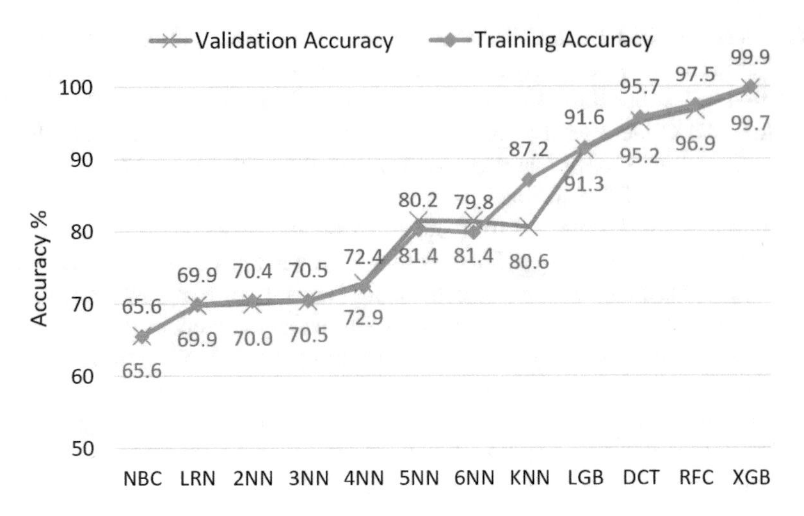

Fig. 10.7 ML model performance comparison using IID 2 training dataset

fold	1	2	3	4	5	6	7	8	9	10
accuracy %	96.84	96.86	96.84	96.88	96.80	96.92	96.84	96.54	96.71	96.90

Fig. 10.8 Validation results of the trained XGBoost model using StratifiedKFold cross validation

the entire dataset.[2]Results for 10 folds of StratifiedKFold cross validation is shown in Fig. 10.8. *The results generated by cross validation confirm that the model is less biased, performing well in unseen data and not overfitting. Since the exploration indicated that XGB performs best in the given scenario, XGB is selected as the ML model for the DoS attack detection method.*

10.5.3 Feature Importance

While using more features can certainly increase model accuracy, extracting redundant features from NoC traffic can lead to unnecessary performance and power overhead. Therefore, features that show the least importance for the decision making process of the ML model-XGB are eliminated. Figure 10.9 shows the feature importance rank of each feature. Since each router runs a model trained from the data extracted at that particular router, the feature importance rank slightly

[2] *KFold* cross validation shuffles the dataset and splits it into k subsets, then trains on $k - 1$ and evaluates on the other set iteratively. In contrast, *StratifiedKFold* cross validation shuffles the dataset and splits it into k subsets by class and uses a subset from each class in the test set emulating a representation of the entire dataset in each fold for both training and validation.

	Router															
Feature	r_0	r_1	r_2	r_3	r_4	r_5	r_6	r_7	r_8	r_9	r_{10}	r_{11}	r_{12}	r_{13}	r_{14}	r_{15}
outport	12	8	8	13	11	9	8	10	11	9	9	10	14	9	9	14
inport	14	12	11	12	10	11	10	14	13	12	11	12	15	11	10	11
cc type	13	15	15	14	15	13	12	16	15	13	13	15	13	14	13	13
flit id	18	18	18	18	20	20	20	19	19	20	20	20	18	18	18	19
flit type	17	17	16	16	19	18	18	18	18	19	18	19	17	17	17	16
vnet	16	19	20	20	18	19	19	20	20	18	19	18	20	20	20	20
vc	10	13	14	10	9	17	15	8	8	16	15	8	10	15	15	10
traversal id	1	1	1	1	1	1	1	1	1	1	1	1	1	1	1	1
hop count	6	7	7	7	7	7	7	7	7	7	7	7	7	7	7	6
current hop	9	16	17	17	14	16	16	12	10	14	16	11	12	16	16	12
hop percentage	15	9	10	11	16	10	11	13	14	10	12	17	11	10	12	15
enqueue time	8	10	9	9	8	8	9	9	9	8	8	9	8	8	8	8
packet count decr	3	4	4	5	4	5	4	4	4	4	4	4	5	5	5	5
packet count incr	5	5	6	6	5	6	6	6	5	6	6	5	6	6	6	7
max packet count	2	2	2	2	2	2	2	2	2	2	2	2	2	2	2	2
packet count index	4	3	3	4	3	3	3	3	3	3	3	3	4	4	4	4
port index	20	14	13	15	13	12	14	11	12	11	10	14	16	12	11	18
traversal index	7	6	5	3	6	4	5	5	6	5	5	6	3	3	3	3
cc vnet index	19	20	19	19	17	15	17	17	17	17	17	16	19	19	19	17
vnet vc cc index	11	11	12	8	12	14	13	15	16	15	14	13	9	13	14	9

Fig. 10.9 Feature importance rank for each feature at each router for IID 2 dataset with least important features highlighted

changes from router to router. However, the overall trend remains consistent where the highlighted features are the least used. Therefore, for the rest of the exploration, the highlighted features are eliminated when training and testing the accuracy of the DoS attack detection mechanism.

10.5.4 DoS Attack Detection Accuracy

With the selected model and features, in this section, the accuracy of the DoS attack detection method is evaluated. As outlined in Sect. 10.4, each model outputs the attack probability independently for a given time window τ_j. The overall attack probability during τ_j ($\mathcal{P}_j$) is calculated according to Eq. 10.1. Figures 10.10 and 10.11 show excerpts from results generated during an attack (IID 2 and test case N-0-15-A-12) and a normal (IID 2 and test case N-0-15) scenario, respectively. The threshold for inferring attacks from $\mathcal{P}_j$ is set to 0.5 ($\lambda = 0.5$) since an attack scenario should give probabilities close to 1, whereas in a normal scenario, the probabilities should be close to 0. Columns "r_0" through "r_{15}" in Figs. 10.10 and 10.11 show the probabilities outputted by models corresponding to each router. Column "$\mathcal{P}_j$"

τ_j	Attack Probability																P_j	Status
	r_0	r_1	r_2	r_3	r_4	r_5	r_6	r_7	r_8	r_9	r_{10}	r_{11}	r_{12}	r_{13}	r_{14}	r_{15}		
τ_1	1	1	1	1	0	1	1	1	1	1	1	1	1	1	1	1	1	ATTACK
τ_2	1	1	1	0.3	1	1	1	1	1	1	1	1	1	1	1	1	1	ATTACK
τ_3	1	1	1	1	1	1	1	1	1	1	1	1	1	1	1	1	1	ATTACK
τ_4	1	1	1	1	1	1	1	1	1	1	1	1	1	1	1	1	1	ATTACK
τ_5	1	1	1	1	1	1	0.7	1	1	1	1	1	1	1	1	1	1	ATTACK
τ_6	1	1	1	1	1	1	1	1	1	1	1	1	1	1	1	1	1	ATTACK
τ_7	0.2	1	0	0	1	1	1	1	1	1	1	1	1	1	1	1	0.9	ATTACK
τ_8	1	1	0.8	1	1	1	0.5	0	1	1	1	1	1	1	1	1	1	ATTACK
τ_9	1	1	1	1	1	1	1	1	1	1	1	1	1	1	1	1	1	ATTACK
τ_{10}	1	1	1	1	1	1	1	1	1	1	1	1	1	1	1	1	1	ATTACK

Fig. 10.10 Results of attack scenario for IID 2 and test N-0-15-A-12 ($\tau_j = 1000$)

τ_j	Attack Probability																P_j	Status
	r_0	r_1	r_2	r_3	r_4	r_5	r_6	r_7	r_8	r_9	r_{10}	r_{11}	r_{12}	r_{13}	r_{14}	r_{15}		
τ_1	0	0	0	0	0	0	0	0	0	0	0	0	0	0	0	0	0.0	NORMAL
τ_2	0	0	0	0.7	0	0	0	0	0	0	0	0	0	0	0	0	0.0	NORMAL
τ_3	0	0	0	0	0	0	0	0	0	0	0	0	0	0	0	0	0.0	NORMAL
τ_4	0	0	0	0	0	0	0	0	0	0	0	0	0	0	0	0	0.0	NORMAL
τ_5	0	0	0	0	0	0	0	0	0	0	0	0	0	0.8	0	0	0.0	NORMAL
τ_6	0	0	0	0	0	0	0	0	0	0	0	0	0	0	0	0	0.0	NORMAL
τ_7	0	0	0	1	0	0	0	0	0	0	0	0	0	0	0	0	0.0	NORMAL
τ_8	0	0	0	0	0	0	0	0	0	0	0	0	0	0	0	0	0.0	NORMAL
τ_9	0	0	0	0	0	0	0	0	0	0	0	0	0	0	0	0	0.0	NORMAL
τ_{10}	0	0	0	0	0	0	0	0	0	0	0	0	0	0	0	0	0.0	NORMAL

Fig. 10.11 Results of normal scenario IID 2 and test N-0-15 ($\tau_j = 1000$)

shows the overall probability for time window τ_j calculated using Eq. (10.1) and the "Status" column indicates the final decision of the ML model for each τ_j. The two excerpts show 100% accuracy since all the time windows are classified accurately. However, each test case consists of more than 3000 such time windows (3280 in the complete table corresponding to Fig. 10.10), which is related to the application execution time. The DoS attack detection accuracy is calculated as the portion of accurately classified time windows.

Figure 10.12 shows DoS attack detection accuracy for all test cases shown in Table 10.2. In IID 1, the model is trained with only two datasets (N-0-15 and N-0-15-A-1) and tested with varying MIP locations (7, 11, and 12). Even though the number of training datasets is low, the ML model still achieves an accuracy of ∼90%. As the number of training datasets is increased, the model achieves very high accuracy (∼99%), even when tested with MIP locations which the model was not trained on. Since a decision is made at the end of each time window, the time taken to detect an attack is τ_j, which is experimentally set to 1000 cycles (1 μs). Attack detection times of previous work that addressed DoS attack detection in real-time systems

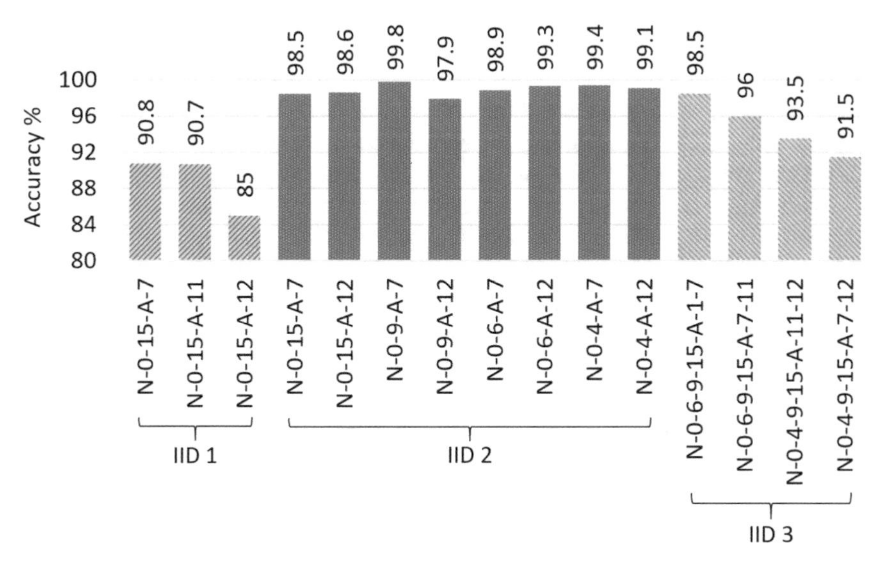

Fig. 10.12 DoS attack detection accuracy for all test cases in Table 10.2

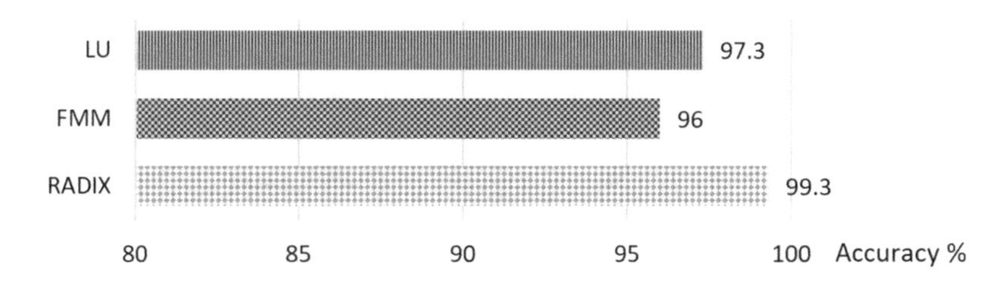

Fig. 10.13 DoS attack detection accuracy across different applications for IID 2, test case N-0-15-A-7

fall in the same range (3–8 μs) [17, 18]. *Results show that the approach is capable of detecting DoS attacks with high accuracy and in real-time, irrespective of the number or the placement of MIPs and the number of applications running on the SoC. High attack detection accuracy is achieved not only if active and malicious IP placements match the training configurations, but also in new MIP placements, which the model has not been trained on.*

To explore the behavior of the method across different applications, the model was trained on IID 1 with the FFT benchmark and tested on test case N-0-15-A-7 with LU, FMM, and RADIX running as application instances. *Results in Fig. 10.13 show that even though the model is not trained on a particular application (traffic pattern), it is capable of detecting attacks with high accuracy.*

10.6 Summary

In this chapter, we discuss about the potential of using a machine learning based DoS attack detection mechanism for NoC-based SoCs. A widely explored threat model was considered for the exploration where a malicious IP floods the NoC with a large number of packets causing deadline violations, performance degradation, or reduced energy efficiency. Unlike the existing DoS attack detection methods that rely on highly predictable NoC traffic patterns and specific use cases, this approach is capable of detecting DoS attacks with high accuracy in real-time, in the presence of unpredictable NoC traffic patterns caused by diverse applications with input variations and application mappings. Experimental results demonstrated that non-linear models, such as gradient boosting, produce the best results for the given architecture and threat models. The observations from ML model performance and feature importance reveal that the key to achieving high accuracy is to carefully craft features out of the data extracted from NoC traffic. This approach is capable of detecting DoS attacks with high accuracy in a wide variety of scenarios.

Acknowledgments This work was partially supported by the National Science Foundation (NSF) grant SaTC-1936040.

References

1. N. Agarwal et al. Garnet: a detailed on-chip network model inside a full-system simulator, in *ISPASS* (2009), pp. 33–42
2. A. Ahmed, F. Farahmandi, Y. Iskander, P. Mishra, Scalable hardware trojan activation by interleaving concrete simulation and symbolic execution, in *Proceedings of the 2018 IEEE International Test Conference (ITC)* (IEEE, New York, 2018), pp. 1–10
3. A. Ahmed, F. Farahmandi, P. Mishra, Directed test generation using concolic testing on RTL models, in *Proceedings of the 2018 Design, Automation and Test in Europe Conference and Exhibition (DATE)* (2018), pp. 1538–1543
4. A. Ahmed, Y. Huang, P. Mishra, Cache reconfiguration using machine learning for vulnerability-aware energy optimization. ACM Trans. Embedded Comput. Syst. (TECS) **18**(2), 1–24 (2019)
5. S. Bhunia et al. *The Hardware Trojan War* (Springer, Berlin, 2018)
6. N. Binkert et al. The gem5 simulator. SIGARCH Comput. Archit. News **39**(2), 1–7 (2011)
7. A.K. Biswas et al. Router attack toward NOC-enabled MPSoC and monitoring countermeasures against such threat. Circuits Syst. Signal Process. **34**(10), 3241–3290 (2015)
8. Bloomberg *The Big Hack: How China Used a Tiny Chip to Infiltrate U.S. Companies*. bloomberg.com/news/features/2018-10-04/the-big-hack-how-china-used-a-tiny-chip-to-infiltrate-america-s-top-companies
9. T. Boraten, A.K. Kodi, Mitigation of Denial of Service Attack with Hardware Trojans in NoC Architectures, in *IPDPS* (2016)
10. T. Boraten et al. Secure model checkers for network-on-chip (noc) architectures, in *GLSVLSI* (2016), pp. 45–50
11. S. Charles, P. Mishra, Reconfigurable network-on-chip security architecture. ACM Trans. Des. Autom. Electron. Syst. (TODAES) **25**(6), 1–25 (2020)

12. S. Charles, P. Mishra, Securing network-on-chip using incremental cryptography, in *IEEE Computer Society Annual Symposium on VLSI (ISVLSI)* (2020)
13. S. Charles, P. Mishra, Lightweight and trust-aware routing in NoC based SoCs, in *IEEE Computer Society Annual Symposium on VLSI (ISVLSI)* (2020)
14. S. Charles et al. Exploration of memory and cluster modes in directory-based many-core CMPS, in *NOCS* (2018), pp. 1–8
15. S. Charles, H. Hajimiri, P. Mishra, Proactive thermal management using memory-based computing in multicore architectures, in *International Green and Sustainable Computing Conference (IGSC)* (2018), pp. 1–8
16. S. Charles, A. Ahmed, U.Y. Ogras, P. Mishra, Efficient cache reconfiguration using machine learning in NOC-based many-core CMPS. ACM Trans. Des. Autom. Electron. Syst. (TODAES) **24**(6), 1–23 (2019)
17. S. Charles et al. Real-time detection and localization of DoS attacks in NoC based SoCs, in *DATE* (2019), pp. 1160–1165
18. S. Charles et al. Real-time detection and localization of distributed DoS attacks in NoC based SoCs, in *IEEE TCAD* (2020)
19. S. Charles et al. Lightweight anonymous routing in NoC based SoCs, in *DATE* (2020)
20. M. Chen, X. Qin, H.-M. Koo, P. Mishra, *System-level Validation: High-level Modeling and Directed Test Generation Techniques* (Springer, Berlin, 2012)
21. C. Ciordas et al. An event-based network-on-chip monitoring service, in *HLDVT* (2004), pp. 149–154
22. F. Farahmandi, P. Mishra, Automated debugging of arithmetic circuits using incremental gröbner basis reduction, in *Proceedings of the 2017 IEEE International Conference on Computer Design (ICCD)* (IEEE, New York, 2017), pp. 193–200
23. F. Farahmandi, P. Mishra, FSM anomaly detection using formal analysis, in *Proceedings of the 2017 IEEE International Conference on Computer Design (ICCD)* (IEEE, New York, 2017), pp. 313–320
24. F. Farahmandi, P. Mishra, Automated test generation for debugging multiple bugs in arithmetic circuits. IEEE Trans. Comput. **68**(2), 182–197 (2018)
25. F. Farahmandi, Y. Huang, P. Mishra, Trojan localization using symbolic algebra, in *Proceedings of the 2017 22nd Asia and South Pacific Design Automation Conference (ASPDAC)* (IEEE, New York, 2017), pp. 591–597
26. F. Farahmandi, Y. Huang, P. Mishra, *System-on-Chip Security: Validation and Verification* (Springer, Berlin, 2019)
27. L. Fiorin et al. A security monitoring service for NoCs, in *CODES+ISSS* (2008), pp. 197–202
28. J. Frey, Q. Yu, A hardened network-on-chip design using runtime hardware trojan mitigation methods. Integration **56**, 15–31 (2017)
29. M. Geller, P. Nair, 5g security innovation with CISCO, in *Whitepaper Cisco Public* (2018), pp. 1–29
30. S. Gupta et al. Vulnerable network analysis using war driving and security intelligence, in *ICICT*, vol. 3 (2016), pp. 1–5
31. U. Gupta, C.A. Patil, G. Bhat, P. Mishra, U.Y. Ogras, DYPO: Dynamic pareto-optimal configuration selection for heterogeneous MPSOCS. ACM Trans. Embedded Comput. Syst. (TECS) **16**(5s), 1–20 (2017)
32. Y. Huang, P. Mishra, Vulnerability-aware energy optimization for reconfigurable caches in multitasking systems. IEEE Trans. Comput. Aided Des. Integr. Circuits Syst. **38**(5), 809–821 (2019)
33. Y. Huang, S. Bhunia, P. Mishra, MERS: statistical test generation for side-channel analysis based trojan detection, in *ACM SIGSAC Conference on Computer and Communications Security (CCS)* (2016), pp. 130–141
34. Y. Huang, S. Bhunia, P. Mishra, Scalable test generation for trojan detection using side channel analysis. IEEE Trans. Inf. Forensics Secur. (TIFS) **13**(11), 2746–2760 (2018)
35. Y. Lyu, P. Mishra, A survey of side-channel attacks on caches and countermeasures. J. Hardw. Syst. Secur. **2**(1), 33–50 (2018)

36. Y. Lyu, P. Mishra, Efficient test generation for trojan detection using side channel analysis, in *Design Automation and Test in Europe Conference (DATE)* (2019), pp. 408–413
37. Y. Lyu, P. Mishra, Automated test generation for activation of assertions in RTL models, in *Proceedings of the 2020 25th Asia and South Pacific Design Automation Conference (ASPDAC)* (IEEE, New York, 2020), pp. 223–228
38. Y. Lyu, P. Mishra, Automated test generation for trojan detection using delay-based side channel analysis, in *Proceedings of the 2020 Design, Automation and Test in Europe Conference and Exhibition (DATE)* (2020), pp. 1031–1036
39. Y. Lyu, P. Mishra, Automated trigger activation by repeated maximal clique sampling, in *Proceedings of the Asia and South Pacific Design Automation Conference (ASPDAC)* (2020), pp. 482–487
40. Y. Lyu, P. Mishra, Scalable activation of rare triggers in hardware trojans by repeated maximal clique sampling. IEEE Trans. Comput. Aided Des. Integr. Circuits Syst. (2020). https://doi.org/10.1109/TCAD.2020.3019984
41. Y. Lyu, P. Mishra, Scalable concolic testing of RTL models. IEEE Trans. Comput. (2020). https://doi.org/10.1109/TC.2020.2997644
42. Y. Lyu, X. Qin, M. Chen, P. Mishra, Directed test generation for validation of cache coherence protocols. IEEE Trans. Comput. Aided Des. Integr. Circuits Syst. **38**(1), 163–176 (2018)
43. Y. Lyu, A. Ahmed, P. Mishra, Automated activation of multiple targets in RTL models using concolic testing, in *Proceedings of the 2019 Design, Automation and Test in Europe Conference and Exhibition (DATE)* (IEEE, New York, 2019), pp. 354–359
44. P. Mishra, F. Farahmandi, *Post-Silicon Validation and Debug* (Springer, Berlin, 2019)
45. P. Mishra, S. Bhunia, M. Tehranipoor, *Hardware IP security and Trust* (Springer, Berlin, 2017)
46. Z. Pan, P. Mishra, Automated test generation for hardware trojan detection using reinforcement learning, in *Proceedings of the Asia and South Pacific Design Automation Conference (ASPDAC)* (2021)
47. Z. Pan, J. Sheldon, P. Mishra, Hardware-assisted malware detection using explainable machine learning, in *Proceedings of the IEEE International Conference on Computer Design (ICCD)* (2020)
48. Z. Pan, J. Sheldon, P. Mishra, Test generation using reinforcement learning for delay-based side channel analysis, in *Proceedings of the IEEE/ACM International Conference on Computer-Aided Design (ICCAD)* (2020)
49. Z. Pan, J. Sheldon, C. Sudusinghe, S. Charles, P. Mishra, Hardware-assisted malware detection using machine learning, In *Design Automation and Test in Europe (DATE)* (2021)
50. N. Prasad et al. Runtime mitigation of illegal packet request attacks in networks-on-chip, in *ISCAS* (2017), pp. 1–4
51. P. Prasse et al. Malware detection by analysing network traffic with neural networks, in *IEEE S and P Workshops (SPW)* (2017), pp. 205–210
52. J.S. Rajesh et al. Runtime detection of a bandwidth denial attack from a rogue network-on-chip, in *NOCS* (2015)
53. J. Sepulveda et al. Towards the formal verification of security properties of a network-on-chip router, in *Proceedings of the 2018 IEEE 23rd European Test Symposium (ETS)* (2018), pp. 1–6
54. A.S. Shekhawat, *Analysis of Encrypted Malicious Traffic* (2018)
55. A. Shoemaker, *How to Identify a Mirai-style DDOS Attack*. https://www.imperva.com/blog/how-to-identify-a-mirai-style-ddos-attack/
56. A. Sodani et al. Knights landing: second-generation intel Xeon phi product. IEEE Micro **36**(2), 34–46 (2016)
57. W. Wang, P. Mishra, S. Ranka, *Dynamic Reconfiguration in Real-Time Systems* (Springer, Berlin, 2012)
58. K. Wang et al. TSA-NOC: learning-based threat detection and mitigation for secure network-on-chip architecture. IEEE Micro **40**(5), 56–63 (2020)
59. D. Wentzlaff et al. On-chip interconnection architecture of the tile processor. IEEE Micro **27**(5), 15–31 (2007)

60. S.C. Woo et al. The splash-2 programs: characterization and methodological considerations, in *ISCA* (1995), pp. 24–36
61. Y.J. Yoon et al. Virtual channels and multiple physical networks: two alternatives to improve NoC performance. TCAD **32**(12), 1906–1919 (2013)

Chapter 11
Trojan Aware Network-on-Chip Routing

Manju Rajan, Abhijit Das, John Jose, and Prabhat Mishra

11.1 Introduction

With the noting growth in Internet-of-Things (IoT) devices and embedded systems, outsourcing of circuit design and fabrication process has significantly accelerated over the years. The race for bringing more devices into the market made semiconductor industries paying less attention to the hardware security of these devices. Due to the reduced emphasis on security standards, new hardware vulnerabilities are uncovered every now and then. For example, one of the recently exposed glitches in modern Intel processors allows an adversary to access kernel memory [46]. The flaw is able to bypass most of the hardware level protections available in the system.

Due to the increasing demand for data and compute-intensive tasks in the era of IoT and embedded systems, design of Multi-Processor System-on-Chips (MPSoCs) gained popularity. The use of packet-based on-chip interconnect technology called Network-on-Chip (NoC) in MPSoCs outperforms the existing bus-based interconnect, thereby circumventing the low wire routing congestion and low operation frequencies of the system. NoC provides separation between computation and communication, supports modularity and Intellectual Property (IP) reuse via standard interfaces and handles synchronisation issues, which in turn improve the performance of the system. However, the long and globally distributed supply chain of hardware IPs makes MPSoC design increasingly vulnerable to diverse trust/integrity issues. For example, MPSoCs built with third party NoCs create more vulnerabilities due to its emphasis on performance and backward compatibility

M. Rajan (✉) · A. Das · J. Jose
Indian Institute of Technology Guwahati, Guwahati, Assam, India
e-mail: manju18@iitg.ac.in; abhijit.das@iitg.ac.in; johnjose@iitg.ac.in

P. Mishra
University of Florida, Gainesville, FL, USA
e-mail: prabhat@ufl.edu

P. Mishra, S. Charles (eds.), *Network-on-Chip Security and Privacy*,
https://doi.org/10.1007/978-3-030-69131-8_11

rather than security [1, 20, 23, 24, 34, 35, 38, 41]. Being the backbone of inter-core communication, NoC has access to all the cores of an MPSoC and hence if malicious, it can wreak havoc. Thus, security aware communication architectures gathered attention and became a major area of recent research [8–14, 16].

A hardware-oriented attack is defined as the exploitation of state values and the manipulation of control signals that can result in the violation of security in a computing platform. The attacker can exploit information like power consumption, branch prediction speculation, execution time, memory access pattern and memory access time to launch an attack. Such hardware orientated attacks become a major concern due to the difficulty in analysing their processing permutations for all the IP cores of an MPSoC, especially in real-time systems. One of the significant hardware-oriented security compromises in MPSoCs is the insertion of a malicious circuit called *Hardware Trojan(HT)* [28], which can alter the functionality of the system to deploy an attack. Example of frequent attacks with HTs includes Denial of Service (DoS), information leakage, high jacking [16], unauthorised memory access [43], etc. These threats in MPSoCs are critical challenges in the hardware security domain.

11.1.1 Overview of Hardware Trojans

One of the biggest HT attacks reported is the failure of the Syrian radar to warn of an incoming air strike [39]. Also, report shows that US NSA embedded HT circuitry into the USB port to steal secret data from all over the world, including the military's network in China and Russia, information about Mexican drug cartels [50], etc.

Figure 11.1 shows the generic structure of an HT. It consists of a trigger logic which initiates the activation of an HT and a payload logic which perform the malicious modification of the circuit or the functionality to deploy the attack [4]. HTs can be broadly classified into two categories based on the mechanism of this trigger/activation and payload/action [4], as shown in Fig. 11.2. They can be further divided based on the types of signal used to initiate the HT, whether analog or digital or anything else. For example, analog HTs can be initiated by conditions like delay, power, temperature, device aging effect, etc. Similarly, digital HTs initiated

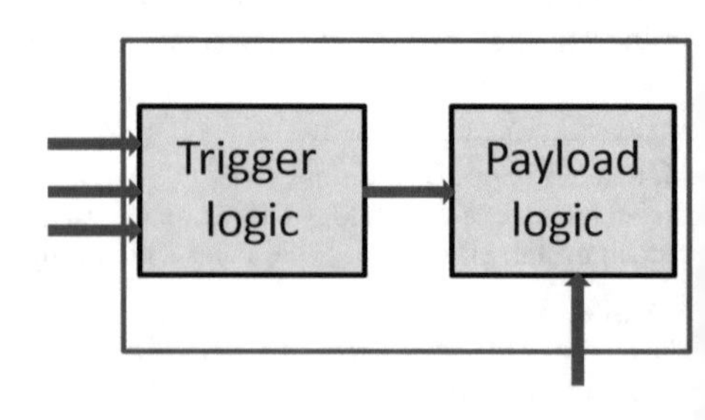

Fig. 11.1 Structure of a Hardware Trojan

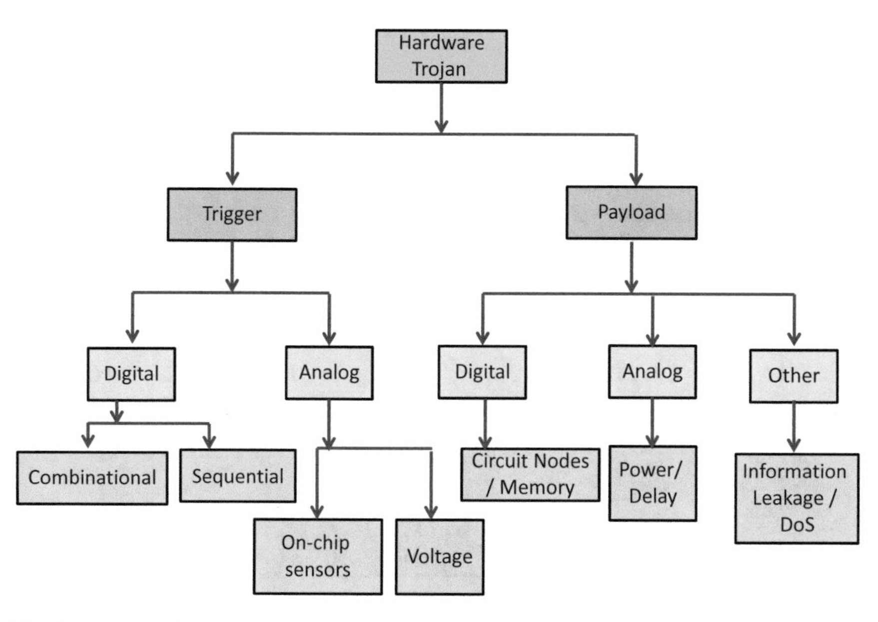

Fig. 11.2 Types of Hardware Trojan

by Boolean logic function use sequential or combinational circuits where the HT gets activated either by the sequence of events that occur in the system or during the outcome of a combination of events which are undetectable during the functional testing phase.

An HT can be injected into a system at any stage of the IP supply chain, like specification, design, fabrication, etc. EDA vendors can alter the logic implementations deduced by their tools, which can do more than required to activate the HT. Third party IP vendors can easily manipulate the RTL and insert malicious codes which can modify the system's behaviour [28]. An untrusted staff who has access to the fabrication process and facility can tweak the system chip [4]. Once activated, the HT payload can create catastrophic effects like DoS, functional failure, information leakage, high jacking [16], unauthorised memory access [43], alteration to the system, etc. Based on the intention and attack model, an adversary can mount HT on locations like processor, interconnect, memory, input-output devices, etc. In modern MPSoCs, HTs can be deployed either in the processor or the interconnect (NoC). Due to the positional advantage, an HT mounted and activated on the NoC can cause severe performance degradation and bring down the entire system to a near halt.

11.1.2 Trojan-Based Attacks on NoC Architectures

As NoC is shared by all the connected heterogeneous cores of an MPSoC, avoiding interference in underlying applications poses a greater challenge. This makes NoC more vulnerable to hardware-oriented security threats, where an attacker can analyse the communication flow, variation in power consumption, etc. to launch the attack. Among other attacks, HTs mounted on NoC of MPSoCs impose unique challenges as they remain hidden until they are triggered. Identifying the rare condition that triggers an HT requires to examine all the possible input patterns to the MPSoC. However, it is not feasible due to the time constrained post-silicon debug and validation [2, 15, 19, 21, 22, 32, 33, 36, 37]. Figure 11.3 lists well known attacks an HT can initiate when mounted on NoC. A brief introduction about each one of them is as follows:

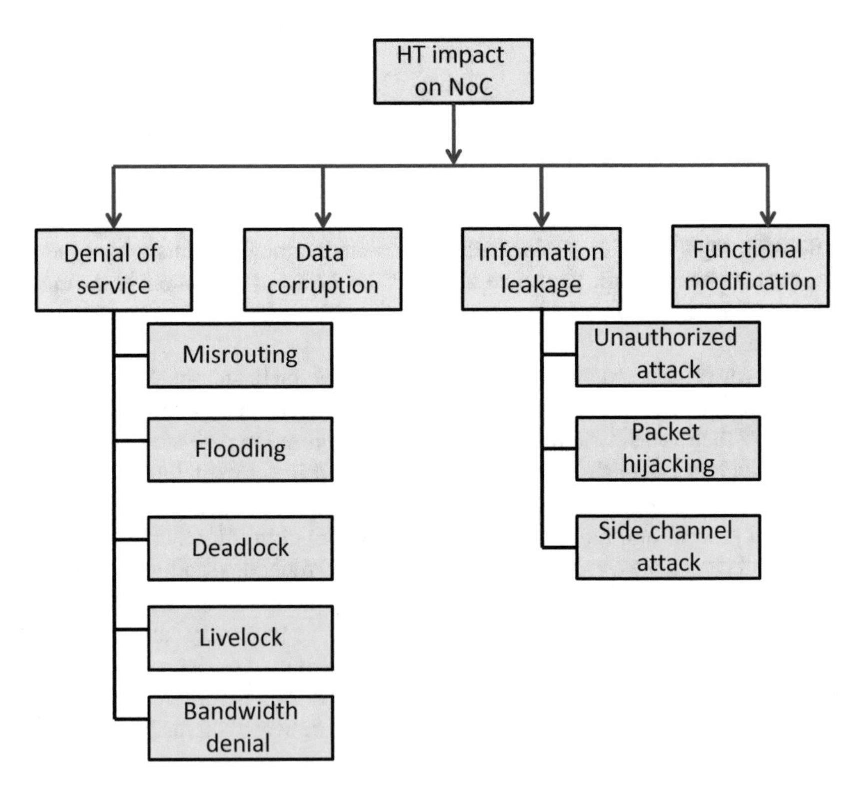

Fig. 11.3 Impact of HT threats on NoC

11.1.2.1 Denial of Service (DoS)

DoS attack by an HT on NoC is mainly aimed towards resource depletion, including bandwidth, being one of the most critical resources in a communication framework. The HT can deploy such an attack by flooding the network with frequent and useless packets. As a consequence, the victim packets suffer from buffer and link unavailability, which eventually leads to a deadlock, halting the entire NoC communication. In some cases, the HT modifies the employed routing algorithm to misroute packets, thus denying them to reach their destination. This behaviour of the HT makes the victim packets suffer from livelock, thus wasting the resources without any productivity.

11.1.2.2 Information Leakage

The objective of a leakage Trojan is to snoop on the ongoing data communication on NoC and steal critical information. These information leakage/data snooping attacks are performed by duplicating the incoming packets from a processing core [3, 45]. In most of the cases, such Trojans get triggered by an accomplice thread that resides in some other core. HT taps the incoming links from the network interface (NI) or even the internal links and watches out for covert signals from a possible accomplice thread. Once an activation signal is received by the HT, it initiates sending snooped data from the link to the accomplice thread, waiting to steal the critical information.

11.1.2.3 Data Corruption

An adversary with the intention of a data corruption attack injects an HT which attempts to learn the key used for packet encryption. When activated, the HT inserts different combination of inputs to the encoder to determine a sequence of errors which can mask a corrupted data into another valid data. With a side channel analysis [26, 27, 29–31, 42], the adversary can also monitor how encoders and decoders respond to different combination of inputs over a period of time. Once the adversary obtains sufficient insights, the HT attempts to decipher the encryption key. Once it gets hold of the key, the whole system running sensitive applications is compromised.

11.1.2.4 Functional Modification

In a functional modification attack, an HT tries to change the operation of a circuit by maliciously modifying its design or by adding new component into the existing design. Since most of the components and the design of a circuit are outsourced, an attacker can be any third party IP vendor or untrustworthy staff who can inject the HT in any one of the phases like logical synthesis, physical P & R, etc. One such attack is the addition of a malicious logic that replaces a circuit's original 'AND' logic with a multiplexer (MUX). The detection of these kind of HTs is very hard as the malicious logic may not be triggered under the normal execution.

The remainder of this chapter is organised as follows. Section 11.2 provides an overview of Trojan placements at four different locations: network interface, network link, input/output buffers and network routers. While there are many possible Trojan placements, Trojans at network routers are the most explored attacks in the literature. Section 11.3 describes a Trojan aware routing algorithm when Trojans are present in the network routers. Section 11.4 presents the experimental evaluation along with the overheads. Finally, Sect. 11.5 concludes the chapter, giving a summary of everything.

11.2 Different Placements of Hardware Trojans

In the previous section, various forms of HT induced attacks are described. Irrespective of the type of attack, an HT can be mounted at different locations within the NoC infrastructure. In this section, three different placements along with the associated Trojan detection and mitigation techniques are discussed. Specifically, HTs placed at major locations within NoC are presented with their abstract/high-level design, detection and mitigation mechanisms. Specific examples are selected from the existing literature so that readers can understand the HTs at a certain depth and can refer to if interested for further details.

11.2.1 Trojan at Network Interface (NI)

Available research shows that an adversary can deploy an HT that can initiate a data snooping attack by modifying the flit queue in the NI of an NoC router [45]. This HT makes use of an accomplice thread that sends messages to initiate the attack. Upon HT activation, packets that reach NI are duplicated with less interference to the standard NI functions and sent to the accomplice thread.

Once activated, the HT starts monitoring the cyclic flit queue associated with the NI. Whenever a head flit is transmitted from the NI queue to the router, HT exploits its location in the queue, as shown in Fig. 11.4. Before the location of the head flit is overwritten in the queue, HT copies the content of the head flit into a new flit and set its destination as the malicious core. The duplicate flit reaches the router like any other flit to travel towards the malicious core where the accomplice thread is residing thereby achieving data snooping.

Unlike the state-of-the-art techniques [3], a snooping detection module (THANOS) that utilises threshold voltage degradation can be used for run-time

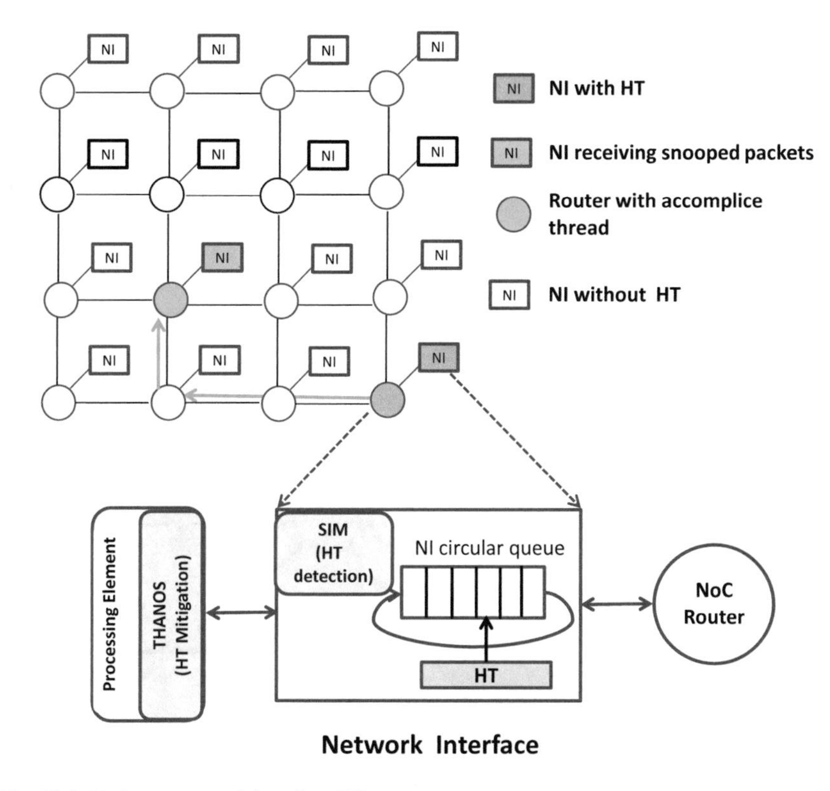

Fig. 11.4 Trojan at network interface (NI)

detection of the core that uses the malicious application for data snooping. A snooping invalidation module (SIM) which uses an encoding-based duplicate packet detection mechanism is employed to prevent malicious data replication by the HT on NoC.

11.2.2 Trojan at Network Link

Existing literature has an attack using HT that snoops on passing packets in the network link to inject a fault into some target packets [7]. The fault corrupts a packet and triggers re-transmission, which in turn creates congestion and then deadlock in the on-chip network. This attack makes use of a kill switch to control the HT activation as well as to avoid HT triggering during the verification process.

Figure 11.5 shows the HT which is named as target-activated sequential payload (TASP). It has a target block which is used to identify the victim packet by checking the information like source, destination, virtual channel (VC), process ID, memory address, etc. It has a payload counter that is used to inject faults at different locations

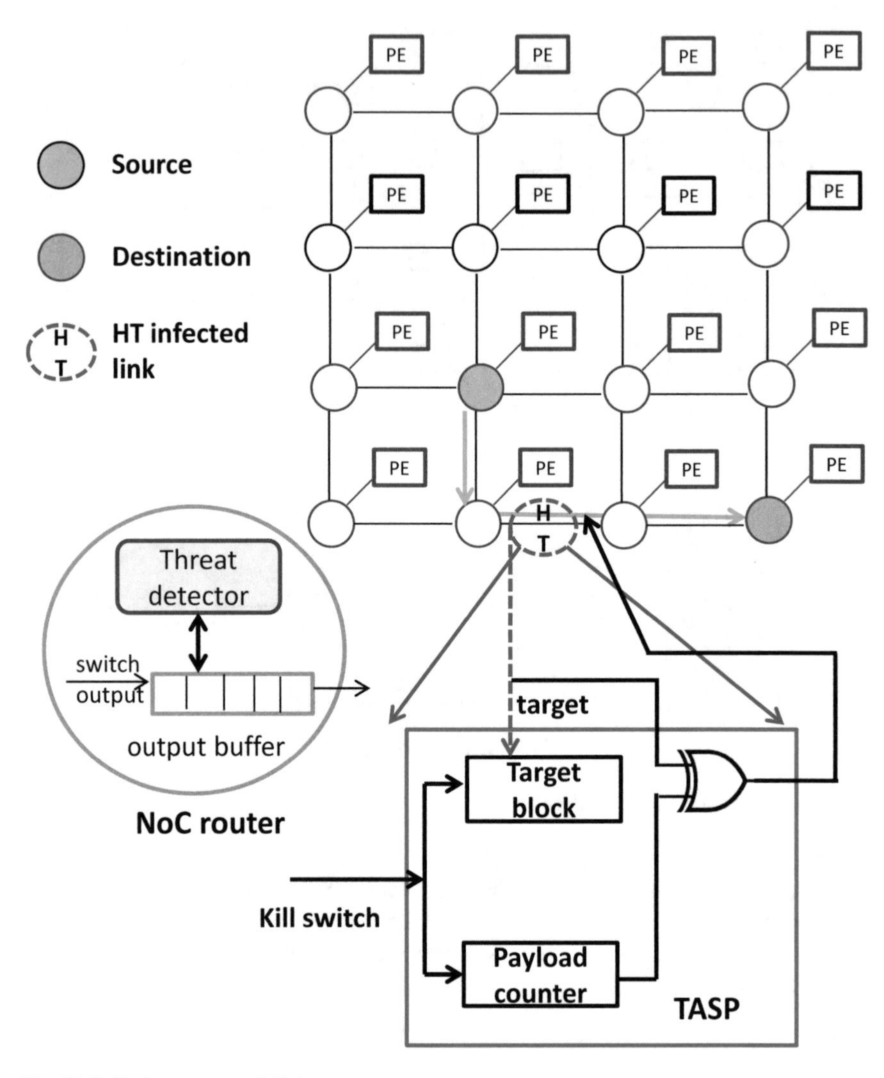

Fig. 11.5 Trojan at network link

and thus avoids getting noticed by fault aware architectures. The HT also has an XOR tree to change the bits on the wires selected within the link during the attack.

A threat detection module integrated into the output buffer of an NoC router is used to monitor the re-transmission of packets and the possibility for transient or permanent faults. These kinds of attacks can be circumvented by a heuristic switch to switch mitigation technique which obfuscates the packets to avoid HT triggering.

11.2.3 Trojan at Input/Output Buffers

A Trojan at input/output buffers, as shown in Fig. 11.6 can initiate an attack that changes the flit type, modify packet's destination and sabotage the integrity of a packet that leaves the NI [25]. In NoC based MPSoCs, FIFO based input/output buffers become a primary target for attackers due to their regularity in implementation and large area footprints. The main consequence of an HT at input/output buffers is resource depletion.

When activated, the HT placed at the input/output buffers modifies the critical bits of a flit. It can change the head/body/tail bit of a flit to change the flit type, thus creating problem in routing and arbitration. The HT can also modify routing path bits to enable misrouting for the victim packets. It can even modify the destination address bits to send the victim packets to a malicious core for data snooping.

To proactively defence the HT's impact on NoC's integrity, the mitigation technique uses three modules HTM1, HTM2 and HTM3. The HTM1 module uses a

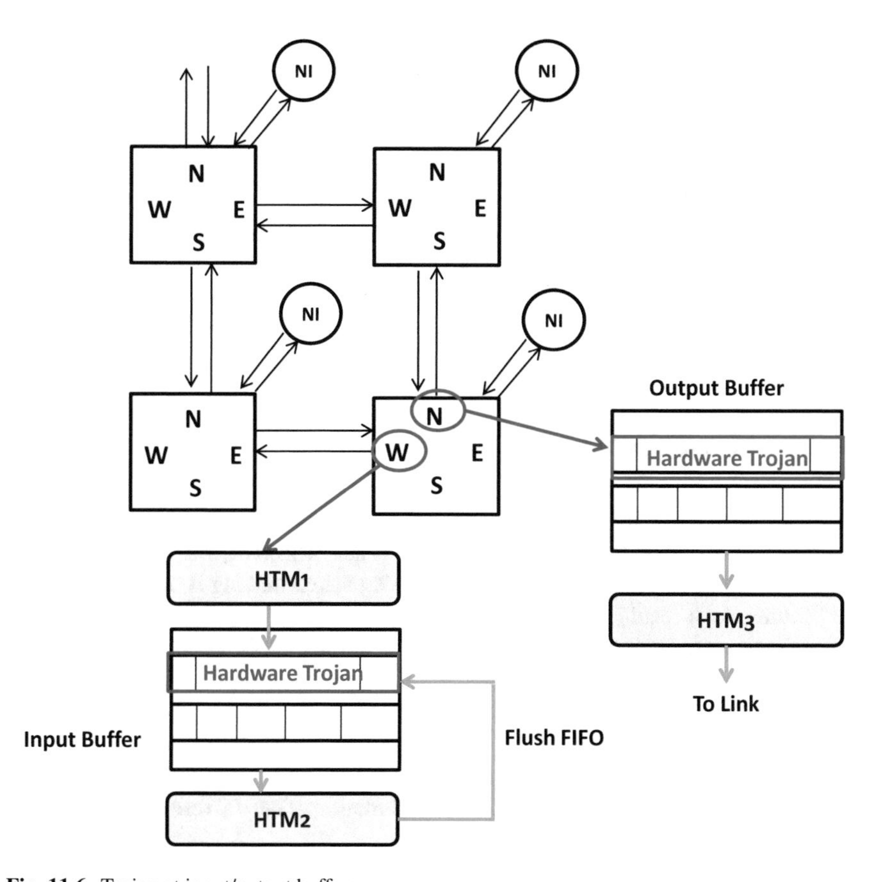

Fig. 11.6 Trojan at input/output buffers

dynamic flit permutation technique to reorder the incoming flits before they are sent to the FIFO based input buffer. HTM2 checks for flit's integrity to flush malicious flits and HTM3 is used to recover the flit to its original form in output buffer.

11.2.4 Trojan at Network Routers

Routers being the most important component in NoC framework are the primary target for attackers. Many popular attacks on NoC place a Trojan in the Routing Computation (RC) unit to manipulate routing information. Such an HT changes the routing algorithm/routing table contents to initiate packet misrouting [6, 17, 44]. Other consequences include DoS, delay of service, information leakage, bandwidth depletion, unauthorised access, etc. Attacks on NoC through an HT at the routers have grown exponentially in recent years due to their ability of creating multiple damages at once. Section 11.3 presents an efficient technique to enable trusted NoC communication in the presence of an HT at the routers [44].

11.3 SECTAR: Secure NoC Using Trojan Aware Routing

SECTAR introduced an intermittent HT threat model that misroutes packets to initiate multiple attacks on NoC based MPSoCs. To deal with a threat of this nature, SECTAR proposed Trojan Aware Routing (TAR), a technique that dynamically detects a misrouting HT, isolates the HT and route packets in the network bypassing the HT. The following subsections describe everything about SECTAR.

11.3.1 Threat Model

SECTAR presented an HT threat model that tampers the routing algorithm employed in RC unit to enable misrouting. When triggered, the HT maliciously assigns a wrong output port to the head flit of a packet, making it travel to a wrong next router. As a result, all the flits of the packet get misrouted (due to wormhole routing) and contribute to one of the attack scenarios: DoS, injection suppression and delay of service. DoS is a scenario where the flits of a packet get indefinitely delayed in the path and never reach their destination. Injection suppression scenario is a by-product of DoS where new flits cannot be injected into the network due to unavailability of router buffers. Delay of service is a scenario similar to DoS except that the flits eventually reach their destination. The HT threat model can be formulated as follows:

An NoC packet P can be represented as:

$$P = \{F^p_{head} \parallel F^p_{body1} \parallel F^p_{body2} \parallel \dots \parallel F^p_{bodyn} \parallel F^p_{tail}\} \tag{11.1}$$

where F^p_i are flits of packet P such that:

$$\begin{aligned} F^p_{head} &= [\{SRC,\ DEST,\ CTRLMSG\}] \\ F^p_{body} &= [\{CTRLMSG\},\ \{Data\}] \\ F^p_{tail} &= [\{CTRLMSG\},\ \{Data\}] \end{aligned}$$

Path of packet P from source to destination can be given as:

$$P = \{R_{src},\ \dots R_{k-1},\ R_k,\ R_{k+1},\ \dots R_{dest}\} \tag{11.2}$$

where R_i denotes router i in the NoC. Let RA_i denote the routing algorithm employed in the RC unit of R_i. Then for packet P, it can be said from Eqs. (11.1) and (11.2) that

$$RA_k(F^p_{head}) = R_{k+1} \tag{11.3}$$

where for the head flit F^P_{head} of packet P, the routing algorithm RA_k employed in router R_k will assign the next router as R_{k+1}.
Let HT denotes the proposed threat model such that

$$\begin{gathered} HT(RA_k) = RA^*_k \text{ and} \\ RA^*_k(F^p_{head}) = R^*_{k+1} \text{ and} \\ R^*_{k+1} \neq R_{k+1} \end{gathered}$$

Packets carry cache miss requests, cache miss replies, evicted cache blocks and coherence messages from source to their destination through the underlying NoC. A router infected with the proposed HT can misroute these packets and degrade the application-level performance of latency-critical applications. Such type of HTs can be added to an NoC IP at any stage of the supply chain, including specification, design and fabrication [28, 47]. To make it hard to get detected, SECTAR assumed that the proposed HT is intermittently malicious and internally triggered [28, 49].

Figure 11.7 shows an illustration of an 8×8 mesh NoC based MPSoC with the proposed HT mounted on router 35. An adversary can insert any number of such HTs in the NoC. However, activating multiple HTs can create an unusual variation in energy and power consumption and hence may be easily noticed (detected). To make it hard to get detected, SECTAR assumed that the adversary had infected only a single router with the proposed HT. Based on the location of HT, the entire 8×8 NoC is divided into eight different regions: N, E, S, W, NE, SE, SW and

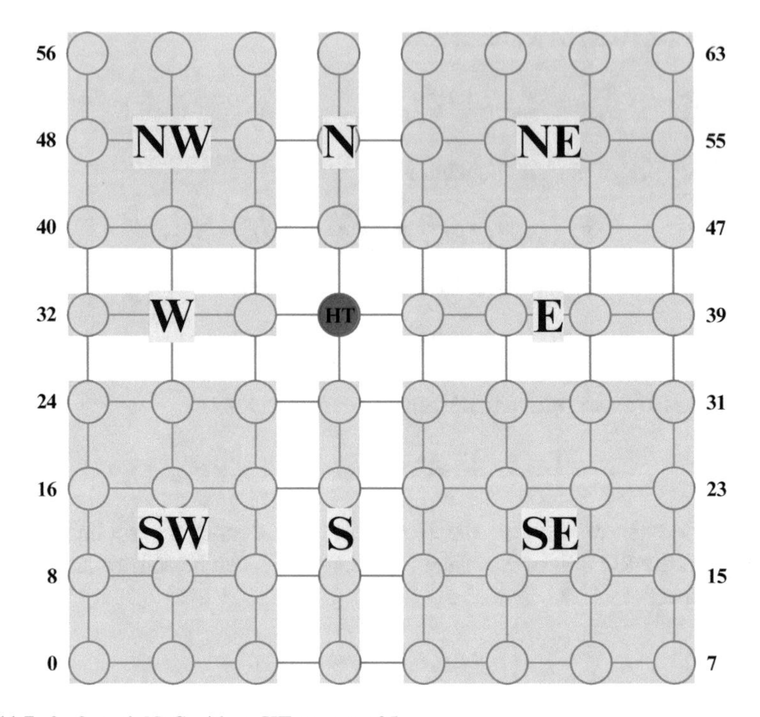

Fig. 11.7 8×8 mesh NoC with an HT at router 35

NW. When triggered, the impact of HT varies across different regions based on their inter-core communication. The compromised MPSoC encounters the following attack scenarios:

11.3.1.1 Attack Scenario: Denial of Service (DoS)

To understand how a misrouting HT can initiate a DoS attack, consider a case shown in Fig. 11.8 with the Trojan hiding at router 35. The underlying NoC employed X-Y dimension order routing algorithm for packet traversal. During inter-core communication, a packet $P1$ with source $S1$ on its way to destination $D1$ reaches router 35. Instead of forwarding $P1$ to router 43 as per X-Y routing, the activated HT at router 35 misroutes $P1$ to router 34. Note that the HT can misroute this packet to any other direction than the valid one (say router 27, router 34 or router 36). When the misrouted packet reaches router 34, following X-Y routing, it will be re-sent to router 35. Destination $D1$ is at router 59, which is on the same column as that of HT infected router 35. So as per X-Y routing, $P1$ can reach destination $D1$ only through router 35, which is currently compromised. Since router 35 always misroutes, $P1$ will never reach its destination $D1$. This is a DoS attack scenario created by the proposed misrouting HT threat model in SECTAR.

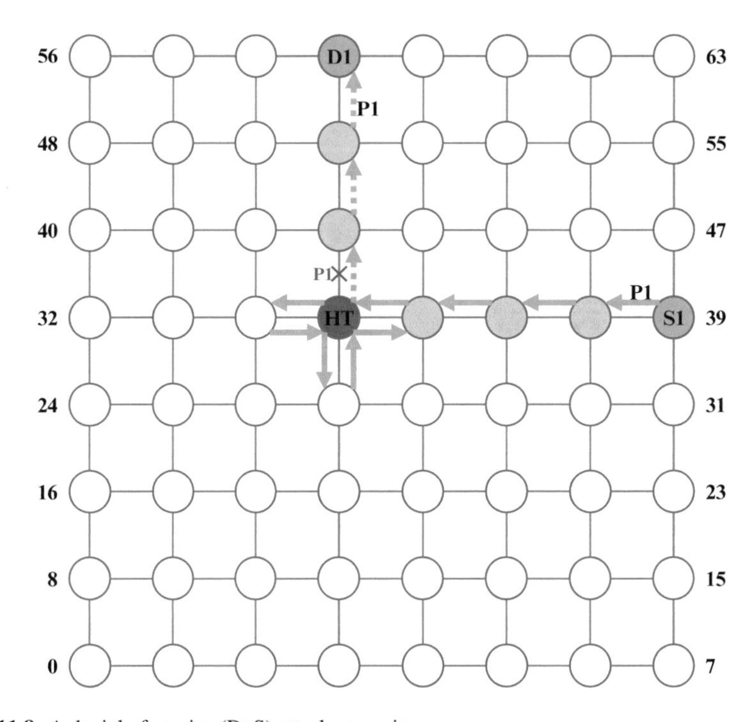

Fig. 11.8 A denial of service (DoS) attack scenario

According to Fig. 11.7, source $S1$ is in region E and destination $D1$ is in region N. With further analysis, it is concluded that an inter-region communication of type $E \rightarrow N$ leads to a DoS attack scenario. In general, for all the inter-region communication where the destination router is on the same column as that of the HT infected router 35, DoS attack like scenario is possible. Hence, any packet traversal between the following regions is susceptible to a DoS attack scenario:
$E \rightarrow N$, $E \rightarrow S$, $W \rightarrow N$, $W \rightarrow S$ $NE \rightarrow S$, $NW \rightarrow S$, $SE \rightarrow N$, $SW \rightarrow N$.

11.3.1.2 Attack Scenario: Injection Suppression

A misrouting HT can also create injection suppression in the network. Most of the time this occurs as the by-product of a DoS attack. SECTAR defined the HT in such a way that the direction of misrouting is non-deterministic. For example, in the case considered in Fig 11.8, the HT infected router 35 can misroute packet $P1$ in different invalid directions at different instances. As a result, $P1$ gets trapped into a ping-pong state between the neighbours of router 35, except with router 43. Packets are buffered in virtual channels (VCs) of routers while taking part in routing and arbitration decisions. Prolonged ping-pong of $P1$ leads to VC

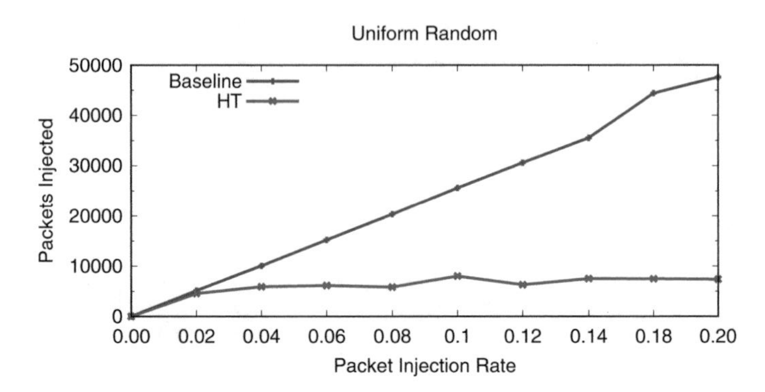

Fig. 11.9 HT triggered injection suppression

unavailability in neighbouring routers and propagates the effect to others by back-pressure. Eventually, a scenario of injection suppression arises in the entire system. When the traffic is high, unavailability of NoC resources due to the ping-pong effect also leads the system into a deadlock.

SECTAR analysed the proposed threat model for injection suppression and its effect on an 8×8 NoC based MPSoC as shown in Fig. 11.9. Simulating a *uniform_random* synthetic traffic reveals that with the increase in packet injection rate, impact of the proposed HT escalates, and the number of packets injected decreases drastically. The increasing injection suppression eventually leads the network into a deadlock.

11.3.1.3 Attack Scenario: Delay of Service

Another attack scenario created by the proposed HT threat model in SECTAR is a delay of service. Consider an inter-core communication, where a packet $P2$ with source $S2$ is travelling towards its destination $D2$, as shown in Fig 11.10. When $P2$ reaches the HT infected router 35, instead of forwarding the packet towards router 36 (as per X-Y routing), the Trojan at router 35 misroutes $P2$. Since the direction of misrouting is random, packet $P2$ can reach any one of the neighbours of router 35, like 27, 34 or 43. If $P2$ reaches either router 27 or 43, following X-Y routing it can reach the destination $D2$ incurring a small delay. However, if the HT misroutes $P2$ towards router 34, then it enters a ping-pong state between router 34 and 35, something similar to the DoS scenario (refer Fig 11.8). But, due to the randomness of misrouting, the ping-pong breaks when $P2$ gets misrouted to either router 27 or 43 in the near future. In that case, packet $P2$ eventually reaches its destination with an indefinite delay. This is a delay of service attack scenario.

Again, according to Fig. 11.7, source $S2$ is in region W and destination $D2$ is in region NE. Thus, an inter-region communication of type $W \rightarrow NE$ creates a delay

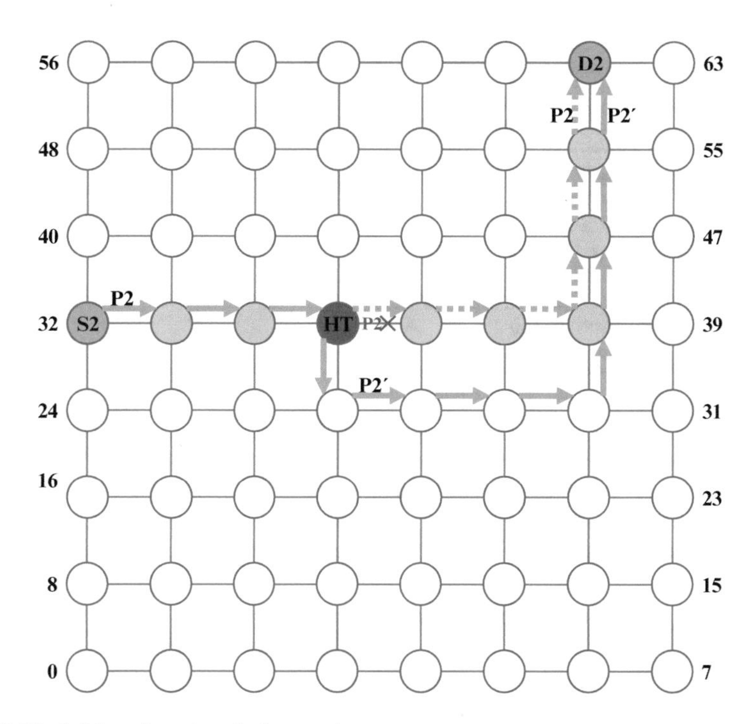

Fig. 11.10 A delay of service attack scenario

of service attack scenario. To generalise, a delay of service attack like scenario is possible when there is a communication between the following regions: $E \rightarrow W,\ E \rightarrow NW,\ E \rightarrow SW,\ W \rightarrow E,\ W \rightarrow NE,\ W \rightarrow SE.$

11.3.2 Trojan Aware Routing

11.3.2.1 Detecting the Trojan

SECTAR employed X-Y routing where a packet travels along the X direction and reaches the same column as that of destination. Then, the packet travels along the Y direction to reach the destination. Let P be a packet with source $S(x1, y1)$ and destination $D(x2, y2)$. As per X-Y routing, when P reaches an intermediate router $R(x, y)$, it is forwarded along the X direction until $(x < x2)$. When P reaches a router where $(x == x2)$, it changes the direction and starts travelling along the Y direction until $(y < y2)$. When P reaches a router where $(y == y2)$, it reaches the destination $D(x2, y2)$. The X-Y dimension order routing algorithm decides the output port for a packet based on the position of destination router with respect to the current router. The routing algorithm does not consider the input port of the packet and its previous router for its routing decisions. The proposed HT threat model in

Algorithm 1: Working of the detection module

Input: Input Direction of flit; in_dir
Output: Violated output direction of a flit: outv_dir
if $x_{diff} < 0$ *and indir is WEST* **then**
| set *outvdir* as WEST
else if $x_{diff} > 0$ *and indir is EAST* **then**
| set *outvdir* as EAST
else if $x_{diff} > 0$ *and indir is NORTH* **then**
| set *outvdir* as NORTH
else if $x_{diff} > 0$ *and indir is SOUTH* **then**
| set *outvdir* as SOUTH
end

SECTAR exploits this feature of the routing algorithm and enables misrouting. Now, even if a packet is misrouted and reaches a router where it should not have reached as per X-Y routing, the employed routing algorithm is not able to detect it. The packet is forwarded to destination without knowing the misrouting that has brought the packet to this router.

To identify packet misrouting and HT infected router, TAR adds a detection module, a 1-bit *alert_flag* and a 3-bit *alert_dir* at every NoC router. The working of the detection module is given in Algorithm 1. *alert_flag* is set only if the neighbour is identified as an HT infected router and reset otherwise. *alert_dir* either denotes no direction or the direction where the HT is detected; north, east, south, or west. In the DoS attack scenario shown in Fig. 11.8, packet $P1$ is forwarded to router 34 because of the misrouting at router 35. With the detection module in place, router 34 knows that $P1$ has entered through east input port from router 35. Analysing the position of destination $D1$ at router 59 with respect to router 35, the detection module concludes that X-Y routing is violated and $P1$ is misrouted. Router 34 sets its *alert_flag* and updates *alert_dir* as east since router 35 misrouted packet $P1$ and hence must be an HT infected router. *alert_flag* and *alert_dir* are also used in the subsequent phases of shielding and bypassing the Trojan.

11.3.2.2 Shielding the Trojan

Once the HT is detected by one of its neighbours (27, 34, 36 or 43), a dynamic shielding protocol is activated. The router that detects the HT generates a special *alert flit* to be sent to its neighbours about the detection of the HT. In TAR, such routers are known as generators. Neighbours upon receiving the *alert flit* propagates the message further by creating a *propagation flit*. In TAR, routers generating the *propagation flits* are called propagators. The structure of these special flits is very similar to normal flits, as shown in Fig. 11.11. *Alert flit* contains a 1-bit *msg_dir* indicating the direction an *alert flit* needs to be forwarded by generators. A 3-bit *DHT_alert_dir* indicates the direction an *alert flit* needs to be forwarded by propagators. The alert message also contains a 3-bit *NHT_alert_dir* which indicates

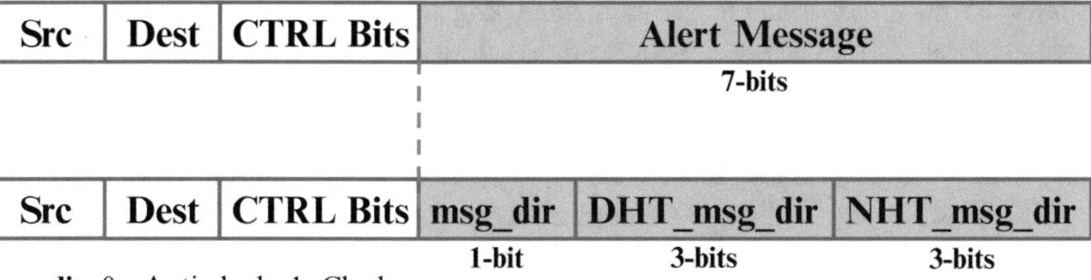

msg_dir: 0: Anti-clock, 1: Clock

DHT_msg_dir/NHT_msg_dir: 000: No Dir, 001: North, 010: West, 011: South, 100: East

Fig. 11.11 Structure of an alert_flit

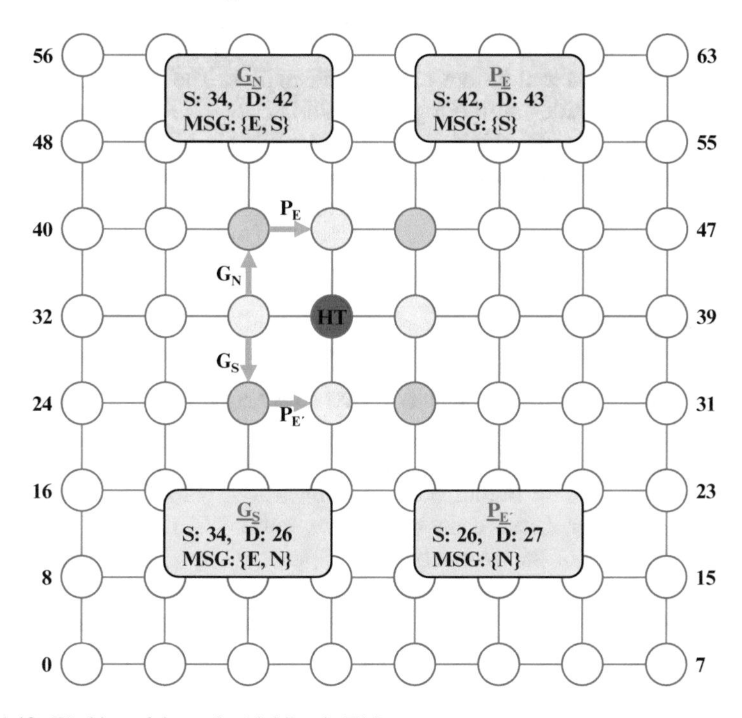

Fig. 11.12 Working of dynamic shielding in TAR

the direction where the HT is detected. Figure 11.11 presents all the possible values for different fields of the *alert flit*. When the message of HT detection is propagated among all the neighbouring routers using *alert* and *propagation flits*, each router accordingly updates its *alert_flag* and *alert_dir*. This results in a shield creation around the HT that successfully isolates the HT infected router from the rest of the network. The third and final phase of TAR uses this shielding to route packets by bypassing the isolated HT infected router.

Working of the dynamic shielding phase in TAR is explained using Fig. 11.12. From the previous phase of HT detection, let us assume that router 34 has identified router 35 as an HT infected router. *alert_flag* in router 34 is now set to 1 and *alert_dir* as 100 (East). As shown in Fig. 11.12, router 34 generates two alert flits, G_N and G_S. With an alert message {*msg_dir* = 0, *DHT_alert_dir* = 100, *NHT_alert_dir* = 011}, alert flit G_N is forwarded from router 34 to router 42, where *msg_dir* = 0 indicates G_N to be forwarded in clockwise direction. *DHT_alert_dir* = 100 (East) in G_N indicates that upon reaching router 42, the message needs to be propagated in East direction. Router 42 generates a propagation flit P_E with an alert message {*msg_dir* = 0, *DHT_alert_dir* = 000, *NHT_alert_dir* = 011} to be forwarded to router 43. When P_E reaches router 43, *NHT_alert_dir* = 011 (South) indicates that the HT is detected in South direction of router 43; which is router 35. *alert_flag* and *alert_dir* are updated as 1 and south, respectively, in router 43 which can be a generator for other neighbours. Similarly, G_S and $P_{E'}$ also propagate the message of HT detection to other neighbours. Here, 27, 34, 43 and 36 are generator routers and 26, 42, 44 and 28 are propagation routers. The message propagation continues from both sides until a logical shield is created around the HT infected router. In this example, the shield is completed when *alert_dir* is set for router 27 as north, router 34 as east, router 43 as south and router 36 as west. After the end of dynamic shielding, the detected HT infected router is isolated from rest of the network.

11.3.2.3 Bypassing the Trojan

The final phase of TAR implements a bypass routing mechanism, as presented in Algorithm 2. When a packet arrives at a router, bypass mechanism checks the *alert_flag* and *alert_dir* of that router. Only if the *alert_flag* is set and *alert_dir* matches with the desired output port direction of the packet, bypass routing is activated. In all other cases, a packet follows normal X-Y routing to reach its destination. Working of the Trojan bypassing phase is explained using Fig. 11.13. Let us consider the same scenarios of DoS and delay of service attacks shown in Figs. 11.8 and 11.10 for the sake of simplicity and continuity. A packet $P1$ with source $S1$ on its way to destination $D1$ reaches router 36. After the completion of shielding in the previous phase, router 36 has its *alert_flag* set and *alert_dir* as west. As per X-Y routing, the desired output port of packet $P1$ at router 36 is west which matches with the *alert_dir* of router 36. Now the Trojan bypass algorithm initiates and reroutes packet $P1$ away from the HT infected router 35 as presented in Part I of Algorithm 2. Packet $P1$ is rerouted from router 36 to router 44, and Part II of Algorithm 2 is initiated since 44 is a propagation router. Now, packet $P1$ is forwarded from router 44 to router 43, and from there it directly reaches destination $D1$ at router 59.

Since destination $D1$ is in the same column as that of HT infected router 35, with HT activated, it becomes impossible for $P1$ to reach $D1$ using the conventional approach and hence it results into a DoS like scenario. With the Trojan bypass algorithm in place, now $P1$ can reach its destination, thus mitigating the impact of DoS. Since packets like $P1$ are not trapped in the network anymore, the proposed

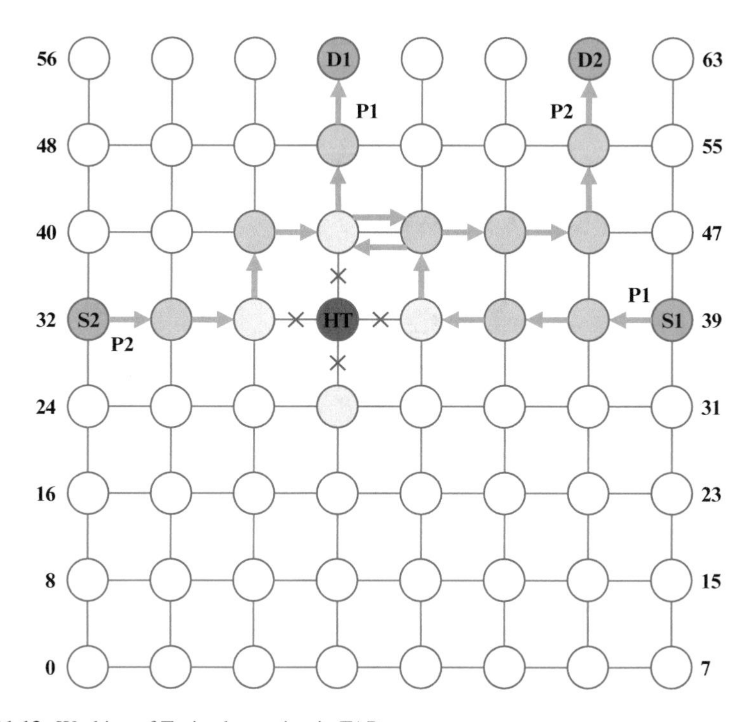

Fig. 11.13 Working of Trojan bypassing in TAR

bypass routing also diminishes the possibility of injection suppression. Similarly, packet $P2$ with source $S2$ on its way to destination $D2$ reaches router 34. Instead of forwarding to router 35 which is HT infected, router 34 reroutes $P2$ towards router 42. The Trojan bypass algorithm rerouted packet $P2$ in such a way that it reaches destination $D2$ without any additional delay. Hence, the delay of service scenario created by the proposed HT threat model in SECTAR is mitigated by intelligent bypassing. Note that router 35 misroutes only those packets that are passing through it. Hence, even after bypassing is activated, the packets whose source or destination is router 35 will continue to come out of/go into router 35, thus not hampering the application executing in the infected core. Due to the nature of run-time detection, when an HT is detected, it might have already misrouted first few flits of some packets while rest of the flits are on the way. Intuitively, it seems that the bypassing algorithm will not allow the rest of the flits to travel to the HT infected router in order to avoid misrouting. However, this situation will not arise since only the head flit takes part in routing and arbitration. Hence, if a head flit is already misrouted before HT detection, all the following flits will go through the same route. After HT detection, when such a misrouted head flit comes out of the HT infected router due to the ping-pong effect, it will never enter the HT again due to the employed bypassing. Hence, even misrouted flits will eventually reach their respective destination.

Rerouting packets using the bypass algorithm violates normal X-Y routing and creates a possibility for network deadlock. To ensure deadlock prevention, TAR employed the concept of intermediate destination [18]. When packet $P2$ is rerouted

Algorithm 2: Trojan bypass

```
Input: Packet header
Output: Output port direction of a flit
/*Part I: Mitigation by generator routers */
if alert flag is SET and current router is generator then
    if x_diff and y_diff is not equal to zero then
        if alertdir is EAST or WEST then
            if y_diff < 0 then
                set outdir as SOUTH
            else
                set outdir as NORTH
            end
    else if x_diff is zero then
        if (y_diff > 0 and alertdir is NORTH) ||
        (y_diff < 0 and alertdir is SOUTH) then
            set outdir as EAST or WEST
    else if y_diff is zero then
        if (x_diff > 0 and alertdir is EAST) ||
        (x_diff < 0 and alertdir is WEST) then
            set outdir as NORTH or SOUTH
    end
/*Part II: Mitigation by propagation routers */
else if alert flag is RESET and current router is propagator then
    if (x_diff < 0 and indir is WEST) ||
    (x_diff > 0 and indir is EAST) then
        if y_diff < 0 then
            set outdir as SOUTH
        else
            set outdir as NORTH
        end
    else if x_diff < 0 and indir is SOUTH then
        set outdir as WEST
    else if x_diff > 0 and indir is NORTH then
        set outdir as EAST
    end
end
```

from router 34 to router 42, it starts travelling in the Y direction. However, when it travels from router 42 to router 43, $P2$ violates X-Y routing, since turning X from Y direction is prohibited. Using the concept of intermediate destination [18], router 42 is made the new destination for packet $P2$. Now, after getting rerouted from router 34, packet $P2$ reaches router 42 and gets ejected into its local output port, since 42 is the new destination. Only after router 42 finds out that $P2$ is meant for destination $D2$ at router 62, it re-injects $P2$ as a new packet destined for $D2$. Packet $P2$ now follows normal X-Y routing like any other packet to reach the destination. The ejection of packet $P2$ and re-injection as a new packet from the intermediate destination 42 makes sure that X-Y routing is not violated thus eliminating deadlock.

11.4 Performance Evaluation

To evaluate the performance of SECTAR, three system models are considered: (1) A baseline system to represent a normal NoC based MPSoC without any embedded Trojan (*Baseline*), (2) an NoC based MPSoC with an HT infected router (*HT*) and (3) an NoC based MPSoC integrated with the Trojan aware routing technique (*TAR*).

11.4.1 Simulation Framework and Workloads

For all the system models to be evaluated (Baseline, HT and TAR), SECTAR considered a 64-core NoC based MPSoC. Each core contains a simple CPU, and 32KB, 4-way set associative, 64B block, private L1 instruction and data caches. L2 cache is shared, and distributed as multiple banks using SNUCA technique. So, each core also has a 256KB, 16-way associative, 64B block, shared L2 cache bank. These MPSoCs use a traditional 8×8 2D mesh NoC with 5 VCs per input port routers and a 128-bit flit channel for inter-core communication. Usually, L1 cache miss triggers the generation of NoC request packets, which travels from source to the appropriate destination core where the corresponding L2 cache sets are mapped. Similarly, reply packets travel back to the requesting core through the underlying NoC. X-Y dimension order routing algorithm is employed for packet traversal, where request packets are 1-flit, and reply packets are of 5-flits. Garnet module in gem5 simulator [5] is modified to implement different router microarchitectures for Baseline, HT and TAR. The HT is modelled in such a way that there exists a single Trojan infected NoC router at specific times.

To evaluate the performance, *Uniform Random* and *Bit Complement* synthetic traffic patterns are used with varying injection rates. SECTAR also used SPEC CPU2006 benchmark based real workloads where one benchmark is assigned to each of the 64 cores of the MPSoC. Different categories of workloads are created by grouping the benchmarks based on their Misses Per Kilo Instructions (MPKIs). These categories are High MPKI (greater than 40), Medium MPKI (less than 40 but greater than 20) and Low MPKI (less than 20). *leslie3d*, *lbm*, *GemsFDTD* and *mcf* fall under High MPKI, *soplex* and *astar* under Medium MPKI and *sjeng*, *bzip2*, *omnetpp* and *sphinx* under Low MPKI, as given in Table 11.1. These categories are used to make six workload mixes, M1, M2, M3, M4, M5 and M6, each having 64 benchmark instances, as presented in Table 11.2.

Table 11.1 Benchmark classification based on MPKI

Category	Benchmarks
High MPKI	leslie3d, lbm, GemsFDTD, mcf
Medium MPKI	soplex, astar
Low MPKI	sjeng , bzip2, omnetpp, sphnix

Table 11.2 Workload characteristics

Workload	Benchmarks
M1	100% High MPKI
M2	100% Low MPKI
M3	100% Medium MPKI
M4	50% High MPKI & 50% Low MPKI
M5	50% Low MPKI & 50% Medium MPKI
M6	50% High MPKI & 50% Medium MPKI

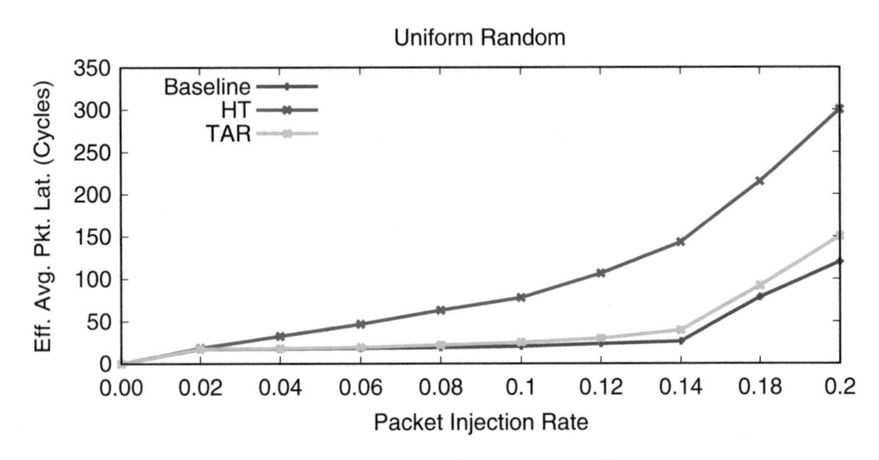

Fig. 11.14 Effective average packet latency: uniform random

11.4.2 Results and Discussion

11.4.2.1 Effective Average Packet Latency

Average Packet Latency (APL) is defined as the number of cycles required for a packet to reach its destination. To understand the effect on packet latency when the HT is triggered, APL is an appropriate metric. However, there are times when an HT infected NoC shows inconsistent latency due to packet loss. Hence, SECTAR used Effective APL (EAPL) to get a consistent measure. EAPL can be defined as:

$$EAPL = APL * \frac{Packets\,Ejected_{without\,HT}}{Packets\,Ejected_{with\,HT}} \tag{11.4}$$

While analysing EAPL using synthetic traffic patterns as shown in Figs. 11.14 and 11.15, it is observed that with increasing injection rate, packet latency also increases in Baseline, HT and TAR. Due to the deflection of packets by the HT router, the MPSoC experiences DoS and delay of service attacks. Hence, there is an escalation in latency in HT system model when compared to the Baseline and TAR models. However, TAR achieves a reduction in EAPL when compared to HT infected NoC model. As TAR uses HT bypassing to secure communication, majority

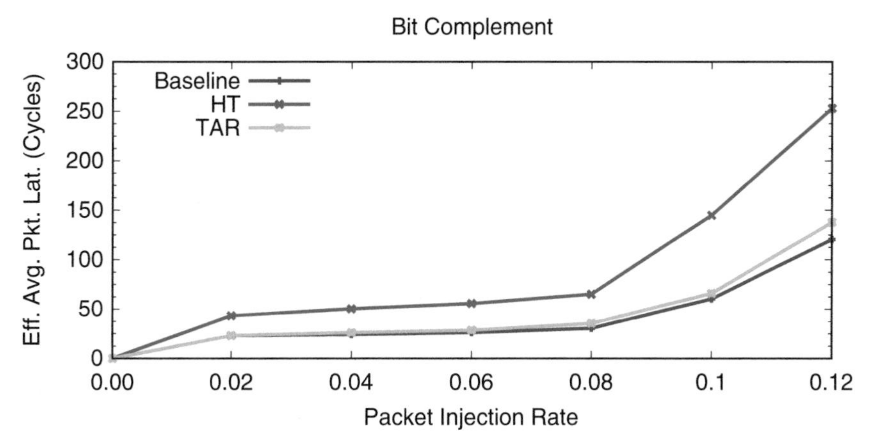

Fig. 11.15 Effective average packet latency: bit complement

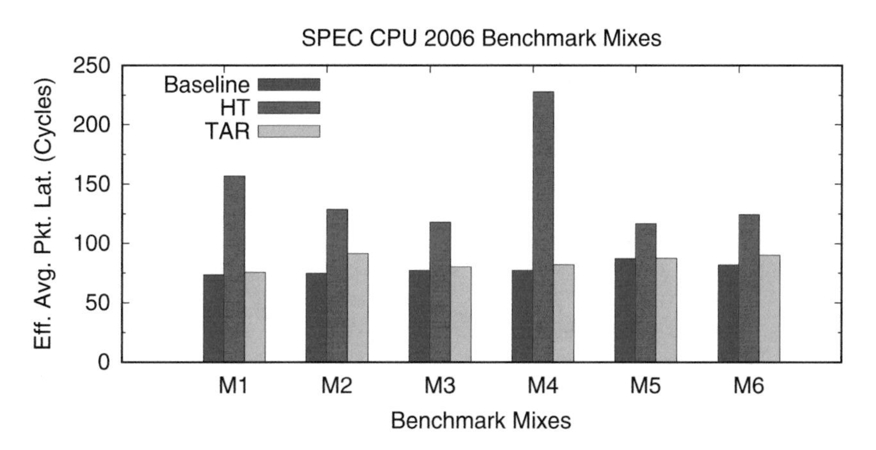

Fig. 11.16 Effective average packet latency: SPEC CPU2006 Workloads

of the packets that are supposed to travel through the HT infected router now take extra few hops through the intermediate destination to reach their actual destination. This leads to an increase in EAPL when compared to the Baseline model. Simulation results show that there is a 16% increase in latency when compared to the Baseline.

While analysing the EAPL using real workloads as shown in Fig. 11.16, it is observed that for all the workload mixes, HT triggering increases packet latency by an average of 87% over the Baseline. However, TAR exhibits a reduction in the EAPL by 38% with respect to HT infected NoC, with 7% increase in latency when compared to the Baseline. This is due to the bypass routing incorporated in the NoC routers.

11.4.2.2 Effective Average Deflected Packet Latency

A variant of Average Packet Latency (APL) called Average Deflected Packet Latency (ADPL) is defined as the APL of only those packets which are meant to travel through the HT infected router. Similar to EAPL, to get realistic measures, SECTAR used Effective ADPL (EADPL) which is given as:

$$EADPL = ADPL * \frac{Deflected\,Packets\,Ejected_{without\,HT}}{Deflected\,Packets\,Ejected_{with\,HT}} \tag{11.5}$$

Analysis using synthetic traffic patterns as given in Figs. 11.17 and 11.18 clearly shows that as injection rate increases, EADPL increases on the HT infected NoC. When HT is active, some of the packets get trapped in ping-pong state between HT's neighbours due to deflection. Because of its intermittent nature, when the HT is not

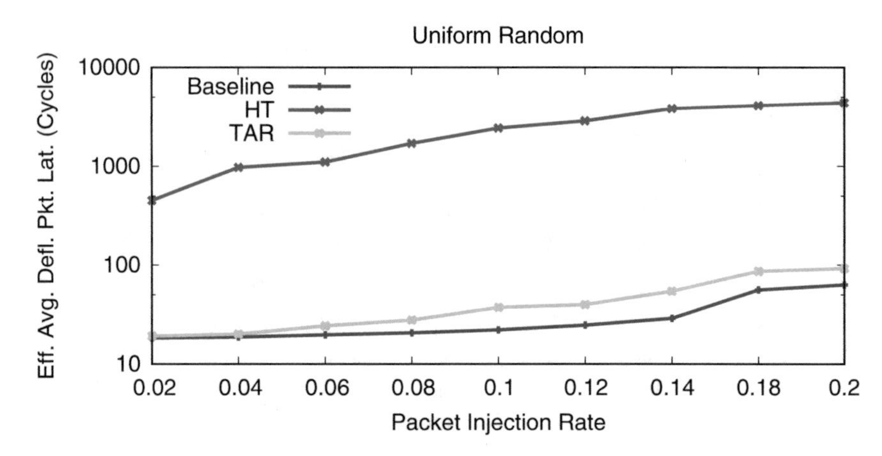

Fig. 11.17 Effective average deflected packet latency: uniform random

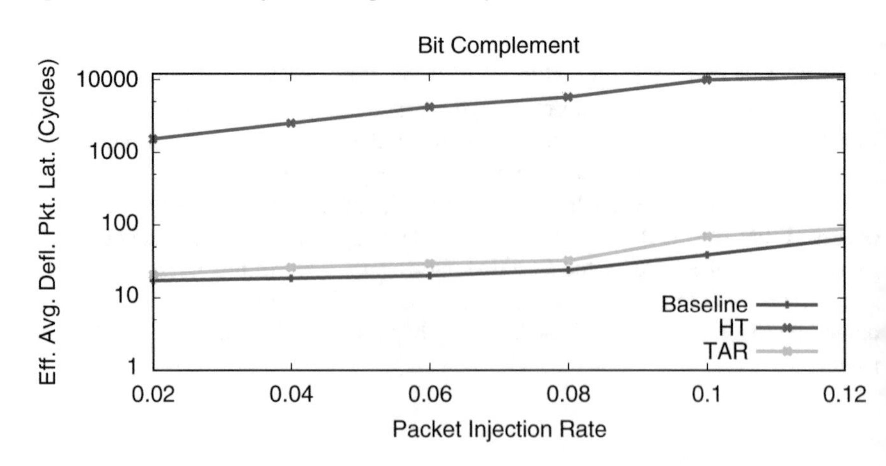

Fig. 11.18 Effective average deflected packet latency: bit complement

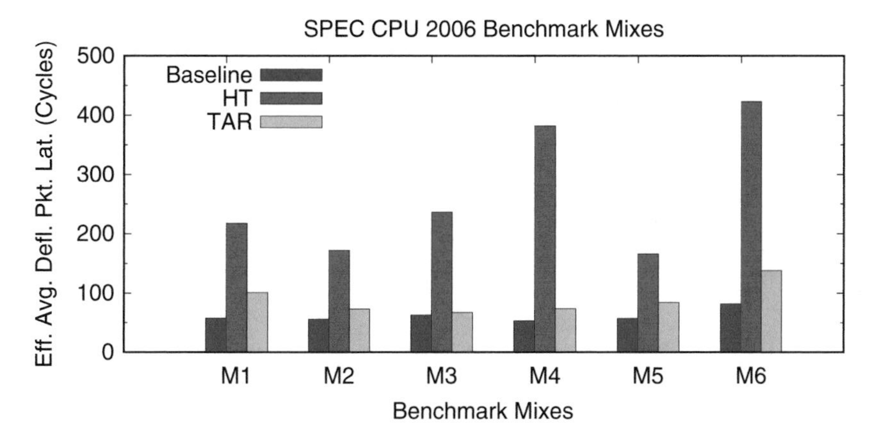

Fig. 11.19 Effective average deflected packet latency: SPEC CPU2006 workloads

active, some of these packets manage to escape the ping-pong state and reach their destination with increased latency. During heavy traffic in NoC (high injection rate), the ping-pong state exhausts router buffers and delays deflected packets even further. With the inclusion of TAR, EADPL reduces significantly as the victim packets are bypassed from the HT. Experimental results show that compared to HT infected NoC, TAR reduces EADPL by 97%. However, due to the rerouting of packets and ejection at intermediate destinations, TAR shows an average increase of 38% in EADPL when compared to the Baseline.

Analysis using SPEC CPU2006 benchmark based real workloads as given in Fig. 11.19 shows that across all workload mixes, TAR achieves 62% reduction in EADPL compared to HT infected NoC model, and 40% increase in EADPL when compared with the Baseline model.

11.4.2.3 Throughput

In every network, throughput plays a vital role in determining the Quality-of-Service (QoS) of the underlying applications. In the context of NoC, throughput is defined as the number of packets that have reached their destination per router per clock cycle. Throughput for the system models when using synthetic traffic patterns are shown in Figs. 11.20 and 11.21. It is clearly evident that the delivery rate of packets is almost similar for both Baseline and TAR models. It was possible due to the efficient implementation of the proposed Trojan aware routing. However, when it comes to HT infected NoC, more than 25% reduction in the delivery rate of packets is observed. This reduction is due to the ping-pong state and induced injection suppression. When analysing the SPEC CPU2006 benchmark based real workloads as shown in Fig. 11.22, the HT infected NoC receives around 80% fewer packets

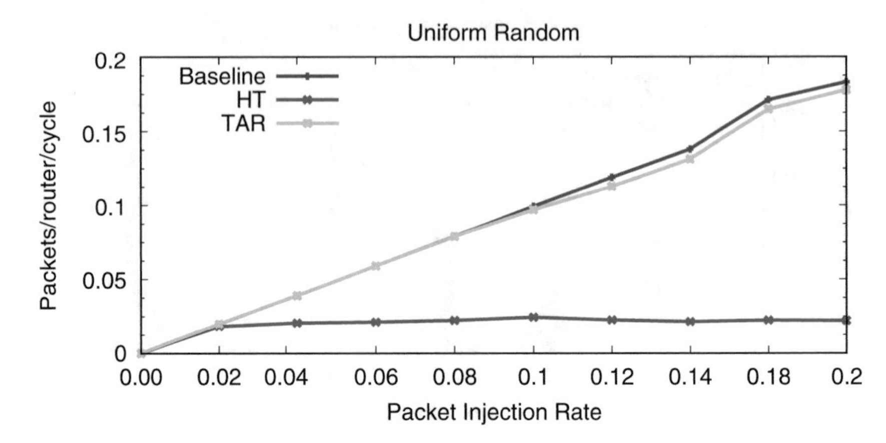

Fig. 11.20 Throughput: uniform random

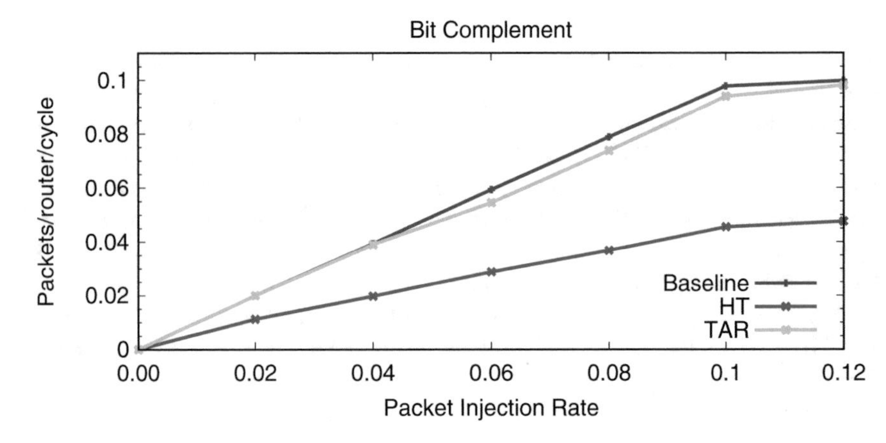

Fig. 11.21 Throughput: bit complement

when compared to the Baseline, while TAR model suffers only 6% throughput reduction.

11.4.2.4 Injection Suppression Avoidance

To understand how Trojan aware routing avoids injection suppression, buffer (VC) availability in NoC routers during the simulation is analysed. Experiments show that the number of packets processed around the HT infected router is very high. This is due to the ping-pong state of the packets created by the HT induced misrouting. The ping-pong state fills up all the VCs in the HT infected router as well as its neighbours. Unavailability of VCs creates back-pressure and eventually leads to injection suppression, as already shown in Fig. 11.9.

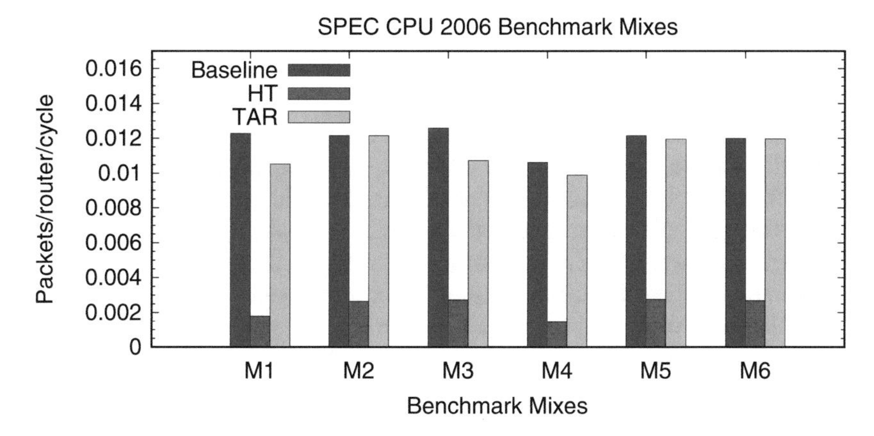

Fig. 11.22 Throughput: SPEC CPU2006 workloads

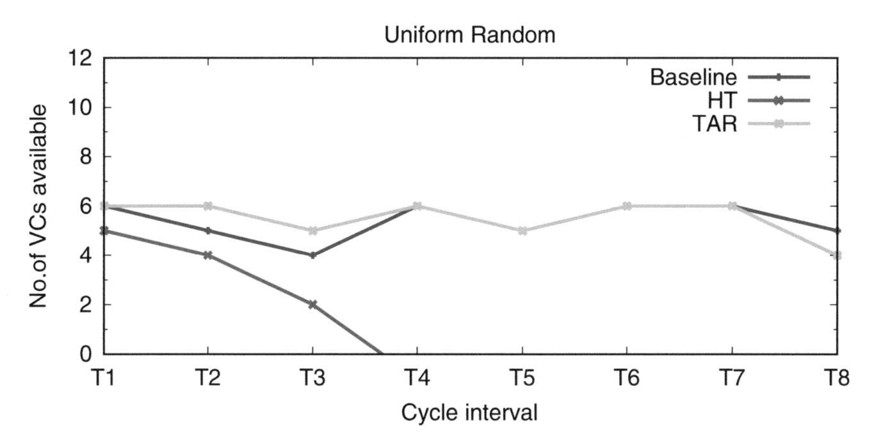

Fig. 11.23 Virtual channel (VC) availability

To analyse VC availability in NoC routers, the number of free VCs available in the routers at different time intervals is calculated, as given in Fig. 11.23. Results show that as the simulation progresses, the number of free VCs in HT infected NoC model reduces drastically. For example, when the simulation reaches time interval T4, VC availability in HT infected NoC model reaches zero. This represents the induced injection suppression in the entire NoC due to packet misrouting. However, the bypassing algorithm employed in the TAR system model ensures that a DoS attack scenario never arises. Since injection suppression is a by-product of DoS, no DoS meaning no injection suppression. Hence, the VC availability in TAR system model is maintained as close as in the Baseline model.

11.4.3 Overhead

SECTAR is implemented using standard 3-stage pipelined input buffered routers, where the stages are (1) buffer write and route computation, (2) VC allocation and switch allocation and (3) switch traversal. The detection module in TAR works in parallel with the route computation stage. Dynamic shielding is entirely independent and works in parallel to the regular router operations. However, the bypass algorithm works within the route computation stage. According to the existing literature [40], VC allocation and switch allocation stage is the slowest and determines the router pipeline latency. As the route compilation stage is comparatively smaller, even with the added bypassing logic, TAR enabled NoC routers can be operated at the same frequency at that of the Baseline model.

The usage of 1-bit *alert_flag* and 3-bit *alert_dir* in NoC routers result in a storage overhead of 4-bits per router. Therefore, in a 64-core NoC based MPSoC, only extra 32B (4-bits × 64-core) is added. To evaluate the area and power overheads, DSENT simulator [48] with 22nm processor technology at 1GHz operating frequency is used. The additional circuitry for detection, shielding and bypassing logic incurs a negligible area overhead of 2.78% and a leakage power overhead of 3% when compared to the Baseline NoC routers.

11.5 Summary

This chapter presented a broad picture of the widely popular hardware Trojans (HTs) in NoC based MPSoCs. It began with an introduction about HTs and different attack scenarios created by such HTs, like DoS, information leakage, data corruption and functional modification. Then, this chapter briefly talked about the placement of HTs at different places in NoC, their impact, detection and mitigation. Finally, it concentrated on a specific state-of-the-art technique called SECTAR that is capable of enabling trusted NoC communication in the presence of HTs at routers. The intermittent HT in SECTAR misroutes packets to initiate DoS, injection suppression and delay of service like attack scenarios. To deal with such an HT, SECTAR proposed Trojan Aware Routing (TAR) that dynamically detects a misrouting HT and isolates the HT by routing packets away from it. Experimental evaluation with synthetic and real workloads validated that TAR is capable of implementing a trusted NoC communication with graceful degradation in overall system performance.

Acknowledgments This work was partially supported by the US National Science Foundation (NSF) grant SaTC-1936040.

References

1. A. Ahmed, F. Farahmandi, Y. Iskander, P. Mishra, Scalable hardware Trojan activation by interleaving concrete simulation and symbolic execution, in *Proceedings of the IEEE International Test Conference (ITC)* (2018)
2. A. Ahmed, F. Farahmandi, P. Mishra, Automated activation of multiple targets in RTL models using concolic testing, in *Proceedings of the Design Automation & Test in Europe (DATE)* (2018), pp. 1538–1543
3. D.M. Ancajas, K. Chakraborty, S. Roy, Fort-NoCs: mitigating the threat of a compromised NoC, in *Proceedings of the Design Automation Conference (DAC)* (2014)
4. S. Bhunia, M.S. Hsiao, M. Banga, S. Narasimhan, Hardware Trojan attacks: threat analysis and countermeasures. Proc. IEEE **102**(8), 1229–1247 (2014)
5. N. Binkert, B. Beckmann, G. Black, S.K. Reinhardt, A. Saidi, A. Basu, J. Hestness, D.R. Hower, T. Krishna, S. Sardashti, R. Sen, K. Sewell, M. Shoaib, N. Vaish, M.D. Hill, D.A. Wood, The gem5 Simulator. ACM SIGARCH Comput. Archit. News **39**(2), 1–7 (2011)
6. A.K. Biswas, S.K. Nandy, R. Narayan, Router attack toward NoC-enabled MPSoC and monitoring countermeasures against such threat. Circ. Syst. Signal Process. **34**(10), 3241–3290 (2015)
7. T. Boraten, A.K. Kodi, Mitigation of denial of service attack with hardware Trojans in NoC architectures, in *Proceedings of the International Parallel and Distributed Processing Symposium (IPDPS)* (2016)
8. S. Charles, P. Mishra, Reconfigurable network-on-chip security architecture. ACM Trans. Des. Autom. Electron. Syst. **25**(6), 1–25 (2020)
9. S. Charles, P. Mishra, Lightweight and trust-aware routing in NoC based SoCs, in *Proceedings of the IEEE Computer Society Annual Symposium on VLSI (ISVLSI)* (2020)
10. S. Charles, P. Mishra, Securing network-on-chip using incremental cryptography, in *Proceedings of the IEEE Computer Society Annual Symposium on VLSI (ISVLSI)* (2020)
11. S. Charles, C. Patil, U. Ogras, P. Mishra, Exploration of memory and cluster modes in directory-based many-core CMPs, in *Proceedings of the IEEE/ACM International Symposium on Networks-on-Chip (NOCS)* (2018)
12. S. Charles, Y. Lyu, P. Mishra, Real-time detection and localization of DoS attacks in NoC based SoCs, in *Proceedings of the Design Automation & Test in Europe (DATE)* (2019), pp. 1160–1165
13. S. Charles, M. Logan, P. Mishra, Lightweight anonymous routing in NoC based SoCs, in *Proceedings of the Design Automation & Test in Europe (DATE)* (2019)
14. S. Charles, Y. Lyu, P. Mishra, Real-time detection and localization of distributed DoS attacks in NoC based SoCs, in *IEEE Transactions on Computer-Aided Design of Integrated Circuits and Systems* (2020)
15. M. Chen, X. Qin, H. Koo, P. Mishra, *System-Level Validation: High-Level Modeling and Directed Test Generation Techniques* (Springer, New York, 2012)
16. J. Coburn, S. Ravi, A. Raghunathan, S. Chakradhar, SECA: Security-Enhanced Communication Architecture, in *Proceedings of the International Conference on Compilers, Architectures and Synthesis for Embedded Systems (CASES)* (2005)
17. L. Daoud, N. Rafla, Routing aware and runtime detection for infected network-on-chip routers, in *Proceedings of the IEEE International Midwest Symposium on Circuits and Systems (MWSCAS)* (2018)
18. A. Das, S. Babu, J. Jose, S. Jose, M. Palesi, Critical packet prioritisation by Slack-aware re-routing in on-chip networks, in *Proceedings of the IEEE/ACM International Symposium on Networks-on-Chip (NOCS)* (2018)
19. F. Farahmandi, P. Mishra, Automated debugging of arithmetic circuits using incremental gröbner basis reduction, in *Proceedings of the IEEE International Conference on Computer Design (ICCD)* (2017), 193–200

20. F. Farahmandi, P. Mishra, FSM anomaly detection using formal analysis, in *Proceedings of the IEEE International Conference on Computer Design (ICCD)* (2017), pp. 313–320
21. F. Farahmandi, P. Mishra, Automated test generation for debugging multiple bugs in arithmetic circuits. IEEE Trans. Comput. **68**(2), 182–197 (2018)
22. F. Farahmandi, P. Mishra, *Post-Silicon Validation and Debug* (Springer, New York, 2019)
23. F. Farahmandi, Y. Huang, P. Mishra, FSM anomaly detection using formal analysis, in *Proceedings of the Asia and South Pacific Design Automation Conference (ASPDAC)* (2017), pp. 591–597
24. F. Farahmandi, Y. Huang, P. Mishra, *System-on-Chip Security: Validation and Verification* (Springer, New York, 2019)
25. J. Frey, Q. Yu, A hardened network-on-chip design using runtime hardware Trojan mitigation methods. Integration **56**(1), 15–31 (2017)
26. Y. Huang, S. Bhunia, P. Mishra, MERS: statistical test generation for side-channel analysis based Trojan detection, in *Proceedings of the ACM SIGSAC Conference on Computer and Communications Security (CCS)* (2016), pp. 130–141
27. Y. Huang, S. Bhunia, P. Mishra, Scalable test generation for Trojan detection using side channel analysis. IEEE Trans. Inf. Forensics Secur. **13**(11), 2746–2760 (2018)
28. H. Li, Q. Liu, J. Zhang, A survey of hardware Trojan threat and defense. Integration **55**(1), 426–437 (2016)
29. Y. Lyu, P. Mishra, A survey of side-channel attacks on caches and countermeasures. Springer J. Hardw. Syst. Secur. **2**(1), 33–50 (2018)
30. Y. Lyu, P. Mishra, Efficient test generation for Trojan detection using side channel analysis, in *Proceedings of the Design, Automation & Test in Europe (DATE)* (2019), pp. 408–413
31. Y. Lyu, P. Mishra, Automated test generation for Trojan detection using delay-based side channel analysis, in *Proceedings of the Design, Automation & Test in Europe (DATE)* (2020), pp. 1031–1036
32. Y. Lyu, P. Mishra, Automated test generation for activation of assertions in RTL models, in *Proceedings of the Asia and South Pacific Design Automation Conference (ASPDAC)* (2020), pp. 223–228
33. Y. Lyu, P. Mishra, Scalable concolic testing of RTL models, in *IEEE Transactions on Computers* (2020)
34. Y. Lyu, P. Mishra, Scalable activation of rare triggers in hardware Trojans by repeated maximal clique sampling, in *IEEE Transactions on Computer-Aided Design of Integrated Circuits and Systems* (2020)
35. Y. Lyu, P. Mishra, Automated trigger activation by repeated maximal clique sampling, in *Proceedings of the Asia and South Pacific Design Automation Conference (ASPDAC)* (2020), pp. 482–487
36. Y. Lyu, X. Qin, M. Chen, P. Mishra, Directed test generation for validation of cache coherence protocols. IEEE Trans. Comput.-Aided Des. Integr. Circ. Syst. **38**(1), 163–176 (2018)
37. Y. Lyu, A. Ahmed, P. Mishra, Automated activation of multiple targets in RTL models using concolic testing, in *Proceedings of the Design Automation & Test in Europe (DATE)* (2019), pp. 254–359
38. P. Mishra, S. Bhunia, P. Mishra, *Hardware IP security and Trust* (Springer, New York, 2017)
39. S. Mitra, H.S.P. Wong, S. Wong, The Trojan-proof chip. IEEE Spectrum **52**(2), 46–51 (2015)
40. C. Nicopoulos, S. Srinivasan, A. Yanamandra, D. Park, V. Narayanan, C.R. Das, M.J. Irwin, On the effects of process variation in network-on-chip architectures. IEEE Trans. Depend. Secure Comput. **7**(3), 240–254 (2008)
41. Z. Pan, P. Mishra, Automated test generation for hardware Trojan detection using reinforcement learning, in *Proceedings of the Asia and South Pacific Design Automation Conference (ASPDAC)* (2021)
42. Z. Pan, J. Sheldon, P. Mishra, Test generation using reinforcement learning for delay-based side channel analysis, in Proceedings of the IEEE/ACM International Conference on Computer-Aided Design (ICCAD) (2020)

43. N. Potlapally, Hardware security in practice: challenges and opportunities, in *Proceedings of the IEEE International Symposium on Hardware Oriented Security and Trust (HOST)* (2011)
44. M. Rajan, A. Das, J. Jose, P. Mishra, SECTAR: secure NoC using Trojan aware routing, in *Proceedings of the IEEE/ACM International Symposium on Networks-on-Chip (NOCS)* (2020)
45. V.Y. Raparti, S. Pasricha, Lightweight mitigation of hardware Trojan attacks in NoC-based manycore computing, in *Proceedings of the Design Automation Conference (DAC)* (2019)
46. Researchers discover troubling new security flaw in all modern Intel processors (2019). https://tinyurl.com/y35n9bzk
47. C. Rooney, A. Seeam, X. Bellekens, Creation and detection of hardware Trojans using non-invasive off-the-shelf technologies. Electronics **7**(7), 124 (2018)
48. C. Sun, C.H.O. Chen, G. Kurian, L. Wei, J. Miller, A. Agarwal, L.S. Peh, V. Stojanovic, A tool connecting emerging photonics with electronics for opto-electronic networks-on-chip modeling, in *Proceedings of the IEEE/ACM International Symposium on Networks-on-Chip (NOCS)* (2012)
49. M. Tehranipoor, F. Koushanfar, A survey of hardware Trojan taxonomy and detection. IEEE Des. Test Comput. **27**(1), 10–25 (2010)
50. The Quantum Program of NSA (2014). http://www.nytimes.com/2014/01/15/us/nsa-effort-pries-open-computers-not-connected-to-internet.html

Part IV
NoC Validation and Verification

Chapter 12
Network-on-Chip Security and Trust Verification

Aruna Jayasena, Subodha Charles, and Prabhat Mishra

12.1 Introduction

Network-on-chip (NoC) provides a scalable on-chip communication architecture that enables energy-efficient communication between hundreds of diverse intellectual property (IP) cores in modern system-on-chip (SoC) designs. Figure 12.1 shows a typical SoC utilizing NoC to communicate between IP cores such as processors, memory, controllers, etc. The complexity of modern SoC designs coupled with time-to-market deadlines have motivated manufacturers to design few IPs in-house and outsource the rest to third-party vendors. While this trend of globally distributed supply chain of designing, manufacturing, and testing has increased manufacturing efficiency, it paved way for security concerns.

In order to meet the performance requirements of different IP cores, NoC design has evolved to be quite complex as different techniques are employed to accommodate high communication bandwidth. As shown in Fig. 12.2, data is transmitted as "packets" that can be divided further into smaller blocks (of fixed length) called "flits" inside NoC. Packets injected by IPs through the network interfaces (NI) are transmitted to their destinations via a network of different routers and communication links, according to a given routing protocol. These routers consist of structures employing various advanced features, such as shared buffers, message prioritization, complex allocation strategies and pipelining methodologies [54]. With these advanced performance features, it has definitely become a

---

A. Jayasena (✉) · P. Mishra
University of Florida, Gainesville, FL, USA
e-mail: arunajayasena@ufl.edu; prabhat@ufl.edu

S. Charles
University of Moratuwa, Colombo, Sri Lanka
e-mail: s.charles@ieee.org

P. Mishra, S. Charles (eds.), *Network-on-Chip Security and Privacy*,
https://doi.org/10.1007/978-3-030-69131-8_12

Fig. 12.1 An example SoC with NoC-based communication fabric to interact with a wide variety of third-party IP cores

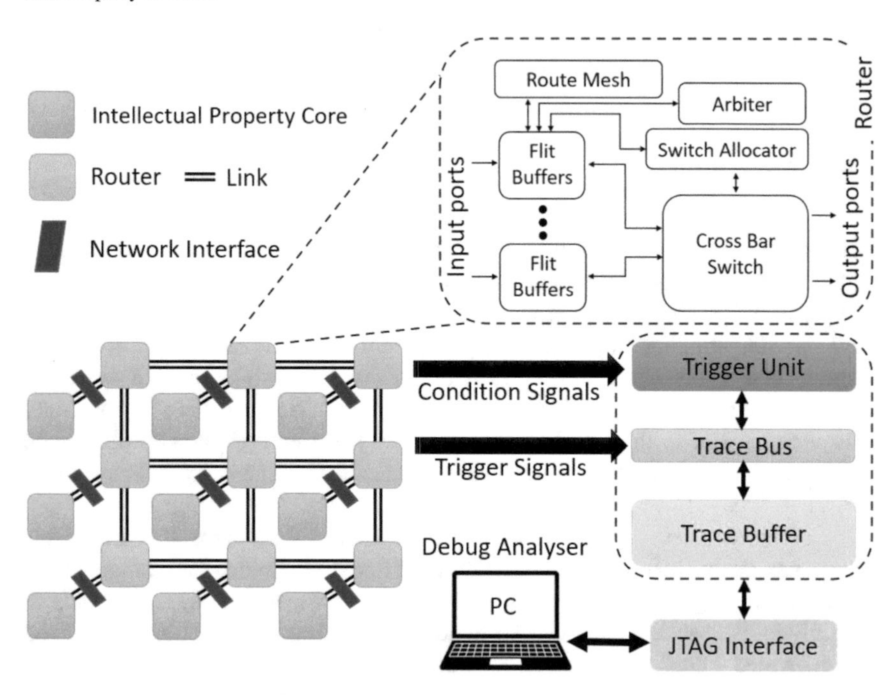

Fig. 12.2 Overview of a Network-on-Chip (NoC) architecture and associated post-silicon debug platform

challenge to ensure correct functionality under all possible scenarios for the entire NoC [1, 5, 27, 54].

Malicious implants such as hardware Trojans can be integrated into the designs during the long IP supply chain [9]. Once integrated, hardware Trojans can contribute to a plethora of attack scenarios such as industrial espionage, device malfunction, and device performance degradation. The distributed nature of the NoC across the SoC makes the impact of attacks even more severe. A survey conducted by an independent market research firm revealed that the FlexNoC on-chip interconnection architecture [5] is used by four out of the top five Chinese fabless semiconductor companies [61]. As a result, Arteris, the company that designed the FlexNoC architecture, has achieved a sales growth of 1002% over three years through IP licensing [8]. Such widespread usage of NoC IPs coupled with the distributed manufacturing process make the NoC a focal point of security attacks. Therefore, developing security countermeasures against potential attacks have been a primary focus of SoC designers [12–14, 16, 18–20]. Recent efforts try to combine the advantages of logic testing and side-channel analysis for effective Trojan detection in integrated circuits [30, 31, 35, 36, 39, 53].

Traditional SoC validation methodology is unlikely to detect malicious implants due to various reasons including: (1) it is infeasible to get 100% coverage of functional scenarios for complex (billion-gate) SoC designs [3, 21, 22, 24, 37, 42–44, 47], (2) Trojans can stay benign and act maliciously when a predefined trigger condition is met, which can be extremely rare [48], and (3) a carefully crafted Trojan has a very low performance and power footprint that can be hidden in typical process variations and environmental noise margins. For example, Ancajas et al. [7] presented an eavesdropping attack based on NoC packet duplication at routers that incur only 4.62%, 0.28%, and 1% area, power, and performance overhead, respectively. Therefore, the likelihood of the Trojan being detected is very small unless suitable security validation mechanism is employed [2, 23, 25, 26, 40, 41, 48, 52]. In this chapter, we describe a NoC security and trust validation framework consisting of assertion-based validation, formal verification, and post-silicon debug of security vulnerabilities.

NoC security validation can be performed during design time (pre-silicon validation) or post fabrication (post-silicon validation and debug). Formal verification and simulation-based validation are two primary techniques for pre-silicon NoC validation. While formal methods can ensure 100% coverage of a design, the complexity of IP designs make the exploration space grow exponentially, making 100% coverage infeasible. On the other hand, simulation-based techniques cannot provide guarantees about the verification completeness. Typically, an effective combination of formal verification, simulation-based validation, and post-silicon debug are used for verifying NoC security vulnerabilities.

The remainder of this chapter is organized as follows. Section 12.2 describes NoC architecture as well as security vulnerabilities. The next three sections cover NoC security validation using formal verification, simulation-based validation, and post-silicon debug. Section 12.6 presents experimental results. Finally, Sect. 12.7 concludes the chapter.

12.2 Network-on-Chip Architectures and Security Vulnerabilities

In this section, we first provide a brief overview of Network-on-Chip (NoC) architectures. Next, we discuss NoC security vulnerabilities.

12.2.1 Network-on-Chip (NoC) Architectures

Figure 12.2 shows an example NoC architecture consisting of several IPs connected together via routers and electrical wires (links). IPs are connected to the routers via a network interface (NI). The combination of an IP, an NI, and a router is referred as a "node" in the NoC. NoC architectures use packets to communicate between IPs. For example, when a memory instruction (Load/Store) is executed by source IP S, the private caches located in the same node are checked first and if it is a miss, the off-chip memory at destination IP D has to be accessed to retrieve the data. Therefore, a memory fetch request message is created and injected in the appropriate virtual network. The message created by the IP is first received by the NI, which converts it to network packets before sending the packets into the network via the local router.[1] The packets are routed through the routers and links according to the routing protocol until the destination node is reached. The NI connected to D re-creates the message from the packets and passes it to D, which initiates the memory access. The response message from memory follows a similar process when going from D to S. Similarly, all IPs integrated in the SoC leverage the resources provided by the NoC to communicate with each other.

12.2.2 NoC Security Vulnerabilities

We consider an attacker who is able to tamper with the NoC IP and implant Trojans in the routers during design time. The malicious implants can be inserted by a rogue designer, buggy design automation/computer-aided design (CAD) tools, or at the foundry via reverse engineering [9]. Once integrated, the Trojans remain hidden (deactivated) in order to avoid detection. Pre-programmed wake times and/or a specific activation logic can be used to fully activate the Trojans. Even when behaving maliciously, Trojans exhibit negligible power and performance overhead. For example, Sepúlveda et al. [59] explored a similar threat model and showed that a malicious router that can corrupt/duplicate/misroute packets incurs 2.2%, 0.2%, and 0.3% area, power, and performance overhead, respectively, when compared to normal operation.

In this chapter, we assume that the Trojans can duplicate, corrupt, drop, misroute, and starve packets when packets pass through routers. Figure 12.3 shows a block diagram of a Trojan architecture that can facilitate the attacks. The capabilities of the Trojan include all possible attacks that can be caused by a Trojan-infected router as outlined in the rest of this section.

---

[1] Most NoC architectures facilitate flits, which is a further breakdown of a packet used for flow control purposes. We stick to the level of packets for the ease of explanation as these methods remain the same at the flit level as well.

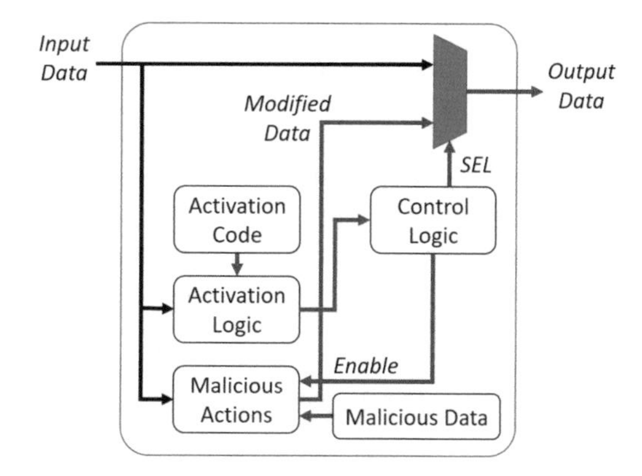

Fig. 12.3 Trojan architecture

12.2.2.1 Packet Duplication

IPs rely on the NoC to ensure secure data communication. An attacker can eavesdrop on the packets in an attempt to leak sensitive information. A common threat model is a hardware-software coalition attack where a Trojan-infected router and an accomplice application work together to eavesdrop. When packets are received at the input buffer of the router, the Trojan copies the packets, modifies the destination address in the header so that the new destination is an IP that runs an accomplice malicious application, and places it back in the input buffer. The NoC then routes the duplicated packets to the malicious application. The same threat model has been widely used to explore eavesdropping attacks in NoC [7, 19].

12.2.2.2 Packet Corruption

SoC relies on the integrity of data communicated through the NoC for correct execution of tasks. If an attacker corrupts data intentionally, it can cause erroneous behavior and/or system failure. Furthermore, since corrupted data can trigger re-transmissions, it can incur significant power and performance overheads leading to denial-of-service attacks. While designers employ a wide variety of techniques to improve energy efficiency in NoC-based SoCs [4, 15, 17, 28, 29, 63], these methods are not suitable here. The Trojan architecture in Fig. 12.3 facilitates data corruption by replacing the packet content with the content in a malicious register. A similar threat model utilized packet corruption at a router to discuss eavesdropping, denial-of-service and illegal packet forwarding based attacks [32].

12.2.2.3 Packet Starvation

The SoC operation and performance guarantees can rely on a few critical components. For example, the response time of a memory controller that provides the interface to off-chip memory can be critical in serving all the memory requests. If an attacker intentionally delays packets originating from such a critical component, the SoC performance can suffer significant degradation. Delays can lead to catastrophic consequences in real-time safety-critical applications. A Trojan can selectively delay packets originating from an IP, which is referred to as "packet starvation." Starvation can be caused by a Trojan-infected router de-prioritizing packets from a particular origin at the arbiter [55]. In other words, packets are treated unfairly such that all the input ports do not get an equal chance of accessing the output.

12.2.2.4 Packet Dropping

Packet dropping is considered as the next step of packet starvation. In starvation, packets are intentionally delayed and can reach the destination at some point of time. However, when the packets are dropped, the destination will not receive the packets unless they are re-transmitted. Similar to the consequences of starvation, packet dropping can cause severe performance degradation and malfunction [46].

12.2.2.5 Packet Misrouting

The NoC uses routing protocols to route packets between the senders and the receivers. A key requirement of routing protocols is to ensure packet routing without causing deadlocks and livelocks. A Trojan that corrupts packet header information and/or routing tables can force some packets to loop around and force deadlocks and livelocks. Such attacks are capable of rendering single application to full chip failures [10]. Re-routing packets are also a critical component in eavesdropping attacks as explained above (see *Packet Duplication*).

12.3 Formal Verification of NoC Security Vulnerabilities

Formal verification is widely used for functional as well as security validation of hardware designs. In this chapter, we focus on how formal verification can be employed for checking the correctness of NoC security properties [60]. Formal verification requires a set of properties that NoC design should satisfy. Figure 12.4 shows an overview of formal verification of NoC architectures that consists of two important steps [60]: property definition and formal verification. We briefly describe these steps. The details are available in [60].

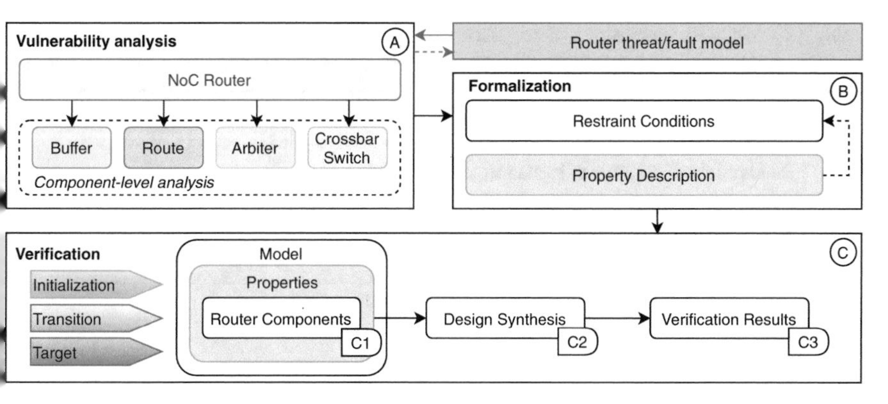

Fig. 12.4 Formal verification of NoC security properties [60]

12.3.1 Definition of NoC Security Properties

The security properties can be defined based on the threat models outlined in Sect. 12.2.2. Table 12.1 elaborates the properties where the first column identifies the property number (P#) and the second column describes the property as well as temporal logic description. Notations used to denote these properties are shown in Table 12.2. Each attack type discussed in Sect. 12.2.2 corresponds to several combinations of properties. The mapping between the attack scenarios and the combined properties is illustrated in the Table 12.3. For example, if we consider the packet duplication attack (discussed in Sect. 12.2.2.1), the combined property corresponding to that is indicated as P6, P11, and P14 from first row in Table 12.3.

12.3.2 Verification of NoC Security Properties

There are many model (property) checking tools for verifying security properties. For example, satisfiability (SAT)-based bounded model checking can accept the NoC design and associated property as inputs, and check whether the NoC design satisfies the property. In case of a failure, the tool produces a counterexample that can be used to localize the vulnerability and fix it.

12.4 Simulation-Based Validation Using Security Assertions

A major challenge in NoC validation is how to increase controllability and observability of the hardware design. The ability to control the internal signal is referred to as controllability, whereas observability refers to the ability to view

Table 12.1 Security properties to detect various NoC attacks

P#	Description of security property
$P1$	Write pointers incremented when wr_en are enabled $wren \wedge (\neg rden \wedge \neg full) \leftrightarrow (X(wr\,ptr) == (wr\,ptr + 1))$ Read pointers incremented when rd_en are enabled $rden \wedge (\neg wren \wedge \neg empty) \leftrightarrow (X(rd\,ptr) == (rd\,ptr + 1))$
$P2$	Age of packet is incremented in each cycle $X(age_{packet}) == (age_{packet} + 1)$
$P3$	Write pointers are not incremented when the buffer is full $(wren \wedge \neg rden \wedge full \rightarrow (wr\,ptr == X(wr\,ptr)))$ Read pointers are not incremented when the buffer is empty $(rden \wedge \neg wren \wedge empty \rightarrow (rd\,ptr == X(rd\,ptr)))$
$P4$	Buffer cannot be both full and empty at the same time $G\neg(empty \wedge full)$
$P5$	Data that was read from the buffer was at some point in time written into the buffer $G(data_{out} == \exists P(data_{in}))$
$P6$	The same number of packets that were written in to the buffer can be read from the buffer $\sum(wren \wedge rden) \wedge (wren \wedge \neg rden \wedge \neg full) ==$ $\sum(wren \wedge rden) \wedge (rden \wedge wren \wedge \neg empty)$
$P7$	Route can issue at most one request $G((\sum_{i=0}^{N_{ports}} req\,port_i) \leq 1)$
$P8$	Route should issue a request whenever a data is valid $G(datavalid \leftrightarrow (\sum_{i=0}^{N_{ports}} req\,port_i) == 1)$
$P9$	Routing algorithm (XY) should be correctly implemented $G((dest_x > current_x \leftrightarrow destport_{next} == EAST) \vee (dest_x < current_x \leftrightarrow$ $destport_{next} == WEST) \vee (dest_y > current_y \leftrightarrow destport_{next} == SOUTH) \vee$ $(dest_y < current_y \leftrightarrow destport_{next} == NORTH) \vee (destport_{next} == LOCAL))$
$P10$	Always at most one grant issued by the arbiter $G((\sum_{i=0}^{N_{ports}} gnt\,port_i) \leq 1)$
$P11$	As long as the request is available, it will eventually be granted by the arbiter within T cycles $(req\,port\ U\ gnt\,port) \rightarrow F(gnt\,port)$
$P12$	No grant can be issued without a request $\neg req\,port \rightarrow X(\neg gnt\,port)$
$P13$	Time between two issued grants is same for all requests $G(\forall i, j \in \{north, west, south, east, local\} \Delta T_i = \Delta T_j)$
$P14$	During multiplexing, output should be equal to input data $G((\sum_{i=0}^{N_{ports}} (select_i \wedge (data_{in}i == data_{out}))) == 1)$
$P15$	Age of the packet leaving the router will be at least T_{min} $G(P2 \leftrightarrow \nexists(age_{dataout} < T_{min}))$
$P16$	Age of the packet leaving the router should not exceed T_{max} $G((P10 \wedge P11 \wedge P12 \wedge P1 \wedge P2 \wedge P4 \wedge P14 \wedge P7 \wedge P8 \wedge P9) \leftrightarrow \nexists(age_{dataout} > T_{max}))$

Table 12.2 Notations used to define properties.

Symbol	Operator	Description
$X\phi$	Next	Property should hold in the next cycle
$G\phi$	Always	Property should always hold
$F\phi$	Eventually	Property will at some point in time (future) hold
$P\phi$	Prev. state	Specifies a state at some point in time in the past
$\phi U \omega$	Hold	ϕ will be the case until a time when ω is the case

Table 12.3 Property combinations developed using properties in Table 12.1 that map to the threat models in Sect. 12.2.2

Vulnerability	Combinations
Packet Duplication	No packet loss inside the router: $G(P6 \wedge P11 \wedge P14)$
Packet Corruption	No packet duplication inside the router: $G(P7 \wedge P10 \wedge P14)$
Packet Starvation	No packet modification inside the router: $G(P1 \wedge P3 \wedge P4 \wedge P5 \wedge P14)$
Packet Dropping	Packet that enters the router will eventually leave the router at some point of time: $G(P1 \wedge P2 \wedge P4 \wedge P7 \wedge P8 \wedge P9 \wedge P10 \wedge P11 \wedge P12 \wedge P14 \wedge P15 \wedge P16)$
Packet Misrouting	Packet is correctly routed to the correct port according to the destination: $G(P7 \wedge P9 \wedge P10 \wedge P14)$

the internal signals by propagating them to observable points (such as primary outputs). Assertion-based validation (ABV) has shown promising results in both of these areas. Assertions can capture unusual behavior and depending on where the assertion is embedded, can give information about the internal state of the design. This increased observability reduces overall hardware validation time significantly. While assertions do not directly improve controllability, a lot of research on ABV have proposed efficient techniques to generate tests that can activate the assertions.

Assertions can be viewed as a check embedded in the design. Failure to adhere to the condition will prompt warnings. For example, assertions can check whether the output of an adder is always equal to the sum of the two inputs. While ABV is widely used for functional validation, there is limited effort in utilizing assertions to detect security vulnerabilities [34]. Note that there is a fundamental difference between the objectives of functional and security assertions—while functional assertions monitor expected behaviors, security assertions are designed to monitor unexpected vulnerabilities.

Figure 12.5 shows an overview of the NoC vulnerability analysis framework using security assertions. It consists of three major tasks. First, we discuss various types of assertions. Next, we elaborate how security assertions can be embedded in the NoC design. Finally, we show how to generate test cases to activate the assertions in order to compute the assertion coverage and prove the validity of the assertions.

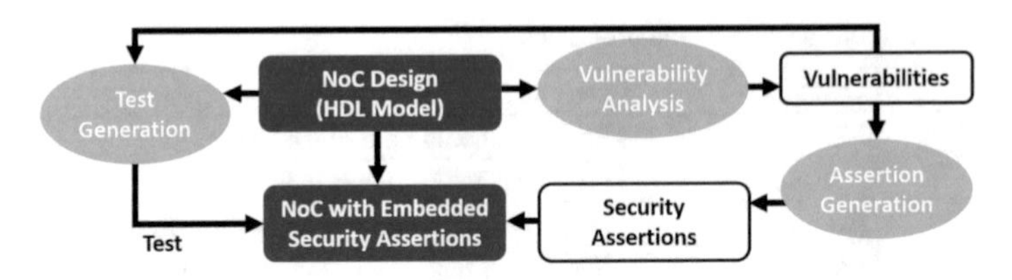

Fig. 12.5 Overview of NoC trust verification framework using security assertions

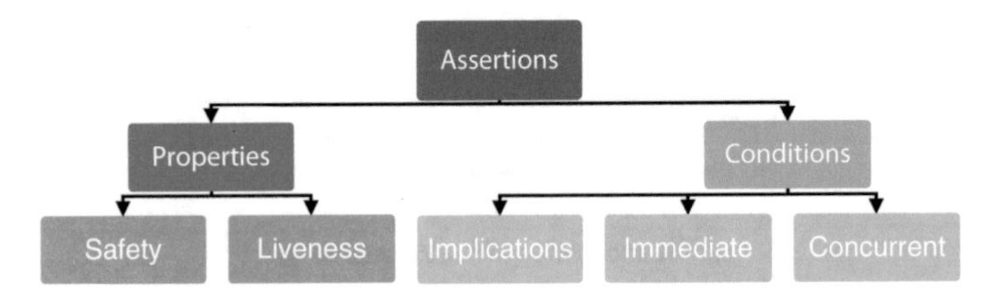

Fig. 12.6 Various types of assertions

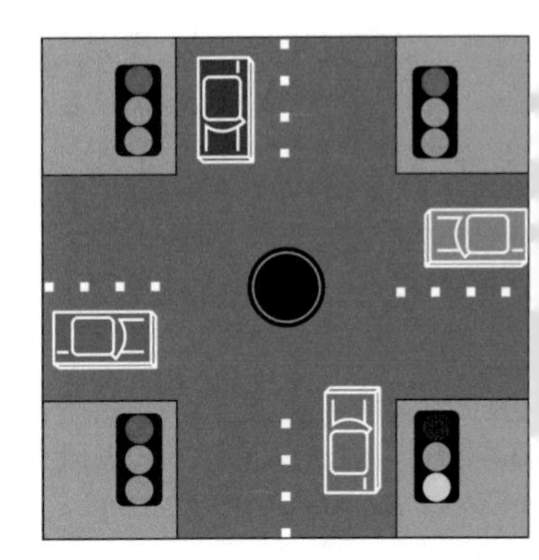

Fig. 12.7 Four-way traffic intersection example

12.4.1 Types of Assertions

Figure 12.6 shows that the assertions can be categorized based on the properties (behaviors) or conditions (sequences). The first category of assertions monitors properties that can be divided as safety or liveness properties. Consider a traffic-light example in Fig. 12.7 to understand safety and liveness properties. A *safety property* implies that *nothing bad will happen*. For example, the green signal should be given to only one direction to avoid accidents. A *liveness property* indicates that *something good will eventually happen*. For example, every direction should eventually get a green light. This prevents starvation (waiting forever) for vehicles from certain direction of the road. Of course, the implementation needs to consider many other aspects such as how long should a vehicle wait for the particular event (green light) to happen.

The second category of assertions monitors conditions and sequences. These assertions can be further divided into three types: *implication assertions*, *immediate assertions*, and *concurrent assertions*. Implication assertions follow the format of $(a \rightarrow b)$ where a and b can be sequence of expressions. Here, a is considered as the antecedent while b is considered as the consequent. Implication assertions simply monitor sequences based on satisfying specific criteria. For example, A1 in Table 12.4 is an implication type security assertion. Immediate assertions check features such as $(a == b)$ that can be embedded inside a sequential code. For example, A6 in Table 12.4 is an immediate type security assertion. Similarly, concurrent assertions are in the format of $\neg(a\&b)$ that has the ability to check relationships between signals from different concurrent blocks. The assertion will fail when both a and b are true at the same time. For example, A5 in Table 12.4 is a concurrent type security assertion.

12.4.2 Generation of Security Assertions

To launch an attack identified in Sect. 12.2.2, the hardware Trojan must change the normal behavior of the NoC. We identify the changes to the normal communication characteristics that happen during an attack and formulate them as security properties that should hold during runtime to detect ongoing attacks. The identified properties are written as assertions and embedded in the NoC design. If the security checks defined by the assertions are not violated during runtime, we can conclude that there are no ongoing attacks. One way to generate security assertions to capture hardware Trojans is to determine the rare nodes (signals) in the design, and generate combinations of these rare nodes as potential triggers for security assertions.

Section 12.3 shows the individual properties in Table 12.1 that should hold during execution. Note that the properties shown in Table 12.1 can be implemented as SystemVerilog assertions (referred with A# in the chapter). This approach is suitable for pre-silicon (design time) trust analysis using security assertions. These security assertions can also be synthesized as checkers to enable post-silicon (run time) security validation as described in Sect. 12.5.

12.4.3 Directed Test Generation to Activate Security Assertions

Given that the security assertions represent unexpected behaviors, they are not expected to be activated during the traditional validation methodology. Therefore, it is important to generate directed tests to activate the security assertions. Once an assertion is activated by a directed test, it indicates that the assertion is valid and it is able to accurately detect a specific security threat. Figure 12.8 shows

Table 12.4 Security assertions to detect various NoC attacks.

A#	Description of security assertions
$A1$	Write pointers incremented when wr_en are set $wren \wedge (\neg rden \wedge \neg full) \leftrightarrow (X(wr\,ptr) == (wr\,ptr + 1))$
$A2$	Read pointers incremented when rd_en are set $rden \wedge (\neg wren \wedge \neg empty) \leftrightarrow (X(rd\,ptr) == (rd\,ptr + 1))$
$A3$	Write pointers are not incremented when the buffer is full $(wren \wedge \neg rden \wedge full \rightarrow (wr\,ptr == X(wr\,ptr)))$
$A4$	Read pointers are not incremented when the buffer is empty $(rden \wedge \neg wren \wedge empty \rightarrow (rd\,ptr == X(rd\,ptr)))$
$A5$	Buffer cannot be both full and empty at the same time $G\neg(empty \wedge full)$
$A6$	Data that was read from the buffer was at some point in time written into the buffer $G(data_{out} == \exists P(data_{in}))$
$A7$	Route can issue at most one request $G((\sum_{i=0}^{N_{ports}} req\,port_i) \leq 1)$
$A8$	Route should issue a request whenever a data is valid $G(datavalid \leftrightarrow (\sum_{i=0}^{N_{ports}} req\,port_i) == 1)$
$A9$	Routing algorithm (XY) should be correctly implemented $G((dest_x > current_x \leftrightarrow destport_{next} == EAST) \vee (dest_x < current_x \leftrightarrow destport_{next} == WEST) \vee (dest_y > current_y \leftrightarrow destport_{next} == SOUTH) \vee (dest_y < current_y \leftrightarrow destport_{next} == NORTH) \vee (destport_{next} == LOCAL))$
$A10$	Always at most one grant issued by the arbiter $G((\sum_{i=0}^{N_{ports}} gnt\,port_i) \leq 1)$
$A11$	No grant can be issued without a request $\neg req\,port \rightarrow X(\neg gnt\,port)$

our test generation framework with two complementary approaches. We use SAT-based bounded model checking (BMC) that accepts the NoC design and assertions (negated properties) as inputs. The counterexamples generated from EBMC model checker [51] can be used as a directed test that is guaranteed to activate the respective security assertion. Unfortunately, EBMC may fail to handle complex properties due to state space explosion. In such cases, we use concolic testing [38] that can effectively utilize concrete simulation and symbolic execution to generate the required test patterns. Concolic testing addresses the state space explosion problem by exploring one path at a time compared to model checking that tries to explore all possible paths. To activate the security assertions non-vacuously, we first convert the security assertions into branch statements and then use concolic testing to activate the specific branches.

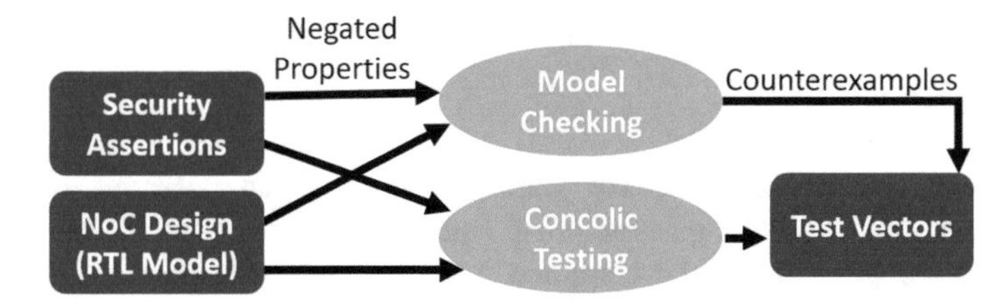

Fig. 12.8 Directed test generation using SAT-based bounded model checking as well as concolic testing

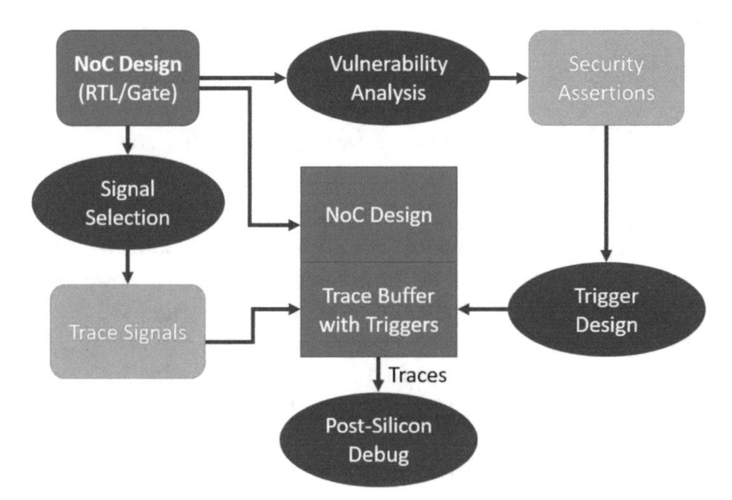

Fig. 12.9 Overview of the post-silicon security validation framework. It consists of four major tasks: vulnerability analysis, signal selection, trigger design, and post-silicon debug

12.5 Post-Silicon NoC Security Validation

Post-silicon validation and debug of SoCs have emerged as a challenging problem [47, 49]. Recent research efforts have addressed post-silicon functional validation issues for NoC architectures [54, 56–58]. Rout et al. [56–58] explored efficient router and trace buffer design for post-silicon validation of NoC-based SoCs. The techniques in [56–58] mainly cater to ascertaining the functional correctness of NoC designs. This chapter describes existing efforts on post-silicon security validation for NoC architectures.

Figure 12.9 provides an overview of the methodology that consists of four major tasks: vulnerability analysis, signal selection, trigger generation, and post-silicon debug. The first task leads to the generation of security assertions. The second task enables automated generation of trigger logic based on security assertions. The third task enables security-aware signal selection without compromising observability requirements. The final task performs trace analysis for post-silicon debug of functional bugs as well as security vulnerabilities. The remainder of this section describes these four tasks in detail.

12.5.1 Vulnerability Analysis for Security Assertion Generation

There are automated approaches for generation of functional assertions [62]. There are some initial efforts in generation of SoC security assertions [45] as well as security properties [60]. We have applied the same basic principles in this work to derive the NoC security assertions based on the threat models outlined in Sect. 12.3. Table 12.2 shows the notations we have used in defining security assertions. Table 12.4 presents the NoC security assertions that we have derived based on different threat models and attacks.

12.5.2 On-Chip Trigger Design Using Security Assertions

Implementation of assertions can assist in checking design correctness [27]. Similarly, security assertions [34] can also be implemented on-chip. Typically, the generation of such assertions can be done automatically [62]. However, as a large number of assertions can be obtained through mining techniques, the implementation of the assertions for on-chip triggers becomes a difficult problem owing to the associated overhead. An approach to obtain the on-chip implementation of assertions (specified in property specification language) is presented in [11] based on the concepts of automata theory. However, these techniques do not specifically cater to the objective of security assertions/properties. Since the enumeration of design behaviors based on specification tend to be typically large, we adopt a threat model-centric approach for obtaining security assertions. The assertions described in the Table 12.4 are defined based on the behavior of the threat models for different types of assertions as outlined in Sect. 12.4.1. Table 12.5 shows illustrative examples of different types of assertions and respective trigger logic. Note that we need to generate one trigger (T_i) for each assertion (A_i) for post-silicon debug. For example, if assertion A2 fails, trigger T2 will be activated. Typically, it is simple (negligible hardware overhead) to implement triggers corresponding to safety properties (e.g., trigger T1 for assertion A1). However, implementation of triggers for liveness checking may introduce significant hardware overhead. For example, in order to implement T6 (trigger for assertion A6), we have implemented check_buffer to store flit data that enters the flit_buffer. If we store the entire flit, it will introduce unacceptable area overhead. We hashed the flit data by converting them from 32 bits to 8 bits using *Knuth Variant on Division* to minimize the overhead. When a packet leaves the flit_buffer, we do a hash calculation on the particular flit and compare it with the check_buffer. If there is no match, the trigger will be activated and the trace data would be dumped to the trace_buffer.

Table 12.5 Illustrative examples of different assertion types and corresponding trigger logic

Assertion type	Example assertion	Trigger logic
Implication	$(a\|->b)$	$a\&\neg b$
Immediate	$(a==b)$	$(a!=b)$
Concurrent	$\neg(a\&b)$	$(a\&b)$

Algorithm 1: $TraceSignalSelection$

Input: $Design, M, SA$
Output: $Tr_{signals}$
1 $sv, dv \leftarrow \emptyset$;
2 $Design$ = RTL description of the $Design$;
3 M = modules of $Design$;
4 $\Omega = \{\Omega_1, \Omega_2 ..... \Omega_M\}$;
5 SA = security assertions;
6 $sv_j \leftarrow$ candidate variables from each A_j in **A** for selection;
7 **for** *each module* Ω_i *in* M **do**
8   **A** $\leftarrow$ assertion(s) related to M_i from SA;
9     $sv \leftarrow sv \cup sv_i$;
10     $dv_i \leftarrow$ variables from Ω_i;
11     $dv \leftarrow dv \cup dv_i$;
12     $DG_i \leftarrow$ construct dependency graph for variables of dv_i in dv;
13 **end**
14 $Edge_i \leftarrow$ calculate edges connected to each dv_i from constructed dependency graph;
15 Rank variables of dv across all Ω_i by $Edge_i$;
16 $sdv \leftarrow dv \cup sv$;
17 $Tr_{signals} \leftarrow$ variables from sdv as per trace-buffer width limit;

12.5.3 Security-Aware Trace Signal Selection

Algorithm 1 outlines the method for selecting trace signals to be stored in on-chip trace buffers to maximize the coverage of security assertions. With the help of dependency graph analysis, the trace signals are selected. During the design execution, the on-chip trace buffers contain the traced signals that are to be off-loaded for further fine-grained analysis (for the purpose of localization).

To enable effective on-chip debug and security validation, trace signals must be selected carefully. It is a major challenge to identify efficient trace signals due to the exponential nature of possible trace signal combinations as well as conflicting requirements such as error detection and internal visibility enhancement [33]. We explain the algorithm using an illustrative example. Consider the following SystemVerilog assertion: $(a == 1)\&\&(b == 0)|-> (c == 0)$. Here, c is the destination signal, and a and b can be register-variables (flip-flops), primary inputs or wires (internal nets). The above assertion basically means that signal c is false when signal a is true and signal b is false. The condition present in the left-hand side (antecedent) can be translated into triggers and the same applies to the signal condition in the right-hand side (consequent). The goal of the signal selection is to trace variables that will be able to infer the values of signals a, b, and c.

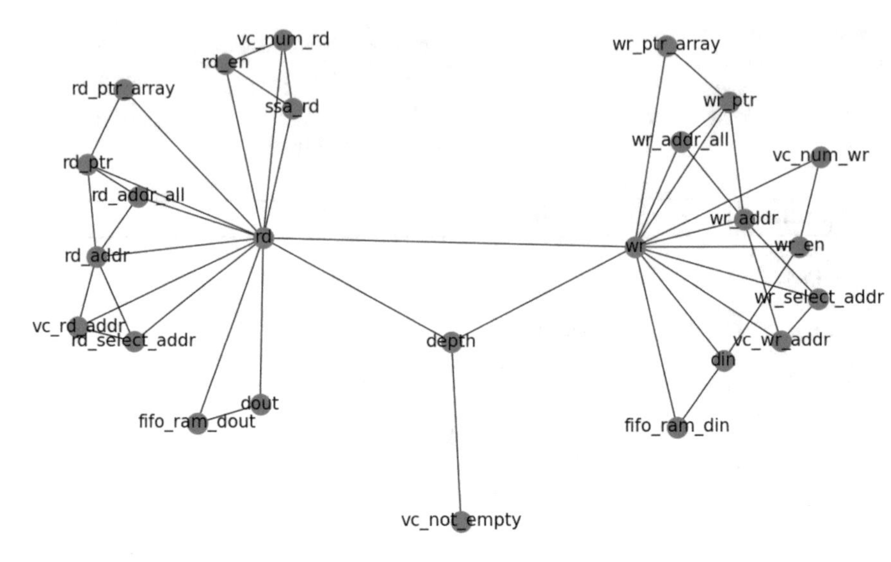

Fig. 12.10 An example signal dependency graph for flit_buffer

One illustration of the dependency graph is shown in Fig. 12.10 where the nodes represent different signals in the design and the edges depict the dependencies between them (inferred from the assignments in the RTL design description). This illustration corresponds to flit_buffer module of the NoC design. Based on these graphs, we select the signals that are maximally connected with other signals. The underlying reasoning is that maximum number of signals needed for functional behavior checking can be obtained for selection. To select variables related to security, we analyzed different types of security assertions developed in Sect. 12.5.1. Thereafter, the variables that were involved in the security assertions are chosen as probable candidates of trace signals. We perform a commonality search between the variables chosen from the security assertions and those chosen from the dependency variable analysis.

As discussed in Sect. 12.6, the trace buffer width is limited to 48 bits and a portion of it is used by the trace header data. Therefore, we cannot select all the signals to the trace buffer. We have to find the most beneficial signals that can be used to regenerate other signals during the offline analysis. For this task, we generate the variable dependency graph for each component of the NoC design. Then we order (sort) all the signals in different modules based on their connectivity (number of edges) with the other signals. This method arranges all the variables in each module in descending order of their restoration capability. Then we select the most relevant signals for a particular trigger giving the priority based on the ordered variables until we reach the trace width limit. For triggers implemented at route_mesh and arbiter, the trace width was enough to fit all the variables. However, for flit_buffer triggers, we applied the above technique to select the most profitable ones in terms of restorability. For example, the selected trace signals for flit_buffer trigger are listed in Table 12.6.

Table 12.6 Selected trace signals for flit_buffer triggers

TID	Signals
T1/T3	depth,wr_ptr,wr_ptr_next,wr_addr,vc_wr_addr,wr_en
T2/T4	depth,rd_ptr,rd_ptr_next,rd_addr,vc_num_rd,vc_rd_addr,rd_en
T5	depth,wr_ptr,rd_ptr,wr_addr,rd_addr,wr_en,rd_en
T6	flit_source(dout),flit_destination(dout),flit_source(check_buffer), flit_destination(check_buffer),Hashed-flit_body(dout), Hashed-flit_body(check_buffer),rd_en

Algorithm 2: $Post - SiliconTraceAnalysis$

```
Input: D_tr, SA
Output: ValidationResult
1  D_tr = data from traced flip-flops/signals;
2  for each SA_i in SA do
3      p_a ← antecedent signals of SA_i;
4      p_c ← consequent signals of SA_i;
5      Check signals values of p_a in D_tr;
6      if p_a signal values as per SA_i then
7          Check signals values of p_c in D_tr;
8          if p_c different from SA_i then
9              SA_i fails;
10         end
11         if p_c signal values as per SA_i then
12             SA_i passes;
13         end
14     end
15     if p_a different from SA_i then
16         SA_i fails;
17     end
18 end
19 ValidationResult ← failed/passed SA_i;
```

12.5.4 Post-Silicon Debug of Security Vulnerabilities

After the activation of on-chip triggers, the fixed number of trace buffers can store certain important information related to the design execution. The contents of these buffers need to be off-loaded and analyzed for several purposes. The primary benefit out of them being the understanding of internal signals after the activation of triggers leading to analysis of the bug (or, the security threat) in a fine-grained manner.

The methodology for offline trace analysis is presented in Algorithm 2. The validation algorithm relies on checking the security assertion on the off-loaded data from the trace buffer. The observed violation of the security property can hint towards a possible attack scenario. Note that because of the on-chip trigger framework, the respective trigger must have been activated. Therefore, the detection of the security attacks is achieved in a quick manner with minimal detection latency.

12.6 Experiments

This section demonstrates the effectiveness of the NoC security validation framework. We first describe the experimental setup. Next, we present the experimental results.

12.6.1 Experimental Setup

We first describe the experimental setup for pre-silicon NoC validation using security assertions. Next, we describe the experimental setup for post-silicon validation and debug.

12.6.1.1 Pre-Silicon NoC Validation Setup

We used the RTL design of an open-source NoC-based SoC generation platform—*ProNoC* [50]. A 2×2 Mesh NoC was configured that interconnects 4 IPs as shown in Fig. 12.11. Each IP was configured with a *mor1k* processor with the parameters outlined in Table 12.7. A simple message passing application and a message gathering application were written in "C" language and compiled using the mor1k tool-chain. The compiled binaries were placed in relevant block-RAMs on respective IPs. The application simulated a scenario where IPs 0, 1, and 2 are sending three types of packets to IP 3 as shown in Fig. 12.11. Debug statements were added to print the intermediate results to the terminal.

The simulation was done using ModelSim to verify the behavior. Once the functional accuracy of the experimental setup was verified, the selected security properties were implemented using SystemVerilog assertions. The assertions were implemented at the corresponding router components in the NoC RTL model.

Table 12.7 Parameters used in the experimental setup

Parameter	Value
Router type	Virtual channel (VC) based router
VCs per port	2
Payload width	32
Switch (SW) allocator arbitration type	Round robin arbitration
Buffer flits per VC	4
Routing algorithm	X-Y routing
VC/SW combination type	Comb-Nonspec: VC allocator combined with non-speculative SW allocator where the validity of speculative requests are checked at the beginning of SW allocation

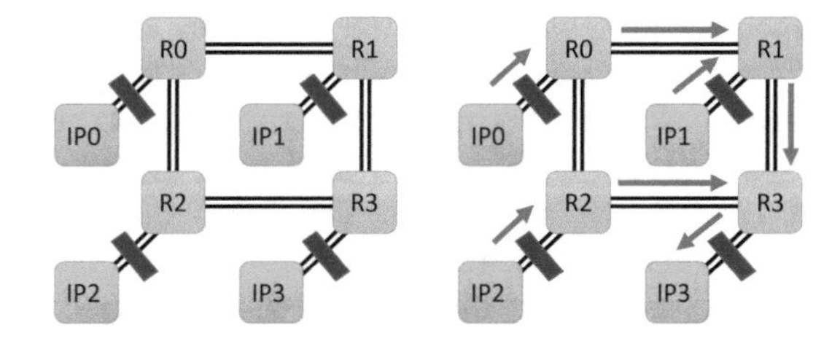

Fig. 12.11 2×2 Mesh NoC-based SoC used for simulation (left) and the message passing scenario (right)

Table 12.8 Assertions implemented at different NoC modules

Module	Implemented assertions
flit_buffer.sv	*A*1, *A*2, *A*3, *A*4, *A*5, *A*6, *A*15, *A*16
route_mesh.sv	*A*7, *A*8, *A*9
arbiter.sv	*A*10, *A*11, *A*12, *A*13
main_comp.sv	*A*14

Listing 12.1 Implementation of A9 assertion in SystemVerilog as well as its equivalent branch representation

```
//Assertion Statement
assert ((dest_x > current_x && destport_next==EAST)
    || (dest_x < current_x && destport_next==WEST)
    || (dest_y > current_y && destport_next==SOUTH)
    || (dest_y < current_y && destport_next==NORTH)
    || (destport_next==LOCAL));
//Branch Statement
if ((dest_x > current_x && destport_next==EAST)
    || (dest_x < current_x && destport_next==WEST)
    || (dest_y > current_y && destport_next==SOUTH)
    || (dest_y < current_y && destport_next==NORTH)
    || (destport_next==LOCAL)) flag_A9<=1'b0;
else flag_A9<=1'b1;
```

Table 12.8 provides information about the security assertion implementation. The first column provides the module name in the ProNoC benchmark. The second column indicates the assertions implemented in that module.

Note that SystemVerilog assertions are not synthesizable as post-silicon checkers. Previous work has proposed several alternatives to address this. Omar et al. [6] proposed a method that generates RTL netlists from assertions. We use a different approach by creating equivalent branch statements corresponding to each assertion. For example, Listing 12.1 shows the SystemVerilog description of assertion A9 as well as its branch equivalent representation. The SoC was simulated again with assertions and the relevant branch statements using ModelSim to verify the assertion implementation.

We considered both safety and liveness related assertions introduced in Sect. 12.4.2. Implementation of liveness assertions is more complex compared to safety assertions since liveness behaviors include the "eventual" operator. For example, assertion *A5* ensures that data that was read from the buffer was at some point in time written into the buffer. This property captures two critical vulnerabilities: i) flit modification and ii) dropped packets. To capture flit modifications, we validate the flit content at the input and the output of the buffer. The straightforward approach would be to keep a copy of the packet data and validate the entire packet at the output. However, this adds unacceptable memory overhead. Alternatively, we used a lightweight hash function called the *Knuth Variant on Division* to store the hashed value of each flit. The header flit, which arrives first, is hashed and stored in a *check buffer*. Subsequently, when the body and tail flits arrive, each flit is hashed and the hashed value is XORed with the existing value in the check buffer. The process is repeated at the buffer output and validated against the value in the check buffer. To capture the behavior of the "eventual" operator, the maximum time each packet spends in the router buffer is required. However, finding an exact upper bound for buffer wait times is not possible using only the RTL design. Therefore, we derived an upper bound using simulations and used in the assertions.

To validate the accuracy of the assertions, the Trojan behavior was implemented according to the threat models introduced in Sect. 12.2.2 by modifying the RTL design. Specific modifications are as follows. *Packet Duplication/Eavesdropping:* Duplicate packets inside the flit buffer and change the destination of the duplicated packet. *Packet Corruption:* Randomly alter the packet data during transmission. *Packet Starvation:* Randomly stop issuing grants through the arbiter to some nodes. *Packet Dropping:* Randomly drop packets. *Packet Misrouting:* Modify the routing algorithm to change the destination.

We ran independent simulations for the above scenarios to observe assertion activation and to verify the accuracy of assertions in Table 12.3. Once design time (pre-silicon) security analysis was complete, the assertion statements were replaced with equivalent branch statements with a dedicated flag for each branch. The design was then synthesized with *Quartus Prime*. The comparison showed that the overhead of added logic for assertions is negligible (approximately 1%) compared to the original *ProNoC* design.

12.6.1.2 Post-Silicon NoC Debug Setup

We created a 4x4 Mesh NoC consisting of 16 "mor1k" processors IPs using the open-source ProNoc tool [50]. Figure 12.12 shows this configuration. This Verilog RTL design has the configuration parameters of the NoC as shown in Table 12.7. A simple message passing scenario was designed to send three packets of data from each IP core to the IP core numbered 10. The designed scenario was implemented in C programming language, and using mor1k tool-chain the required binaries were created for each IP core. The binary files were placed in the RAM modules of relevant IP cores. Then the design was simulated in Modelsim to validate the functionality. Next, we injected vulnerabilities and performed different debug experiments using Modelsim simulator.

A centralized trace buffer was created, with a buffer length of 48 bits. Individual trigger circuits were designed and implemented to convert each of the security assertion (discussed in Section 12.5.1) to a synthesizable trigger logic. The Design-

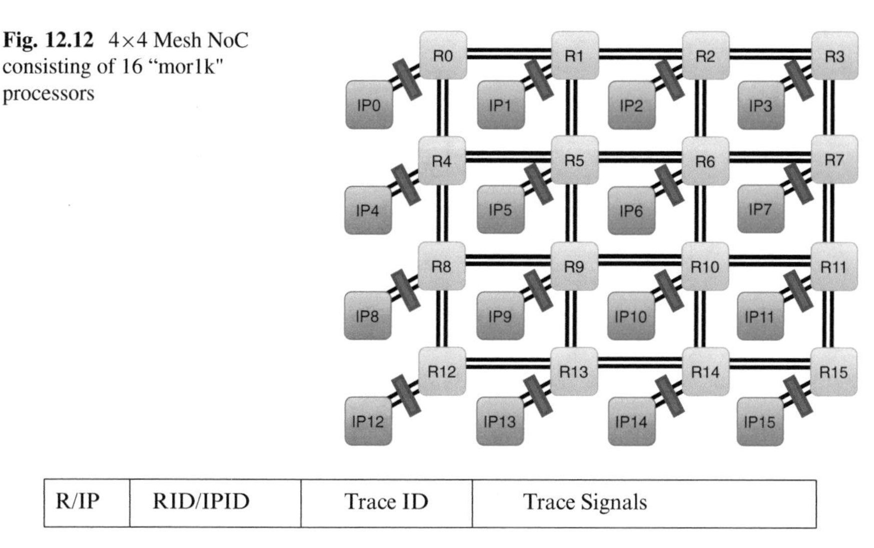

Fig. 12.12 4×4 Mesh NoC consisting of 16 "mor1k" processors

Fig. 12.13 Abstract trace packet structure

for-Debug (DfD) circuit was created with a dedicated packet structure for the trace. Figure 12.13 illustrates the bit structure for the trace packet. The first bit (R/IP) is dedicated to identifying the source of the trace, whether it was originated from the NoC or an IP. The next four bits represent the routerID or IP-ID (16 routers/IPs in this instance). The next four bits represent the trace ID. The remaining bits store the selected trace signals.

12.6.2 Pre-Silicon Validation Utilizing Security Assertions

The previous section presented the approach for creating assertions to monitor vulnerabilities. In this section, we discuss the assertion validation results using the test generation framework outlined in Sect. 12.4.3. We used the unbounded model checker EBMC [51] to generate directed tests. The generated tests were used to verify whether the added assertions are valid by activating these assertions. Table 12.9 presents the time (s) and space (MB) taken by EBMC [51] to activate the assertions. All the experiments were performed on a machine with an Intel i7-10510U CPU @ 1.80 GHz CPU with 16 GB RAM.

As shown in Table 12.9, we are able to activate all but three assertions (A5, A6 and A13) using EBMC [51]. These three assertions exceeded the capability of the model checker due to state space explosion (insufficient memory). We used concolic testing [38] to activate A5, A6, and A13. Results in Table 12.9 show the time and memory requirement for test generation using concolic testing. Overall, we generated tests for activating all the assertions mentioned in Table 12.4 using either

Table 12.9 Test generation time and memory requirement when using EBMC and concolic testing to activate assertions

Framework	EBMC [51]													Concolic [38]		
Assertion	A1	A2	A3	A4	A7	A8	A9	A10	A11	A12	A14	A15	A16	A5	A6	A13
Time (s)	0.08	0.08	0.09	0.08	0.02	0.02	0.02	0.01	0.02	0.02	0.01	0.08	0.09	1.15	1.89	8.78
Memory (MB)	11.7	12.0	11.8	11.2	4.5	3.9	4.6	4.8	5.2	5.1	4.5	11.6	11.8	64.2	65.8	38.3

EBMC or concolic testing. Therefore, we can confirm that the security assertions are valid and satisfies the design requirements.

12.6.3 Post-Silicon Debug of Injected Vulnerabilities

In order to mimic real-life debug context, several attack scenarios were created by closely following the threat models without considering the assertions or triggers. These bugs and attacks were inserted in different routers (selected randomly) to be activated at random clock cycles. Separate Modelsim simulations were carried out for each of these attack scenarios (one randomly inserted vulnerability at a time). Table 12.10 presents different cases of evaluation with the debug framework. The first column provides different types of vulnerabilities. The second column indicates whether it is a functional bug or a security vulnerability. The third column indicates the activated trigger. The last column shows the average latency for activating each vulnerability. From the off-loaded trace contents, we checked for possible violation of the security assertions. As expected, trigger T6 requires different number of cycles since it captures liveness behavior, whereas the activation of the remaining triggers can be detected in one clock cycle.

Table 12.10 Results of bug injections with average time to activate the trigger from the activation of the vulnerability

Vulnerability	Type	Activated trigger	Latency (cycles)
Eavesdropping attack	Security	T6	446
Packet corruption (FIFO input)	Security	T6	7
Packet missing(FIFO input)	Security	T1/T6	1/330
Packet missing(FIFO output)	Security	T2/T6	1/336
Starvation (Arbiter)	Security	T11	1
Wr/rd pointer fails when buffer is not full/empty	Functional	T1/T2	1
Wr/rd pointer increments when buffer is full/empty	Functional	T3/T4	1
Packet destination changing (Flit buffer)	Security	T6	450
Packet misrouting (Algorithm bug)	Functional and Security	T9	1
Invalid destination ports from route	Functional	T8	1
Multiple destination port selection (route)	Functional	T7	1
Invalid grants (arbiter)	Functional and Security	T11	1
Multiple grants (arbiter)	Functional	T10	1

12.7 Summary

Network-on-Chip (NoC) is widely used as a scalable solution to provide communication between a large number of Intellectual Property (IP) cores in modern System-on-Chip (SoC) designs. It is critical to protect NoC against security threats in order to design trustworthy systems. In this chapter, we defined a set of vulnerabilities for NoC architectures, and described security assertions to monitor these vulnerabilities. We also described a test generation framework to activate the security assertions. We have also described an efficient post-silicon debug framework for NoC designs utilizing security assertions. On-chip triggers derived from these security assertions provide an opportunity to enable the debugging features. Experimental results using an NoC benchmark demonstrated that existing validation methods are effective in NoC vulnerability analysis using security assertions.

Acknowledgments This work was partially supported by the National Science Foundation (NSF) grant SaTC-1936040.

References

1. R. Abdel-Khalek, R. Parikh, A. DeOrio, V. Bertacco, Functional correctness for CMP interconnects, in *2011 IEEE 29th International Conference on Computer Design (ICCD)* (2011), pp. 352–359
2. A. Ahmed, F. Farahmandi, Y. Iskander, P. Mishra, Scalable hardware trojan activation by interleaving concrete simulation and symbolic execution, in *2018 IEEE International Test Conference (ITC)* (IEEE, New York, 2018), pp. 1–10
3. A. Ahmed, F. Farahmandi, P. Mishra, Directed test generation using concolic testing on RTL models, in *2018 Design, Automation & Test in Europe Conference & Exhibition (DATE)* (2018)
4. A. Ahmed, Y. Huang, P. Mishra, Cache reconfiguration using machine learning for vulnerability-aware energy optimization. ACM Trans. Embed. Comput. Syst. **18**(2) (2019). https://doi.org/10.1145/3309762
5. Alteris FlexNoC Resilience Package. www.arteris.com/flexnoc-resilience-package-functional-safety [Online]
6. O. Amin et al., System Verilog assertions synthesis based compiler, in *MTV* (2016)
7. D.M. Ancajas, K. Chakraborty, S. Roy, Fort-NoCs: mitigating the threat of a compromised NoC, in *2014 51st ACM/EDAC/IEEE Design Automation Conference (DAC)* (2014), pp. 1–6
8. Arteris makes big gains on inc. 500 list of America's fastest-growing private companies. www.arteris.com/Inc-500-Arteris-pr-2013-august-20, August 2013. [Online]
9. S. Bhunia, M. Tehranipoor, *The Hardware Trojan War* (Springer, New York, 2018)
10. A.K. Biswas, S.K. Nandy, R. Narayan, Router attack toward NoC-enabled MpSoC and monitoring countermeasures against such threat. Circ. Syst. Signal Process. **34**(10), 3241–3290 (2015)
11. M. Boule, Z. Zilic, Incorporating efficient assertion checkers into hardware emulation, in *2005 International Conference on Computer Design*, October 2005, pp. 221–228
12. S. Charles, P. Mishra, Lightweight and trust-aware routing in NoC based SoCs, in *IEEE Computer Society Annual Symposium on VLSI (ISVLSI)* (2020)

13. S. Charles, P. Mishra, Reconfigurable network-on-chip security architecture. ACM Trans. Des. Autom. Electron. Syst. **25**(6), 1–25 (2020). https://doi.org/10.1145/3406661
14. S. Charles, P. Mishra, Securing network-on-chip using incremental cryptography, in *ISVLSI* (2020)
15. S. Charles, H. Hajimiri, P. Mishra, Proactive thermal management using memory-based computing in multicore architectures, in *2018 Ninth International Green and Sustainable Computing Conference (IGSC)* (2018), pp. 1–8
16. S. Charles, C.A. Patil, U.Y. Ogras, P. Mishra, Exploration of memory and cluster modes in directory-based many-core CMPs, in *2018 Twelfth IEEE/ACM International Symposium on Networks-on-Chip (NOCS)* (2018), pp. 1–8
17. S. Charles, A. Ahmed, U.Y. Ogras, P. Mishra, Efficient cache reconfiguration using machine learning in NoC-based many-core CMPs. ACM Trans. Des. Autom. Electron. Syst. **24**(6), 1–23 (2019)
18. S. Charles et al., Real-time detection and localization of DoS attacks in NoC based SoCs, in *DATE* (2019)
19. S. Charles, M. Logan, P. Mishra, Lightweight anonymous routing in NoC based SoCs, in *2020 Design, Automation & Test in Europe Conference & Exhibition (DATE)* (IEEE, New York, 2020)
20. S. Charles, Y. Lyu, P. Mishra, Real-time detection and localization of distributed DoS attacks in NoC based SoCs, in *IEEE Transactions on Computer-Aided Design of Integrated Circuits and Systems* (2020)
21. M. Chen, X. Qin, H.-M. Koo, P. Mishra, *System-Level Validation: High-Level Modeling and Directed Test Generation Techniques* (Springer, New York, 2012)
22. F. Farahmandi, P. Mishra, Automated debugging of arithmetic circuits using incremental gröbner basis reduction, in *2017 IEEE International Conference on Computer Design (ICCD)*, pp. 193–200 (IEEE, New York, 2017)
23. F. Farahmandi, P. Mishra, FSM anomaly detection using formal analysis, in *2017 IEEE International Conference on Computer Design (ICCD)* (IEEE, New York, 2017), pp. 313–320
24. F. Farahmandi, P. Mishra, Automated test generation for debugging multiple bugs in arithmetic circuits. IEEE Trans. Comput. **68**(2), 182–197 (2018)
25. F. Farahmandi, Y. Huang, P. Mishra, Trojan localization using symbolic algebra, in *2017 22nd Asia and South Pacific Design Automation Conference (ASP-DAC)* (2017), pp. 591–597
26. F. Farahmandi, Y. Huang, P. Mishra, *System-on-Chip Security: Validation and Verification* (Springer Nature, New York, 2019)
27. H. Foster, D. Lacey, A. Krolnik, *Assertion-Based Design*, 2nd edn. (Kluwer Academic Publishers, New York, 2003)
28. U. Gupta et al., DyPo: Dynamic pareto-optimal configuration selection for heterogeneous MpSoCs. TECS **16**(5s), 1–20 (2017)
29. Y. Huang, P. Mishra, Vulnerability-aware energy optimization for reconfigurable caches in multitasking systems. IEEE Trans. Comput.-Aided Des. Integr. Circ. Syst. **38**(5), 809–821 (2019)
30. Y. Huang, S. Bhunia, P. Mishra, MERS: statistical test generation for side-channel analysis based Trojan detection, in *Proceedings of the 2016 ACM SIGSAC Conference on Computer and Communications Security* (2016), pp. 130–141
31. Y. Huang, S. Bhunia, P. Mishra, Scalable test generation for Trojan detection using side channel analysis. IEEE Trans. Inf. Forensics Secur. **13**(11), 2746–2760 (2018)
32. M. Hussain, A. Malekpour, H. Guo, S. Parameswaran, EETD: an energy efficient design for runtime hardware Trojan detection in untrusted network-on-chip, in *2018 IEEE Computer Society Annual Symposium on VLSI (ISVLSI)* (2018), pp. 345–350
33. B. Kumar, K. Basu, M. Fujita, V. Singh, Post-silicon gate-level error localization with effective and combined trace signal selection. IEEE Trans. Comput.-Aided Des. Integr. Circ. Syst. **39**(1), 248–261 (2020)
34. Y. Lyu, P. Mishra, System-on-chip security assertions. Accessed on March 2020 from https://arxiv.org/pdf/2001.06719.pdf

35. Y. Lyu, P. Mishra, A survey of side-channel attacks on caches and countermeasures. J. Hardw. Syst. Secur. **2**(1), 33–50 (2018)
36. Y. Lyu, P. Mishra, Efficient test generation for trojan detection using side channel analysis, in *2019 Design, Automation & Test in Europe Conference & Exhibition (DATE)*, pp. 408–413 (IEEE, New York, 2019)
37. Y. Lyu, P. Mishra, Automated test generation for activation of assertions in RTL models, in *2020 25th Asia and South Pacific Design Automation Conference (ASPDAC)* (IEEE, New York, 2020), pp. 223–228
38. Y. Lyu, P. Mishra, Automated test generation for activation of assertions in RTL models, in *ASP-DAC* (2020)
39. Y. Lyu, P. Mishra, Automated test generation for Trojan detection using delay-based side channel analysis, in *2020 Design, Automation & Test in Europe Conference & Exhibition (DATE)* (2020), pp. 1031–1036
40. Y. Lyu, P. Mishra, Automated trigger activation by repeated maximal clique sampling, in *Asia and South Pacific Design Automation Conference (ASPDAC)* (2020), pp. 482–487
41. Y. Lyu, P. Mishra, Scalable activation of rare triggers in hardware Trojans by repeated maximal clique sampling, in *IEEE Transactions on Computer-Aided Design of Integrated Circuits and Systems* (2020)
42. Y. Lyu, P. Mishra, Scalable concolic testing of RTL models, in *IEEE Transactions on Computers* (2020)
43. Y. Lyu, A. Ahmed, P. Mishra, Automated activation of multiple targets in RTL models using concolic testing, in *2019 Design, Automation & Test in Europe Conference & Exhibition (DATE)* (IEEE, New York, 2019), pp. 354–359
44. Y. Lyu, X. Qin, M. Chen, P. Mishra, Directed test generation for validation of cache coherence protocols. IEEE Trans. Comput.-Aided Des. Integr. Circ. Syst. **38**(1), 163–176 (2018)
45. Y. Lyu, P. Mishra, System-on-chip security assertions (2020). Preprint. arXiv:2001.06719
46. J.Y.V. Manoj Kumar, A.K. Swain, S. Kumar, S.R. Sahoo, K. Mahapatra, Run time mitigation of performance degradation hardware Trojan attacks in network on chip, in *2018 IEEE Computer Society Annual Symposium on VLSI (ISVLSI)* (2018), pp. 738–743
47. P. Mishra, F. Farahmandi, *Post-Silicon Validation and Debug*(Springer, New York, 2019)
48. P. Mishra, S. Bhunia, M. Tehranipoor, *Hardware IP security and Trust* (Springer, New York, 2017)
49. P. Mishra, R. Morad, A. Ziv, S. Ray, Post-silicon validation in the SoC era: a tutorial introduction. IEEE Des. Test **34**(3), 68–92 (2017)
50. A. Monemi, J.W. Tang, M. Palesi, M.N. Marsono, Pronoc: a low latency network-on-chip based many-core system-on-chip prototyping platform. Microprocess. Microsyst. **54**, 60–74 (2017)
51. R. Mukherjee et al., Hardware verification using software analyzers, in *ISVLSI* (2015)
52. Z. Pan, P. Mishra, Automated test generation for hardware trojan detection using reinforcement learning, in *Asia and South Pacific Design Automation Conference (ASPDAC)* (2021)
53. Z. Pan, J. Sheldon, P. Mishra, Test generation using reinforcement learning for delay-based side channel analysis, in *IEEE/ACM International Conference on Computer-Aided Design (ICCAD)* (2020)
54. R. Parikh, V. Bertacco, Forever: a complementary formal and runtime verification approach to correct NoC functionality. ACM Trans. Embed. Comput. Syst. **13**(3s), 104:1–104:30 (2014)
55. S. Pasricha, N. Dutt, *On-Chip Communication Architectures: System on Chip Interconnect* (Morgan Kaufmann, Burlington, 2010)
56. S. Rout, K. Basu, S. Deb, Efficient post-silicon validation of network-on-chip using wireless links, in *International Conference on VLSI Design* (2019), pp. 371–376
57. S. Rout, S.B. Patil, V.I. Chaudhari, S. Deb, Efficient router architecture for trace reduction during NoC post-silicon validation, in *IEEE International System-on-Chip Conference (SOCC)* (2019), pp. 230–235
58. S. Rout, M. Badri, S. Deb, Reutilization of trace buffers for performance enhancement of NOC based MpSoCs, in *Asia and South Pacific Design Automation Conference (ASP-DAC)* (2020), pp. 97–102

59. J. Sepúlveda, A. Zankl, D. Flórez, G. Sigl, Towards protected MPSoC communication for information protection against a malicious NoC. Proc. Comput. Sci. **108**, 1103–1112 (2017). *International Conference on Computational Science, ICCS 2017*, 12–14 June 2017, Zurich
60. J. Sepúlveda, D. Aboul-Hassan, G. Sigl, B. Becker, M. Sauer, Towards the formal verification of security properties of a network-on-chip router, in *2018 IEEE 23rd European Test Symposium (ETS)* (IEEE, New York, 2018)
61. K. Shuler, Majority of leading china semiconductor companies rely on Arteris network-on-chip interconnect IP (2013)
62. S. Vasudevan, D. Sheridan, S.J. Patel, D. Tcheng, W. Tuohy, D.R. Johnson, Goldmine: automatic assertion generation using data mining and static analysis, in *Design, Automation and Test in Europe, DATE 2010, Dresden, March 8–12, 2010* (IEEE Computer Society, San Francisco, 2010), pp. 626–629
63. W. Wang, P. Mishra, A. Gordon-Ross, Dynamic cache reconfiguration for soft real-time systems. ACM Trans. Embed. Comput. Syst. **11**(2), (2012). https://doi.org/10.1145/2220336.2220340

Chapter 13
NoC Post-Silicon Validation and Debug

Sidhartha Sankar Rout, Mitali Sinha, and Sujay Deb

13.1 Introduction

Network-on-Chip (NoC) is considered as a suitable and scalable interconnect solution for modern multi-core Systems-on-Chip (SoC) [1]. NoC is primarily comprised of routers, which are responsible for taking data routing decisions. Routers along with the wired links establish the whole data communication network. All the routers are connected to their neighboring nodes through the wired links. Each processing core (PC) is attached to a router through a network interface (NI) and can communicate with other on-chip modules through router switch paths. The shared and distributed nature of NoC allows multiple cores to transfer data simultaneously.

The contemporary SoCs integrating multiple heterogeneous cores in a single chip demand efficient and reliable communication among on-chip modules. Performance and power efficiency along with a high level of Quality of Service (QoS) are desired during on-chip data communication. To achieve these goals, the NoC module is supported with multiple advanced features like power management, speculation, redundancy, run-time controllability, fault tolerance, etc. Although these features result in efficient communication, it makes the contemporary NoCs extremely complex. Due to design intricacy, it is very difficult and time-consuming to detect all the functional bugs during the pre-silicon verification stage [2]. As a result, multiple

S. S. Rout · S. Deb (✉)
Department of Electronics and Communication Engineering, Indraprastha Institute of Information Technology Delhi, Delhi, India
e-mail: sidharthas@iiitd.ac.in; sdeb@iiitd.ac.in

M. Sinha
Department of Computer Science Engineering, Indraprastha Institute of Information Technology Delhi, Delhi, India
e-mail: mitalis@iiitd.ac.in

P. Mishra, S. Charles (eds.), *Network-on-Chip Security and Privacy*,
https://doi.org/10.1007/978-3-030-69131-8_13

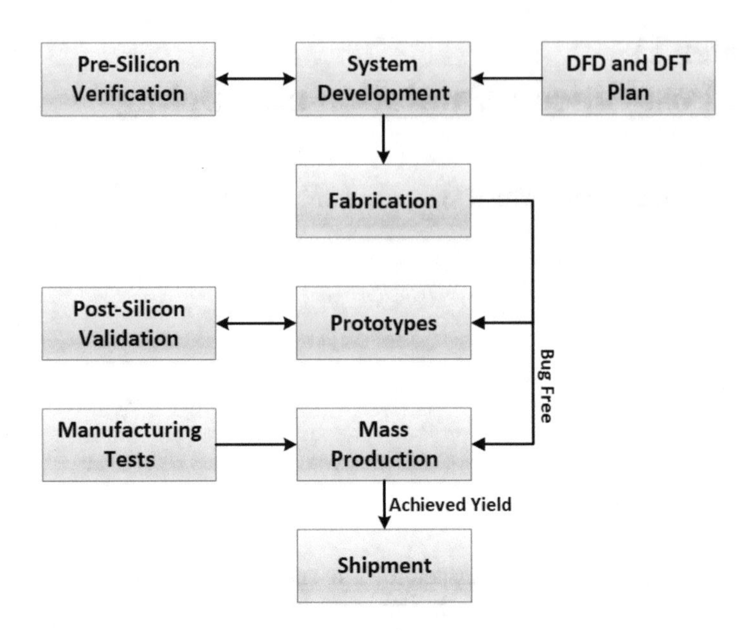

Fig. 13.1 Broad verification steps in a system development cycle

errors may slip on to fabricated interconnect module. A survey conducted by [3] shows that a significant percentage of total design errors in multi-core architectures come from the interconnection networks. The survey summarizes the design errors found in different components of recent multi-core architectures collected from their respective errata sheets. It finds around 48%, 26%, 16.07%, and 13% errors appearing in the interconnect modules of ARM MX6, Intel Xeon E5, Intel Xeon Phi, and AMD Opteron, respectively. The errors on interconnect may introduce several data communication faults, and lead to complete system failure. Therefore, efficient post-silicon validation is required to capture the escaped errors and provide a bug-free interconnect system.

Figure 13.1 shows typical verification steps in a system development process. Pre-silicon verification is performed before fabrication, whereas post-silicon validation and manufacturing tests are performed after the fabrication. Post-silicon validation is executed on the initial prototype chips to find functional and electrical bugs in the design. However, manufacturing tests are carried out on each of the fabricated chips after the mass production to detect any manufacturing defect. Finally, the good chips are shipped, and the bad ones are discarded. Though post-silicon validation and manufacturing tests are performed after fabrication, it can be seen that planning for them starts from the very beginning during the system development phase. Design for Debug (DFD) and Design for Test (DFT) infrastructures are augmented to the actual design, which facilitates observable and controllable points within the design during the post-fabrication validation steps. This particular chapter focuses on state-of-the-art post-silicon validation methods for NoC.

Post-silicon validation is performed on a few fabricated prototypes of the design. The prototype chip is mounted on a test board with all the peripherals connected, and test stimuli are applied for validation. It is expected that the escaped design errors should be detected during this phase. Pre-silicon verification is performed through simulators. The slow execution rate of such simulators results in longer testing time. In contrast, post-silicon validation is a faster process as it can run at the speed of system clock frequency. A significant increase in test speed allows a debug engineer to validate the design for more test cases within a short period of time. Another advantage of post-silicon validation is that the design can be tested for electrical errors that originate due to voltage fluctuation, noise, hotspot, process variation, etc. Detecting electrical errors during pre-silicon verification is challenging as it is cumbersome to model such operating environments on the simulator.

While post-silicon validation provides an efficient platform to remove elusive design bugs, it suffers from very poor system observability and controllability, which is limited to the I/O pins of the chip. As a result, it becomes very difficult to debug and localize a bug upon the detection of a fault. To enhance the system's internal observability during validation, DFD structures are instrumented to the original design during the development phase. Such structures include on-chip trace buffer, trigger unit, trace bus, etc. During the DFD plan, observable points and trigger conditions are decided. Through the validation phase, whenever trigger conditions are satisfied, the run-time traces of the observable points are captured and stored in the trace buffer. These traces improve the internal observability of the system and help to detect as well as find the root-cause of the bugs during fault analysis.

The remainder of the chapter is organized as follows. Section 13.2 discusses different data communication faults in NoC. A generalized NoC validation platform is presented in Sect. 13.3. Packet trace collection, trace data transfer, and fault analysis are three major steps of NoC post-silicon validation, which are discussed in Sects. 13.4, 13.5, and 13.6, respectively. An example NoC debug framework is discussed in Sect. 13.7. To optimize the debug hardware overhead cost, reuse of NoC debug infrastructure is discussed in Sect. 13.8. Finally, Sect. 13.9 concludes the work along with insights into the future research opportunities in this domain.

13.2 NoC Fault Model

NoC being an interconnect module, reliable and efficient data transfer is its main functionality. So, to evaluate the functional correctness of NoC, communication-centric validation methods are popularly adopted [4]. This section presents a fault model that illustrates commonly found data communication faults on an NoC [5, 6]. These faults can be of two different types such as short-lived fault and permanent fault. Moreover, this section also indicates the probable buggy component(s) within the NoC router for which a particular fault may occur.

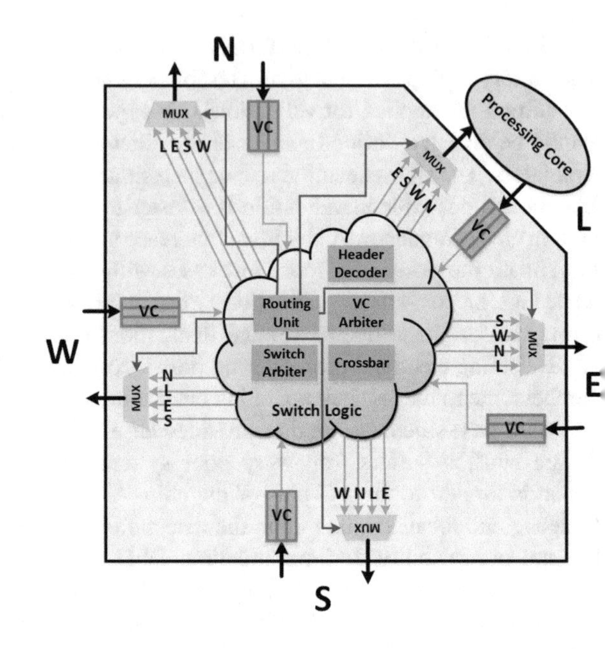

Fig. 13.2 Basic architecture of NoC router

Figure 13.2 shows a basic architecture of NoC router. Before discussing different data communication faults, an overview of the data routing operation inside a router is presented. Here, we have considered *wormhole* flow control routing [7] for our discussion. Whenever a header flit of a packet arrives at an input port of a router, the Header Decoder (HD) reads the destination and VCID (Virtual Channel Identifier) of the packet. According to VCID, the flit is placed in a particular VC. The Routing Computation (RC) within the Routing Unit (RU) decides the output port for the flit based on its destination. VC Arbiter (VA) allocates a VC in the downstream router to the flit after the arbitration among the competing flits. Similarly, Switch Arbiter (SA) assigns the Crossbar inputs to different flits. Arbitration in both VA and SA are performed in a *round-robin* or *oldest-first* manner. Many a time, priority logics are implemented in these arbiters for early allocation of resources to critical data. As soon as the reserved output port and VC in the downstream router are available, RU removes the flit from the VC and asserts the corresponding select signal at the output multiplexer. The flit is forwarded to the allocated Crossbar input, and the Crossbar unit establishes the routing path to the desired output port multiplexer. Finally, the multiplexer transfers the data flit available at its selected input to the output link. Till the time the reserved resources are not available, the flit has to wait in the VC. Once the header flit is forwarded, the remaining flits of the packet follow the same path. Anomalies in packet routing operation introduce several data communication faults to NoC, which are illustrated in Fig. 13.3. The following sub-sections provide a brief discussion on these network faults.

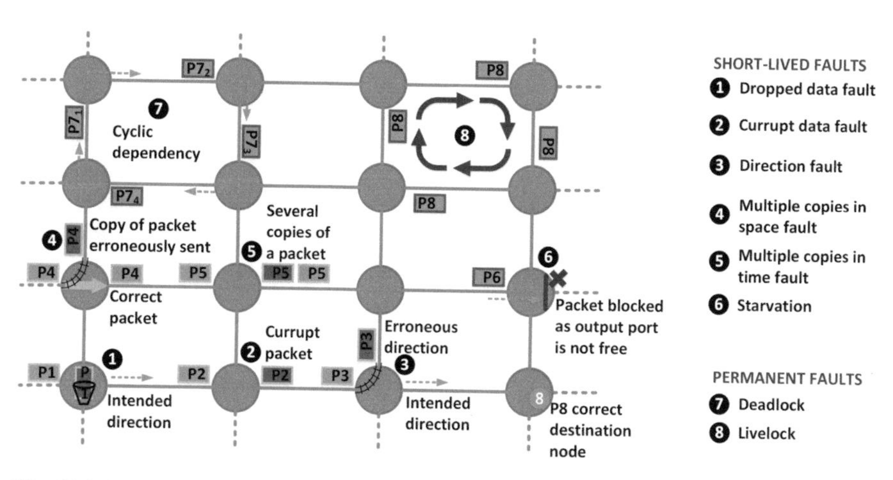

Fig. 13.3 Different data communication faults on an NoC

13.2.1 Short-lived Faults

Short-lived faults exist on the network for a short span of time. These faults may occur due to bugs in different functional units of NoC. Even though the visibility period of such faults is small on the network, they can badly impact system performance. Moreover, if the functional bugs that are the root-cause of such faults remain unaddressed, the fault may affect a critical packet or multiple data packets leading to catastrophic situation. As an example, data drop is a short-lived fault. Let us say a particular input port of an NoC router is buggy, and the result is that any packet coming to that port is getting dropped. This may happen that memory access packets for a critical application may get dropped leading to significant performance degradation. Therefore, the root-cause of such faults should be carefully identified and debugged. This sub-section discusses the commonly observable short-lived network faults.

13.2.1.1 Dropped Data Fault (DDF)

Whenever a packet drops midway before reaching its destination, the error is known as dropped data fault (DDF). In this case, the packet is received at an input port of a router but is never forwarded out of its intended output port. Such fault may occur in VC buffer or RU. For instance, a buggy head counter pointing to the first location of a VC buffer would increase by 2 or more each time a flit is removed. This will lead to multiple flit-drops. Moreover, an erroneous RU may remove the flit from the VC but would not issue an appropriate select signal for output multiplexer. In this case, the flit will never come out of the output port and will be lost inside the router. DDF is visible on the network until the actual arrival time of the packet at its destination.

13.2.1.2 Corrupt Data Fault (CDF)

Whenever the actual information carried by the packet is altered, the fault is known as a corrupt data fault (CDF). Such fault may occur due to error in the VC buffer. Error in one or more bits of buffer memory would change the data value from "0" to "1" or vice versa. This would change the payload information if any of the body flits is corrupted or may change the destination if the head flit is corrupted. Such fault is active till the time the destination node consumes the packet.

13.2.1.3 Direction Fault (DF)

If a packet is sent to an output port other than the intended one, then the fault is known as direction fault (DF). Such fault occurs due to bug in RU or Crossbar component. If the RC module inside the RU decodes a wrong output port from the destination address, the packet flits would be transferred to an unintended output port. Similarly, a fault in the Crossbar module may forward a flit in a wrong direction. This type of fault exists in the network until the packet traverses on the unintended path.

13.2.1.4 Multiple Copies in Space Fault (MCSF)

Due to a buggy RU, it may happen that select signal of multiple output multiplexers may get activated. Thereby a flit would be forwarded to multiple output ports. Such fault is known as multiple copies in space fault (MCSF). The fault can be experienced on the network till the time all the flits taking the wrong paths are live.

13.2.1.5 Multiple Copies in Time Fault (MCTF)

Multiple copies in time fault (MCTF) may originate from the VC buffer in the router. For instance, if the *empty* signal of a VC is faulty, it may happen that multiple copies of the same flit are being received from an upstream router. Though all the copies of the flit are taking the correct path, all of them are not intended to be on the network. Such a fault can be realized until the unwanted packets are on the network.

13.2.1.6 Starvation

A flit experiences starvation, when it does not get an NoC resource to move forward for a long time. Such type of fault may appear due to faulty arbitration mechanism in VA and/or SA modules. In general, whenever the VC waiting time of a flit exceeds beyond an allowable threshold, the anomaly is considered as starvation. This fault exists until the flit gets access of the resource to move forward.

13.2.2 *Permanent Faults*

A permanent fault exists on the NoC forever once it occurs, and continues to thwart the normal operation of the network. We discuss two commonly observed permanent faults on NoC such as deadlock and livelock.

13.2.2.1 Deadlock

Deadlock occurs whenever a cyclic dependency among packets is developed as shown in Fig. 13.3. This happens when all the packets wait for each other to release the resource in a cyclic manner, but none of them proceeds. Such type of scenario may occur due to error in VC allocator. Deadlock is a permanent fault and the affected packets keep on waiting in their respective VC forever.

13.2.2.2 Livelock

Livelock is another permanent fault, where the packet keeps on traversing across network routers, but never reaches its destination as demonstrated in Fig. 13.3. This may happen due to error in RU, or Crossbar module.

From the above discussions, we came to know about the probable erroneous NoC component(s) that may result in a data communication fault. Therefore, in a communication-centric validation framework, whenever such a fault is detected, the root-cause of the bug can be found by evaluating the suspected NoC components.

13.3 NoC Post-Silicon Validation Framework

Post-silicon validation framework enables capturing signal states during system run, and analyzing the same for fault detection and localization. A trace-based debug platform inherited from [8, 9] is demonstrated as debug framework for NoC in Fig. 13.4. As shown in the figure, an NoC-based multi-core system is considered as the Circuit Under Debug (CUD). The integrated DFD hardware provides the suitable infrastructure to observe and capture the NoC behavior during normal operation. The off-chip debug analyzer examines the behavioral traces of the interconnect by running debugger software, and detects any existing network fault. Upon any fault detection, the localization process is carried out. Finally, the root-cause of the fault is diagnosed and fixed.

In Fig. 13.4, it can be observed that the DFD hardware includes an embedded trace buffer, a trigger unit, trace bus, a trace port, and a JTAG interface. Trace buffer is an on-chip memory, where the run-time signal traces can be stored. Trigger unit invokes the tracing operation and trace bus provides the communication medium for

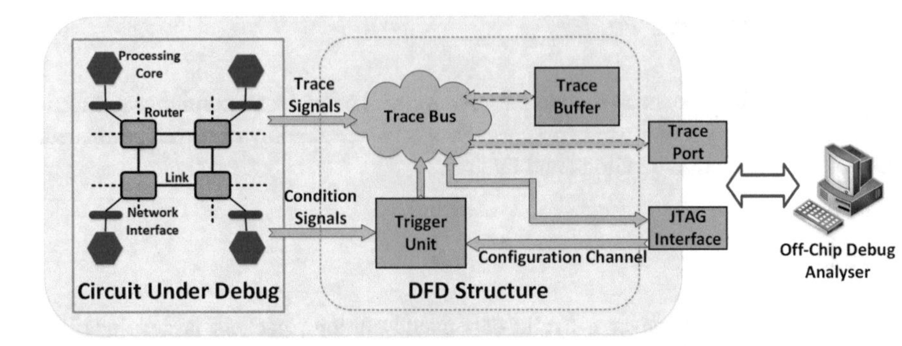

Fig. 13.4 Trace-based NoC debug platform

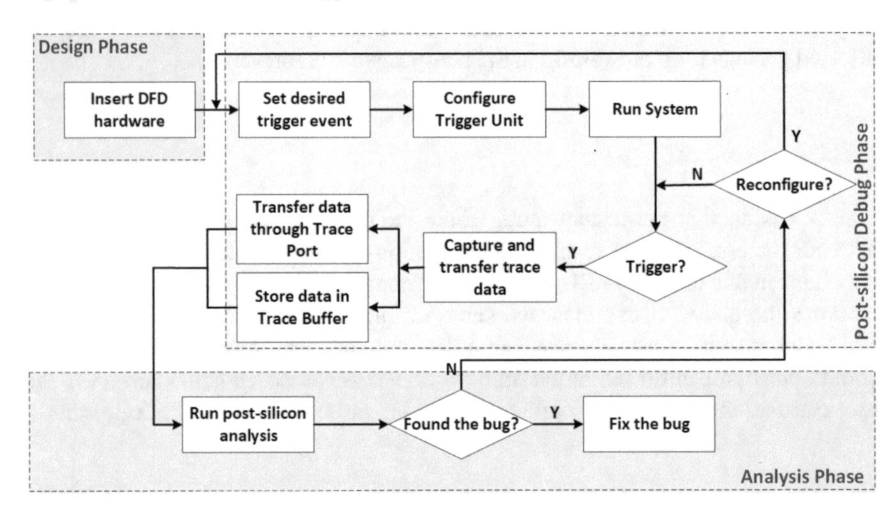

Fig. 13.5 Debug flow during traced-based post-silicon validation

trace transfer. Trace port and JTAG interface export the traces to the external debug analyzer.

The cost of debug hardware is majorly dependent on the size of the trace buffer [10]. This restricts the number of observable points, frequency of tracing, and the width of observation window during the debug phase. Trace buffer width decides the number of observable points, and its depth decides the width of the observation window. By varying the frequency of trace capture, the width of the observation window can also be varied. As the observable points are limited, it is a challenging task to decide which signals to be traced among the plethora of signals in a design. This is known as *Signal Selection*, and is performed during design time. Multiple researchers have proposed several efficient signal selection techniques that can provide greater visibility to the system's internal states [9, 11].

Figure 13.5 presents a generalized flow of trace-based post-silicon validation process. At the beginning of the debug phase, the trigger unit is configured through a debug access port (DAP). Generally, the JTAG interface is used as the DAP. The trigger unit is configured so that signal traces can be captured whenever pre-set trigger events are detected. A trigger event can be expressed in terms of certain state conditions that are transferred by the condition signals from the DUT to the trigger unit. The efficiency of bug localization depends upon the event detection capability and amount of traced data [12]. The event detection capability depends on how well a trigger event is specified, and its relevance to a particular fault. Whereas, the amount of traced data is driven by the size of the trace buffer. The captured traces are either streamed to the off-chip analyzer via the trace port during run-time or stored in the on-chip trace buffer for later analysis. Due to the trace bandwidth limitation and low-speed I/O interface, most of the debug framework follows the store and forward trace mechanism. In this mechanism, the stored traces in the trace buffer are accessed and transferred to the debug analyzer via the JTAG interface later. During post-silicon debug analysis, the traces are checked for any anomalous behavior. Upon the detection of any fault, the root-cause is investigated and fixed. If no fault is detected, then more traces are collected for the same trigger setup, or the trigger event and/or the trigger unit are reconfigured.

As discussed earlier, a communication-centric debug is adopted for NoC post-silicon validation. Therefore, the packet movement is traced during NoC debug. The trigger unit is configured based on trace capture frequency. For short-lived faults, the capture frequency is kept high, whereas it can be kept low for permanent faults. The NoC validation process can broadly be divided into three steps, namely Packet Trace Collection (PTC), Trace Data Transfer (TDT), and Fault Analysis (FA).

13.4 Packet Trace Collection

Trace collection is the most important step in post-silicon validation. In an NoC, payload data is transported in the form of packets. Thus, capturing packet traces can provide a globally consistent view of data communication across the interconnect. The following sub-sections discuss about monitoring infrastructure for NoC-based system, and the process of packet trace collection.

13.4.1 NoC Monitoring Infrastructure

To capture packet movement activities, monitoring units are traditionally deployed at different interface levels of NoC as a significant part of debug infrastructure. Such monitors collect transaction-level packet traces, which are then analyzed to detect any anomaly in packet communication. Many research works propose such monitoring infrastructure for NoC that enhances the debug process of whole SoC [13–17]. Figure 13.6 shows a basic monitoring infrastructure for transaction-based communication-centric system debug. The *read* and *write* transactions of

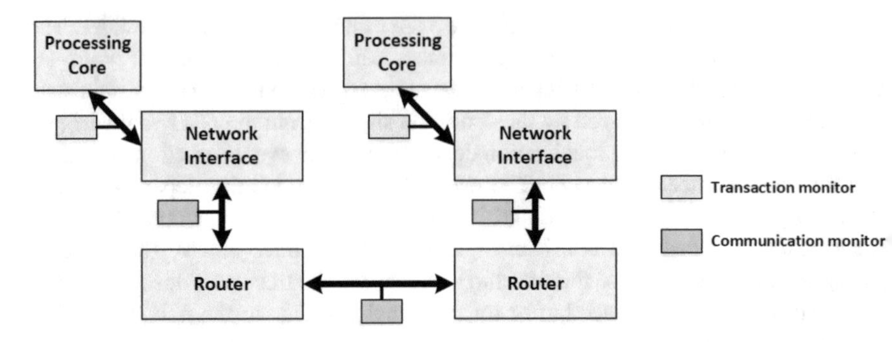

Fig. 13.6 Monitoring infrastructure for transaction-based communication-centric NoC debug

processing cores (PCs) initiate activities on interconnect. These transactions are fragmented into packets that communicate through the network. Based on the placement in the NoC architecture, the monitors can be broadly classified as transaction monitors and communication monitors. Transaction monitors are placed in between the PC and network interface (NI). It inspects the transactions and detects either missing transactions or transactions with incorrect attributes. On the other hand, communication monitors residing between the routers observe the packet communication to detect faulty packets and paths.

The authors in [13] present a monitoring setup for system debug that uses structural and temporal abstraction techniques along with debug data interpretation to visualize an SoC's state at the logical communication level. A monitoring infrastructure is deployed in [14] to carry out the debugging of the interactions among the embedded processors, along with system performance analysis. An event-based NoC monitoring framework is proposed in [15] that integrates hardware probes to NoC's NI ports. The infrastructure provides run-time observability of NoC behavior and supports system-level and application debugging. In another work [16], a system-level debug agent is provided along with the hardware probes integrated between the PCs and the NIs. The debug agent delivers several in-depth analysis features of an NoC-based system such as NoC transaction analysis, multi-core cross-triggering, and global synchronized timestamping for more effective debugging. The authors in [17] propose to integrate monitors and filters to NoC, which observe and filter transactions at run-time. This work also implements programmable finite state machines in the debug unit to validate the correct relation of transactions at run-time.

The above-mentioned proposals integrate monitoring infrastructure to NoC interfaces for providing SoC level debug solutions. Additionally, there are proposals [6, 18, 19], which embed monitors inside each NoC routers to provide dedicated NoC debug solutions. Such an in-router monitoring setup is shown in Fig. 13.7 that can observe and capture the packet traces within the router. These packet traces can be analyzed to find any data communication fault discussed in Sect. 13.2. Upon the detection of any fault, the packet traversal path can be reconstructed using the

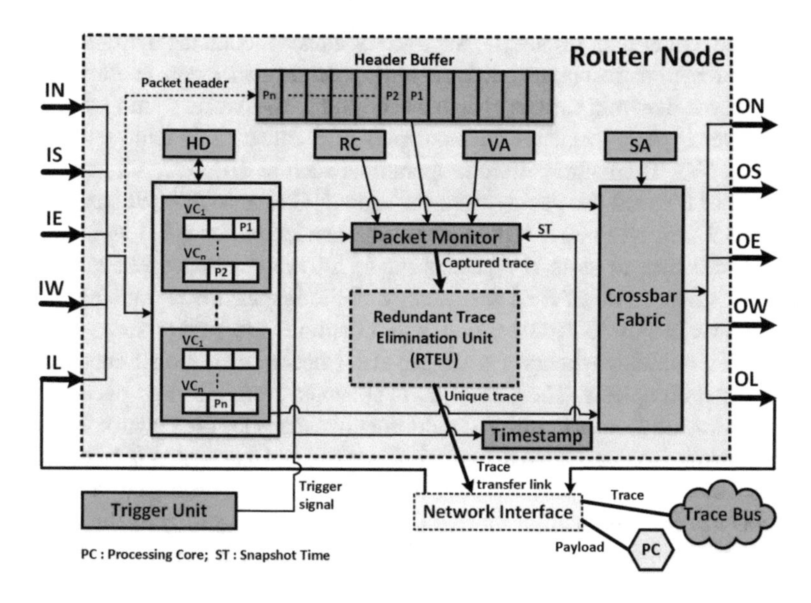

Fig. 13.7 Packet monitoring infrastructure inside NoC router [19]

Fig. 13.8 Individual packet trace pattern

Packet ID	Current Node	Input Port	Input VC	Output Port	Output VC	Timestamp
	Packet routing status					

Packet ID : {Packet Number, Source, Destination}

corresponding packet traces, and thereby the buggy functional unit of the NoC can be localized. The following sub-section discusses the process of packet trace collection using the in-router monitoring infrastructure.

13.4.2 *Process of Trace Collection*

During the packet trace collection phase, routing status of all the packets inside each router node is monitored. Periodic snapshots of packet states are captured and stored as traces in the trace buffer. The trigger unit is configured such that it can repeatedly generate the trigger signal for trace capture after a fixed time unit known as snapshot interval (SI). Packet trace is constructed by concatenating packet information extracted from its header flit, packet's routing status, and timestamp value. Figure 13.8 shows the trace pattern of individual packet trace. Packet ID in the trace is a piece of static information and acts as the identifier of the corresponding packet. Packet number along with source and destination information from the header flit constitute the unique packet ID for each packet. At every core, a packet number is produced and embedded into the header flit whenever a packet is generated. Other than packet ID, the trace of a packet contains multiple dynamic information such as instantaneous timestamp, current node, port number, and VC number. These dynamic entities change according to the packet routing status.

Whenever a packet reaches an input port of a router, it is temporarily stored in an input VC. Thereafter different operations such as HD, RC, VA, and SA are performed to forward the packet from the input VC to an output VC as discussed in Sect. 13.2. An additional buffer known as header buffer (HB) is included in the DFD structure to store the header flit of all the existing packets inside the router. With the arrival of a packet, a copy of the header flit is transferred to the HB as shown in Fig. 13.7. The monitor unit captures the packet traces from HB, RC, and VA modules whenever a trigger event occurs or a global snapshot time (GST) period completes. The packet ID is collected from HB, and packet routing status is extracted from RC and VA. Additionally, snapshot time is provided by the timestamp unit. GST is a large enough time period, and a snapshot is captured after each GST that would help in detecting permanent faults as discussed in Sect. 13.6. A redundant trace elimination unit (RTEU) is provided to discard the redundant traces, which is discussed in Sect. 13.5.3. Once the packet traces are collected, they are transferred via trace bus to the trace buffer for storage.

13.5 Trace Data Transfer and Storage

Both interconnection fabric and trace buffer are important components of DFD structure. DFD interconnection allows the transfer of trace data, whereas trace buffer stores these trace data on-chip for further analysis.

13.5.1 Trace Transfer

Trace bus provides a dedicated routing path for trace data transfer to trace buffer and/or trace port. Standard solutions typically use pipelined multiplexer (MUX) trees to transfer traces [20, 21]. But, such MUX tree-based interconnection incurs high area overhead to provide increased trace bandwidth. To overcome this limitation, the authors in [22] propose a two-level interconnection network consisting of a MUX network and a non-blocking concentration network for trace transfer. In the case of NoC-based system debug, there is an opportunity to reuse the same NoC infrastructure for transferring the traces. Both [18] and [6] have reused the same NoC (which is the CUD) for on-chip trace transfer. Conventionally, a JTAG interface is used to transfer the trace data to an off-chip debug analyzer for trace analysis [23]. The authors in [18] propose to use augmented wireless interface (WI) on NoC for this purpose to increase the off-chip trace transfer speed.

13.5.2 Trace Storage

Trace buffer is an on-chip memory, which stores run-time traces for debugging purpose. The amount of trace data that can be collected during a single post-silicon validation run is limited by the storage capacity of the on-chip trace buffer. To efficiently use the limited storage space in the trace buffer, the authors in [24] have proposed distributed as well as dynamic allocation of trace buffers at run-time. Moreover, the authors in [18] and [6] have proposed debug structure for NoC sub-system, where L2 cache of each processing core is effectively reused as local trace buffer of corresponding router node. To deal with the trace buffer storage constraint while maintaining the system's internal observability, several trace reduction techniques are proposed in the literature, which is highlighted in the following sub-section.

13.5.3 Trace Reduction

Size of trace buffer is one of the major controlling factors of on-chip DFD cost. Therefore, different trace reduction mechanisms are used to maintain observability with minimal overhead. State restoration and trace compression are the two major techniques that are used for trace reduction. Special state restoration mechanisms [9, 11, 25] can enhance the amount of traced information by data expansion. This allows less number of observable signals to be selected without any visibility loss, which leads to a smaller trace buffer requirement. Moreover, trace compression techniques can reduce the total amount of trace by 20–30% [26–28].

In the case of an NoC-based system, there is another opportunity to reduce trace amount driven by an inherent nature of NoC operation. During NoC debug, packet traces are captured periodically once after each SI. If the state of a particular packet does not change frequently, then with small SI, multiple redundant traces get generated for the same packet. This scenario can easily be visualized for the congested network, where packets in routers get stuck at different pipeline stages for multiple additional cycles than expected. Based on such observations, a mechanism is presented in [19] that can distinguish the redundant traces and keep only one copy from each redundant group. As a result, one instance of all unique traces is collected and thus the amount of trace data reduces while maintaining the system observability. The scheme is called Redundant Trace Elimination (RTE) that is able to save a significant amount of trace buffer space in an NoC-based system by discarding the redundant traces. Figure 13.9 demonstrates the structure and operation of both periodic trace collection, and redundant trace elimination. Whenever the packet monitor unit collects new packet trace *Snapshot(t)* (Fig. 13.9a), it is compared with the corresponding packet trace *Snapshot(t-SI)* previously stored in the Snapshot Buffer (SB) (Fig. 13.9b). If both the traces are different, then the new trace *Snapshot(t)* would be overwritten in the SB. The output of the SB would

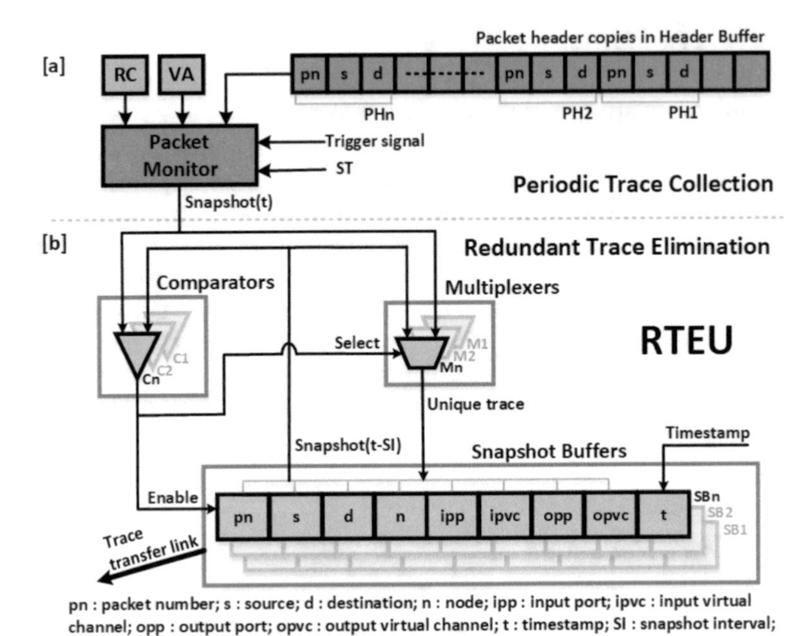

Fig. 13.9 Periodic trace collection and redundant trace elimination [19]

be enabled to export the new trace through the trace transfer link to the trace bus. Otherwise, if both the traces are the same then the previously stored trace *Snapshot(t-SI)* will be retained in the SB. No trace will be transferred to the trace bus. This enables the system to discard all the redundant traces and only stores the unique ones. Thus, the RTE mechanism reduces the total trace amount without degrading the network internal observability.

13.6 Fault Analysis

During the fault analysis phase, the collected traces are analyzed for fault detection, identification, root-causing, and localization. The root-cause of faulty behavior needs to be located both in space and time. Bug localization being the most important step in debug [10], it requires extensive packet path reconstruction. In the case of an NoC debug process, the individual packet paths are reconstructed using packet ID and timestamp values once the traces are collected in the debug analyzer. All the traces related to a particular packet are extracted using the unique packet ID assigned to it. Then the traces corresponding to each packet are stitched together according to the increasing timestamp values. Thus the paths traversed by every packet are reconstructed. The percentage of packet path reconstruction depends upon the amount of traces present for the respective packet. Each packet

Algorithm 1: Short-lived fault detection algorithm

```
1: while (curtime − starttime) < GST do
2:     if curtime == prevtime then
3:         if curnode, prevnode on path(src, dest) then
4:             fault ← MCTF
5:         else
6:             fault ← MCSF
7:         end if
8:     else if curnode ! on path(src, dest) then
9:         fault ← DF
10:    else if endtrace && (curnode ! = dest) then
11:        fault ← DDF
12:    end if
13: end while
```

Algorithm 2: Permanent fault detection algorithm

```
1: while (curtime − starttime) > GST do
2:     if curtrace == prevtrace then
3:         fault ← Deadlock
4:     else
5:         fault ← Livelock
6:     end if
7: end while
```

path is examined by several fault detection algorithms based on the faults for which the design is getting evaluated. This section discusses two such algorithms presented in [19]; one for short-lived faults (Algorithm 1), and another for permanent faults (Algorithm 2). If the time difference between the current trace and the packet start trace is less than GST, the packet path is evaluated for short-lived faults, else the path is evaluated for permanent faults.

When more than one instance of the same packet is present on the network, the fault can be MCTF or MCSF. In such cases, if all the instances are taking the correct path one after another, then the fault is known to be MCTF. Otherwise, if all the instances are taking different paths then the fault is called MCSF. These faults are described from line numbers 2–6 in Algorithm 1. Line numbers 7 and 8 represent DF that indicates the packet is routed to a wrong output port than the desired one. DDF is a case of a lost packet inside a router, which is shown in line numbers 9 and 10. Deadlock condition arises when a cyclic dependency is formed and packets keep on waiting for each other to release the network resources but none of them proceed. So, the packet state never changes, which can be realized from the line numbers 2 and 3 of Algorithm 2. It must be noted that the traces collected at each GST (discussed in Sect. 13.5.3) are directly sent to the trace buffer and do not pass through the RTE unit. Hence, in case of a deadlock, the packet trace collected at a GST that is the same as the previous trace would not be discarded. This will help in detecting deadlock during the analysis phase. Similarly, livelock displayed in line numbers 4 and 5 happens when a packet continues to move on the network but it never reaches its destination.

Once a fault is detected, the fault location can be traced back from the constructed path of the corresponding packet. After fault localization, the functional modules of the concerned router node are examined to find the root-cause of the fault and identify the exact functional bug. Thereafter the bug is fixed, and once all such bugs are identified and fixed the design is sent for mass production.

13.7 NoC Validation Framework using Wireless Links

This section discusses an efficient post-silicon validation framework for NoC using augmented wireless links [18]. The framework facilitates better observability of the system in case of short-lived packet faults like direction fault and packet drop. This is achieved without any additional overhead in terms of trace buffer size and trace bandwidth requirement. The wireless interfaces (WIs) on the network are used to efficiently transport the traces to the external debug analyzer. This results in eliminating the need for additional trace bus while elevating the speed of trace communication.

The average hop count of packet communication reduces notably in wireless augmented NoC in comparison to wired NoC [29]. This creates an opportunity to capture traces either for more packets at the same capture frequency or for the same number of packets at the increased capture frequency without increasing the trace buffer size. As a result, it enables the debug system to detect the network short-lived faults (that demand more frequent trace collection) more efficiently. Augmented on-chip wireless links are used to accelerate the trace data transfer. The traditional low bandwidth JTAG interface used for transferring traces to the external debug analyzer is replaced by WIs. This significantly increases the speed of trace data transfer.

13.7.1 Debug Operation using WIs

An NoC validation framework using wireless links is shown in Fig. 13.10. For an illustration purpose, a 36-node system is shown in the figure that is augmented with 4 WIs. The whole network is divided into four clusters, and each cluster is assigned with a WI. The framework also uses a wireless enabled off-chip debug analyzer. The WIs transfer long-range on-chip test payload during packet trace collection (PTC), and perform off-chip trace communication during trace data transfer (TDT) phase. Each WI is responsible for transferring traces of the corresponding cluster. The framework provides trace buffer at each router node (distributed trace buffer) that stores the traces of the packets traversing across that node. To save the on-chip area and routing overhead, the proposed method reuses L2 cache of each core as local trace buffer of the corresponding router node. A specified portion of the

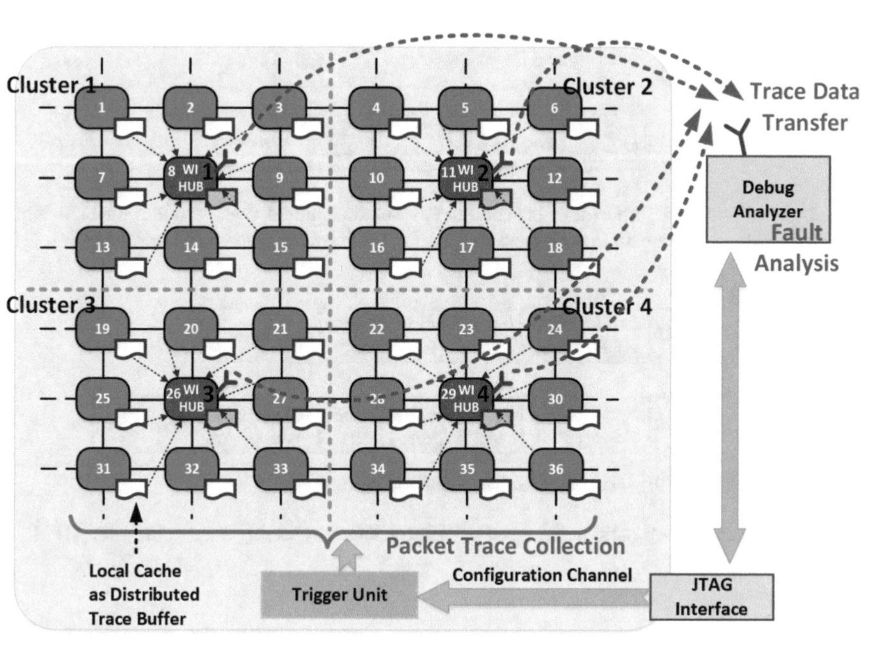

Fig. 13.10 NoC validation framework using wireless links [18]

local caches are temporarily used to store packet traces during the debug phase and are released for normal operations after the post-silicon validation. Dedicated trace buffers are provided to the nodes that are not connected to a processing core.

During the PTC phase, traces are captured for all the packets present inside a router as discussed in Sect. 13.4.2. The router architecture can be referred from Fig. 13.7. The tracing operation is performed after every snapshot interval (SI), and the collected traces in each router are stored in the reserved space of the attached L2 cache (local trace buffer). The SI value can be controlled by the trigger unit depending on the type of suspected fault and the type of network node. The SI is kept low for short-lived faults and high for permanent faults. To ensure the correctness of the wireless communication, SI of WI hubs are fixed at one cycle. In contrast, the SI of wired nodes is decided based upon the size of the local trace buffer.

The packet traces are built from the packet header information and the packet routing status. Figure 13.11 shows the format of packet and header flit transmitted over wireless enabled NoC, where each WI is assigned with a unique WI Address. The packet header holds the WI Address of both source and destination WI along with the actual source and destination node addresses of the corresponding packet. The pattern of packet trace at both wired node and WI hub is shown in Fig. 13.12.

Once the trace buffer space is filled, the TDT phase is initiated. Trace data need to be communicated to the off-chip debug analyzer for Fault Analysis (FA). Conventionally, a dedicated trace bus is used for trace communication. In the proposed method, this is achieved by the existing network resources. This further

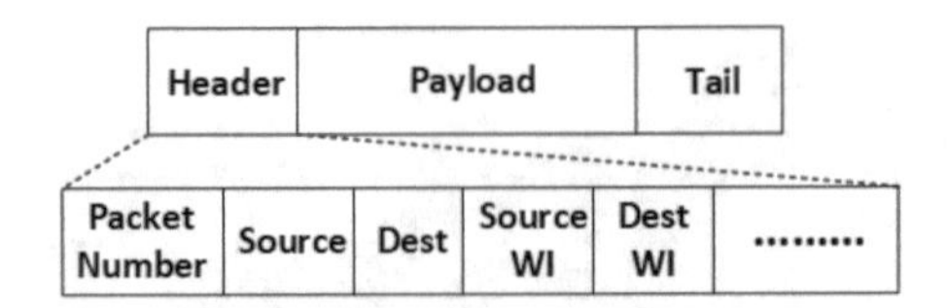

Fig. 13.11 Format of packet and header flit content transmitted over wireless enabled NoC. Unused space in header flit can be used to hold Packet Number, Source WI and Dest WI. [18]

[a]

Packet ID	Current Node	Input Port	Input VC	Output Port	Output VC	Timestamp

[b]

Packet ID	Current Node	Input Port	Input VC	Output Port	Output VC	Source WI	Dest WI	Timestamp

Packet ID: {Packet Number, Source, Destination}

Fig. 13.12 Packet trace pattern. (**a**) trace pattern for wired node, (**b**) trace pattern for WI HUB [18]

reduces the on-chip area and routing overhead. During the TDT phase, the normal execution of the network is stopped. Trace data stored in the local trace buffer of each router gets transferred to the WI hub of the corresponding cluster using the wired NoC path. Then, the accumulated traces at each WI hub are sent to the off-chip debug analyzer for post-processing. This communication is conventionally accomplished using a low bandwidth JTAG interface [23]. In the proposed approach, high bandwidth inter-chip wireless communication is used to transport the trace data to the analyzer as shown in Fig. 13.10. This enhances the speed of trace data transfer during the debug phase.

Upon the collection of packet traces, the debug analyzer processes the data and reconstructs the packet traversal paths. The detection algorithm in the analyzer can take these paths as input to make a decision on the occurrence of any functional fault as discussed in Sect. 13.6. The reconstructed path also indicates the location of the fault. Once the fault is localized, its root-cause and concerned buggy functional units are identified and fixed.

13.7.2 Wireless Interface

In the proposed validation framework, WIs are used for trace as well as test payload communication. This wireless enabled NoC topology is comprised of Base Routers (BRs) and few Hybrid Routers (HRs) shown as WI hubs in Fig. 13.10. HR integrates WI with BR components and can convert digital packet data to the RF domain and vice versa. This is composed of serializer/deserializer, modulator/demodulator, Power Amplifier (PA), and Low-Noise Amplifier (LNA) components. An omni-directional antenna is used for both transmission and reception of wireless data.

Table 13.1 Network topology and simulation setup

Component	Configuration
Topology	8×8 wireless enabled NoC (Mesh augmented with WIs), 4 WI hubs
Router	5 I/O wired ports, 2 virtual channels, 8 flit buffers, 8 flit packets, 32 bit flits
Wireless Link	60 GHz carrier, 16 Gbps bandwidth, single cycle latency

For low overhead implementation, on-off keying (OOK) based wireless transceivers are used, and a single wireless channel is shared between all the WIs [30]. A variable gain PA [31] is used, which can provide different data transfer amplification levels depending on whether the WI is used for normal on-chip test payload communication or is used for off-chip trace transfer. The amplifier pumps more power to transfer traces to the off-chip debug analyzer. Power gated WIs are used to reduce the energy consumption [32].

13.7.3 Results and Analysis

This section discusses the experimental setup for WI-based debug platform. The multi-core interconnect fabric is modeled on a cycle-accurate network simulator, Noxim [33]. The details of the network topology and the simulation setup are shown in Table 13.1. All cores and wired network modules are operated at a clock frequency of 1 GHz. Fault detection and path reconstruction capability of the proposed framework is evaluated by injecting several faults into the network. To evaluate the efficacy of the proposed framework, the authors have compared the performance of the wireless platform against the wired platform.

Figure 13.13 shows the fault detection and path reconstruction ability of the proposed platform with different SI and for *Random* traffic workload. The results are shown for short-lived network faults. From the results, it can be observed that with increasing SI, the percentage of fault detection, as well as path reconstruction, degrades significantly. This is because snapshots after large SI are unable to capture most of the traversal information of short-lived faulty packets. Hence, such faults demand very frequent trace collection for efficient debugging.

13.7.3.1 Trace Buffer Size

Smaller trace buffer size is always desired as it saves silicon area as well as cost. Though this work reuses a small portion of each L2 cache memory as trace buffer, it is required to assign a large portion of the cache for normal operation even during the debug phase. In this work, the size of a packet trace from a wired node is 52 bits. This includes 22 bits of packet ID, 14 bits of packet routing status, and 16 bits

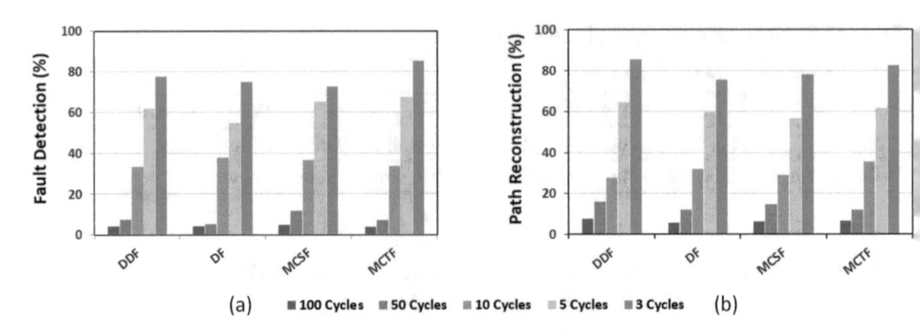

Fig. 13.13 (**a**) Percentage fault detection, (**b**) Percentage path reconstruction for different network short-lived faults with varying SI [18]

Table 13.2 Comparison of wireless and wired post-silicon validation platforms

% Improvement	DDF	DF	MCSF	MCTF
Fault detection	25	37	11	26
Path reconstruction	33	26	38	35

of timestamp value. Trace of a packet from WI hub adds 4 more bits of WI address values on top of the wired packet trace and consumes a total of 56 bits.

This work considers an 8×8 2D mesh network with 4 WI hubs. For a *Random* traffic workload, the wireless enabled network results average hop count of 4 in comparison to 6 per packet communication with a wired-only network. If packet trace is captured once at each hop during a packet transmission, in the case of a wired network, total of 3.81 KB of trace buffer size is needed to accommodate traces for 100 packets. In case of a wireless network, two hops involved in wireless transmission would require 56 bits each to store the trace while the remaining hops require 52 bits only. So, with around 10% of test payload transferred through wireless medium, the network requires a 2.54 KB of trace buffer which is around 67% of wired case. Hence, the proposed wireless post-silicon validation platform can show a considerable amount of improvement both in the case of short-lived fault detection and path reconstruction in comparison to a wired post-silicon validation platform. This can be observed in Table 13.2 for a *Random* traffic workload for different short-lived faults.

13.7.3.2 Efficient Trace Data Transfer

In the proposed post-silicon validation framework, trace data is transferred to the external debug analyzer using the wireless medium. An average of 54% packet latency reduction in the case of wireless inter-chip communication has been sown in [34] compared to wire-based silicon interposer communication. In terms of bit-rate capacity, the proposed wireless interface can export traces with 16 Gbps data rate. Whereas a JTAG based ARM CoreSight debug and trace connector (ULINK*pro*) provides up to 800 Mbps of trace communication speed [35]. This ensures that the proposed framework provides fast off-chip trace transfer during NoC post-silicon validation.

In summary, the framework proposed in this work augments WIs to the NoC debug platform that provides high-speed trace transfer and better short-lived fault analysis during post-silicon validation of the NoC sub-system.

13.8 Reuse of NoC Debug Infrastructure

On-chip DFD structures are used to capture escaped faults during post-silicon debug. Most of the DFD modules are left idle after the debug process. Reuse of such structures can compensate for the area overhead introduced by them. The work in [36] proposes to re-utilize the trace buffers as extended virtual channels (VCs) for the router nodes of an NoC during in-field execution. The addition of VCs to routers can significantly improve network throughput [7, 37]. Therefore, extending the number of VCs by reusing the trace buffers can improve the average network throughput and latency considerably.

System internal visibility and DFD overhead are competing in nature. Trace buffer is one of the significant components of DFD structure. As the size and complexity of NoC-based multi-core SoCs are increasing, the trace buffer footprint is also growing to maximize the system observability during the debug process. The area overhead introduced by trace buffers is considered as a major design concern as the DFD hardware becomes non-functional once the SoC goes into production. To address this issue, reuse of architectural component as DFD component or vice versa has been proposed in several research fronts [6, 18, 38–42]. The work in [38], instead of using a dedicated trace buffer, uses data cache to store both traces and data during debug. The authors in [39] propose to use L1 and L2 caches to store memory operation activity logs required for memory consistency and coherence validation. A dedicated portion of L2 cache of each node is used as trace buffer for NoC validation in [6, 18]. Similarly, there are few efforts that demonstrate the reuse of DFD hardware for some architectural enhancements. The authors in [40] have re-employed the debug structure for online monitoring during run-time verification. Embedded trace buffers are re-purposed for malware prevention in [41], and reused as victim cache to enhance cache performance in [42].

The work in [36] proposes to reuse the NoC trace buffer as extended VCs of network routers to improve the throughput and performance of NoC-based multi-core SoCs. Total trace buffer can be distributed among all the routers to store corresponding packet traces during debug phase as shown in Fig. 13.14a. Whereas the same trace buffers can be reused as extended VCs of router input channels to improve network throughput during in-field execution as shown in Fig. 13.14b.

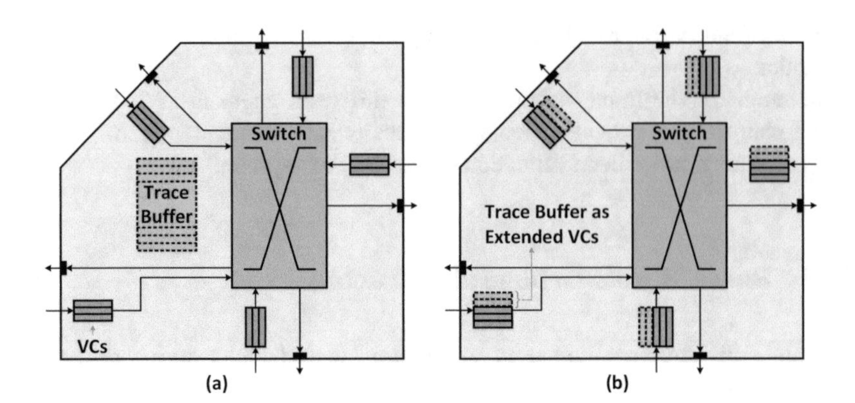

Fig. 13.14 (**a**) Trace buffer used for storing traces during debug, (**b**) Trace buffer used as extended VCs during in-field execution [36]

13.8.1 Trace Buffer Distribution

In this work, the trace buffers are distributed among all the nodes present in the network and are utilized as extended VCs during in-field normal operation. Distribution of trace buffers are performed using a *Fair Division* (FD) algorithm. A proportional fair division method says every agent receives at least its due share of a set of resources according to its own value function [43]. In this context, routing nodes are agents, trace buffers are resources, and profile index (PI) of each node based on traffic condition are considered to be respective value function $f(v)$. The Following two sub-sections present the details about the calculation of PI of each node, and accordingly fair distribution of available trace buffer resources among all the nodes.

13.8.1.1 Profiling the Router Nodes

Each of the network routing nodes is profiled based on the amount of traffic passing through it for an application. Separate profiling of individual node is done for all the applications meant for the corresponding multi-core SoC. Finally, each node is assigned with a PI, which is the arithmetic mean of all of its profiling values. For a network with n nodes $\{N_1, N_2, \ldots N_n\}$ and m applications $\{A_1, A_2, \ldots A_m\}$, PI of each node can be realized as below in Eq. (13.1):

$$PI_{N_i}|_{i\in\{1\ to\ N\}} = \left\{\sum_{j=1}^{m} p_{i,j}\right\}/j \tag{13.1}$$

Here $p_{i,j}$ represents profiling of ith node for jth application, and PI_{N_i} represents the final profile index value of ith node. Value of $p_{i,j}$ becomes high if large number of packets traverse through the ith node while jth application is executed. The PI value indicates an average occupancy level of the respective node by the payload packets. It can be concluded that the trace buffer requirement of each node is proportional to its PI value. So, value function $f(v)$ of each routing node is calculated based upon the corresponding normalized PI as shown in Eq. (13.2).

$$f_i(v) = PI_{N_i} / \left\{ \sum_{i=1}^{n} PI_{N_i} \right\} \tag{13.2}$$

These value functions are used in the FD method for the optimal distribution of the available trace buffers.

13.8.1.2 Fair Division of Trace Buffers

The objective of fair division is to divide and allocate resource share to each candidate so that everyone would feel that it has got its due share according to its value function. In the proposed work *Dubins-Spanier Moving-Knife Procedure* [44] is adopted for the fair distribution of trace buffers. The *Moving-Knife Procedure* is originally proposed to solve the fair division of cake problem among n people, where $n > 2$. A knife is slowly moved across a cake from its extreme left position, and whenever a person feels that the knife has moved $1/n$th of the total cake according to his measure, he calls "cut." Then he takes that piece of cake and exits. If two persons call at the same time, then the piece is given randomly to any one of them. The process is repeated for $n - 1$ participants.

In the context of trace buffer distribution, a modified FD algorithm is illustrated in Algorithm 3. This is shown for a NoC comprised of n nodes with individual node value function $f_i(v)$ and total trace buffer size of S_{tb}. The cake problem deals with the division of continuous resource, whereas trace buffer distribution is a discrete resource division problem. The smallest unit of trace buffer that can be assigned to a particular port of a router node should have the size equal to one VC. In the case of a 2D mesh topology (considered for experiments in the work), each router having 5 ports results into minimum trace buffer share per router node ($tb_{N(min)}$) as 5 times a VC. Each time the trace buffer share of a node according to its value function is found not to be a whole number multiple of $tb_{N(min)}$, the final share is rounded up to either the next or previous whole number multiple of $tb_{N(min)}$. Final adjustment of trace buffer share to meet the size limit of available trace buffer is done on the top ranked nodes claiming the largest portions of it, as shown in Algorithm 3.

Algorithm 3: Modified fair division algorithm for trace buffer distribution

assumption: $S_{tb} = p * tb_{N(min)}$; ▷ where S_{tb} is the trace buffer size, $tb_{N(min)}$ is minim trace buffer share per router, and p is a whole number

initial: $m = n$; $j = 1$; ▷ total n number of nodes in the network $N = \{N_1, N_2 \ldots N_n$ represents node number

while $m > 0$ **do** ▷ trace buffer distribu
  call function ***tb_fair_div***
  i = Node number that asked for the trace buffer partition
  if $tb_{N_i} \leqslant tb_{N(min)}$ **then**
    $tb_{N_i} = tb_{N(min)}$
  else
    if $(tb_{N_i} \% \, tb_{N(min)}) \geqslant tb_{N(min)}/2$ **then**
      $tb_{N_i} = \lceil tb_{N_i} \rceil$
    else
      $tb_{N_i} = \lfloor tb_{N_i} \rfloor$
  end if
    ▷ $\lceil tb_{N_i} \rceil$ and $\lfloor tb_{N_i} \rfloor$ returns the next and previous whole number divisible by $tb_{N(}$ respectively
  end if
  $S_{tb} = S_{tb} - tb_{N_i}$
  make $i \notin N$
  $m - -$
end while

if $(q = \sum_{j=1}^{n} tb_{N_j} \; / \; tb_{N(min)}) > p$ **then**
  For top q-p nodes assigned with maximum trace buffer share are reduced by $tb_{N(min)}$ *ea*
else if $q < p$ **then**
  For top q-p nodes assigned with maximum trace buffer share are increased by $tb_{N(min)}$ *e*
end if

function ***tb_fair_div*** definition
  Start the S_{tb} *division*
  Wait till any node asks for the trace buffer partition according to its value function f(v)
  If more than one node ask for the same partition then assign randomly to any one of th

13.8.2 Network Operation

This section discusses the network operation for the proposed solution in both debug as well as in-field execution modes. Figure 13.15a shows a 64-node network and (b) illustrates the supporting router structure. Each router node gets a portion of the total trace buffer based on its value function as discussed in Sect. 13.8.1.2. The following sub-sections highlight the use of trace buffer for different purposes during different operation modes. A mode signal coming from the debug support unit (DSU) decides the mode of operation of the network.

13.8.2.1 During Debug Mode

During post-silicon debug, the trace buffer is used for packet trace storage. Trace buffers corresponding to a particular router can equally be distributed among all the input ports during physical implementation (Fig. 13.15b). Though distributed among ports, the trace buffer controller (TBUF Ctrl) considers all the trace buffers inside the router as a lumped trace storage space as can be seen in Fig. 13.14a. The packet monitor unit collects the packet state information (packet ID, current node, input port, output port, and VC number) from HD, RC, and VA and builds the respective packet trace by adding the corresponding timestamp value. The TBUF Ctrl generates the trace buffer address (*tbaddr*) and forwards the generated packet trace (*tdata_in*) to the corresponding *tbaddr* location. Generated *tbaddr* is usually a trace buffer location associated with the input port receiving the packet flit. If no trace buffer space is available at a particular input port, the TBUF Ctrl looks for empty trace buffer at other input ports of the same router and generates the corresponding *tbaddr* of a free trace buffer space. TBUF Ctrl asserts a *tb_full* signal when it finds no free trace buffer space inside a router. This necessitates exporting the trace buffer content (*tdata_out*) of the filled router to the external debug analyzer.

Trace transfer from the network routers is performed in two stages, namely local trace transfer (LTT), and global trace transfer (GTT). Whenever a particular router node runs out of trace buffer space, the LTT phase starts for the corresponding router. As an example, node A in Fig. 13.15a is a busy node, and its trace buffer fills up quickly. During the LTT phase, traffic switching in router A is paused and packet traces stored in its trace buffer are transferred to the trace bus through the existing network trace port (NTP). The network can be divided into multiple sub networks and each of them can have an NTP. While transferring the packet traces to the corresponding NTP, the intermediate routers (routers on the path from A to B) give higher priority to the traces over the normal payload while switching, and thus

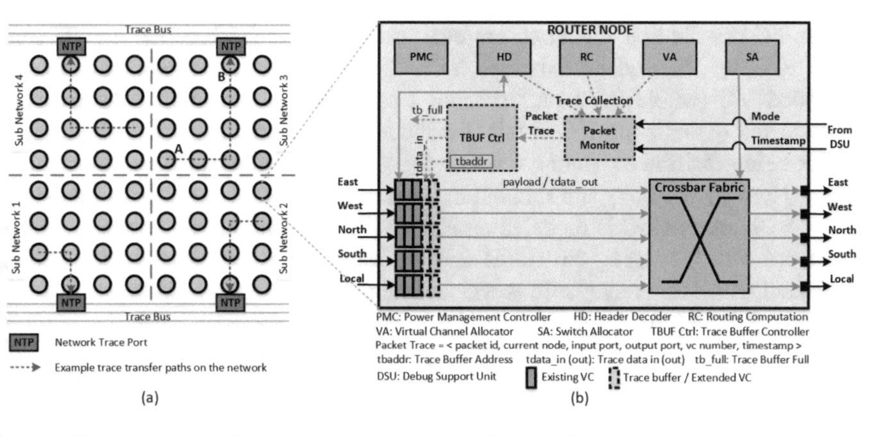

Fig. 13.15 Debug and in-field operations with trace buffer reused as NoC VCs [36]. (**a**) A 64 node network, (**b**) Router node architecture

provide a faster trace transfer. The remaining network exhibits normal operation during LTT. After the completion of the LTT phase, node A resumes with its normal operation. A GTT phase starts periodically after a pre-decided (during the design phase) large time period. During this phase, the normal operation of the whole network is paused, and traces from all router nodes are collected through the NTP ports. This is performed to collect complete network traces, so that trace analysis can be started in the debug analyzer. The network gets back to its normal operation after the GTT phase.

13.8.2.2 During In-field Execution Mode

During in-field execution mode, the trace buffers are no more used for debug purposes. In the proposed method, the router trace buffer is re-utilized as extended VCs (Fig. 13.15b) to enhance the network performance. The incoming payload flits now find additional buffer space at each router input ports, leading to reduced chance of packet drop and deadlock conditions. During this mode, the TBUF Ctrl and packet monitor units are power gated by the Power Management Controller (PMC). The VA unit is designed according to the total VC (existing + extended), so that it can perform the desired arbitration during in-field execution mode. VA module keeps on snooping the available VC credit information of the downstream routers to assign the output channel to a particular incoming flit. The extended VC scenario in the proposed method increases the possibility of VC credit having a positive value in most of the time, and thus increases the throughput of the network.

13.8.3 Experimental Results

This section describes the simulation setup and discusses the results generated. Three VC distribution scenarios are evaluated such as: (1) baseline architecture (only existing VC and no extended VC), (2) existing VC + equally distributed extended VC (whole trace buffer being equally distributed among the existing nodes), and (3) existing VC + fair distribution based extended VC (whole trace buffer being distributed among the existing nodes according to their value functions). Network topology and simulation setup is listed in Table 13.3. For simulation, a 8×8 2D mesh NoC is modeled on Noxim [33]. The router module is modified to incorporate the debug structures and VC extension. The proposed scheme is evaluated using three synthetic workloads and three workloads from SPLASH-2 benchmark suite [45].

Table 13.3 Network topology and simulation setup

Component	Configuration
Topology	8×8 baseline 2D mesh NoC, XY routing
Router	5 I/O ports, 4 existing VCs per port, extended VCs based upon value function, 2 flit buffers, 8 flit packets, 32-bit flits, 1 GHz operating clock
Debug Setup	8 KB trace buffer size, single cycle snapshot, 32 bits packet trace size
Workload	Synthetic—*Random, Transpose, Butterfly*; SPLASH-2—*Barnes, FFT, Radix*

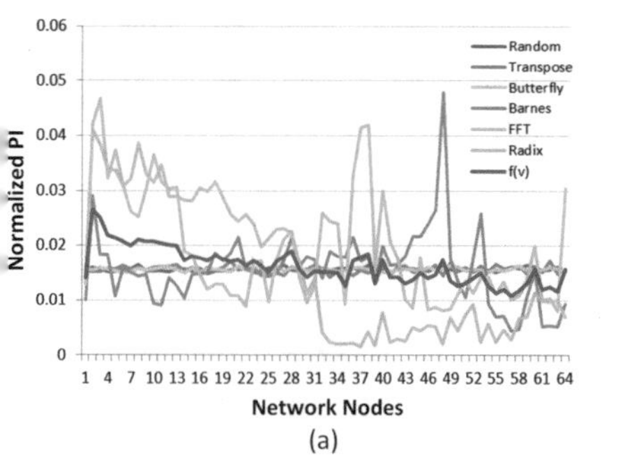

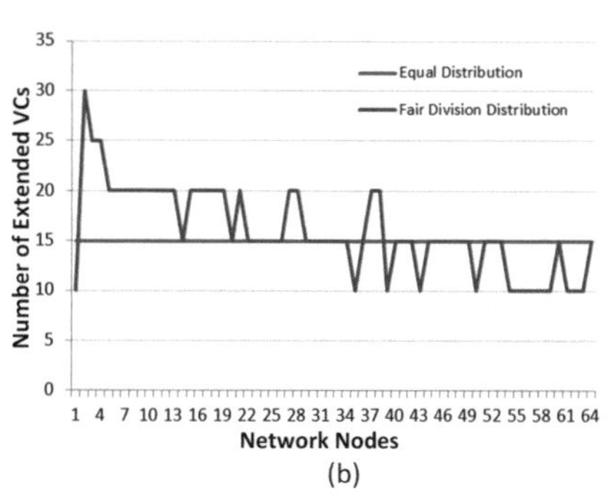

Fig. 13.16 (**a**) Normalized PI and value function of each node, (**b**) Trace buffer distribution in terms of extended VCs for both equal and Fair Division distribution [36]

13.8.3.1 Value Function Calculation and Trace Buffer Distribution

Profiling of each NoC node is performed based on traffic conditions for the calculation of value function $f(v)$. All the 6 workload patterns are executed on the 64 nodes NoC and profile index $p_{i,j}$ of each node for the individual workload is calculated. Figure 13.16a shows the graphs of normalized $p_{i,j}$ for all the workload patterns. The $f(v)$ of each node is calculated from Eq. (13.2) mentioned in Sect. 13.8.1.1 and is plotted in the same figure. Based on these $f(v)$ values, each node gets its due share of the trace buffer. Figure 13.16b shows the distribution of trace buffer as a measure of extended VCs at each node of the NoC for both equal distribution and FD based distribution. The figure shows that for an 8 KB of total trace buffer size, an equal distribution provides 15 additional VCs per router node while FD allocates as high as 30 additional VCs and as low as 10 additional VCs, which is proportional to the corresponding value function.

13.8.3.2 Trace Buffer Overflow

Whenever the local trace buffer of a router node gets filled, the local trace transfer (LTT) phase starts as discussed in Sect. 13.8.2. This pauses the concerned router operation and degrades the intermediate routers' operation while transferring the

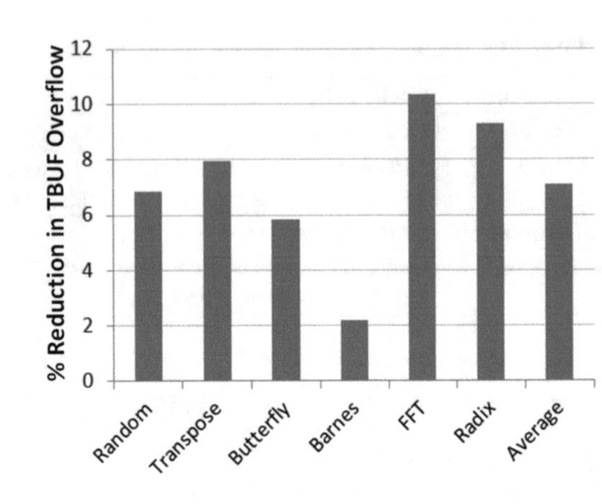

Fig. 13.17 Percentage reduction in local trace buffer overflow in case of Fair Division over equal distribution [36]

traces to the network trace port (NTP). Therefore, it can be concluded that as the number of local trace buffer overflow increases, the time period of the debug process also gets extended. Figure 13.17 shows an average 7% reduction in local trace buffer overflow in case of FD distribution over equal trace buffer distribution. This is because, fair distribution of trace buffers encourages parallel filling of buffer space in all routers, while in case of equal distribution, trace buffers in few routers fill up quickly, and in few others, they are mostly unused. This shows that FD based trace buffer distribution would considerably speed up the debug process than the equal distribution case.

13.8.3.3 Network Performance

Network performance has been evaluated in terms of throughput and latency. The proposed scheme provides a few additional VCs to each port of every router. As a result, network congestion reduces, and thereby network throughput increases, and global average packet delay decreases. Results shown in Fig. 13.18a, b demonstrate that equal trace buffer distribution provides an average of 8.36% throughput improvement and 9.25% average delay reduction over baseline architecture. Whereas the proposed FD scheme provides an average of 11.36% increase in throughput and 13.97% decrease in average packet delay. The enhancement in performance parameters in the case of the FD scheme over equal distribution is achieved as the router nodes are allocated with additional VCs according to their requirement.

In summary, proposed trace buffer reuse as extended VCs of NoC improves the network performance considerably, and thereby notably reduces the overhead associated with the DFD structure.

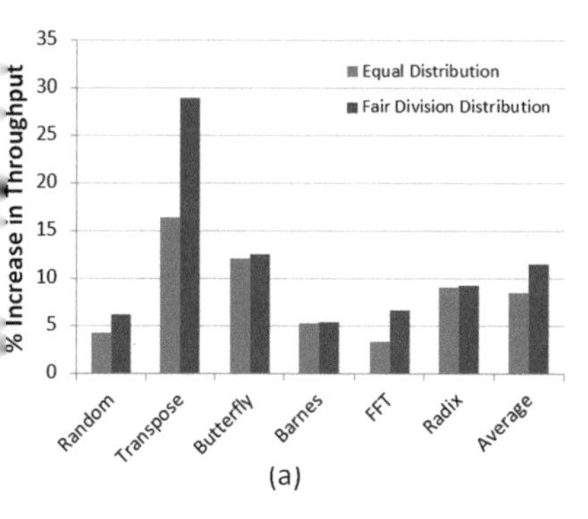

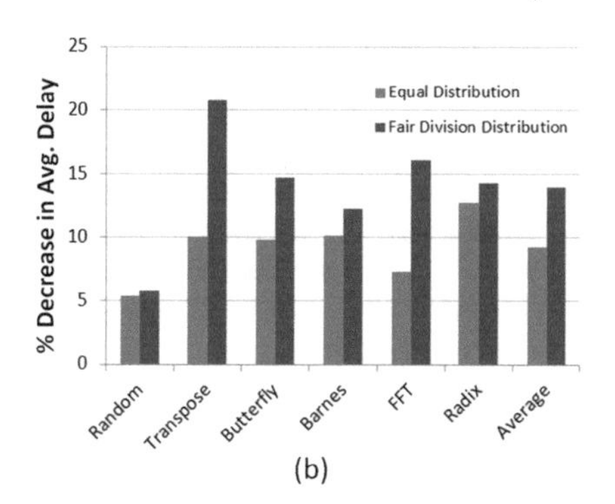

Fig. 13.18 (**a**) % increase in network throughput (**b**) % decrease in network average delay for equal and Fair Division trace buffer distribution in comparison to the baseline architecture [36]

13.9 Conclusion and Future Work

This chapter provides an overview of the NoC post-silicon debug framework. It discusses about communication-centric trace-based debug platform for NoC, and the steps as well as infrastructure involved in it. This chapter presents two detailed example works on NoC validation; the first one illustrates a wireless enabled efficient debug structure, and the second one shows the reuse of trace buffer as extended VCs for the improvement of NoC performance.

Post-silicon debug of NoC has a tremendous amount of future research scope. This is because NoC is an emerging interconnect solution for future multi-core SoCs. Moreover, due to multi-fold complexity growth in embedded systems, post-silicon debug is the ultimate validation step followed by the industries before mass production. Secured NoC debug structure and reuse of debug components for architectural purpose are two less explored areas that can be pursued for new findings.

Acknowledgments This work was supported in part by Scheme for Promotion of Academic and Research Collaboration (SPARC), India under sanction ID SPARC/2018-2019/P541/SL.

References

1. L. Benini, G. De Micheli, Networks on chips: a new SoC paradigm. Computer **35**(1), 70–78 (2002)
2. P. Mishra, R. Morad, A. Ziv, S. Ray Post-silicon validation in the soc era: a tutorial introduction. IEEE Des. Test **34**(3), 68–92 (2017)
3. P. Jayaraman, R. Parthasarathi, A survey on post-silicon functional validation for multicore architectures. ACM Comput. Surv. (CSUR) **50**(4), 1–30 (2017)

4. K. Goossens, B. Vermeulen, R. Van Steeden, M. Bennebroek, Transaction-based communication-centric debug, in Proceedings of the First International Symposium on Networks-on-Chip (NOCS'07) (IEEE, New York, 2007), pp. 95–106
5. N. Karimi, A. Alaghi, M. Sedghi, Z. Navabi, Online network-on-chip switch fault detection and diagnosis using functional switch faults. JUCS **14**(22), 3716–3736 (2008)
6. R. Abdel-Khalek, V. Bertacco, Post-silicon platform for the functional diagnosis and debug of networks-on-chip. ACM Trans. Embedded Comput. Syst. (TECS) **13**(3s), 1–25 (2014)
7. W.J. Dally, B.P. Towles, *Principles and Practices of Interconnection Networks* (Elsevier, Amsterdam, 2004)
8. Q. Xu, X. Liu, On signal tracing in post-silicon validation, in *Proceedings of the 2010 15th Asia and South Pacific Design Automation Conference (ASP-DAC)* (IEEE, New York, 2010), pp. 262–267
9. H.F. Ko, N. Nicolici, Algorithms for state restoration and trace-signal selection for data acquisition in silicon debug. IEEE Trans. Comput. Aided Des. Integr. Circuits Syst. **28**(2), 285–297 (2009)
10. S. Mitra, S.A. Seshia, N. Nicolici, Post-silicon validation opportunities, challenges and recent advances, in *Design Automation Conference* (IEEE, New York, 2010), pp. 12–17
11. K. Basu, P. Mishra, Rats: restoration-aware trace signal selection for post-silicon validation. IEEE Trans. Very Large Scale Integr. VLSI Syst. **21**(4), 605–613 (2012)
12. H.F. Ko, N. Nicolici, On automated trigger event generation in post-silicon validation, in *Proceedings of the conference on Design, automation and test in Europe* (2008), pp. 256–259
13. K. Goossens, B. Vermeulen, A.B. Nejad. A high-level debug environment for communication-centric debug, in *Proceedings of the 2009 Design, Automation and Test in Europe Conference and Exhibition* (IEEE, New York, 2009), pp. 202–207
14. B. Vermeulen, K. Goossens, A network-on-chip monitoring infrastructure for communication-centric debug of embedded multi-processor SoCs, in *Proceedings of the 2009 International Symposium on VLSI Design, Automation and Test* (IEEE, New York, 2009), pp. 183–186
15. C. Ciordas, T. Basten, A. Radulescu, K. Goossens, J. Meerbergen, An event-based network-on-chip monitoring service, in *Proceedings of the Ninth IEEE International High-Level Design Validation and Test Workshop (IEEE Cat. No. 04EX940)* (IEEE, New York, 2004), pp. 149–154
16. S. Tang, Q. Xu, A multi-core debug platform for NOC-based systems, in *Proceedings of the 2007 Design, Automation and Test in Europe Conference and Exhibition* (IEEE, New York, 2007), pp. 1–6
17. M. Dehbashi, G. Fey, Transaction-based online debug for noc-based multiprocessor socs. Microprocess. Microsyst. **39**(3), 157–166 (2015)
18. S.S. Rout, K. Basu, S. Deb, Efficient post-silicon validation of network-on-chip using wireless links, in *Proceedings of the 2019 32nd International Conference on VLSI Design and 2019 18th International Conference on Embedded Systems (VLSID)* (IEEE, New York, 2019), pp. 371–376
19. S.S. Rout, S.B. Patil, V.I. Chaudhari, S. Deb, Efficient router architecture for trace reduction during NOC post-silicon validation, in *Proceedings of the 2019 32nd IEEE International System-on-Chip Conference (SOCC)* (IEEE, New York, 2019), pp. 230–235
20. M. Abramovici, In-system silicon validation and debug. IEEE Des. Test Comput. **25**(3), 216–223 (2008)
21. A.B. Gomes, F.A. Alves, R.S. Ferreira, J.A.M. Nacif, Vericonn: a tool to generate efficient interconnection networks for post-silicon debug, in *Proceedings of the 2015 16th Latin-American Test Symposium (LATS)* (IEEE, New York, 2015), pp. 1–6
22. X. Liu, Q. Xu, Interconnection fabric design for tracing signals in post-silicon validation, in *Proceedings of the 2009 46th ACM/IEEE Design Automation Conference* (IEEE, New York, 2009), pp. 352–357
23. K. Morris, *On-chip Debugging–built-in Logic Analyzers on your FPGA* (2004)
24. H.F. Ko, A.B. Kinsman, N. Nicolici, Design-for-debug architecture for distributed embedded logic analysis. IEEE Trans. Very Large Scale Integr. VLSI Syst. **19**(8), 1380–1393 (2010)

25. K. Rahmani, S. Ray, P. Mishra, Postsilicon trace signal selection using machine learning techniques. IEEE Trans. Very Large Scale Integr. VLSI Syst. **25**(2), 570–580 (2016)
26. E. Anis, N. Nicolici, On using lossless compression of debug data in embedded logic analysis, in *Proceedings of the 2007 IEEE International Test Conference* (IEEE, New York, 2007), pp. 1–10
27. E. Anis, N. Nicolici, Low cost debug architecture using lossy compression for silicon debug, in *Proceedings of the 2007 Design, Automation and Test in Europe Conference and Exhibition* (IEEE, New York, 2007), pp. 1–6
28. K. Basu, P. Mishra, Test data compression using efficient bitmask and dictionary selection methods. IEEE Trans. Very Large Scale Integr. VLSI Syst. **18**(9), 1277–1286 (2009)
29. S. Deb, A. Ganguly, P.P. Pande, B. Belzer, D. Heo, Wireless NoC as interconnection backbone for multicore chips: promises and challenges. IEEE J. Emerging Sel. Top. Circuits Syst. **2**(2), 228–239 (2012)
30. S.H. Gade, S.S. Rout, M. Sinha, H.K. Mondal, W. Singh, S. Deb, A utilization aware robust channel access mechanism for wireless NOCs, in *Proceedings of the 2018 IEEE International Symposium on Circuits and Systems (ISCAS)* (IEEE, New York, 2018), pp. 1–5
31. S. Kaushik, M. Agrawal, H.K. Mondal, S.H. Gade, S. Deb, Path loss-aware adaptive transmission power control scheme for energy-efficient wireless NOC, in *Proceedings of the 2017 IEEE 60th International Midwest Symposium on Circuits and Systems (MWSCAS)* (IEEE, New York, 2017), pp. 132–135
32. H.K. Mondal, S. Kaushik, S.H. Gade, S. Deb, Energy-efficient transceiver for wireless NoC, in *Proceedings of the 2017 30th International Conference on VLSI Design and 2017 16th International Conference on Embedded Systems (VLSID)* (IEEE, New York, 2017), pp. 87–92
33. V. Catania, A. Mineo, S. Monteleone, M. Palesi, D. Patti, Cycle-accurate network on chip simulation with Noxim. ACM Trans. Model. Comput. Simul. (TOMACS) **27**(1), 1–25 (2016)
34. A. Ganguly, M.M. Ahmed, R. Singh Narde, A. Vashist, M.S. Shamim, N. Mansoor, T. Shinde, S. Subramaniam, S. Saxena, J. Venkataraman, et al. The advances, challenges and future possibilities of millimeter-wave chip-to-chip interconnections for multi-chip systems. J. Low Power Electron. Appl. **8**(1), 5 (2018)
35. (Accessed on October 10, 2020) armKEIL, CoreSight^{TM} Technology. http://www2.keil.com/coresight/
36. S.S. Rout, M. Badri, S. Deb, Reutilization of trace buffers for performance enhancement of NOC based MPSoCs, in *Proceedings of the 2020 25th Asia and South Pacific Design Automation Conference (ASP-DAC)* (IEEE, New York, 2020), pp. 97–102
37. W.J. Dally, C.L. Seitz, *Deadlock-free Message Routing in Multiprocessor Interconnection Networks* (1988)
38. C.H. Lai, Y.C. Yang, J. Huang, A versatile data cache for trace buffer support. IEEE Trans. Circuits Syst. I Regul. Pap. **61**(11), 3145–3154 (2014)
39. A. DeOrio, I. Wagner, V. Bertacco, Dacota: post-silicon validation of the memory subsystem in multi-core designs, in *Proceedings of the 2009 IEEE 15th International Symposium on High Performance Computer Architecture* (IEEE, New York, 2009), pp. 405–416
40. N. Jindal, S. Chandran, P.R. Panda, S. Prasad, A. Mitra, K. Singhal, S. Gupta, S. Tuli, Dhoom: reusing design-for-debug hardware for online monitoring, in *Proceedings of the 2019 56th ACM/IEEE Design Automation Conference (DAC)* (IEEE, New York, 2019), pp. 1–6
41. K. Basu, R. Elnaggar, K. Chakrabarty, R. Karri, Preempt: preempting malware by examining embedded processor traces, in *Proceedings of the 2019 56th ACM/IEEE Design Automation Conference (DAC)* (IEEE, New York, 2019), pp. 1–6
42. N. Jindal, P.R. Panda, S.R. Sarangi, Reusing trace buffers as victim caches. IEEE Trans. Very Large Scale Integr. (VLSI) Syst. **26**(9), 1699–1712 (2018)
43. S.J. Brams, A.D. Taylor, *Fair Division: From Cake-cutting to Dispute Resolution* (Cambridge University, Cambridge, 1996)
44. L.E. Dubins, E.H. Spanier, How to cut a cake fairly. Am. Math. Mon. **68**(1P1), 1–17 (1961)
45. S.C. Woo, M. Ohara, E. Torrie, J.P. Singh, A. Gupta, The splash-2 programs: characterization and methodological considerations. ACM SIGARCH Comput. Archit. News **23**(2), 24–36 (1995)

Chapter 14
Design of Reliable NoC Architectures

Noel Daniel Gundi, Prabal Basu, Sanghamitra Roy, and Koushik Chakraborty

14.1 Introduction

Rapid technology scaling has fueled a seismic growth in the number of on-chip resources. To procure an efficient performance throughput, effective communication between the hundreds of cores proves to be a very vital feature. Communication delay in System-on-Chips is a massive determinant in the overall system performance. In order to facilitate the ongoing communication needs between hundreds of cores, Network-on-Chip has been embraced as the *de facto* standard for the on-chip communication, owing to their performance, scalability, and flexibility advantages.

Providing a reliable NoC design has been a challenging task, as the performance of an NoC is primarily based on the network topology and routing algorithm. As the NoC plays an important role in the performance and energy efficiency of the system, addressing the factors affecting the NoC reliability appears to be of prime importance. This chapter focuses on the enhanced design techniques for an NoC architecture with prime stress on addressing factors affecting the reliability of an NoC. Section 14.2 discusses the challenges posing a threat on the NoC reliability. Section 14.3 elaborates the various schemes to tackle the NoC reliability issues. Section 14.4 summarizes the promising design solutions discussed in this chapter.

N. D. Gundi (✉) · P. Basu · S. Roy · K. Chakraborty
Utah State University, Logan, UT, USA
e-mail: noeldaniel@aggiemail.usu.edu; prabalb@aggiemail.usu.edu; sanghamitra.roy@usu.edu; koushik.chakraborty@usu.edu

P. Mishra, S. Charles (eds.), *Network-on-Chip Security and Privacy*,
https://doi.org/10.1007/978-3-030-69131-8_14

14.2 Factors Affecting NoC Reliability

As an NoC is deployed across the parallel computing environment, multiple issues emerge, which questions the credibility of an NoC design. Reliability of NOC is affected by various factors ranging from the problems arising due to device aging to unbalanced utilization of NoC components. Sections 14.2.1–14.2.6 address the various issues which degrade the performance of the NoC thereby, affecting the entire system performance.

14.2.1 Negative Bias Temperature Instability and Electromigration

Negative Bias Temperature Instability (NBTI) occurs due to the negative bias voltages at higher temperatures creating traps between layers of MOSFETs [1]. NBTI causes a degradation in drain current and absolute increase in the threshold voltage. On the other hand, Electromigration is the process of the transportation of metallic atoms by the electron current flow.

Table 14.1 shows the different schemes considering the varying impact of NBTI and Electromigration on the NoC routers and links. Figure 14.1 depicts the increase of latency with time due to the individual and combined effect of NBTI and Electromigration.

14.2.2 Asymmetric Traffic Utilization

Asymmetric utilization of NoC components significantly exacerbates the aging degradation. Higher utilization in particular NoC components manifests in a power-performance degradation due to rapid aging of these NoC components. Mishra et al. [2] observed that there is up to $2\times$ utilization in the centralized routers in comparison to the peripheral routers. Increase in utilization symmetry in the centralized routers is demonstrated in Fig. 14.2.

Table 14.1 Different degradation schemes

Scheme	Degradation in routers	Degradation in links
A	NBTI	NONE
B	NBTI	NBTI
C	NBTI	Electromigration
D	NBTI	NBTI and Electromigration

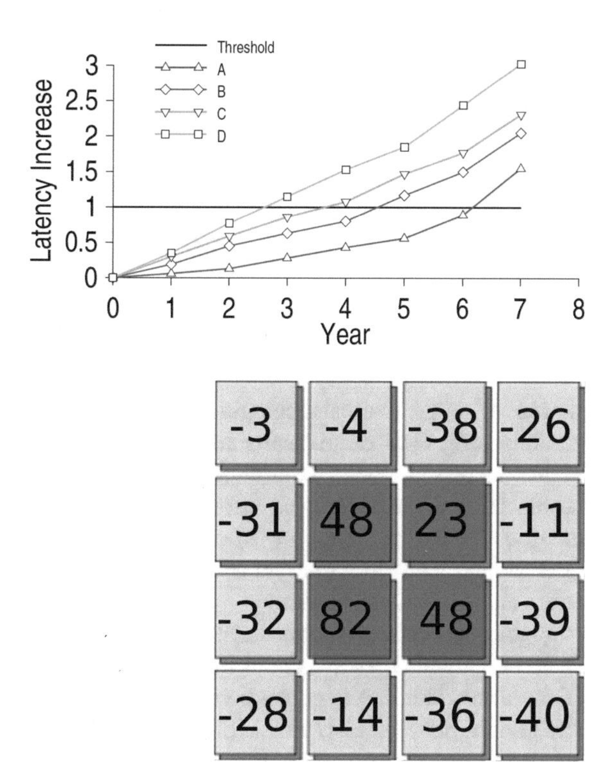

Fig. 14.1 Time taken for the network to become faulty under various aging models (high injection rate)

Fig. 14.2 Percentage traffic increase of each router using Buffered-Router Aware Routing (average across PARSEC benchmarks). This utilization difference leads to more than 2× divergence in NBTI induced performance degradation

14.2.3 *Hot Carrier Injection*

The phenomenon of Hot Carrier Injection (HCI) occurs when a carrier leaves the channel overcoming the potential barrier between the silicon and the gate oxide [3]. Carriers leaving the channel are deposited in the gate oxide region of the transistor. Over a period of time, the conductive properties of the transistor are altered due to the deposited carriers leading to an overall degradation in the threshold voltage, drain saturation current, and transconductance [4–6]. HCI degradation is majorly dependent on the switching activity of the transistors.

Figure 14.3 depicts the switching activity for the gates across an NoC architecture. From Fig. 14.3 it is evident that only 25% of the gates are responsible for 75% of the switching activity. The resulting asymmetry leads to an unbalanced HCI degradation across the NoC architecture leading to an early failure of an NoC.

14.2.4 *Quality-of-Service (QoS) Policies*

Enforcement of Quality-of-Service (QoS) Policies becomes quintessential to ensure fairness among different users/programs when limited number of resources are

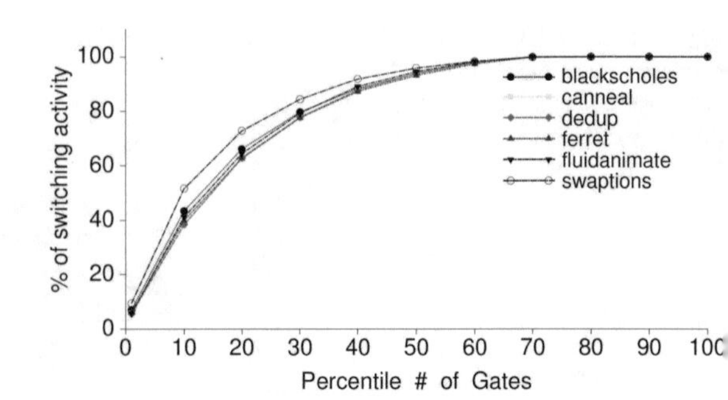

Fig. 14.3 Cumulative distribution function of the switching activity vs gate count

shared by large exascale computing system [7]. However, as NoC is scaled administering QoS dramatically lowers its Mean Time To Failure (MTTF) due to the increased power consumption and raised thermal profile. The elevated power/thermal characteristics arises due to the balanced resource management provided by the QoS support [8], rather than an increase in performance. Hence QoS support leads to a wearout acceleration and shortened lifetime even though it offers an identical bandwidth.

Figure 14.4a demonstrates the three nodes A, B, and E attempting to send flits to D. Nodes A and B receive unfair treatment without QoS as they receive only 1/4th of the bandwidth due to contention. Fair distribution of the link bandwidth for all three nodes between E–D link is provided by the QoS support. However, the risen network activity results in an increase in the power consumption which results in wearout acceleration for NoC devices. Figure 14.4b and c demonstrate the effects of QoS support on the power and MTTF of an NoC.

14.2.5 Voltage Emergencies

Voltage Emergencies in an NoC (VEN) arise due to the collaboration of various technology trends. A substantial increase in energy savings can be observed in computation than in communication due to technology scaling. NoCs consume a remarkable proportion (i.e., 36%) of chip power [9]. NoC draws a large current in its circuit components due to its rising power footprint. VENs emerge in the system resulting in timing errors, due to the variations in the current drawn by the NoC. Timing errors[1] generated by VEN can be mitigated by voltage guardbands. However, using guardbands alone can significantly deteriorate the energy efficiency. Timing errors in an NoC router pipeline presents a distinct challenge in comparison to the processor pipeline [10], as pipeline flush and recovery mechanisms cannot be used in the NoC pipeline.

[1]A timing error is observed when the pipe stage delay in exceeds the clock period.

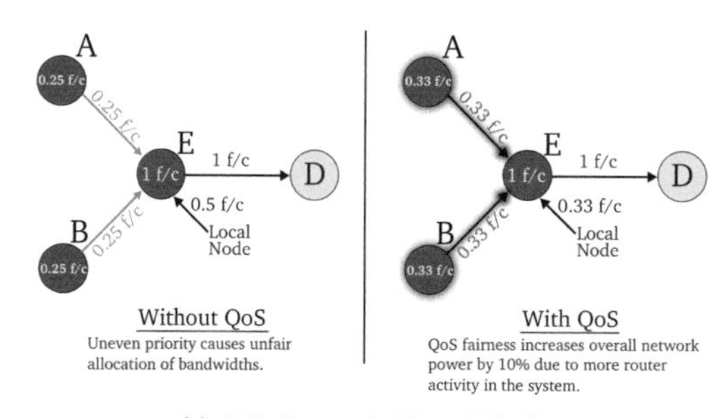

(a) QoS effect on the Network Traffic.

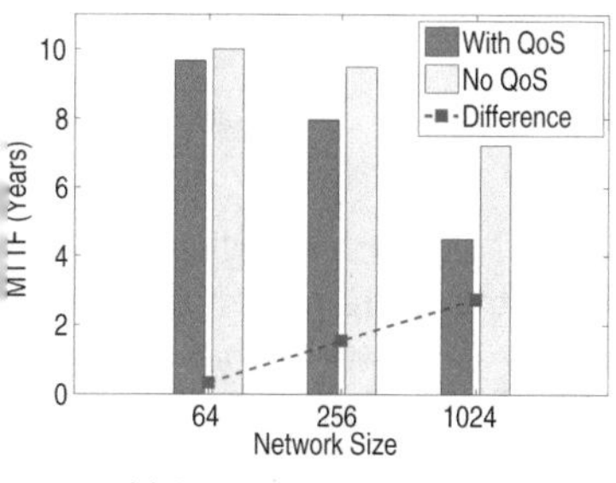

(b) MTTF Impact of QoS Support.

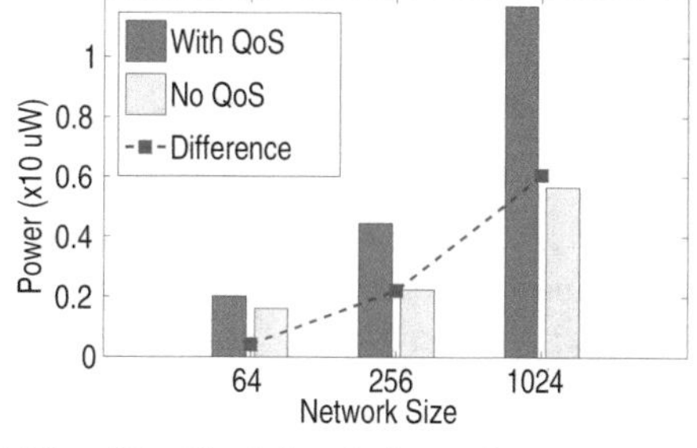

(c) Effect of Providing QoS on the Average Router Power Consumption.

Fig. 14.4 Figure (**a**) and (**b**) shows the conflicting goals of QoS support and sustainability: although the bandwidth offered by the NoC remains unchanged, different resource usage under QoS causes an accelerated wearout and a shortened lifetime. Figure (**c**) shows the effect of providing QoS on the average router power consumption

Fig. 14.5 Frequency of timing errors in the routers of a 8 × 8 NoC for real world applications

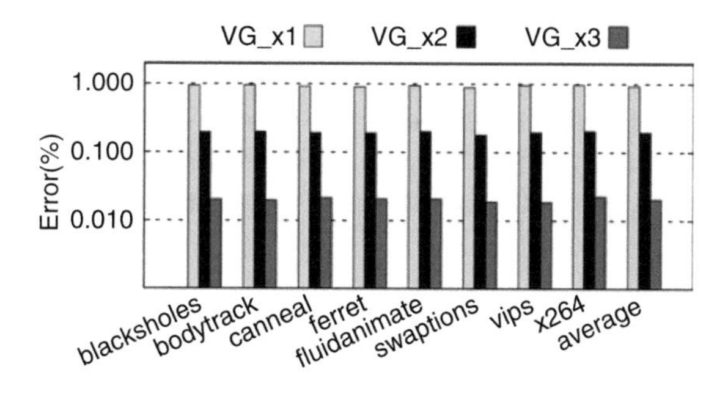

Figure 14.5 depicts the frequency of timing errors in the routers of a 8 × 8 NoC for the voltage guardbands (VG_x1, VG_x2, VG_x3) set at (22%, 26%, and 30%) above the nominal supply voltage. Timing errors induced from VEN lead to data corruption, flit redirection, and other functional errors. Hence, it is crucial to design energy efficient techniques to handle VEN induced timing errors.

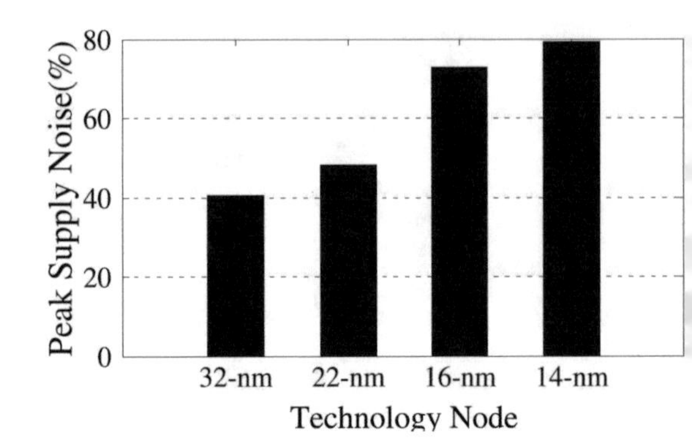

Fig. 14.6 Result is normalized to the corresponding 32-nm technology values. Figure highlights the variation of interconnect circuit parameters per unit length

14.2.6 Power Supply Noise

Modern multiprocessor system-on-chips (MPSoCs) encounter a rising concern due to the integrity of supply voltage. Switching of logic devices due to the uneven distribution of current results in the emergence of noise in Power Delivery Network (PDN), leading to a drop in the supply voltage. The performance and energy efficiency of the system components is severely affected by the Power Supply Noise (PSN). Additionally, scaling of technology node further exacerbates the problem due to the decreasing size and higher device density.

Sources of voltage noise in a PDN are: resistive drop (IR) and inductive drop ($L(\Delta i/\Delta t)$). Voltage drop across the resistances of the power delivery wires causes IR drop, which is proportional to the current (I) in the circuit. Inductive drop, on the other hand, is caused by the wire inductance (L) of the power grid and is proportional to the rate of change of current through the inductance. Figure 14.6 depicts the trend of RLC parameters at smaller technology nodes. Figure 14.6 shows that, the peak noise increases from 40% of the supply voltage at the 32-nm technology node to about 80% of the supply voltage at the 14-nm technology node, if the power distribution strategy remains unchanged.

14.3 Reliable NoC Design Methodologies

Overcoming the reliability problems requires a profound understanding of the intrinsic architecture details, which in turn can be utilized to procure a feasible solution. Additionally, understanding whether the problem can be mitigated or whether the effects of the problem can be delayed proves vital in the direction of developing a reliable design. For example, NBTI (Sect. 14.2.1) is critical, but a recoverable device aging mechanism. However, HCI (Sect. 14.2.3) is an unrecoverable aging phenomenon [11]. To restore the impacts of the factors affecting an NoC design discussed in Sect. 14.2, variety of strategies based on the investigations from innovative research [12–17] will be explored in this section, in addition to various concurrent research works (Sect. 14.3.7) in this field of work.

14.3.1 Overcoming NBTI and Electromigration

To tackle the problem of NBTI and Electromigration, balancing of the network traffic is essential. Balancing of the asymmetric network utilization can be achieved using a reliability metric and utilizing this metric in an aging-aware adaptive routing algorithm.

The reliability metric is determined based on the intensity of traffic a stressed router/link can handle. Hence the reliability metric *TTPE* is defined as the fraction of the nominal traffic that a stressed router/link should accept during a particular epoch [12]. Significance of the *TTPE* for an aging-stressed NoC design is based on the following facts:

1. *TTPE* determines an upper limit on the amount of traffic that a router or link should accept so as to keep the variation in network latency below a pre-defined threshold for a particular aging period.
2. *TTPE* is derived from continuous monitoring of the traffic, and is used to adapt the routing policies for every epoch to mitigate the long-term degradation in the NoC.

TTPE varies over the runtime with different values during different epochs for each stressed router and link.

The calculation of *TTPE* involves the following stages:

- **Threshold calculation:** The congestion-aware routing algorithm that routes the flits based on both local and global congestion information is profiled. The total time taken to route these flits is then divided into several epochs. The significance of adding epochs lies in the fact that an application's communication characteristics may change during the runtime and therefore the traffic must be monitored continuously. This process keeps track of the link and the router utilization during runtime and takes additional measures if the utilization reaches *TTPE* for the epoch under consideration. For each epoch, the n most stressed links and routers are considered based on their utilization. Based on the NBTI and electromigration of these stressed links and routers, the *TTPE* is calculated.
- **Using *TTPE* Estimation in Routing:** The computed *TTPE* for different epochs is stored in the form of lookup tables (SL_{set}) in each router. The router at runtime can then select the appropriate *TTPE* depending on the epoch. During this stage, the routing tables for each router are computed. In order to minimize network latency and communication energy, only the deadlock-free shortest paths for each flow are selected.

The routing algorithm involves the following two stages (Algorithm 1):

Algorithm 1: Aging_Adaptive

For each flow,
1. Select the best shortest path from the routing table which:
   a) suffers from least delay variation due to aging (sc_{age} is minimum).
   b) is least congested based on global and local congestion information (sc_{cong} is minimum).
2. For each stressed link in SL_{set} of each epoch:
   a) Check if the link meets its $TTPE$:
      - If the link has already reached its $TTPE$, keep the link idle for the rest of the epoch (insert recovery cycles).
      - If link utilization is safely below its $TTPE$ then there is no need for inserting recovery cycles.

1. **Congestion and aging-aware routing:** For each flow at runtime, the routing algorithm selects the best shortest path from the routing table that (i) suffers from least aging degradation i.e. the path that suffers from least delay variation due to aging (1-a); and (ii) is least congested (1-b). Higher priority is given to a path that least degraded as compared to a path with the least congestion.
2. **Honoring *TTPE* by employing recovery cycles:** During the execution of the routing algorithm, each stressed link in SLset is checked to see if it meets its respective *TTPE* for every epoch (2-a). There can be two possible cases: (i) In the epoch, if the link has already reached its *TTPE*, then the link must be kept idle for the rest of the epoch so that its utilization does not exceed its *TTPE*; and (ii) If the link operates safely inside its *TTPE* for that epoch, then there is no need for inserting idle cycles. The physical significance of inserting these idle cycles is that they provide additional time to the links and routers to recover from the aging stress. Therefore, these additional idle cycles are called as recovery cycles. This procedure also avoids unnecessary insertion of recovery cycles in the epoch and thus keeps the network latency in check.

14.3.2 Balancing Traffic Utilization

Balancing of the traffic utilization can be achieved by exploiting the criticality of the various flits in the NoCs [13]. The health of the routers in the network is tracked using a Wearout Monitoring System (WMS). The WMS and the criticality information are used to implement an aging-aware routing schemes.

14.3.2.1 Criticality of Different Flits in NoCs

The latencies of various packets transmitted through an NoC can have varied effects on performance. Previous works have exploited this criticality to improve system performance [18, 19].

Criticality Classification

In general, precise estimation of the packet criticality at the NoC router is hard as it merely has information about source–destination and the packet type. A thorough criticality estimation may require information about the relative performance of running program threads [18, 20], detailed cache coherence transitions, and so forth. To mitigate this complexity, a low-complexity approach is employed, which requires no change in existing interfaces. This involves identifying criticality based on packet type and source–destination. Table 14.2 shows the summary of classification. Using this policy, data packet transmitted from L1 to L2 (destination) is tagged as non-critical in a shared two level cache hierarchy. A vast majority of these packets are writebacks because of cache eviction, and thus the system performance is insensitive to their network latency. Some of these packets are also a result of data sharing among on-chip cores, but these are expected to be a much smaller component due to the predominance of private data even in multi-threaded programs [21].

Figure 14.7 shows the percentage of non-critical packets of PARSEC benchmarks averaged across all the buffered routers. An average of 49% of packets traversing through the buffered routers are non-critical and can actually take a different routing path with minimal performance degradation. Moreover, all benchmarks show substantial opportunity, ranging from 44% to 51% in these benchmarks. By redirecting non-critical traffic to the bufferless routers, the utilization of the buffered routers is minimized, thereby mitigating the aging effects in the buffered routers.

Table 14.2 Packet criticality classification

Data messages		
Source	**Destination**	**Classification**
L1 Cache	L2 Cache	Non-critical
L2 Cache	L1 Cache	Critical
Memory	L2 Cache	Critical
L2 Cache	Memory	Non-critical
Control Messages		
Source	**Destination**	**Classification**
L1 Cache	L2 Cache	Critical
L2 Cache	L1 Cache	Critical
Memory	L2 Cache	Critical
L2 Cache	Memory	Critical

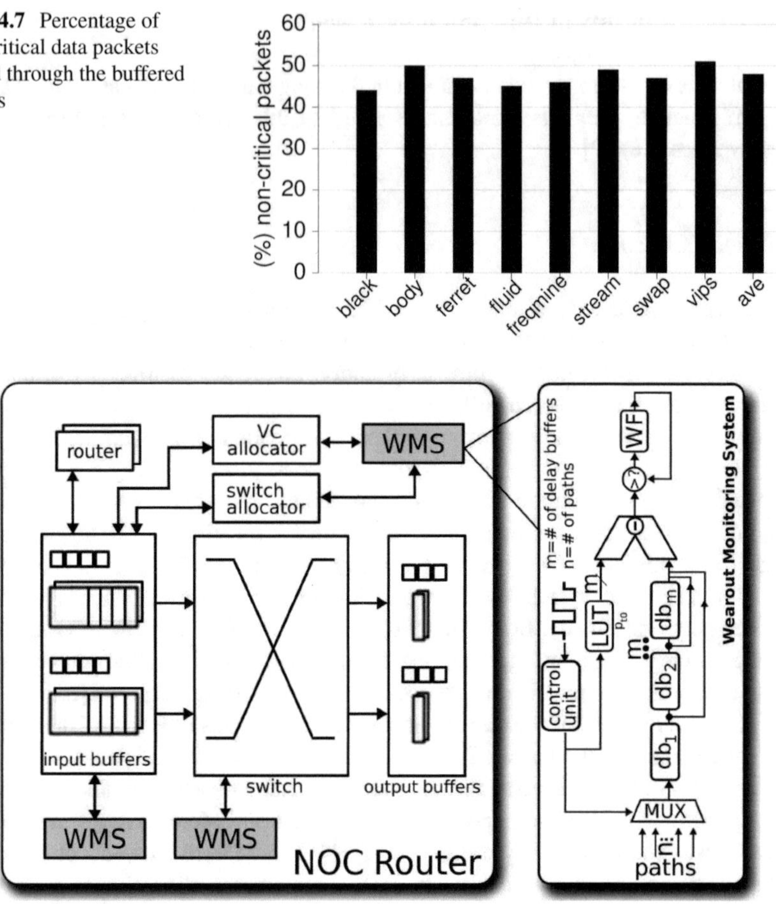

Fig. 14.7 Percentage of non-critical data packets routed through the buffered routers

Fig. 14.8 WMS circuit. Each path delay is sampled through a buffer sequence and compared with the reference delay to calculate the WF

14.3.2.2 Wearout Monitoring System (WMS) for NoC Routers

To be able to guide the aging-aware routing algorithm, the WMS profiles the extent of degradation in each router. The WMS circuit shown in Fig. 14.8 augments all pipeline stages of a router. As the performance degradation of a router is dictated by the worst case delay degradation in any pipeline stage, the monitoring system measures the maximum delay degradation across all paths in different pipeline stages. Within a stage, the WMS uses a multiplexer to estimate the delay of all n paths in a combinational logic. The control unit in Fig. 14.8 alters the multiplexer select signal in each cycle to choose which path to measure. Then, a series of m cascaded delay buffers (db_1, db_2, ..., db_m) sample the signal at equal time intervals. The state transition captured at the output of each delay buffer provides an estimate of the delay of the path. Finally, the comparator selects the maximum delay degradation among the n paths over a span of n cycles. The WMS measures the Wearout Factor (WF) as follows:

$$WF_{router} = \max(wf_1, wf_2, \ldots, wf_N) \quad (14.1)$$

$$wf_i = \max(wf_{p1}, wf_{p2}, \ldots, wf_{pn}) \quad (14.2)$$

where, $wf_1, wf_2, \ldots, wf_N$ are the wearout factors for N stages of the router micro-architecture, and $wf_{p1}, wf_{p2}, \ldots., wf_{pn}$ are the wearout factors of the n paths in a single stage i.

14.3.2.3 Criticality-Driven Path Selection

The criticality-driven routing incorporates two major design considerations:

1. Criticality of the incoming packet.
2. WF that dictates the current aging.

The maximum threshold for deflecting non-critical packets is defined as DFL_{Max}. Subsequently, based on the aging degradation in a router, the defection rate is pro-rated in that router.

Integrating Criticality in Routing

To drive the deflection logic in the routing path selection, the source router adds a single bit to store the criticality in the header flit of every packet. All intermediate routers peek into this criticality bit to select different routing paths based on criticality.

Integrating Wearout Monitoring

Different routers can undergo different aging degradation based on their utilization history. In a given router, the WF provides its current aging degradation. Table 14.3 shows the pro-rating scheme used in this work. For example, a router with a WF

Table 14.3 WF based deflection estimation

Wearout factor range	Scheme
0.00–0.50	$\frac{1}{8} \times DFL_{max}$
0.50–0.75	$\frac{1}{4} \times DFL_{max}$
0.75–1.00	$\frac{1}{2} \times DFL_{max}$
1.00– $+\infty$	$1 \times DFL_{max}$

of 0.8 will deflect 25% of all non-critical packets, assuming DFL_{Max} is 0.5. At every sampling interval of the WMS, the WF will be sent to adjacent routers to communicate the degradation of a particular router and a corresponding link. Each router stores the WF of four adjacent routers (North, South, East, West) in dedicated WF registers.

Deflecting Non-critical Packets

For every incoming flit in a router, the deflection logic uses the WF and packet criticality information to determine whether the packet will be sent in the direction of the pre-established path or deflected away from the buffered router. For a bufferless router, this task is accomplished by using a multiplexer and a selection logic. For a buffered router, an additional entry is added in the routing table corresponding to the possible deflection paths for each output port. For instance, an output in the North direction can be deflected to East or West if it is coming from the South input. This logic is accomplished using a 4-bit XOR of the number of ports (N,S,E,W) and the ports used for input and the desired output. Since there can be multiple deflection paths, the one that has no pending flits in the output buffer is used. For ties, the first port using a standard priority encoder is utilized.

14.3.3 Tackling HCI

HCI degradation can be handled by distributing the switching activity across the NoC. The following four techniques are explored in the router micro-architecture: Bit Cruising (BC); Distributed Cycle Mode (DCM); Crossbar Lane Switching (CLS); and BCCLS that is a combination of schemes BC and CLS [14].

14.3.3.1 Bit Cruising (BC)

Bit Cruising interchanges the different portions of the data being transmitted in the crossbar. Bit Cruising is largely motivated by two properties of the programs.

1. Most data in the cache line are aggregated at the lower bits. Hence, most data traversing through the NoC does not occupy the complete channel width of the network. In some cases, all data bits are actually zero.
2. Control requests sent as a single flit do not store information in the most significant portions of the channel as routing information can fit in the first few bytes of the whole channel. The control flit only utilizes 25% of the channel width, leaving the remaining 75% constant [14]. These two characteristics radically lower the switching activity in certain bits while emphasizing others.

To prevent the asymmetry in HCI degradation, the data being sent across the network must be such that the switching activity across the channel is distributed. Passing different data values each time a gate is used will balance the switching activity and uniformly degrade all gates. Hence, the highly changing bits are being circulated around the channel. The Bit Cruiser circuit will be situated in the Network Interface (NI) and does not add any overhead in the critical path of the pipeline of an NoC.

14.3.3.2 Distributed Cycle Mode (DCM)

The Distributed Cycle Mode balances out degradation of transistors by latching an input value in the crossbar during idle times such that unswitched transistors in previous cycles will transition and experience equivalent aging. This scheme does not relieve any HCI aging compared to other schemes but can be beneficial as equally aged transistors have smaller leakage power.

14.3.3.3 Crossbar Lane Switching (CLS)

Another asymmetrical degradation also occurs in the crossbar lanes that are immune to techniques applied in the channel level. This type of asymmetric degradation arises when some input–output pairs are used more than others. This occurrence is demonstrated with an example in Fig. 14.9 where there are two paths (p0 and p1) that both use the same East output port. If path $p0$ is used more than $p1$, then the transistors along the path $p0$ will be sensitized more and hence, experience more HCI degradation. CLS is situated at the frontend of the router pipeline and aims to balance the usage of the crossbar lanes. In the canonical router model, an input port directly forwards flits to the output ports by establishing a physical connection between the two via the crossbar switch. As such, flits coming from the same input port will always use the same crossbar lane to connect to different output

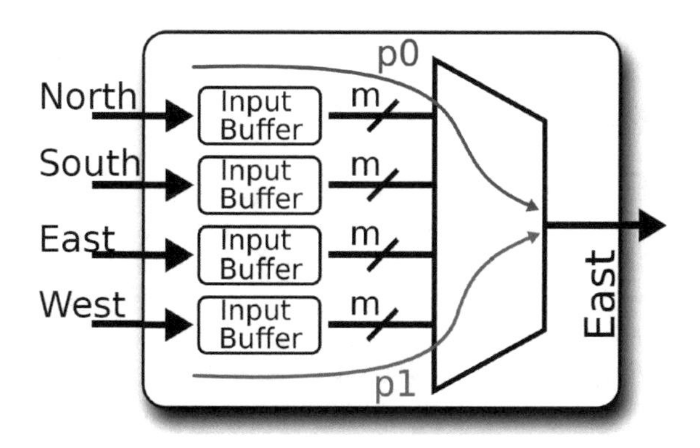

Fig. 14.9 East section of A crossbar switch. CLS works on the inter-lane (by changing the path of the data) level while BC works only on the intra-lane level (by changing the bit ordering within a path)

ports. However, the introduction of Input Buffers (IB) and Virtual Channels (VC) in modern router architectures decouples this one-to-one association because the flits are first stored in the IB before being transmitted to the output ports. With trivial modifications in the VC allocator and the Route Calculation part of the pipeline, it is possible to control the crossbar lane, which an input port will utilize at any given time. This new allocation and routing policy will now cause the crossbar circuit to use a different path and activation circuit, but still send the same data as if it were coming from the original input port. Thus, the correctness of the flit and the route is preserved. Similar to the Bit Cruising technique's cruise setting, CLS will need a knob input to indicate the new mapping between input ports and crossbar lanes.

14.3.3.4 Bit Cruising and Crossbar Lane Switching (BCCLS)

Bit Cruising and Crossbar Lane Switching (BCCLS) is a combination of the BC and CLS schemes. BCCLS combines both the benefit of switching distribution inside a channel (BC scheme) and the distribution of activity across many channels (CLS scheme). The implementation of BCCLS comes naturally because both BC and CLS tackle different portions of the router circuit. BC reshuffles the data sent through the network while CLS effectively changes the port a flit is coming from by modifying the VC allocation and route calculation.

14.3.4 Managing QoS support

Wearout degradation due to a QoS support in an NoC can be managed [15] using a three-step approach as follows:

1. Device level wearout of routers and links is monitored using NoC Health Meter.
2. The wearout information is communicated across the NoC.
3. The wearout information is utilized during NoC routing to dynamically mitigate the effects of aging.

14.3.4.1 NoC Health Meter (NHM)

The NHM profiles the level of degradation in each router and incoming links. The pipe stages of a router is augmented by the NHM circuit as shown in Fig. 14.10. NHM measures the delay degradation in the combinational circuit between two pipeline registers by measuring the slack in each stage. A high resolution all-digital, self-calibrating time-to-digital converter (HR-TDC) consisting of a Vernier Chain (VChain) circuit that has a measurement resolution of 5 ps [22] is used by the NHM to measure the slack. HR-TDC is an in situ delay-slack monitor consisting of a Vernier Chain circuit with an overall measurement window of 150 ps, which

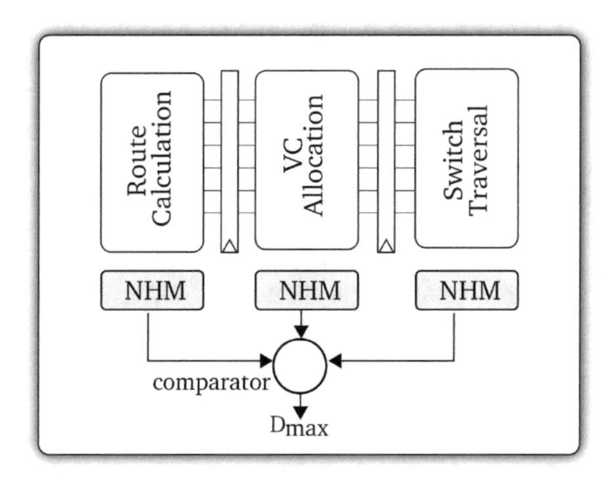

Fig. 14.10 NoC router augmented with NHM

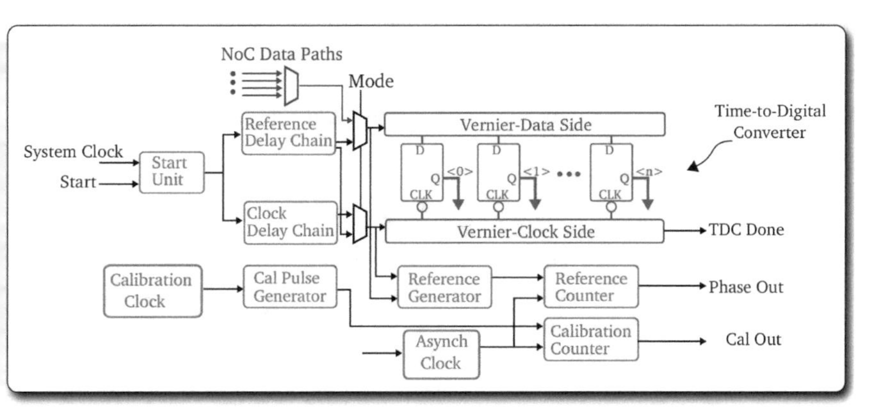

Fig. 14.11 High resolution in situ delay-slack measurement from Fick et al. [22]

is sufficient for timing slack measurements in 2 Ghz+ systems. After measuring the delay degradation of each stage, D_{max}: the maximum degradation among all pipe stages is estimated. Fick et al. has demonstrated that a complete full self-calibration of an entire TDC implemented on a 64-bit Alpha processor can take only 5 min [22].

HR-TDC in NoCs

Usage of HR-TDC circuits to measure the slack or propagation delay of each pipeline stage in an NoC is important because exascale chips with thousands of nodes can experience both global and local Process–Voltage–Temperature (PVT) variability. HR-TDC operates in three modes:

1. **Normal operation**: HR-TDC is measuring the delay fed from the NoC Data Path. Delays of only 30% of the top most critical paths are measured, as measuring all paths is expensive [23]. Data for the Time-to-Digital converter will be aggregated by the NHM to decide the maximum delay among all the pipeline stages.
2. **Reference Delay Chain (RDC) Calibration**: HR-TDC measures the delay of the "Reference Delay Chain" using statistical sampling. Before VChain calibration starts, calibration of the RDC has to be completed.
3. **Vernier Chain Calibration**: HR-TDC calibrates the Vernier Chain in order to maintain a delay of 5ps in each stage of the chain. Eight firmware-controlled capacitor loads are used to make a stage in the VChain tunable, with each load designed to introduce 1 ps shifts in the delay.

Vernier Chain (i.e. red portion of Fig. 14.11) is responsible for measuring the slacks from the NoC data paths in each pipeline stage and converting it to a digital code.

14.3.4.2 Propagating Delay Information and Routing Table Update

The encoded delay information is estimated and propagated through the firmware during the system boot-up, once a month by performing the following three steps :

1. All nodes estimate their D_{max} in parallel throughout the system.
2. D_{max} is broadcasted through the flit link network. To avoid extreme flooding, the network is divided into small equally sized regions. Then, one node from each region broadcasts its D_{max} throughout the system.
3. The routing tables in each node are updated using this D_{max} information.

14.3.4.3 Routing Algorithm

The routing algorithm profiles all two-turn minimal paths of all source–destination pairs. The paths are chosen based on a particular metric such as average router degradation or maximum router degradation. The path for a particular source–destination pair is updated once per month. Figure 14.12 shows an example of our routing algorithm in action. The firmware has already decided which turns to make for a flit with a source–destination of 0 and 11, respectively. The turns are made on nodes 2 and 10. Additionally, a single bit in the head flit is used to indicate which direction the flit should first go, X or Y direction. Whether it is up/down or left/right will be decided by the algorithmic routing based on the relative address of the source and the turning points. Once the flit hits one of the turning nodes, it is going to turn towards the direction of the destination. The algorithm is very scalable because no matter what the size of the exascale NoC is, the routing information stored in a flit (i.e. address of turning points) to be sent from a node to another will only grow by log(n) with *n* being the number of nodes.

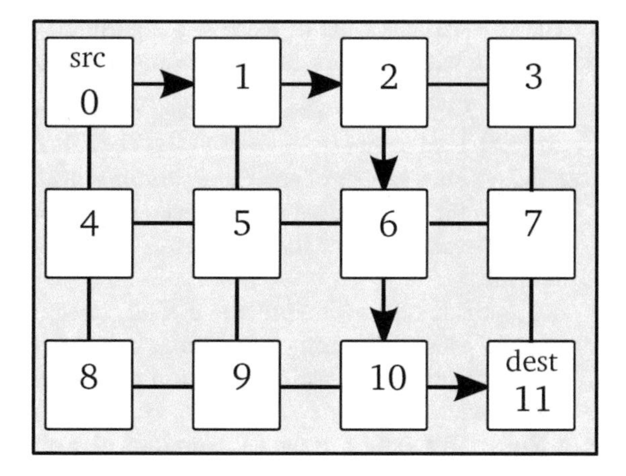

Fig. 14.12 Two-turn path routing

Deadlock Avoidance

Routing packets using various two-turn path configurations can lead to protocol deadlock when cyclic resource dependencies exist. One Virtual Channel (VC) is allocated in each port as an escape channel only to be used when avoiding a deadlock. Normally, when there is no contention, the flits will be routed on the non-escape channels. However, when all non-escape VCs from all routers are occupied for a certain period of time, a cyclic dependency could exist. This is possible because the flits are not restricted to use the same VC ID in each hop in order to maximize the bandwidth of the network. This cyclic dependency is broken by halting further injection in the NoC and allowing in-flight flits to arrive at their destination using deterministic routing via the escape channels.

14.3.4.4 Applying NoC Health Meter in Dynamic Wearout Resilient Routing

NoC health meter can be harnessed by the routing algorithm in two unique ways to dampen QoS-induced traffic stress in NoC routers. Duato's theory is used to restrict virtual channels to specific packet classes to avoid deadlocks [24]. The two algorithms are explained below:

1. **Fresh Routing (FR)**: This algorithm always routes the flits using the least degraded path. This path is constructed by considering several minimal paths and comparing the average wearout information in each path.
2. **Latency Reclamation routing (LR)**: This algorithm seeks to balance congestion and reliability objectives by using dynamic runtime information when deciding a path. LR first compares the number of available credits—a metric quantifying the level of congestion in a node of neighboring routers. If the least degraded path is congested, LR will choose the non-congested path.

The two variants each of these two algorithms are elaborated considering the routing path with p routers, having maximum delays $D_1, D_2, \ldots, D_p$, respectively.

- FR_{Avg}: This scheme uses the average wearout of all routers in a path to select the least-aged path. ($D_{path} = avg(D_1, D_2, \ldots, D_p)$).
- FR_{Max}: This variant of the FR algorithm selects the least-aged path using the maximum router wearout of each path. ($D_{path} = max(D_1, D_2, \ldots, D_p)$). This scheme seeks to limit the wearout of the most degraded router at any time interval.
- LR_{Avg}: This scheme is similar to FR_{Avg}, selecting the least-aged path based on average. However, during congestion, it avoids queuing delay by sending flits in the direction with more credits at times, when the least-aged path is overly congested.
- LR_{Max}: This variant of the LR algorithm also allows credit-based exceptions to the least-aged path. However, like the FR_{Max} scheme, it determines the least aged path using the maximum router delay in each path.

14.3.5 Voltage Emergencies

A reliable design to tackle Voltage Emergencies [16] will comprise of two key parts

- Error detection and confinement system.
- Recovery mechanisms used to recover corrupted flits.

14.3.5.1 Error Detection and Confinement

VEN induced timing errors are detected at the NoC router pipeline registers using shadow flip-flops [10]. The shadow flip-flops use a delayed clock, allowing double sampling of the combinational logic output. A discrepancy between the sample data in the regular flip-flop and the shadow flip-flop indicates a timing error. Inserting shadow flip-flop is relatively straightforward in an NoC router, as the circuit path in a router pipeline is more uniform in comparison to a typical processor pipeline. Figure 14.13 outlines the circuit-level modifications in an NoC router with 4 pipe stages: input buffer/route calculation, VC allocation, switch traversal, and output buffer. Once an error is detected, restoring error-free communication can only proceed after the error is confined within the router pipeline. On the detection of error, the error has to be confined within the route pipeline, to restore the NoC to error-free communication state. As a traditional NoC pipeline cannot stop a flit from transmission after it has reached the switch traversal stage, two strategies for error confinement based on the error location are explored:

1. ***Error before switch traversal***: Mark the VC as free and increase the credit for the specific port to block the flit before switch traversal. The corrupted flit is

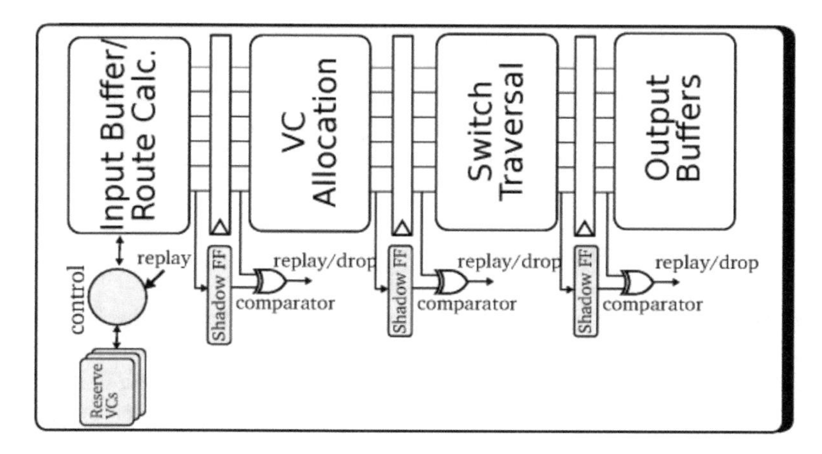

Fig. 14.13 Error detection, confinement, and SRE

overridden, as the new flip is allocated to the free VC entry in the subsequent cycle.

2. ***Error during switch traversal***: Add a poison bit to every output buffer entry. Poison bit is set, when an error is detected on a flit during switch traversal. Therefore, the link traversal is revoked for the particular flit in the next cycle and the buffer and poison bit are cleared to reclaim that entry.

14.3.5.2 Recovery Mechanisms

Two variants of the design based on the tradeoff in performance and complexity overhead are explored.

1. **Router Temporization (RT)** is a low-complexity source-based recovery technique that relies on flit re-transmission.
2. **Selective Router Echo (SRE)** is an in situ dynamic recovery mechanism with a low performance overhead.

Router Temporization (RT)

Router Temporization uses a combination of flit re-transmission and temporary frequency scaling to implement error-free communication in the presence of VEN.

- **Re-Transmission**: The NoC router checks the source for the acknowledgment (ACK) packet to verify the receipt of the data at the destination. The router assumes that the flit has been dropped if the ACK packet is not received after a set amount of time and sends the same flip again until an ACK packet is received.

- **Frequency Scaling**: As the threshold of dropped flits is exceeded, the frequency is lowered (i.e. frequency is halved) to prevent the continuous corruption of flits VEN typically lasts for a short time span [25]. If the errors persist, the frequency will be consequently lowered until the errors stop. Once the errors stop, the original frequency will be restored using an exponential back-off algorithm.

Selective Router Echo (SRE)

Selective Router Echo is an error recovery system embedded in the NoC router pipeline. In SRE, the router micro-architecture is augmented to mimic a processor pipeline. Figure 14.13 shows the pipeline for the SRE-enabled router. Extra virtual channels are added in the router, called Reserve VCs (RVCs) to keep a record of all in-flights flits which have crossed the input buffer stage. RVCs will replay the erroneous flits in the pipeline in the event of a VEN.

The steps involved in the recovery mechanism are:

- **Stall**: In the case of a VEN induced timing error, the router is stalled and incoming flits to the router are temporarily delayed.
- **Restart**: The router is restarted after stall completion. The delayed flits in the input buffers are permitted to pass through, as the input buffers are cleared to enable the recovery of flits from the RVCs.
- **Restore**: The entries from the RVCs are restored to the input buffers thereby restoring the router to an earlier state.
- **Resume**: The credit restrictions are lifted and the flits in the input buffer are sent to the targeted output buffers thereby, resuming the normal operation of the router.

14.3.6 *Power Supply Noise*

PSN can be tackled using flow-control protocols and routing algorithms. The design of a PSN-aware flow-control (PAF) involves a hierarchical approach to dictate the Maximum Current Load (MCL) across the NoC, while ensuring a minimal performance impact [17]. The flow-control information is then utilized in a PAF-aware routing algorithm to tackle PSN.

14.3.6.1 Hierarchical MCL Allocation

High concurrent switching of proximal regions is avoided by carefully adjusting the MCL allocated to each region. To realize the MCL allocation principles at different granularities, a metric Flit Acceptance Potential (FLAP) is defined. For a given input channel of a router, the FLAP is set to 1 when it can receive an incoming flit (otherwise it is set to 0). For a router, the FLAP indicates the aggregate FLAP of its input channels. Similarly, the FLAP of a particular region represents the aggregate FLAP of the routers in that region. At any given time, the FLAP of a router employing wormhole flow control in a 2-D mesh with four input channels is 4, when all of its input channels can receive at least one flit. The PAF allocates variable MCL to each region by dynamically throttling their FLAPs, irrespective of the space availability in the input channel's buffers. MCL allocation is a hierarchical process that can be applied at multiple spatial granularities. For example, a large region consists of many smaller subregions. The allocated MCL for the large region is distributed among the subregions, ensuring that proximal subregions are not simultaneously allocated with high MCLs. At the lowest granularity, each router's FLAP is managed in a manner that is consistent with the MCL allocation of the entire subregion.

14.3.6.2 Optimizations of PAF

The generic PAF technique needs multiple optimizations to efficiently tackle the design challenges.

Minimizing Performance Impact

Complementary approaches are explored to retain a high performance in the PAF.

- **Judicious FLAP management**: To avoid a large flit delay in a given region, the PAF allows intermittent high and low FLAPs in a router.
- **Topological awareness**: The PAF can be adapted based on the network topology and expected traffic pattern. For example, central routers in a mesh typically experience a high resource demand. This demand can be met by allocating greater FLAPs to the central routers.
- **Congestion awareness**: Two broad classifications of the PAF are explored (Sects. 14.3.6.2 and 14.3.6.2).

Congestion-Agnostic PAF

This variant of PAF statically allocates high and low FLAPs to the regional routers based on a round-robin fairness scheme. The FLAP allocation policy is not influenced by the network buffer occupancy.

Congestion-Aware PAF

This variant of the PAF manages the FLAP allocation based on the relative congestion of the network buffers. The following two congestion awareness at different granularities are considered.

1. **Channel granularity**: The FLAP of the least congested channel of a router is set to 1, so that it can always receive an incoming flit. The other channels' FLAPs are dictated by the aggregate FLAP of the router.
2. **Router granularity**: The least congested router of a region is allocated with a high FLAP. However, the other routers are allocated with low FLAPs to avoid high simultaneous switching. The aggregate FLAPs of the routers are consistent with the allocated MCL of the region.

Avoiding Starvation

Repeated blocking of the flits at the same input channel of a router in successive cycles can cause a starvation. To avoid starvation, the PAF adopts a round-robin fairness scheme to restrict flit reception across all the input channels of a router. Moreover, the PAF uses deterministically routed escape VCs, allowing all the possible turns without a deadlock situation.

Scalability

The PAF is a hierarchical technique that uses local network information at the smallest regional granularity to ascertain the FLAPs of the routers. As the size of the smallest region remains the same even for a larger NoC, the PAF can scale efficiently with the network size.

14.3.6.3 PAF-Aware Adaptive Routing Algorithm

Dynamically throttling the FLAP of a router may cause an intermittent upsurge in the local PSN due to an increased resource contention. This upsurge is circumvented using a PAF cognizant routing algorithm—PAR, which steers the flit toward an unthrottled downstream path. Figure 14.14 depicts the conceptual overview of the PAR. PAR primarily makes the routing decision based on the relative regional congestion information, aggregated solely along the minimal paths. If the chosen output channel has a throttled FLAP, the PAR reroutes the flit to an orthogonal output channel, strictly maintaining the minimal path constraint. This strategy reduces local current spike and the PSN by relieving router contention, but may occasionally increase the network latency by routing some flits toward more congested downstream paths. In a scenario, where both the minimal paths are

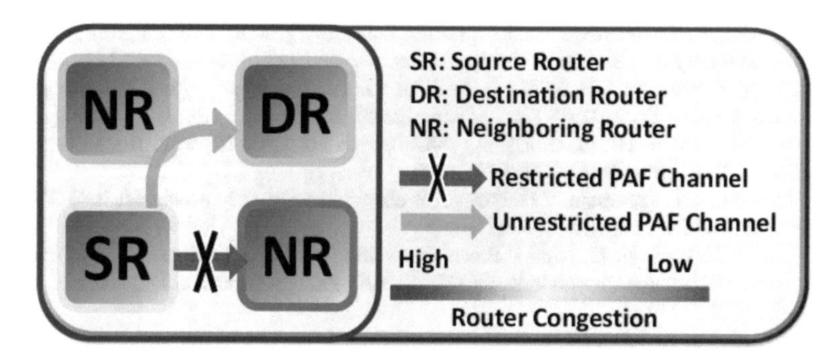

Fig. 14.14 PAR algorithm

blocked due to throttled FLAPs, the flit adheres to the initial channel assignment and waits in the upstream router for another cycle. The PAR incurs no additional circuit overhead as it utilizes the same information required for the PAF.

14.3.7 Concurrent Research Works

In addition to the methodologies discussed through Sects. 14.3.1–14.3.6, cutting-edge research contributions have also been made towards achieving an enhanced NoC design which further stresses on the impact of the reliability threat posed by the issues addressed in Sect. 14.2 [26–35].

14.4 Summary

Increasing performance needs have led to a rapid deterioration of the components in the communication network (NoC). A major cause of this degradation has been the asymmetric utilization of the network components due to device characteristics, resource allocation policies, and uneven traffic flow. In order to restore the reliability of an NoC infrastructure, unique solutions have been explored to mitigate the impending issues. The in situ solutions aid in increasing the lifetime of an NoC and contribute towards the overall system performance.

References

1. D.K. Schroder, Negative bias temperature instability: what do we understand? Microelectron. Reliab. **47**(6), 841–852 (2007)
2. A.K. Mishra, N. Vijaykrishnan, C.R. Das, A case for heterogeneous on-chip interconnects for CMPs, in *ACM SIGARCH Computer Architecture News* (2011), pp. 389–400

3. K. Bhardwaj, K. Chakraborty, S. Roy, An MILP based aging aware routing algorithm for NoCs in *Proceedings of the IEEE/ACM Design Automation and Test in Europe* (2012), pp. 326–331
4. E. Takeda, Y. Nakagome, H. Kume, S. Asai, New hot-carrier injection and device degradatio in submicron MOSFETs. IEEE Proc. I (Solid-State Electron Dev.) **130**(3), 144–150 (1983)
5. T. Ning, C. Osburn, H. Yu, Emission probability of hot electrons from silicon into silico dioxide. J. Appl. Phys. **48**(1), 286–293 (1977)
6. P.E. Cottrell, R.R. Troutman, T.H. Ning, Hot-electron emission in n-channel IGFET's. IEE Trans. Electron Dev. **26**(4), 520–533 (1979)
7. B. Grot, S.W. Keckler, O. Mutlu, Preemptive virtual clock: a flexible, efficient, and cost effective QOS scheme for networks-on-chip, in *EEE/ACM International Symposium on 200* (2009), pp. 268–279
8. J. Lee, M.C. Ng, K. Asanovic, Globally-synchronized frames for guaranteed quality-of-servic in on-chip networks, in *ISCA'08: Proceedings of the 35th Annual International Symposium o Computer Architecture* (2008), pp. 89–100
9. Y. Hoskote, S.R. Vangal, A. Singh, N. Borkar, S. Borkar, A 5-GHz mesh interconnect for teraflops processor. IEEE Micro **27**(5), 51–61 (2007)
10. D. Ernst, S. Das, S. Lee, D. Blaauw, T.M. Austin, T.N. Mudge, N.S. Kim, K. Flautner, Razor circuit-level correction of timing errors for low-power operation. IEEE Micro **24**(6), 10–2 (2004)
11. H. Kufluoglu, Mosfet Degradation due to NBTI and HCI and Its Implications for Reliability Aware VLSI Design, Ph.D. dissertation, Purdue University, West Lafayette, IN (2007)
12. K. Bhardwaj, K. Chakraborty, S. Roy, Towards graceful aging degradation in NoCs through a adaptive routing algorithm, in *DAC Design Automation Conference 2012* (2012), pp. 382–391
13. D.M. Ancajas, K. Chakraborty, S. Roy, Proactive aging management in heterogeneous NoC through a criticality-driven routing approach, 2013, pp. 1032–1037
14. D.M. Ancajas, J.M. Nickerson, K. Chakraborty, S. Roy, HCI-tolerant NoC router microarchi tecture, in *2013 50th ACM/EDAC/IEEE Design Automation Conference (DAC)* (IEEE, Ne York, 2013), pp. 1–10
15. D.M. Ancajas, K. Chakraborty, S. Roy, J.M. Allred, Tackling QoS-induced aging in exascal systems through agile path selection, in *2014 International Conference on Hardware/Softwar Codesign and System Synthesis (CODES+ISSS)* (2014), pp. 1–10
16. R.J. Shridevi, D.M. Ancajas, K. Chakraborty, S. Roy, Tackling voltage emergencies in No through timing error resilience, in *2015 IEEE/ACM International Symposium on Low Powe Electronics and Design (ISLPED)* (2015), pp. 104–109
17. P. Basu, R.J. Shridevi, K. Chakraborty, S. Roy, Iconoclast: tackling voltage noise in the No power supply through flow-control and routing algorithms. IEEE Trans. VLSI Syst. **25**(7) 2035–2044 (2017)
18. R. Das, O. Mutlu, T. Moscibroda, C.R. Das, Application-aware prioritization mechanisms fo on-chip networks, in *2009 42nd Annual IEEE/ACM International Symposium on Microarchi tecture (MICRO)* (2009), pp. 280–291
19. Z. Li, J. Wu, L. Shang, R.P. Dick, Y. Sun, Latency criticality aware on-chip communication, i *2009 Design, Automation & Test in Europe Conference & Exhibition* (2009), pp. 1052–1057
20. B. Datta, W. Burleson, Analysis and mitigation of NBTI-impact on PVT variability in repeate global interconnect performance, in *GLSVLSI '10: Proceedings of the 20th symposium o Great lakes symposium on VLSI* 2010, pp. 341–346
21. J.F. Cantin, J.E. Smith, M.H. Lipasti, A. Moshovos, B. Falsafi, Coarse-grain coherenc tracking: RegionScout and region coherence arrays. IEEE Micro **26**(1), 70–79 (2006)
22. D. Fick, N. Liu, Z. Foo, M. Fojtik, J. sun Seo, D. Sylvester, D. Blaauw, In situ delay-slac monitor for high-performance processors using an all-digital self-calibrating 5 ps resolutio time-to-digital converter, in *2010 IEEE International Solid-State Circuits Conference - (ISSCC* (2010), pp. 188–189
23. S. Das, C. Tokunaga, S. Pant, W.-H. Ma, S. Kalaiselvan, K. Lai, D. Bull, D. Blaauw, RazorII: i situ error detection and correction for PVT and SER tolerance. IEEE J. Solid-State Circ. **44**(1 32–48 (2009)

24. W.J. Dally, B. Towles, *Principles and Practices of Interconnection Networks* (Morgan Kaufmann, San Francisco, CA, 2004)
25. Y. Kim, L.K. John, S. Pant, S. Manne, M.J. Schulte, W.L. Bircher, M.S.S. Govindan, Audit: stress testing the automatic way, in *2012 45th Annual IEEE/ACM International Symposium on Microarchitecture* (2012), pp. 212–223
26. A.K. Kodi, A. Sarathy, A. Louri, J. Wang, Adaptive inter-router links for low-power, area-efficient and reliable Network-on-Chip (NoC) architectures, in *2009 Asia and South Pacific Design Automation Conference* (IEEE, New York, 2009), pp. 1–6
27. D. Zoni, W. Fornaciari, NBTI-aware design of NoC buffers, in *Proceedings of the 2013 Interconnection Network Architecture: On-Chip, Multi-Chip* (2013), pp. 25–28
28. J. Alshraiedeh, A. Kodi, An adaptive routing algorithm to improve lifetime reliability in NoCs architecture, in *2016 IEEE International Symposium on Defect and Fault Tolerance in VLSI and Nanotechnology Systems (DFT)* (IEEE, New York, 2016), pp. 127–130
29. L. Wang, X. Wang, T. Mak, Dynamic programming-based lifetime aware adaptive routing algorithm for network-on-chip, in *2014 22nd International Conference on Very Large Scale Integration (VLSI-SoC)* (IEEE, New York, 2014), pp. 1–6
30. J. Heißwolf, R. König, J. Becker, A scalable NoC router design providing QoS support using weighted round robin scheduling, in *2012 IEEE 10th International Symposium on Parallel and Distributed Processing with Applications* (IEEE, New York, 2012), pp. 625–632
31. S. Avramenko, S.P. Azad, S. Esposito, B. Niazmand, M. Violante, J. Raik, M. Jenihhin, QoSinNoC: analysis of QoS-aware NoC architectures for mixed-criticality applications, in *2018 IEEE 21st International Symposium on Design and Diagnostics of Electronic Circuits & Systems (DDECS)* (IEEE, New York, 2018), pp. 67–72
32. R. Tamhankar, S. Murali, S. Stergiou, A. Pullini, F. Angiolini, L. Benini, G. De Micheli, Timing-error-tolerant network-on-chip design methodology. IEEE Trans. Comput.-Aided Des. Integr. Circ. Syst. **26**(7), 1297–1310 (2007)
33. D. DiTomaso, T. Boraten, A. Kodi, A. Louri, Dynamic error mitigation in NoCs using intelligent prediction techniques, in *2016 49th Annual IEEE/ACM International Symposium on Microarchitecture (MICRO)* (IEEE, New York, 2016), pp. 1–12
34. V.Y. Raparti, S. Pasricha, PARM: power supply noise aware resource management for NoC based multicore systems in the dark silicon era, in *Proceedings of the 55th Annual Design Automation Conference*, 2018, pp. 1–6
35. N. Dahir, T. Mak, F. Xia, A. Yakovlev, Minimizing power supply noise through harmonic mappings in networks-on-chip, in *Proceedings of the Eighth IEEE/ACM/IFIP International Conference on Hardware/software Codesign and System Synthesis* (2012), pp. 113–122

Part V
Emerging NoC Technologies

Chapter 15
Securing Silicon Photonic NoCs Against Hardware Attacks

Ishan G. Thakkar, Sai Vineel Reddy Chittamuru, Varun Bhat, Sairam Sri Vatsavai, and Sudeep Pasricha

15.1 Introduction

Since the end of Dennard scaling in mid-2000s, the furtherance of CMOS transistor scaling has allowed continued increase in transistor count per microprocessor chip. This growing number of transistors, which is already in a few billions today, have been utilized to integrate increasingly greater number of cores per microprocessor chip. Such microprocessor chips with multiple integrated cores are often referred to as chip-multiprocessors (CMPs). To enable efficient communication between these cores within a CMP, conventional bus-based interconnects have been replaced with network-on-chips (NoCs) (e.g., [4, 36]). In NoCs, processing cores use routers connected by segmented links for efficient and scalable communication devoid of global wire delays.

I. G. Thakkar
Electrical and Computer Engineering Department, University of Kentucky, Lexington, KY, USA
e-mail: igthakkar@uky.edu

S. V. R. Chittamuru
Micron Technology, Inc, Austin, TX, USA
e-mail: schittamuru@micron.com

V. Bhat
Qualcomm, San Diego, CA, USA

S. S. Vatsavai
University of Kentucky, Lexington, KY, USA
e-mail: ssr226@uky.edu

S. Pasricha (✉)
Electrical and Computer Engineering Department, Colorado State University, Fort Collins, CO, USA
e-mail: sudeep@colostate.edu

P. Mishra, S. Charles (eds.), *Network-on-Chip Security and Privacy*,
https://doi.org/10.1007/978-3-030-69131-8_15

15.2 State of the Art in NoCS

In CMPs, electrical NoCs (ENoCs) have emerged as the standard communication fabric, as they are more scalable and modular compared to the traditional bus-based interconnects. Many CMPs have been implemented with ENoCs, such as Intel's 48-core SCC [33], Tilera's 72-core TILE-Gx [34], Kalray's 256-core MPPA [35], Intel's 80-core TeraFlops chip [37], Sun's Niagara [38], and MIT's RAW chip [39]. Moreover, a 496-core CMP mesh topology ENoC has been recently implemented in [40]. Increasing core count and decreasing transistor size elevates the concerns associated with reliability, power consumption, performance and security in ENoCs. This also reflects in the ongoing research efforts in the field of NoCs. With shrinking technology node size, ENoCs become prone to faults that arise due to manufacturing defects, increased demand for traffic, and aging of various components. Several techniques have been proposed to address fault tolerance in ENoCs (e.g., [41–45]). For instance, in [41], a bypass path around faulty nodes is set up to improve fault tolerance in ENoCs.

In addition to faults, avoiding congestion, deadlock, and livelock in ENoCs is also important, as the performance of ENoCs depends on that. Congestion is caused in an ENoC when majority of data packets traverse through a common path, deadlock is caused because of a cyclic buffer dependency, and livelock occurs when the packets keep spinning around the network without progressing to their destination. To avoid congestion, deadlock, and livelock conditions in ENoCs, several effective routing algorithms have been proposed in prior works (e.g., [46–50]). For instance, [46] uses the Hamiltonian routing strategy to adaptively route the packets through deadlock-free paths in a diametrical 2D mesh ENoC.

Furthermore, the increase in number of cores and shrinking feature size also increases power density in CMPs, which causes electromigration, negative bias temperature instability, hot carrier injection, and time-dependent dielectric breakdown. All of these effects accelerate aging in CMPs that degrades reliability and lifetime of CMPs. Substantial research has been done to address the reliability and low lifetime related issues in ENoCs (e.g., [51–53]). For example, an adaptive routing algorithm is utilized in [51] to improve lifetime reliability of ENoCs. Moreover, high dynamic energy consumption in ENoCs is also a critical challenge, as ENoCs consume significant portion of overall CMP power budget [54, 55]. Several prior works have focused on minimizing the dynamic energy consumption in ENoCs (e.g., [55–59]). For instance, [59] manages the dynamic energy consumption in ENOCs by combining thread migration with dynamic voltage and frequency scaling techniques. Further, a good amount of research is also being conducted in using CMPs with NoCs for developing application specific processors for datacenters (e.g., SmarCo [60]), cloud based computing (e.g., Piton [61]), and reconfigurable accelerator systems (e.g., MITRACA [62]).

Another critical challenge for ENoCs is their poor performance scalability. With increasing core count and the resultant increase in communication distances, the achievable throughput and latency for data transfers with ENoCs substantially degrade. With the motivation of providing better communication for longer distances, the use of wireless networks-on-chip (WiNoCs) has been proposed in place of ENoCs [63]. A WiNoC architecture is proposed in [64] that uses on-chip antennas to transfer data across long distances, while an arbitration mechanism to ensure high-average and guaranteed performance for WiNoCs had been developed in [65]. In [66], a WiNoC uses an adaptable algorithm that works in the background along with a token sharing scheme to fully utilize the wireless bandwidth efficiently. Furthermore, the recent advancements in emerging technologies like silicon nanophotonics have made it possible to realize photonics with NoCs (PNoCs). Therefore, several PNoC architectures have been proposed for CMPs, e.g., Corona [67], FireFly [7], LumiNoC [68], CAMON [69]. These PNoC architectures leverage the high bandwidth density and low dynamic power consumption of photonics to address the scalability issues of ENoCs.

Another crucial challenge faced by ENoCs is security. ENoCs are vulnerable to various types of security attacks such as Hardware Trojans, hijacking, extraction of sensitive information, and Denial of Service (DoS). Physical hardware security concerns due to Hardware Trojan (HT) attacks are especially important. An HT is realized through malicious modification of an integrated circuit (IC) during its design or fabrication phase. Such malicious ICs are generally third-party IPs that are used for reducing the design time of CMPs. Such HT-related security risks need to be detected and mitigated in order to ensure trusted functionality. Therefore, several recent works have addressed these security issues in ENoCs (e.g., [70–74]). To mitigate the DoS and timing attacks in ENoCs, [70] proposed the use of a non-Interference based adaptive routing. Similarly, to improve hardware security in WiNoCs, [71] employed a machine learning based engine to protect WiNoCs from DoS, spoofing, and eavesdropping attacks.

15.3 Photonic NoCS (PNoCS) and Related Security Challenges

Due to the ever-increasing core count to meet the growing performance demands of modern Big Data and cloud computing applications, ENoCs suffer from high power dissipation and severely reduced performance [31]. The crosstalk and electromagnetic interference are also increasing with technology scaling, which further reduces the performance and reliability of electrical NoCs [32]. Thus, there is a crucial need of a new viable interconnect technology that can address the shortcomings of ENoCs. Recent developments in silicon photonics have enabled the integration of photonic components and interconnects with CMOS circuits on a chip. The ability to communicate at near light speed, larger bandwidth density, and lower dynamic power dissipation are the prolific advantages that Photonic NoCs (PNoCs) provide over their metallic counterparts [5]. These advantages motivate the use of PNoCs for inter-core communication in modern CMPs [6]. Several PNoC architectures have been proposed to date (e.g., [19–21]). These architectures employ on-chip photonic links, each of which connects two or more gateway interfaces. A cluster of processing cores are connected to PNoC by a gateway interface (GI). Each photonic link comprises one or more photonic waveguides and each waveguide can support a large number of dense-wavelength-division-multiplexed (DWDM) wavelengths. A data signal is carried by a wavelength. Typically, Each source GI generates multiple data signals in the electrical domain (as sequences of logical 1 and 0 voltage levels) that are modulated onto the multiple DWDM carrier wavelengths simultaneously using a bank of modulator MRs at the source GI [9].The data-modulated carrier wavelengths traverse a link to a destination GI, where an array of detector MRs filter them and drop them on photodetectors to regenerate electrical data signals.

In general, each GI in a PNoC is able to send and receive data in the optical domain on all of the utilized carrier wavelengths. Therefore, a bank of modulator MRs (i.e., modulator bank) and a bank of detector MRs (i.e., detector bank) are present at each GI. Each MR in a bank resonates with and operates on a specific carrier wavelength. The high bandwidth parallel data transfers is enabled in PNoCs because of the excellent wavelength selectivity of MRs and DWDM capability of waveguides. However, the excellent wavelength selectivity of MRs and DWDM capability of waveguides also impose serious hardware security threats in PNoCs. The hardware security issues in PNoCs are especially exacerbated due to the complexity of hardware in modern CMPs. This is because to meet the growing performance demands of modern Big Data and cloud computing applications, the complexity of hardware in modern CMPs has increased. To reduce the hardware design time of these complex CMPs, third-party hardware IPs are frequently used. But these third party IPs can introduce security risks [1, 2]. For instance, the presence of Hardware Trojans (HTs) in the third-party IPs can lead to leakage of critical and sensitive in-formation from modern CMPs [3]. Thus, security researchers are now increasingly interested in overcoming hardware-level security risks in addition to traditionally focused software-level security.

Similarly, the CMPs with PNoCs are also expected to use several third party IPs similar to ENoCs and therefore, are vulnerable to security risks [10]. For instance, if the entire PNoC used within a CMP is a third-party IP, then this PNoC with HTs within the control units of its GIs can snoop on packets in the network. Sensitive in-formation can be determined by malicious core (a core running a malicious program) in CMP using these transferred packets.

Unfortunately, MRs of PNoCs are especially susceptible to security threatening manipulations from HTs. In particular, *the MR tuning circuits that are essential for supporting data broadcasts and to counteract MR resonance shifts due to process variations (PV) make it easy for HTs to retune MRs and initiate snooping attacks*. The tuning circuits of detector MRs partially detune them from their resonance wavelengths to enable data broadcast, [7, 11–12], such that a significant portion of the photonic signal energy in the data carrying wavelengths continues to propagate in the waveguide to be absorbed in the subsequent detector MRs. On the other hand, resonance wavelengths shifts in MRs due to process variations (PV) [13]. MR tuning circuits are used to counteract PV-induced resonance shifts in MRs by retuning the resonance wavelengths using carrier injection/depletion or thermal tuning [6]. These tuning circuits of detector MRs can be manipulated by the HT to partially tune the detector MR to a passing wavelength in the waveguide, which enables snooping of the data that is modulated on the passing wavelength. *Such covert data snooping is a serious security risk in PNoCs.*

In this work, we present a framework that improves the hardware security in PNoCs by protecting data from snooping attacks. Our framework can be easily implemented in any existing DWDM-based PNoC without major changes to the architecture and it has low overhead. To the best of our knowledge, this is the first work that attempts to improve hardware security for PNoCs. Our novel contributions are:

- We analyze security risks in photonic devices and extend this analysis to linklevel, to determine the impact of these risks on PNoCs;
- We propose a circuit-level PV-based security enhancement scheme that uses PV-based authentication signatures to protect data from snooping attacks in photonic waveguides.
- We propose an architecture-level reservation-assisted security enhancement scheme to improve security in DWDM-based PNoCs;
- We combine the circuit-level and architecture-level schemes into a holistic framework called *SOTERIA*; and analyze it on the Firefly [7] and Flexishare [8] crossbar-based PNoC architectures.

15.4 Related Work

Several prior works [10, 15, 16] discuss the presence of security threats in ENoCs and have proposed solutions to mitigate them. Data scrambling, packet certification, and node obfuscation were used to present three-layer security system approach [10] to enable protection against data snooping attacks. A symmetric-key based cryptography design was presented in [15] for securing the NoC. In [16], a framework was presented to use permanent keys and temporary session keys for NoC transfers between secure and non-secure cores. *However, no prior work has analyzed security risks in photonic devices and links; or considered the impact of these risks on PNoCs.*

Fabrication-induced PV impact the cross-section, i.e., width and height, of photonic devices, such as MRs and waveguides. Thermal tuning or localized trimming are techniques used at device level to counteract the drifts in resonant wavelength because of PV in MRs [6]. Trimming can induce blue shifts in the resonance wavelengths of MRs using carrier injection into MRs, whereas thermal tuning can induce red shifts in MR resonances through heating of MRs using integrated heaters. *Device level trimming/tuning techniques are inevitable to remedy PV; but their use also enables partial detuning of MRs that can be used to snoop data from a shared photonic waveguide.* In addition,the impact of PV-remedial techniques on crosstalk noise and proposed techniques to mitigate it were discussed in prior works [17–18]. *None of the prior works analyze the impact of PV-remedial techniques on hardware security in PNoCs.*

Our proposed framework in this chapter is novel as it enables security against snooping attacks in PNoCs for the first time. Our framework improves security for any DWDM-based PNoC architecture being network agnostic, mitigating PV, and with minimal overhead.

15.5 Hardware Security Concerns in PNoCS

15.5.1 Device-Level Security Concerns

Undesirable changes in MR widths and heights due to Process variation (PV) causes "shifts" in MR resonance wavelengths, which can be remedied using localized trimming and thermal tuning methods. The localized trimming method injects (or depletes) free carriers into (or from) the Si core of an MR using an electrical tuning circuit, which reduces (or increases) the MR's refractive index owing to the electro-optic effect, thereby remedying the PV-induced red (or blue) shift in the MR's resonance wavelength. An integrated micro-heater is employed in thermal tuning to adjust the temperature and refractive index of an MR (owing to the thermo-optic effect) for PV remedy. Typically, the modulator MRs and detectors use the same electro-optic effect (i.e., carrier injection/depletion) implemented through the same electrical tuning circuit as used for localized trimming, to move in and out of resonance (i.e., switch ON/OFF) with a wavelength [19]. *A HT can manipulate this electrical tuning circuit, which may lead to malicious operation of modulator and detector MRs, as discussed next.*

Figure 15.1 (left) shows the malicious operation of a modulator MR. A malicious modulator MR is partially tuned to a data-carrying wavelength (shown in purple) that is passing by in the waveguide. The malicious modulator MR draws some power from the data-carrying wavelength, which can ultimately lead to data corruption as optical '1's in the data can lose significant power to be altered into '0's. Alternatively, a *partially* tuned malicious detector (Fig. 15.1 (right)) to a passing data-carrying wavelength can filter only a small amount of its power and drop it on a photodetector for data duplication. This small amount of filtered power does not alter the data in the waveguide so that it continues to travel to its target detector for legitimate communication [11]. Further, a malicious detector MR can also cause data corruption (by partially tuning to a wavelength) and denial of communication (by fully tuning to a wavelength). Thus, both malicious modulator and detector MRs can corrupt data (which can be detected and corrected) or cause Denial of Service (DoS) type of security attacks. In addition, malicious detector MRs can also snoop data from the waveguide without altering it. Thus, major security threat in photonic links is malicious detector MRs snooping data from the waveguide without altering it. Note that malicious modulator MRs only corrupt data (which can be detected) and do not covertly duplicate it and are thus not a major security risk.

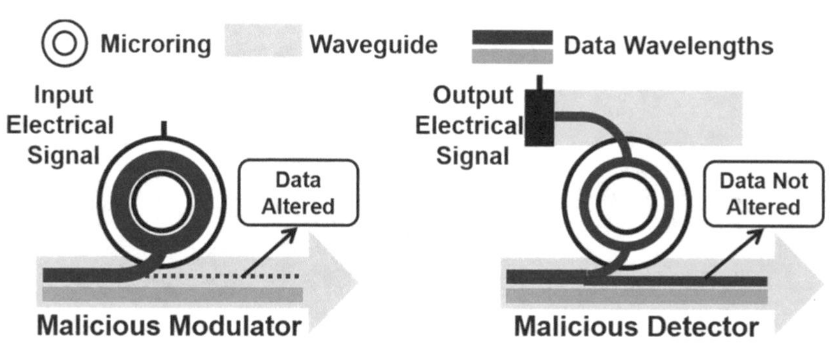

Fig. 15.1 Data transfer in a DWDM-based photonic waveguide, with (**left**) a malicious modulator MR leading to data corruption, and (**right**) a malicious detector MR leading to data snooping [75]

15.5.2 *Link-Level Security Concerns*

Typically, one or more DWDM-based photonic waveguides are present in a photonic link. A modulator bank (a series of modulator MRs) at the source GI and a detector bank (a series of detector MRs) at the destination GI are used in a DWDM based photonic waveguide. DWDM-based waveguides can be broadly classified into four types: single-writer-single-reader (SWSR), single-writer-multiple-reader (SWMR), multiple-writer-single-reader (MWSR), and multiple-writer-multiple-reader (MWMR). We restricted our link-level analysis to MWMR waveguides as SWSR,SWMR and MWSR waveguides are subsets of MWMR.

An MWMR waveguide typically passes through multiple GIs, connecting the modulator banks of some GIs to the detector banks of the remaining GIs. Multiple GIs (referred to as source GIs) can send data using their modulator banks and multiple GIs (referred to as destination GIs) can receive (read) data using their detector banks in an MWMR waveguide. An MWMR waveguide with two source GIs and two destination GIs are shown in Fig. 15.2 as an example. The impact of malicious source and destination GIs on this MWMR waveguide is presented in Fig. 15.2a, b, respectively. The modulator bank of source GI S_1 is sending data to the detector bank of destination GI D_2. When source GI S_2, which is in the communication path, becomes malicious with an HT in its control logic, it can manipulate its modular bank to modify the existing '1's in the data to '0's leading to data corruption. For example, in Fig. 15.2a, S_1 is supposed to send '0110' to D_2, but due to data corruption by malicious GI S_2, '0010' is received by D_2. However, using parity or error correction code (ECC) bits in the data can be used to detect and correct this type of data corruption. Thus, malicious source GIs do not cause major security risks in DWDM-based MWMR waveguides.

Let us consider another scenario for the same data communication path (i.e., from S_1 to S_2). When destination GI D_1, which is in the communication path, becomes malicious with an HT in its control logic, the detector bank of D_1 can be partially tuned to the utilized wavelength channels to snoop data. In the example shown in

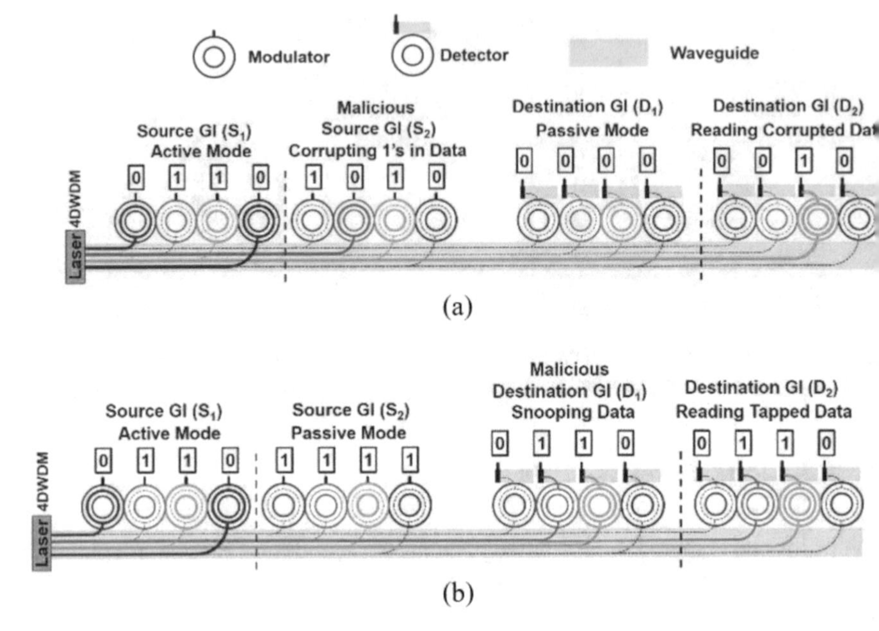

Fig. 15.2 Impact of (**a**) malicious modulator (source) bank, (**b**) malicious detector bank on data in DWDM-based photonic waveguides [30]

Fig. 15.2b, D_1 snoops '0110' from the wavelength channels that are destined to D_2. The sensitive information can be determined by transferring snooped data from D_1 to a malicious core within the CMP. Since, the intended communication among CMP cores is not disrupted, this type of snooping attack from malicious destination GIs are hard to detect. Therefore, there is a pressing need to address the security risks imposed by snooping GIs in DWDM-based PNoC architectures. To address this need, we propose a novel framework SOTERIA that improves hardware security in DWDM-based PNoC architectures.

15.6 *SOTERIA* Framework:Overview

Our proposed multi-layer *SOTERIA* framework enables secure communication in DWDM-based PNoC architectures by integrating circuit-level and architecture-level enhancements. Figure 15.3 gives a high-level overview of this framework. The Privy Data Encipherment Scheme (PDES) uses the PV profile of the destination GIs' detector MRs to encrypt data before it is transmitted via the photonic waveguide. This scheme is sufficient to protect data from snooping GIs, if they not aware of target destination GI. However, a snooping GI can decipher the encrypted data if the target destination GI information is known. Many PNoC architectures (e.g.,

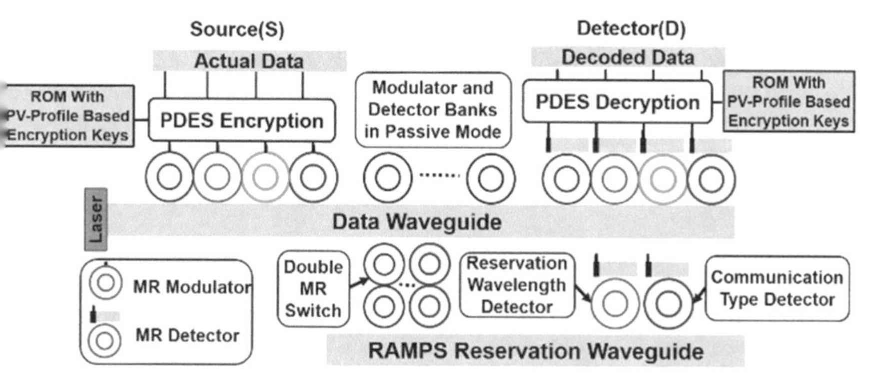

Fig. 15.3 Overview of proposed *SOTERIA* framework that integrates a circuit-level Privy Data Encipherment Scheme (*PDES*) and an architecture-level Reservation-Assisted Metadata Protection Scheme (*RAMPS*) [75]

10, 26]) use the same waveguide to transmit both the destination GI information and actual data, making them vulnerable to data snooping attacks despite using PDES. We devise an architecture-level reservation-assisted metadata protection scheme (RAMPS) that uses a secure reservation waveguide to avoid the stealing of destination GI information by snooping GIs to enhance the security of these PNoCs. The next two sections present details of our PDES and RAMPS schemes.

15.7 Privy Data Encipherment Scheme (PDES)

As discussed earlier (Sect. 1.3.2), malicious destination GIs can snoop data from a shared waveguide. Data encryption can be used to address this security concern so that the malicious destination GIs cannot decipher the snooped data. The encryption key used for data encryption should be kept secret from the snooping GIs for the encrypted data to be truly undecipherable, which can be challenging as the identity of the snooping GIs in a PNoC is not known. Therefore, it becomes very difficult to decide whether or not to share the encryption key with a destination GI (that can be malicious) for data decryption. Each destination GI can have a different key to resolve this conundrum so that a key that is specific to a secure destination GI does not need to be shared with a malicious destination GI for decryption purpose. Moreover, to keep these destination specific keys secure, the malicious GIs in a PNoC must not be able to clone the algorithm (or method) used to generate these keys.

To generate unclonable encryption keys, the PV profiles of the destination GIs' detector MRs are used in our Privy Data-Encipherment scheme (*PDES*) scheme. As discussed in [13], PV induces random shifts in the resonance wavelengths of the MRs used in a PNoC. These resonance shifts can be in the range from 3 nm to 3 nm

[13]. PV profiles are different for the MRs that belong to different GIs in a PNoC. In fact, the MRs that belong to different MR banks of the same GI also have different PV profiles. Due to their random nature, these MR PV profiles cannot be cloned by the malicious GIs, which makes the encryption keys generated using these PV profiles truly unclonable. PDES uses the PV profiles of detector MRs to generate a unique encryption key for each detector bank of every MWMR waveguide in a PNoC.

Our PDES scheme generates encryption keys during the testing phase of the CMP chip, by using a dithering signal based in-situ method [14] to generate an antisymmetric analog error signal for each detector MR of every detector bank that is proportional to the PV-induced resonance shift in the detector MR. Then, it converts the analog error signal into a 64-bit digital signal. Thus, a 64-bit digital error signal is generated for every detector MR of each detector bank. We consider 64 DWDM wavelengths per waveguide, and hence, we have 64 detector MRs in every detector bank and 64 modulator MRs in every modulator bank. For each detector bank, our *PDES* scheme XORs the 64 digital error signals (of 64 bits each) from each of the 64 detector MRs to create a unique 64-bit encryption key. Note that our *PDES* scheme also uses the same anti-symmetric error signals to control the carrier injection and heating of the MRs to remedy the PV-induced shifts in their resonances.

To understand how the 64-bit encryption key is utilized to encrypt data in photonic links, consider Fig. 15.4 which depicts an example photonic link that has one MWMR waveguide and connects the modulator banks of two source GIs (S_1 and S_2) with the detector banks of two destination GIs (D_1 and D_2). *PDES* creates two 64-bit encryption keys corresponding to two destination GIs on the link, and stores them at the source GIs. When data is to be transmitted by a source GI, the key for the appropriate destination is used to encrypt data at the flit-level granularity, by performing an XOR between the key and the data flit. The encryption key matching

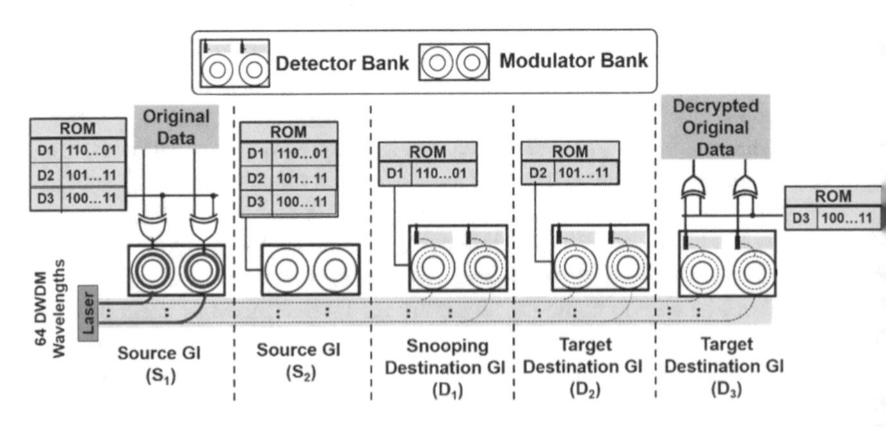

Fig. 15.4 Overview of proposed PV-based security enhancing Privy Data Encipherment Scheme (*PDES*) [75]

the data flit size is required. We consider the size of data flits to be 512 bits. Therefore, the 64-bit encryption key is appended eight times to generate a 512-bit encryption key. In Fig. 15.4, 512-bit encryption keys (for destination GIs D_1 and D_2) are stored in the source GI local ROM, whereas every destination GI stores only its corresponding 512-bit key in its ROM. To eliminate the latency overhead of affixing 64-bit keys we store 512-bit key to generate 512-bit keys, at the cost of a reasonable area/energy overhead in the ROM. As an example, if S_1 wants to send a data flit to D_2, then S_1 first accesses the 512-bit encryption key corresponding to D_2 from its local ROM and XORs the data flit with this key in one cycle, and then transmits the encrypted data flit over the link. As the link employs only one waveguide with 64 DWDM wavelengths, therefore, the encrypted 512-bit data flit is transferred on the link to D_2 in eight cycles. At D_2, the data flit is decrypted by XORing it with the 512-bit key corresponding to D_2 from the local ROM. In this scheme,D_1 cannot decipher the data even if D_1 snoops the data intended for D_2, as it does not have access to the correct key (corresponding to D_2) for decryption. Thus, our PDES encryption scheme protects data against snooping attacks in DWDM-based PNoCs.

Limitations of PDES: The PDES scheme can protect data from being deciphered by a snooping GI, if the following two conditions about the underlying PNoC architecture hold true: (i) the snooping GI does not know the target destination GI for the snooped data, (ii) the snooping GI cannot access the encryption key corresponding to the target destination GI. As discussed earlier, only all source GIs have an encryption key stored and at the corresponding destination GI making it physically inaccessible to a snooping destination GI. However, if more than one GIs in a PNoC are compromised due to HTs in their control units and if these HTs launch a coordinated snooping attack, then it may be possible for the snooping GI to access the encryption key corresponding to the target destination GI.

For instance, consider the photonic link in Fig. 15.4. If both S_1 and D_1 are compromised, then the HT in S_1's control unit can access the encryption keys corresponding to both D_1 and D_2 from its ROM and transfer them to a malicious core (a core running a malicious program). Moreover, the data intended for D_2 can be snooped by the HT in D_1's control unit and transfer it to the malicious core. Thus, the malicious core may have access to the snooped data as well as the encryption keys stored at the source GIs. Nevertheless, to decipher the snooped data accessing the encryption keys stored at the source GIs is not sufficient for the malicious GI (or core). This is because the compromised ROM typically has multiple encryption keys corresponding to multiple destination GIs, and choosing a correct key that can decipher data requires the knowledge of the target destination GI. Thus, our PDES encryption scheme can secure data communication in PNoCs as long as the malicious GIs (or cores) do not know the target destinations of the snooped data.

Unfortunately, many PNoC architectures, e.g., [10, 26], that employ photonic links with multiple destination GIs utilize the same waveguide to transmit both the target destination information and actual data. In such PNoCs, from shared waveguide malicious GI can manage to tap the target destination information, then it can access the correct encryption key from the compromised ROM to decipher the snooped data. Thus, there is a need to conceal the target destination information from malicious GIs (cores). This motivates us to propose an architecture-level solution, as discussed next.

15.8 Reservation-Assisted Metadata Protection Scheme

In PNoCs that use photonic links with multiple destination GIs, data is typically transferred in two time-division-multiplexed (TDM) slots called reservation slot and data slot [10, 26]. Figure 15.5a shows PNoCs using the same waveguide to transfer both slots to minimize photonic hardware. To enable reservation of the waveguide, each destination is assigned a reservation selection wavelength. In Fig. 15.5a, λ_1 and λ_2 are the reservation selection wavelengths corresponding to destination GIs D_1 and D_2, respectively. Ideally, detector switches ON its detector bank to receive data in the next data slot when a destination GI detects its reservation selection wavelength in the reservation slot. But in the presence of an HT, a malicious GI can snoop signals from the reservation slot using the same detector bank that is used for data reception. For example, in Fig. 15.5a, malicious GI D_1 is using one of its detectors to snoop λ_2 from the reservation slot. By snooping λ_2, D_1 can identify that

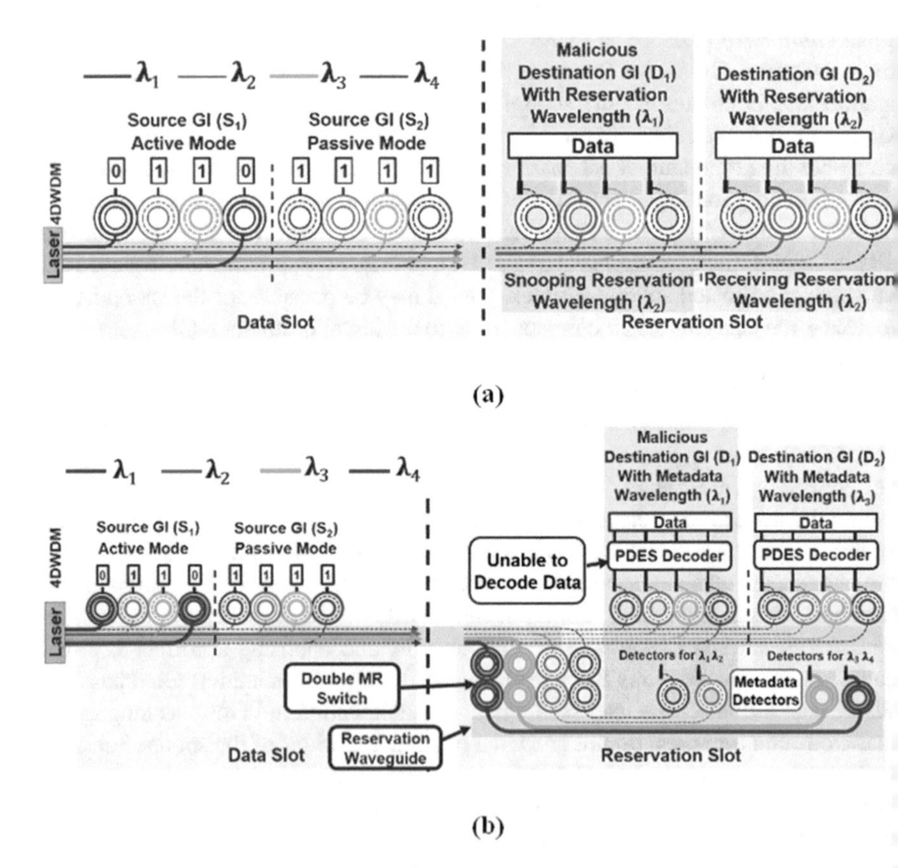

Fig. 15.5 Reservation-assisted data transmission in DWDM-based photonic waveguides (**a**) without RAMPS, (**b**) with RAMPS [30]

the data it will snoop in the subsequent data slot will be intended for destination D_2. Thus, D_1 can decipher its snooped data now by choosing the correct encryption key from the compromised.

To address this security risk, we propose an architecture-level reservation-assisted metadata protection scheme (RAMPS) scheme. In RAMPS, a reservation waveguide is added, whose main function is to carry reservation slots, whereas data slots are carried by data waveguide. We use double MRs to switch the signals of reservation slots from the data waveguide to the reservation waveguide, as shown in Fig. 15.5b. Double MRs are used instead of single MRs for switching to ensure that the switched signals do not reverse their propagation direction after switching [28].Double MRs also have lower signal loss due to steeper roll-off of their filter responses [28] compared to single MRs.

In a photonic link the double MRs are switched ON only in a reservation slot, otherwise they are switched OFF to let the signals of the data slot pass by in the data waveguide. Furthermore, in RAMPS, each destination GI has only one detector on the reservation waveguide, which corresponds to its receiver selection wavelength. For example, in Fig. 15.5b, D_I and D_2 will have detectors corresponding to their reservation selection wavelengths λ_1 and λ_2, respectively, on the reservation waveguide. Figure 15.5b shows this making it difficult for the malicious GI D_I to snoop λ_2 from the reservation slot, as D_I does not have a detector corresponding to λ_2 on the reservation waveguide. However, the HT in D_I's control unit may still attempt to snoop other reservation wavelengths (e.g., λ_2) in the reservation slot by retuning D_I's λ_1 detector. The HT would required to perfect the timing and target wavelength of its snooping attack to succeed in these attempts, which is very difficult due to the large number of utilized reservation wavelengths. Thus, the correct encryption key cannot be identified by D_I to decipher the snooped data. In summary, RAMPS enhances security in PNoCs by protecting data from snooping attacks, even if the encryption keys used to secure data are compromised.

15.9 Implementing *SOTERIA* Framework on PNoCS

We characterize the impact of SOTERIA on two popular PNoC architectures: Firefly [7] and Flexishare [8], both of which use DWDM-based photonic waveguides for data communication. We consider Firefly PNoC with 8 × 8 SWMR crossbar [7] and a Flexishare PNoC with 32 × 32 MWMR crossbar [8] with 2-pass token stream arbitration. We adapt the analytical equations from [28] to model the signal power loss and required laser power in the SOTERIA-enhanced Firefly and Flexishare PNoCs. XOR gates are required to enable parallel encryption and decryption of 512-bit data flits at each source and destination GI of the SOTERIA-enhanced Firefly and Flexishare PNoCs. The overhead for encryption and decryption of every data flit was as 1 cycle delay. The overall laser power and delay overheads for both PNoCs are quantified in the results section.

Firefly PNoC: Firefly PNoC [7], for a 256-core system, has 8 clusters (C1–C8) with 32 cores in each cluster. Firefly uses reservation-assisted SWMR data channels in its 8x8 crossbar for inter-cluster communication. Each data channel consists of 8 SWMR waveguides, with 64 DWDM wavelengths in each waveguide. A reservation waveguide was added to every SWMR channel to integrate SOTERIA with Firefly PNoC. This reservation waveguide has 7 detector MRs to detect reservation selection wavelengths corresponding to 7 destination GIs. Furthermore, 64 double MRs (corresponding to 64 DWDM wavelengths) are used at each reservation waveguide to implement RAMPS. To enable PDES, each source GI has a ROM with seven entries of 512 bits each to store seven 512-bit encryption keys corresponding to seven destination GIs. In addition, each destination GI requires a 512-bit ROM to store its own encryption key.

Flexishare PNoC: We also integrate SOTERIA with the Flexishare PNoC architecture [8] with 256 cores. We considered a 64-radix 64-cluster Flexishare PNoC with four cores in each cluster and 32 data channels for inter-cluster communication. Each data channel has four MWMR waveguides with each having 64 DWDM wavelengths. In SOTERIA-enhanced Flexishare, we added a reservation waveguide to each MWMR channel. Each reservation waveguide has 16 detector MRs to detect reservation selection wavelengths corresponding to 16 destination GIs. A ROM with 16 entries of 512 bits each to store the encryption keys at each source GI is required, whereas each destination GI requires a 512-bit ROM to enable PDES.

15.10 Evaluations

15.10.1 Evaluation Setup

To evaluate our proposed SOTERIA (PDES+RAMPS) security enhancement framework for DWDM-based PNoCs, we integrate it with the Firefly [7] and Flexishare [9] PNoCs, as explained in Sect. 1.7. We modeled and performed simulation based analysis of the SOTERIA-enhanced Firefly and Flexishare PNoCs using a cycle-accurate SystemC based NoC simulator, for a 256-core single-chip architecture at 22 nm. The power dissipation and energy consumption were validated from the DSENT tool [21]. We used real-world traffic from the PARSEC benchmark suite [22]. GEM5 full-system simulation [23] of parallelized PARSEC applications were used to generate traces that were fed into our NoC simulator. We set a "warmup" period of 100 million instructions and then captured traces for the subsequent 1 billion instructions. These traces are extracted from parallel regions of execution of PARSEC applications. We performed geometric calculations for a 20 mm × 20 mm chip size, to determine lengths of SWMR and MWMR waveguides in Firefly and Flexishare. Based on this analysis, the time needed for light to travel from the first to the last node was estimated as 8 cycles at 5 GHz clock frequency [12]. We use a 512-bit packet size, as advocated in the Firefly and Flexishare PNoCs.

Similar to [28], we adapt the VARIUS tool [19] to model random and systematic die-to-die (D2D) as well as within-die (WID) process variations in MRs for the Firefly and Flexishare PNoCs.

The static and dynamic energy consumption values for electrical routers and concentrators in Firefly and Flexishare PNoCs are based on results from DSENT [21]. We model and consider the area, power, and performance overheads for our framework implemented with the Firefly and Flexishare PNoCs as follows. SOTERIA with Firefly and Flexishare PNoCs has an electrical area overhead of 12.7mm2 and 3.4mm2, respectively, and power overhead of 0.44 W and 0.36 W, respectively, using gate-level analysis and CACTI 6.5 [24] tool for memory and buffers. The photonic area of Firefly and Flexishare PNoCs is 19.83mm2 and 5.2mm2, respectively, based on the physical dimensions [20] of their waveguides, MRs, and splitters. For energy consumption of photonic devices, we adapt model parameters from recent work [25, 27] with 0.42pJ/bit for every modulation and detection event and 0.18pJ/bit for the tuning circuits of modulators and photodetectors. The MR trimming power is 130μW/nm [29] for current injection and tuning power is 240μW/nm [29] for heating.

15.10.2 *Overhead Analysis of SOTERIA on PNoCs*

Our first set of experiments compare the baseline (without any security enhancements) Firefly and Flexishare PNoCs with their SOTERIA enhanced variants. From Sect. 1.7, all 8 SWMR waveguide groups of the Firefly PNoC and all 32 MWMR waveguide groups of the Flexishare PNoC are equipped with PDES encryption/decryption and reservation waveguides for the RAMPS scheme.

The total signal loss at the detectors of the worst-case power loss node (N_{WCPL}) were calculated by adapting the analytical models from [28], which corresponds to router C4R0 for the Firefly PNoC [7] and node R63 for the Flexishare PNoC [8]. Figure 15.6a summarizes the worst-case signal loss results for the baseline and SOTERIA configurations for the two PNoC architectures. The loss is increased by 1.6 dB for Firefly PNoC with SOTERIA and Flexishare PNoC with SOTERIA increased by 1.2 dB on average, compared to their respective baselines. Compared to the baseline PNoCs that have no single or double MRs to switch the signals of the reservation slots, the double MRs used in the SOTERIA-enhanced PNoCs to switch the wavelength signals of the reservation slots increase through losses in the waveguides, which ultimately increases the worst-case signal losses in the SOTERIA-enhanced PNoCs. Using the worst-case signal losses shown in Fig. 15.6a, we determine the total photonic laser power and corresponding electrical laser power for the baseline and SOTERIA-enhanced variants of Firefly and Flexishare PNoCs, shown in Fig. 15.6b. From this figure, the laser power overheads are 44.7% and 31.40% for the Firefly and Flexishare PNoCs with SOTERIA on average, compared to their baselines.

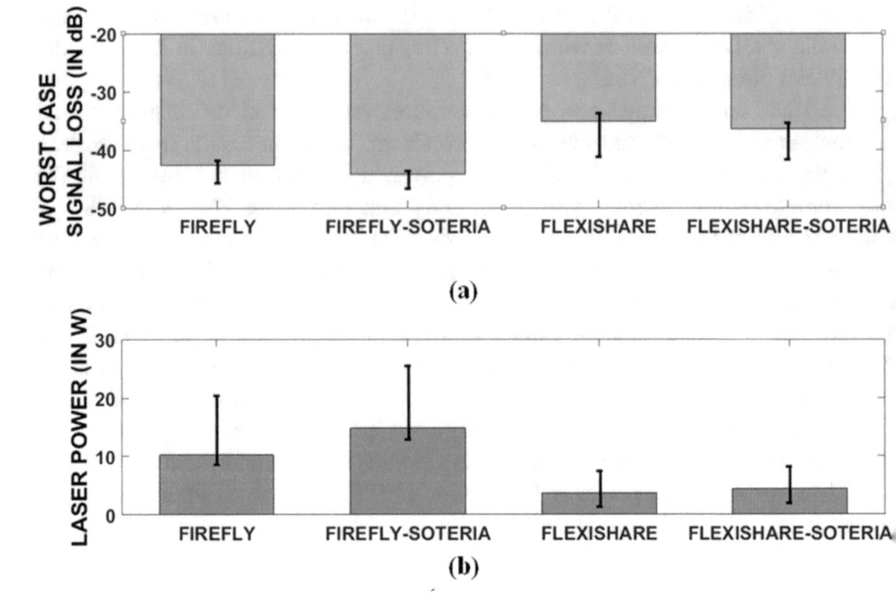

Fig. 15.6 Comparison of (**a**) worst-case signal loss and (**b**) laser power dissipation of SOTERIA framework on Firefly and Flexishare PNoCs with their respective baselines considering 100 process variation maps [30]

Figure 15.7 presents detailed simulation results that quantify the average packet latency and energy-delay product (EDP) for the two configurations of the Firefly and Flexishare PNoCs. Results are shown for twelve multi-threaded PARSEC benchmarks. From Fig. 15.7a, Firefly with *SOTERIA* has 5.2% and Flexishare with *SOTERIA* has 10.6% higher latency on average compared to their respective baselines. The increase in average latency is due to the additional delay due to encryption and decryption of data (Sect. 1.7.1) with *PDES*.

From the results for EDP shown in Fig. 15.7b, Firefly with *SOTERIA* has 4.9% and Flexishare with *SOTERIA* has 13.3% higher EDP on average compared to their respective baselines. The increase in their average packet latency and the presence of additional RVSC reservation waveguides leads to increase in EDP for the *SOTERIA*-enhanced PNoCs, which increases the required photonic hardware (e.g., more number of MRs) in the SOTERIA-enhanced PNoCs. This in turn increases static energy consumption (i.e., laser energy and trimming/tuning energy), ultimately increasing the EDP.

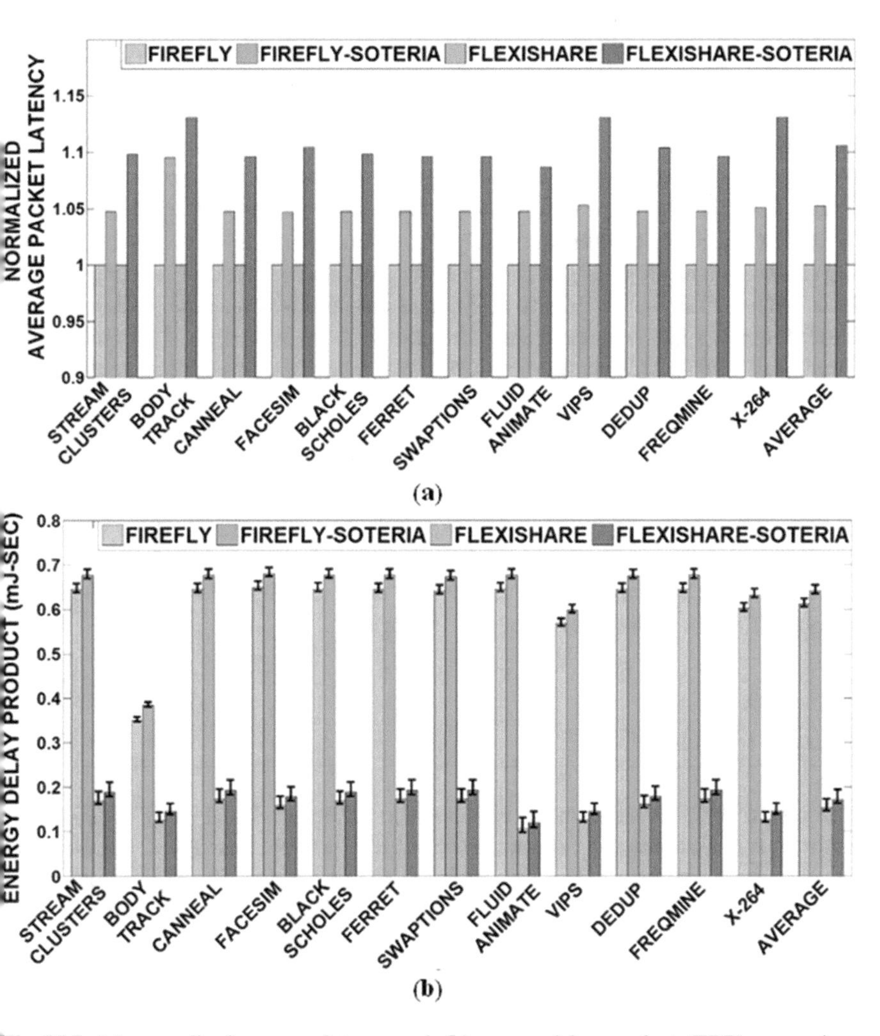

Fig. 15.7 (**a**) normalized average latency and (**b**) energy-delay product (EDP) comparison between different variants of Firefly and Flexishare PNoCs that include their baselines and their variant with SOTERIA framework, for PARSEC benchmarks. Latency results are normalized with their respective baseline architecture results. Bars represent mean values of average latency and EDP for 100 PV maps; confidence intervals show variation in average latency and EDP across PARSEC benchmarks [30]

15.10.3 Analysis of Overhead Sensitivity

Our last set of evaluations explore how the overhead of SOTERIA changes with varying levels of security in the network. Typically, in a manycore system, sensitive information (i.e., keys) is present only at certain portion of the data and hence only a certain number of communication links need to be secure. Therefore, we secure only a certain number channels using SOTERIA for our analysis in this section, instead of securing all data channels of the Flexishare PNoC. Out of the total 32 MWMR channels in the Flexishare PNoC, we secure 4 (FLEX-ST-4), 8 (FLEX-ST-8), 16 (FLEX-ST-16), and 24 (FLEX-ST-24) channels, and evaluate the average packet latency and EDP for these variants of the SOTERIA-enhanced Flexishare PNoC.

In Fig. 15.8, we present average packet latency and EDP values for the five SOTERIA-enhanced configurations of the Flexishare PNoC. From Fig. 15.8a, FLEX-ST-4, FLEX-ST-8, FLEX-ST-16, and FLEX-ST-24 have 1.8%, 3.5%, 6.7%, and 9.5%higher latency on average compared to the baseline Flexishare. Increase in number of SOTERIA enhanced MWMR waveguides increases number of packets that are transferred through the PVSC encryption scheme, which contributes to the increase in average packet latency across these variants. From the results for EDP shown in Fig. 15.8b, FLEX-ST-4, FLEX-ST-8, FLEX-ST-16, and FLEX-ST-24 have 2%, 4%, 7.6%, and 10.8% higher EDP on average compared to the baseline Flexishare. EDP in Flexishare PNoC increases with increase in number of SOTERIA enhanced MWMR waveguides. Overall EDP across these variants is increased due to increase in average packet latency and signal loss due to the higher number of reservation waveguides and double MRs.

15.11 Conclusion

We presented a novel security enhancement framework called SOTERIA that secures data during unicast communications in DWDM-based PNoC architectures from snooping attacks. Our proposed SOTERIA framework shows interesting trade-offs between security, performance, and energy overhead for the Firefly and Flexishare PNoC architectures. Our analysis shows that SOTERIA enables hardware security in crossbar based PNoCs with minimal overheads of up to 10.6% in average latency and of up to 13.3% in EDP compared to the baseline PNoCs. Thus, an attractive solution to enhance hardware security in emerging DWDM-based PNoCs is presented as SOTERIA.

Acknowledgments This research is supported by grants from the University of Kentucky and NSF (CCF-1813370).

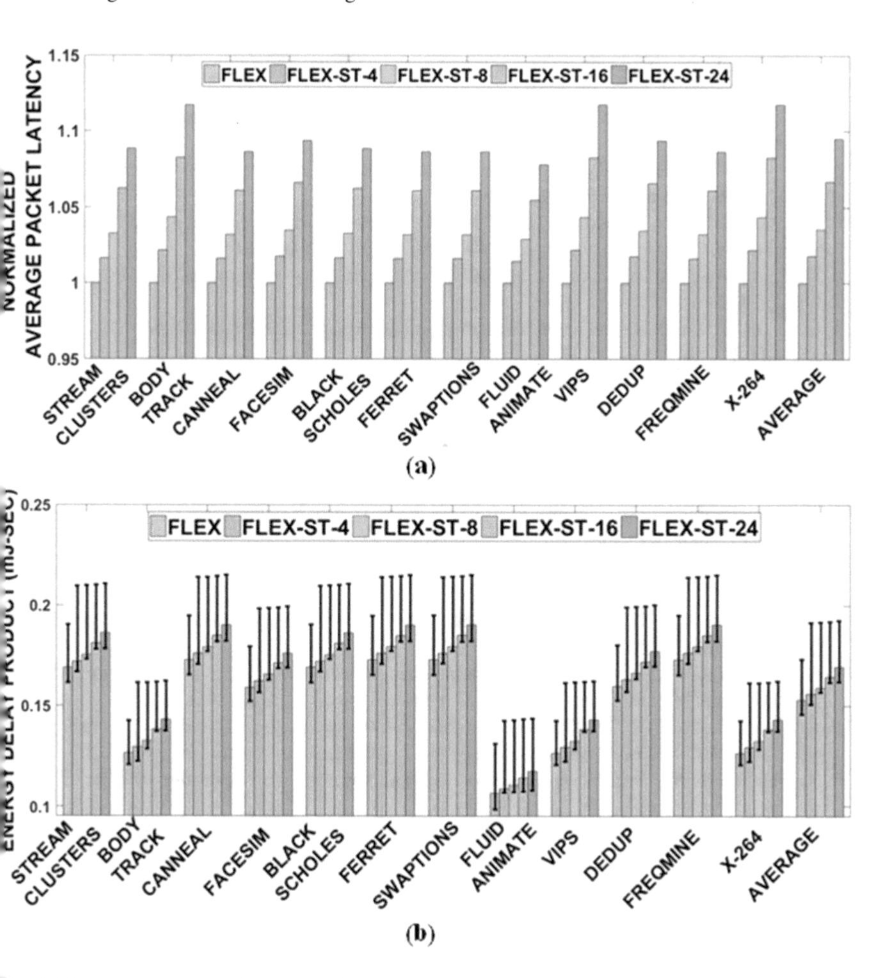

Fig. 15.8 (**a**) normalized latency and (**b**) energy-delay product (EDP) comparison between Flexishare baseline and Flexishare with 4, 8, 16, and 24 SOTERIA enhanced MWMR waveguide groups, for PARSEC benchmarks. Latency results are normalized to the baseline Flexishare results [30]

References

1. R. Chakraborty, S. Narasimhan, S. Bhunia, *Hardware trojan: threats and emerging solutions*. Proceedings of HLDVT, (2009), pp. 166–171
2. M. Tehranipoor, F. Koushanfar, A survey of hardware trojan taxonomy and detection. IEEE Des. Test **25**, 10–25 (2009)
3. S. Skorobogatov, C. Woods, *Breakthrough silicon scanning discovers backdoor in military chip*. Proceedings of CHES, (2012), pp. 23–40
4. W.J. Dally, B. Towles, *Route packets, not wires*. Proceedings of DAC, (2001)
5. D.A.B. Miller, Device requirements for optical interconnects to silicon chips. JPROC **97**(7), 1166–1185 (2009)

6. C. Batten et al., *Building manycore processor-to-dram networks with monolithic silicon photonics*, HotI, (2008), pp. 21–30
7. Y. Pan et al., *Firefly: Illuminating future network-on-chip with nanophotonics*. Proceedings of ISCA, (2009)
8. Y. Pan, J. Kim, G.Memik, *Flexishare: Channel sharing for an energy efficient nanophotonic crossbar*. Proceedings of HPCA, (2010)
9. S.V.R. Chittamuru, S. Pasricha, *SPECTRA: a framework for thermal reliability man-agement in silicon-photonic networks-on-chip*. Proceedings of VLSID, (2016)
10. D.M. Ancajas, et al., *Fort-NoCs: mitigating the threat of a compromised NoC*. Proceedings of DAC, (2014)
11. C. Li, et al., *Energy-efficient optical broadcast for nanophotonic networks-on-chip*. Proceedings of OIC, (2012), pp. 64–65
12. S.V.R. Chittamuru, S. Desai, S. Pasricha, SWIFTNoC: a reconfigurable silicon photonic network with multicast enabled channel sharing for multicore architectures. ACM JETC **13**(4), 58 (2017)
13. S.K. Selvaraja, Wafer-scale fabrication technology for silicon photonic integrated circuits. PhD thesis, Ghent University, 2011
14. K. Padmaraju et al., Wavelength locking and thermally stabilizing microring resonators using dithering signals. JLT **32**(3), 505–512 (2013)
15. C.H. Gebotys, et al., *A framework for security on NoC technologies*, Proceedings of ISVLSI, (2003)
16. H.K. Kapoor et al., A security framework for NoC using authenticated encryption and session keys. CSSP **32**, 2605 (2013)
17. S.V.R. Chittamuru, I. Thakkar, S. Pasricha, *Process variation aware crosstalk mitigation for DWDM based photonic noc architectures*. Proceedings of ISQED, (2016)
18. S.V.R. Chittamuru, I. Thakkar, S. Pasricha, *PICO: mitigating heterodyne crosstalk due to process variations and intermodulation effects in photonic NoCs*, Proceedings of DAC, (2016)
19. S. Sarangi et al., Varius: A model of process variation and resulting timing errors for microarchitects. IEEE TSM **21**(1), 3–13 (2008)
20. S. Xiao, M.H. Khan, H. Shen, M. Qi, Modeling and measurement of losses in silicon-on-insulator resonators and bends. Optics Express **15**(17), 10553–10561 (2007)
21. C. Sun et al., *DSENT – a tool connecting emerging photonics with electronics for opto-electronic networks-on-chip modeling*, NOCS, (2012)
22. C. Bienia et al., *The PARSEC benchmark suit: characterization and architectural implica-tions*, PACT, (2008)
23. N. Binkert et al., The gem5 Simulator, CA News, 2011
24. CACTI 6.5, http://www.hpl.hp.com/research/cacti/
25. S.V.R. Chittamuru, I. Thakkar, S. Pasricha, *Analyzing voltage bias and temperature induced aging effects in photonic inter-connects for manycore computing*, Proceedings of SLIP, (2017)
26. C. Chen A. Joshi, *Runtime management of laser power in silicon-photonic multibus NoC architecture*, Proceedings of IEEE JQE, (2013)
27. I. Thakkar, S.V.R. Chittamuru, S. Pasricha, *Mitigation of homodyne crosstalk noise in silicon photonic NoC architectures with tunable decoupling*, Proceedings of CODES+ISSS, (2016)
28. S.V.R. Chittamuru, I. Thakkar, S. Pasricha, HYDRA: Heterodyne crosstalk mitigation with double microring resonators and data encoding for photonic NoCs. TVLSI **26**(1), 168–181 (2018)
29. D. Dang, S.V.R. Chittamuru, R. Mahapatra, S. Pasricha, *Islands of heaters: a novel thermal management framework for photonic NoCs*, Proceedings of ASPDAC, (2017)
30. S.V. Reddy Chittamuru, I. G Thakkar, V. Bhat, S. Pasricha, *SOTERIA: Exploiting Process Variations to Enhance Hardware Security with Photonic NoC Architectures*, 2018 55th ACM/ESDA/IEEE Design Automation Conference (DAC), (San Francisco, CA, 2018)
31. L. Zhou A.K. Kodi, *PROBE: Prediction-based optical bandwidth scaling for energy-efficient NoCs*, Proceedings of the IEEE/ACM International Symposium on Networks-on-Chip (NOCS), (2016)

32. S. Pasricha N. Dutt. *On-chip communication architectures*. (Morgan Kauffman, 2008b)
33. J. Held, Single-chip cloud computer: An experimental many-core processor from intel labs. Presented at Intel Labs Single-chip Cloud Computer Symposium, Santa Clara, California, 2010
34. M. Mattina, Architecture and performance of the tile-gx processor family, White Paper, Tilera Corporation
35. B. de Dinechin, R. Ayrignac, P.-E. Beaucamps, P. Couvert, B. Ganne, P. de Massas, F. Jacquet, S. Jones, N. Chaisemartin, F. Riss, T. Strudel, *A clustered manycore processor architecture for embedded and accelerated applications*, High Performance Extreme Computing Conference (HPEC), 2013
36. L. Benini, G.D. Micheli, Networks on chips: A new soc paradigm. IEEE Comput. **35**, 70–78 (2002)
37. Y. Hoskote, S. Vangal, A. Singh, N. Borkar, S. Borkar, A 5-ghz mesh interconnect for a teraflops processor. IEEE Micro **27**, 51–61 (2007)
38. H. McGhan. Niagara 2. Microprocessor Report, 2006
39. J.S. Kim, M.B. Taylor, J. Miller, D. Wentzla. Energy characterization of a tiled architecture processor with on-chip networks. (ISLPED ’03, New York, NY, USA, 2003). ACM
40. A. Rovinski et al., A 1.4 GHz 695 Giga Risc-V Inst/s 496-Core Manycore processor with mesh on-chip network and an all-digital synthesized PLL in 16nm CMOS, 2019 Symposium on VLSI Circuits, (Kyoto, Japan, 2019), pp. C30-C31
41. S. Priya, S. Agarwal, H.K. Kapoor, Fault tolerance in network on chip using bypass path establishing packets. 2018 31st international conference on VLSI design and 2018 17th international conference on embedded systems (VLSID), (Pune, 2018), pp. 457–458
42. A.C. Pinheiro, J.A.N. Silveira, D.A.B. Tavares, F.G.A. Silva, C.A.M. Marcon, Optimized fault-tolerant buffer design for network-on-chip applications. 2019 IEEE 10th Latin American symposium on circuits & systems (LASCAS), (Armenia, Colombia, 2019), pp. 217–220
43. T. Boraten, A.K. Kodi, Runtime techniques to mitigate soft errors in network-on-chip (NoC) architectures. IEEE Trans. Comput. Aided Des. Integr. Circuits Syst. **37**(3), 682–695 (2018)
44. H. Kim, A. Vitkovskiy, P.V. Gratz V. Soteriou, Use it or lose it: Wear-out and lifetime in future chip multiprocessors. 2013 46th Annual IEEE/ACM international symposium on microarchitecture (MICRO), (Davis, CA, 2013), pp. 136–147
45. D. DiTomaso, T. Boraten, A. Kodi, A. Louri, Dynamic error mitigation in NoCs using intelligent prediction techniques. 2016 49th Annual IEEE/ACM international symposium on microarchitecture (MICRO), (Taipei, 2016), pp. 1–12
46. P. Bahrebar, D. Stroobandt, Hamiltonian path strategy for deadlock-free and adaptive routing in diametrical 2D Mesh NoCs. 2015 15th IEEE/ACM international symposium on cluster, cloud and grid computing, (Shenzhen, 2015), pp. 1209–1212
47. M. Parasar, A. Sinha, T. Krishna, Brownian bubble router: enabling deadlock freedom via guaranteed forward progress. 2018 Twelfth IEEE/ACM international symposium on networks-on-chip (NOCS), (Turin, 2018), pp. 1–8
48. A. Ramrakhyani, T. Krishna, Static bubble: a framework for deadlock-free irregular on-chip topologies. 2017 IEEE international symposium on high performance computer architecture (HPCA), (Austin, TX, 2017), pp. 253–264
49. A. Ramrakhyani, P.V. Gratz, T. Krishna, Synchronized progress in interconnection networks (SPIN): A new theory for deadlock freedom. IEEE Micro **39**(3), 110–117 (2019)
50. L. Wang, X. Wang, T. Mak, Adaptive routing algorithms for lifetime reliability optimization in network-on-chip. IEEE Trans. Comput. **65**(9), 2896–2902 (2016)
51. J. S. Kim, J. Beom Hong, J. Y. Kang, T. Hee Han, Lifetime improvement method using threshold-based partial data compression in NoC. 2018 International SoC Design Conference (ISOCC), (Daegu, Korea (South), 2018), pp. 269–270
52. L. Huang et al., A lifetime-aware mapping algorithm to extend MTTF of networks-on-chip. 2018 23rd Asia and South Pacific Design Automation Conference, ((ASP-DAC), Jeju, 2018), pp. 147–152

53. V. Rathore, V. Chaturvedi, A. K. Singh, T. Srikanthan, M. Shafique, Towards scalable lifetime reliability management for dark silicon manycore systems, 2019 IEEE 25th international symposium on on-line testing and robust system design (IOLTS), (Rhodes, Greece, 2019), pp. 204–207
54. H. Kim, P. Ghoshal, B. Grot, P. V. Gratz, Reducing network-onchip energy consumption through spatial locality speculation. Proceedings of international symposium on networks-on-chip (NOCS'11), (2011), pp. 233–240
55. G. B. P. Bezerra, S. Forrest, M. Forrest, A. Davis, P. ZarkeshHa, Modeling NoC traÿc locality and energy consumption with rent's communication probability distribution, Proceedings of international workshop on system level interconnect prediction (SLIP'10), (2010), pp.3–8
56. A. Samih, R. Wang, A. Krishna, C. Maciocco, C. Tai, Y. Solihin, Energy-eÿcient inter-connect via Router Parking, 2013 IEEE 19th international symposium on high performance computer architecture (HPCA), (Shenzhen, 2013), pp. 508–519
57. J. Sun, Y. Zhang, An energy-aware mapping algorithm for mesh-based network-on-chip architectures, 2017 international conference on progress in informatics and computing (PIC), (Nanjing, 2017), pp. 357–361
58. G. Ascia, V. Catania, S. Monteleone, M. Palesi, D. Patti, J. Jose, Improving energy consumption of NoC based architectures through approximate communication, 2018 7th mediterranean conference on embedded computing (MECO), (Budva, 2018), pp. 1–4
59. M. G. Moghaddam, Dynamic energy and reliability management in network-on-chip based chip multiprocessors. 2017 eighth international green and sustainable computing conference (IGSC), (Orlando, FL, 2017), pp. 1–4
60. D. Fan et al., SmarCo: An Eÿcient many-core processor for high-throughput applications in datacenters. 2018 IEEE international symposium on high performance computer architecture (HPCA), (Vienna, 2018), pp. 596–607
61. M. McKeown et al., Piton: a manycore processor for multitenant clouds. IEEE Micro **37**(2), 70–80 (Mar.-Apr. 2017)
62. R. Ben Abdelhamid, Y. Yamaguchi, T. Boku, MITRACA: Manycore interlinked torus reconfigurable accelerator architecture. 2019 IEEE 30th international conference on application-specific systems, architectures and processors (ASAP), (New York, NY, USA, 2019), pp. 38–38
63. H. Elmiligi, F. Gebali, M. Watheq El-Kharashi, A. A. Morgan, Traÿc analysis of multi-core body sensor networks based on wireless NoC infrastructure. 2015 IEEE pacific rim conference on communications, computers and signal processing (PACRIM), (Victoria, BC, 2015), pp. 201–204
64. C. Wang, W. Hu, N. Bagherzadeh, A wireless network-on-chip design for multi-core platforms. 2011 19th international euromicro conference on parallel, distributed and network-based processing, (Ayia Napa, 2011), pp. 409–416
65. M. Baharloo, A. Khonsari, P. Shiri, I. Namdari, D. Rahmati, High-average and guaranteed performance for wireless networks-on-chip architectures. 2018 IEEE computer society annual symposium on VLSI (ISVLSI), (Hong Kong, 2018), pp. 226–231
66. D. DiTomaso, A. Kodi, D. Matolak, S. Kaya, S. Laha, W. Rayess, A-WiNoC: adaptive wireless network-on-chip architecture for chip multiprocessors. IEEE Trans. Parallel Distrib. Syst. **26**(12), 3289–3302 (2015)
67. D. Vantrease et al., Corona: system implications of emerging nanophotonic technology. 2008 international symposium on computer architecture, (Beijing, 2008), pp. 153–164
68. Cheng Li, M. Browning, P. V. Gratz, S. Palermo, LumiNOC: A power-efficient, high-performance, photonic network-on-chip for future parallel architectures. 2012 21st international conference on parallel architectures and compilation techniques (PACT), (Minneapolis, MN, 2012), pp. 421–422
69. Z. Wang et al., CAMON: low-cost silicon photonic chiplet for manycore processors, IEEE transactions on computer-aided design of integrated circuits and systems
70. T.H. Boraten, A.K. Kodi, Securing NoCs against timing attacks with non-interference based adaptive routing, 2018 Twelft

1. K. Madden, J. Harkin, L. McDaid, C. Nugent, Adding security to networks-on-chip using neural networks. 2018 ieee symposium series on computational intelligence (SSCI), (Bangalore, India, 2018), pp. 1299–1306
2. B. Lebiednik, S. Abadal, H. Kwon, T. Krishna, Architecting a secure wireless network-on-chip. 2018 Twelfth IEEE/ACM international symposium on networks-on-chip (NOCS), (Turin, 2018), pp. 1–8
3. A.K. Biswas, Efficient timing channel protection for hybrid (packet/circuit-switched) network-on-chip. IEEE Trans. Parallel Distrib. Syst. **29**(5), 1044–1057 (2018)
4. S. Das, K. Basu, J.R. Doppa, P.P. Pande, R. Karri, K. Chakrabarty, Abetting planned obsolescence by aging 3d networks-on-chip. 2018 Twelfth IEEE/ACM international symposium on networks-on-chip (NOCS), (Turin, 2018), pp. 1–8
5. S.V.R. Chittamuru, I. Thakkar, S. Pasricha, S.S. Vatsavai, V. Bhat, Exploiting process variations to secure photonic NoC architectures from snooping attacks. IEEE transactions on computer-aided design of integrated circuits and systems, (TCAD), 2020

Chapter 16
Security Frameworks for Intra and Inter-Chip Wireless Interconnection Networks

M. Meraj Ahmed, Abhishek Vashist, Andrew Keats, Amlan Ganguly, and Sai Manoj Pudukotai Dinakarrao

16.1 Introduction

With the advent of the multi or many-core paradigm towards enhanced performance, traditional bus-based interconnect mechanisms were found to be non-scalable from a design perspective. This led to the adoption of the Network-on-Chip (NoC) paradigm for interconnecting tens to hundreds of cores on a single die [1]. Regular NoC architectures such as mesh or torus-based architectures have shown a reduced design complexity and provided the benefits such as easy-to-replicate, verify and reduced time-to-market [2]. However, such regular architectures resulted in non-scalable performance with increase in number of cores due to long multi-hop paths over wired links [3]. Along with other emerging interconnect technologies such as silicon photonics or Through-Silicon-Vias (TSVs) for 3D NoCs, wireless interconnects were envisioned to enable scalable communication fabrics in multicore chips [4]. Though emerging interconnects such as silicon photonics and 3D TSVs provide high bandwidth communication, the design and overhead costs, and concerns of reliability limits their adoptability [5, 6].

In contrast, advancements in low-power Millimeter-Wave (mm-wave) wireless transceivers, efficient on-die miniature antennas, and smart designs of hybrid architectures with wired as well as single hop wireless links resulted in lower packet latency and energy consumption in on-chip communication and facilitated investigation of wireless NoCs (WiNoCs) in emerging many-core systems [7] as well as in multichip computing systems [8].

M. M. Ahmed · A. Vashist · A. Keats · A. Ganguly (✉)
Rochester Institute of Technology, Rochester, NY, USA
e-mail: ma9205@rit.edu; av8911@rit.edu; axk7655@rit.edu; axgeec@rit.edu

S. M. Pudukotai Dinakarrao
George Mason University, Fairfax, VA, USA
e-mail: spudukot@gmu.edu

P. Mishra, S. Charles (eds.), *Network-on-Chip Security and Privacy*,
https://doi.org/10.1007/978-3-030-69131-8_16

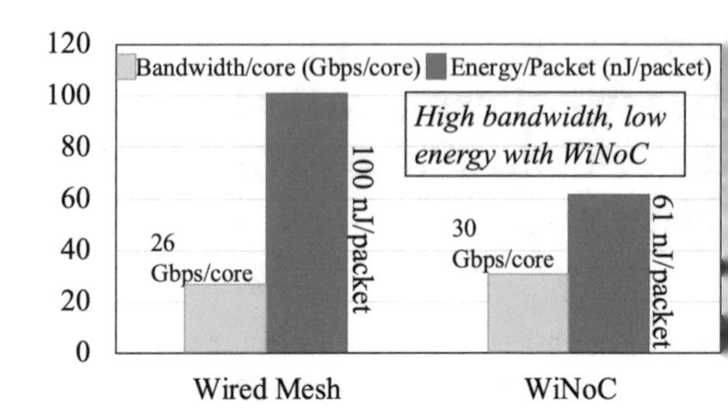

Fig. 16.1 Bandwidth and energy consumption comparison between wired and WiNoC

A case study is performed to compare a wired NoC and a WiNoC for a 64 core system with four wireless interfaces (WIs) overlaid on a mesh in 65nm technology node over a 20 mm×20 mm die. The size of each packet is set to 2 Kb. More details on experimental setup are presented in Sect. 16.5.1. One can observe from Fig. 16.1 that the WiNoC improves the bandwidth per core and energy per packet by 15% and 39%. This demonstrates the potential benefits of adopting WiNoCs for interconnecting multicore or many-core processors.

On the other hand, high-performance computing nodes such as blade servers and embedded systems have been undergoing a massive paradigm shift from single core, single chip architecture to multicore-multichip (MCMC) architecture. This paradigm shift is justified as follows, for a large single chip, different factors such as sub-wavelength lithography, line edge roughness, and random dopant fluctuations can cause wide process variations, which can result in higher fault density and hence, reduces the manufacturing yield.

Performance of the MCMC system is mostly limited by the high latency and power-hungry off-chip I/Os. Conventionally, C4 bumps coupled with in-package transmission lines or flip-chip packaging [9] is used to interconnect chips within a MCMC system. However, signal quality deterioration due to microwave effects, crosstalk coupling effects, and frequency-dependent line losses in the transmission line limit the number of concurrent, high-speed inter-chip I/O and hence chip-to-chip bandwidth. In recent literature it has been shown that WIs operating at GigaHertz (GHz) bandwidth in mm-wave bands can mask off-chip I/O delay by establishing single hop, energy-efficient chip-to-chip communication links [8, 10]. In this chapter, we refer such MCMC systems with in-package mm-wave wireless interconnect as Wireless Network-in-Package (WiNiP).

Although extensive research has been carried out towards improving performance and energy dissipation in both WiNoCs and WiNiPs, relatively little attention has been given to the information integrity and security or privacy aspects of the on and off-chip wireless communication. While security of traditional wired NoCs against various kinds of attacks such as hardware Trojans (HT) [11], eavesdropping (ED) has resulted in appropriate defense mechanisms, the additional threats that unguided wireless interconnects can engender have not received the necessary attention. Wireless interconnects are vulnerable to attacks, similar to those encountered in other wireless networks such as sensor networks or mobile networks. Furthermore, conventional defenses against persistent jamming attacks such as frequency or channel hopping [12] are not applicable in a WiNoC/WiNiP, as the WIs have access to a single shared channel and extremely limited resources. This calls for an embedded defense mechanism for current and emerging mm-wave based on and off-chip interconnection architectures.

Wireless being an unguided, shared transmission medium is vulnerable to many attacks such as Denial-of-Service (DoS), ED, and spoofing. Although each of these attacks requires its own detection and defense mechanism, in this chapter we focus on persistent jamming-based DoS attack as well as ED attacks as they are some of the most common, simple, and yet powerful attack on wireless systems. To replicate such an attack, we consider an external attacker that produces a high energy electromagnetic (EM) radiation that causes interference in the ongoing wireless transmission.

Moreover, it is also possible that a HT planted in the system from a vulnerable design and manufacturing process can cause a WI to transmit persistent jamming signals to cause DoS for other WIs. In this case, one of the WIs infected by a HT will send data over the wireless channel irrespective of whether it is enabled by the adopted Medium Access Control (MAC) mechanism. This will cause contention or interference with legitimate transmissions causing DoS on the remaining WIs. While well-known defenses exist against DoS attacks in large-scale wireless networks [12], those techniques are not directly applicable to the WiNoC or WiNiP scenario due to specific architecture and MAC constraints for wireless interconnection architecture. Similarly, for ED attack, we assume an external receiver can be tuned to the wireless channel used for on and off-chip wireless communication, resulting in information leakage or that an internal HT passes packets received over the wireless channel downstream to a malicious node even when the packet is not addressed to the particular receiver. Such an attack can lead to leakage of sensitive information either directly via an external eavesdropper or through an internal malicious node.

In this chapter, we propose a mechanism to detect and recover from persistent jamming-based DoS attacks that can disable the wireless communication in both WiNoCs and WiNiPs. We re-use the existing Design for Testability (DFT) hardware and deploy a Machine Learning (ML) classifier to detect and defend against persistent jamming attack. To handle more intelligently crafted jamming attacks and ensure a robust, accurate detection and defense mechanism we utilize an Adversarial Machine Learning (AML) and adversarial training for the deployed ML classifiers. Moreover, under such jamming attack, especially for WiNiP architectures, it is non-trivial to synchronize and inform all other WIs about the presence of an adversary as inter-chip communication happens through only wireless medium which is itself vulnerable to the attack. To address this issue, we also develop a MCMC wireless communication protocol along with a reconfigurable MAC that can ensure robust and secure communication under internal and external persistent jamming attack. In addition to DoS attack, we propose to equip each wireless transceiver with a mechanism to prevent information leakage through ED. To achieve this, each wireless transceiver will encode sensitive data using a secret code to minimize the overheads as well as maintain a high throughput.

The chapter has been organized as follows. Section 16.2 describes the contemporary works in NoC security. The attack model for wireless interconnect architectures considered in this chapter has been described in Sect. 16.3. Section 16.4 elaborates the security architecture for WiNoCs and Sect. 16.6 expands the security methodology for WiNiP communication. Results for secure WiNoC and WiNiP interconnection architectures are presented in Sects. 16.5 and 16.7, respectively. Section 16.8 concludes the chapter.

16.2 Contemporary Works

Considerable research has been done on developing techniques for securing conventional NoCs and NoC based multicore processors [13–16]. However, these security measures are confined to wired NoCs and not applicable to wireless interconnections. Very little attention has been dedicated to this important problem of securing on and off-chip wireless communication, although it has been identified as an important challenge to be overcome to make intra and inter-chip wireless communication a reality [7]. In [17], a small-world graph based WiNoC architecture was proposed to mitigate DoS attacks. On the other hand, hash based authentication to prevent eavesdropping has been proposed in [18]. In [19] a secure WiNoC architecture has been proposed that can protect against DoS, eavesdropping and spoofing but engages the Operating System (OS) to block DoS attacks in a WiNoC with contention-free channel access. However, this work does not address the issue of jamming attack from an external attacker assuming that the packaging will protect against such attacks. This may not be true for all kinds of chips or packaging materials and can create DoS attack on WiNiPs. Persistent jamming-based DoS attack for on-chip wireless interconnect has been addressed in [20]. In case of external jamming, the authors in [20] utilized the underlying wired NoC to sustain communication. However, such solution cannot be adopted for WiNiP, as in WiNiP, off-chip communication happens only through wireless interconnect In addition, there exist techniques such as Signature-based attack detection [21], event-based attack detection, which are primarily carried out in software, as the hardware to support such techniques will incur overheads. Though such techniques can detect anomalies, they are hampered by its large latencies and processing overheads, which might not suit a multicore NoC. Furthermore, threshold-based attack detection [22] can be seen as another viable approach for attack detection with low complexity. However, if one utilizes the recoverable error rate as a threshold to distinguish burst and jamming induced errors, the chances of false negatives could be high, as unrecoverable burst errors need not always be caused by jamming.

Similarly, ED poses another threat in securing the communication information. Encryption has been widely proposed in order to secure the communication information. Some of the encryption techniques such as asymmetric key encryption though efficient, poses large overheads, especially in the case of on-chip communications due to utilization of hash tables and computational complexities. This necessitates adoption of symmetric key encryption techniques such as Advanced Encryption Standard (AES) [23]. Though AES has a proven robustness against side-channels in the networking domain, adapting it for WiNoC communications adds large processing overheads and thus is not feasible. In contrast, the work presented in this chapter performs encoding to scramble or mask the data to protect against external eavesdropping and an embedded functional block to detect and prevent internal eavesdropping, leading to lower overheads with sophisticated detection schemes for different kinds of attacks.

In [24], the authors developed a spoofing detection and defense mechanism based on received signal power for on-chip wireless interconnect. However, the proposed mechanism in [24] imposes placement restrictions for WI nodes to distinctly identify the senders that are equidistant from the receiver. Such WI placement restrictions can have significant performance impacts and placement challenges. Moreover, such mechanism cannot be extended for WiNiP systems specially in the presence of an internal or external jammer. Persistent jamming-based DoS attack in WiNiP context has been addressed in [25] which forms the basis of WiNiP security mechanism described in Sect. 16.6. Though persistent jamming attacks are less studied in WiNiPs, a vast amount of research is performed in Wireless Sensor Networks (WSN) for potential solutions. For instance, frequency hopping has been traditionally employed in order to overcome the presence of a jammer [26]. However, multiple jamming devices operating on different bands can effectively block the entire spectrum.

On other hand, although ML has been used in the context of NoC systems for congestion-aware routing [27], it is not used for securing NoC, especially against DoS attacks due to resource constraints. However, there exist works on detecting DoS attacks on cloud or IoT systems. We review some of them and outline the differences here. In [28], a decision tree (DT) based algorithm is devised for detecting DoS attacks in cloud environment. Further, it is combined with signature detection techniques for improving efficiency. Similar works using radial basis function (RBF) neural networks (RBFNNs) [29], artificial NNs (ANNs) [30], are proposed, and [31] presents a comparison of different ML algorithms when detecting Distributed DOS (DDoS) attacks in cloud and IoT devices. The work in [32] employs 23 features to detect the DDoS attacks using different ML classifiers. Despite having the similar objective of detecting DoS/DDoS attacks, the constraints, protocols, and traffic flow are different for miniature NoC systems.

Thus, the main differences and challenges compared to existing works using ML for security against DoS/ED attacks can be outlined as follows: in the existing works, the detection is carried out in a cloud or resource-ample environment, where complex computations can be afforded. However, on a NoC like miniature system that is considered in this work, the overhead and processing resources are limited and play a pivotal role. As such, a direct adoption is inefficient and leads to large overhead and performance penalties.

16.3 Attack Model for WiNoCs and WiNiPs

Here, we discuss the attacks and their manifestations on the adopted WiNoC and WiNiP architectures considered in this chapter. Several security and privacy attacks have emerged in the recent times on multicore processors [33]. In this chapter, we consider persistent jamming-based DoS attacks and ED (arising internally as well as externally) on the wireless interconnection architecture used in WiNoC and WiNiPs. In the presence of a persistent DoS jamming attack either from an external or internal attacker, there will be interference among the attacker and the legitimate transmitter. This interference will cause high error rates due to interference noise. Moreover, as the attack is over a relatively long period of time, it will cause errors in contiguous bits of flits resulting in burst errors. Over the duration of attack, these errors will span multiple flits and therefore, cause burst errors in multiple consecutive flits of a packet. On the other hand, burst errors in both wired and wireless NoC links can happen as a random event as well. Burst errors can also be a result of power source fluctuations, ground bounce or crosstalk [34]. However, the burst errors due to random events such as crosstalk will be relatively short lived, typically, a single clock cycle, due to the data transition pattern in that cycle. On the other hand, burst errors resulting from persistent jamming could be sustained for longer duration, as a short DoS attack is not an effective attack.

A few burst errors caused by a short-lived DoS can be corrected/detected by a burst error correction/detection (BEC/BED) code depending on its correction capability. In the absence of such a BEC mechanism, a request for retransmission can be sent in case of erroneous flits from the upper layers of the NoC protocol stack. Therefore, to be truly effective as an attack, the jamming has to last for relatively long duration to cause enough flits to be in error such that the existing BEC mechanism either cannot correct it or retransmission requests are prohibitively expensive due to a potentially large number of requests. Therefore, we need a mechanism to detect jamming-based DoS attacks and distinguish it from a random burst error. We consider attacks either from a single external attacker or a single internal hardware trojan based attacker which affect one or more WIs in the communication infrastructure. The jamming signal can be caused by an external source equipped with a RF transmitter tuned to the spectral band used in the on- and off-chip wireless communication. Another likely scenario is when a particular WI already existing in the WiNoC/WiNiP is affected by a hardware trojan which forces the WI to ignore the contention-free MAC mechanism and continue to inject traffic from the transmitter of the WI. This constitutes an internal attack. We do not assume that an additional WI is placed as a trojan in the chip as that would be relatively easy to detect. Rather, one of the existing WIs is infected by the trojan and ignore the MAC rules and create jamming even when it is not supposed to transmit over the shared wireless medium.

Despite ML being robust to random noises, it has been shown that ML techniques are vulnerable to crafted threats, termed as *adversarial samples* [35]. Adversarial samples exploit the sensitive features in the input or the ML model, adding noise o which can lead to misleading the output of the ML model [36, 37]. In similar nanner, in this work, we introduce an adversarial attacker who can attack the system y cognitively crafting the attack.

The first step to launch such an adversarial threat is to determine the model and/or parameters). This is performed through reverse engineering process by teratively sending in the data and obtaining the responses, similar to that in [38]. Once the reverse engineered model is built, then, the attacker tries to estimate the nodel and introduce the perturbations by incrementally increasing the noise to the nput features that are sensitive similar to [39] to evade detection or to induce false larms. In this chapter, we utilize Fast Gradient Sign Method (FGSM) attack [35] to raft such an adversarial attack. However, it needs to be noted that direct application f FGSM is not feasible, as it does not have a notion of relativity between individual eatures when crafting an adversarial sample. To combat such scenario, we introduce he relationship between different features such as number of errors not more than he total number of packets sent in the form of constraints.

In similar manner, we consider the ED attack can arise either from an external r internal attacker. In both cases, we assume that the attacker is passive, thus hard-o-detect and can receive any information communicated between different nodes n the wireless interconnection fabric by tuning into the unguided and unprotected vireless channel. For external eavesdropper, we consider a passive external receiver uned to the band used in the WIs with enough sensitivity capable of receiving the lata transmitted over the wireless channel. In the case of internal eavesdropper, the assive attacker receives data that is not addressed for it and routes it downstream o a malicious agent.

6.4 Security Framework for On-Chip Wireless NoC (WiNoC)

Iere, we describe the elements and design of a secure WiNoC interconnec-ion architecture focused on mm-wave wireless communication between various ores/modules within a multicore single chip environment. This adopted WiNoC rchitecture is generic and adopts elements from various designs over the past few ears.

6.4.1 WiNoC Topology

Ve consider each core in the multicore chip is connected to a NoC switch via a Jetwork Interface (NI). The switches are then connected by wired links forming a nesh topology. We adopt a mesh architecture for the wired NoC topology due to its ow complexity, ease to verify and manufacture due to uniformity of link lengths.

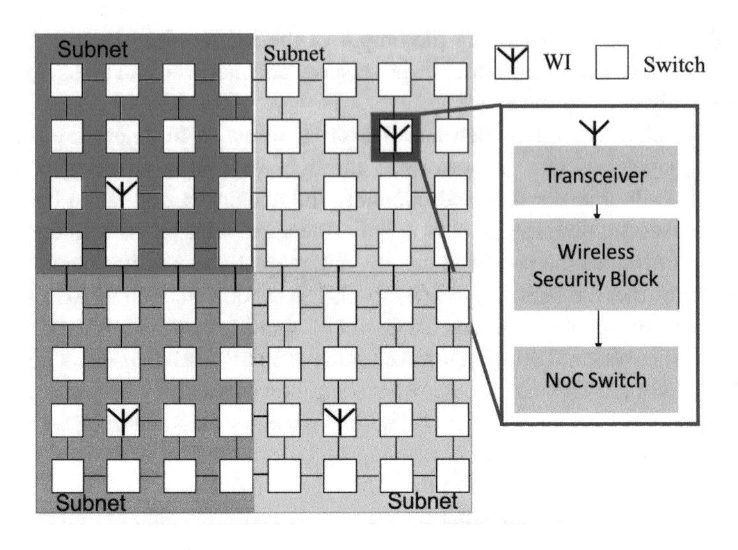

Fig. 16.2 WiNoC topology and WI components (inset)

However, other topologies such as torus or small-world can be chosen if require by the system design constraints. In addition to the wired links, a few NoC switche are equipped with an additional port connected to a wireless transceiver to access t the mm-wave channel, thus forming a hybrid WiNoC architecture. These switche are referred to as Wireless Interfaces (WIs).

Based on several previous works such as [40], we partition the mesh into multipl subnets to deploy the WIs among the NoC switches, as shown in Fig. 16.2. A central switch in each subnet is a WI to facilitate access to the wireless medium The selection of subnet size (or the number of WIs) offers a trade-off betwee performance of the WiNoC and area overhead of the WIs, which can be designe with system-level simulations. The underlying cores are not shown for the purpos of brevity. To ensure WiNoC security, we consider equipping the WIs with Wireless Security Unit (WSU), which can detect and protect against persisten jamming and ED attacks from both internal and external attackers. The WSU i embedded in the WIs so that it can process the data and detect the attack before th data passes downstream to other NoC switches. More details on WSU are presente in the next section.

16.4.2 Wireless Interconnect Overview

We consider using of on-chip embedded miniature antennas operating in the 60 GH mm-wave band unlicensed by Federal Communications Commission (FCC), whic

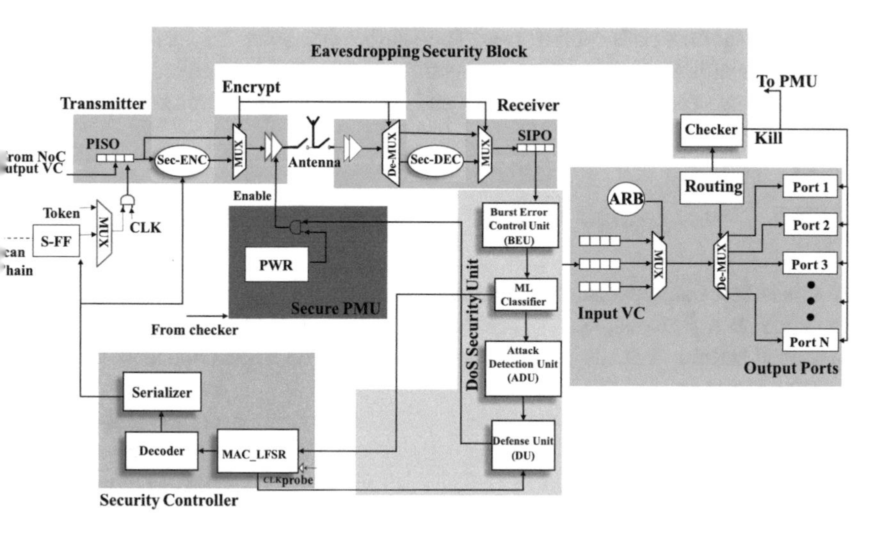

ig. 16.3 WSU architecture

an establish direct communication channels between the WIs. We intend that ne chosen antenna to be compact as well as non-directional, so that they can ommunicate with other WIs in all directions. We adopt the 60 GHz zig-zag antenna vith these characteristics from [40].

To ensure high bandwidth and energy efficiency, we adopt a transceiver design vhere low-power design considerations are taken into account [41, 42]. Non-oherent On-Off Keying (OOK) modulation is chosen, as it allows relatively simple nd low-power circuit implementation without the need for power-hungry carrier ecovery and high-frequency synchronization circuitry. Each WI is a combined ransceiver with a single antenna enabling half-duplex communication. Parallel data rom a NoC switch is serialized using a Parallel In Serial Out (PISO) register before ransmission and vice-versa after reception, where they are received into a Serial In 'arallel Out (SIPO) buffer. The PISO buffer receives data from the output virtual hannel (VC) of the transmitting WI while the SIPO sends the received data to the nput VCs of the receiving WI, as shown in Fig. 16.3.

To avoid non-scalable central arbitrations and power-hungry synchronization cross the chip and facilitate contention-free wireless channel access, we adopt a istributed wireless token passing MAC mechanism to grant access of the shared vireless channel to only the WI possessing the token. Each WI can only occupy the oken for a pre-determined maximum time that is optimized based on system-level imulations.

We consider using a forwarding-table based routing algorithm over pre-computed hortest paths along a Minimum Spanning Tree (MST) determined by Dijk-tra's algorithm. Consequently, deadlock is avoided by transferring flits along the extracted shortest path routing tree. The routing decisions are made locally based on the forwarding table for determining the next hop and is done only for the heade flit, reducing computing requirements and maintaining global routing information.

16.4.3 WSU Design for Secure Wireless Communication

In this subsection, we discuss the mechanism to secure the adopted WiNoC agains the DoS and ED attacks. To enable the proposed secure WiNoC, each WI i equipped with a WSU to sustain functionality of the interconnection fabric even under jamming-based DoS attack.

The proposed WSU shown in Fig. 16.3 has two main components, the DoS Security Block and the ED Security Block. The DoS Security Block consists of a Linear Feedback Shift Register (LFSR) called MAC-LFSR, a Burst Error Contro Unit (BEU), an Attack Detection Unit (ADU), and a Defense Unit (DU). In the normal mode of operation, the data flits are received at the SIPO buffer of a NoC switch equipped with a WI. Upon reception of flits at the receiver's SIPO buffer flits are sent to the BEU. The BEU then detects a burst error and sends its output to the ADU. The BEU employs the BEC proposed in [34] to detect burst errors. The corrected flits after burst error correction are sent to the input VCs of the NoC switch to be routed downstream in parallel to the error related information (as discussed in the next subsection) being sent to the ML Classifier. This removes the DoS detection mechanism from the critical path of the data transfer. The ADU further comprises of an intelligent unit which uses an ML classifier, and an attacker detection unit. The ML classifier is responsible for detecting if the system is under attack based on the input it receives from the BEU. More details of the ML classifier are presented in the next subsection. If the ML classifier detects an attack as opposed to a random burs error, it asserts a flag to the ADU. The ADU receives the input from the ML classifie and determines if the attack is internal or external as discussed in Sect. 16.4.4.3 Based on the kind of attack, corresponding security measure is chosen.

The Eavesdropping Security Block consists of a Linear Feedback Shift Registe (LFSR) called code-LFSR, security encoder (Sec-ENC), and security decode (Sec-DEC) as shown in Fig. 16.3. The code-LFSR generates code words pseudo randomly which are used to encode the data flits using parallel bitwise XOR gates in the Sec-ENC. The Sec-DEC in the receiver uses the same code to XOR the received flits to recover the original data. To protect against brute-force eavesdropping the code-LFSR is clocked periodically to generate a new code. This will protect agains a suspected external ED. To protect from internal ED, each packet that accesse the wireless interconnect will have the address of the intermediate WIs embedded in a header field in addition to its final destination. A rule-based checker in the W compares the address of the target WI of the received packet headers with that of the WI to verify if it is a legitimate packet or if it is being eavesdropped. The operation of these components are elaborated in the next subsections.

6.4.4 *DoS Attack Detection and Defense Mechanism*

n order to detect a DoS attack, various ML classifiers have been deployed. First, we resent the details of ML classifier and modeling of DoS attack, followed by DoS ttack detection and activated defense in the event of DoS attack detection.

6.4.4.1 Machine Learning for Attack Detection

s aforementioned, the considered attacks in this chapter primarily result in causing ontinuous sustained burst errors in the flits (data corruption). In the proposed ViNoC, the output of BEU, which is the number of burst errors within a block, s fed to an ML classifier to detect and differentiate attacks. We experimented vith multiple ML classifiers to evaluate the robustness and efficiency of attack letection in the proposed system. The different ML classifiers considered here are: nulti-layer perceptron (MLP), support vector machine (SVM), k-nearest neighbors KNN), Decision tree (DT), and J48 classifiers. The rationale for experimenting vith different classifiers are: (a) there exist no unique classifier that has "perfect" ield; (b) different classifiers have different resource requirements and performance accuracy, and latency) and (c) the chosen classifiers represent different branch of AL, thus representing a wide spectrum of ML classifiers. The ML classifiers in the DU utilizes an offline learning with runtime inference to alleviate the complexity, rocessing overheads and facilitate faster inference (attack detection). The ML lassifiers output a flag signal when an attack is detected. The ML classifier does not end any data to the switch buffers. This prevents ML classifier from creating any DoS attack. In addition, we also assume that the detector unit along with all other ecurity blocks is designed, verified, and tested in a secure environment, similar to ecure IC design, thereby preventing any HT insertion in the security blocks.

In order to train the ML classifier, the attacks mentioned in Sect. 16.3 are leployed on a WiNoC (shown in Fig. 16.2) with no security mechanisms deployed. cycle-accurate NoC simulator was modeled to operate in one of the three modes: ormal, random burst errors, and DoS attack. In the normal mode, the wireless nterconnects are assumed to work with the reliability level determined by the peration of the transceiver and their operating thermal noise. This type of noise is hown to result in a random Bit Error Rate (BER) of 10^{-10} or less [42]. The second node (random burst errors) is modeled with higher BERs as the burst errors are ontiguous bits of flits. BERs of 10^{-5} are used in this case [34]. Lastly, under DoS ttack, a high BER of 0.5 is assumed as for identically and independently distributed iid) data bits even a very high power jamming signal can cause errors only half f the time on an average. This is because the adopted modulation mechanism in hese wireless interconnects is OOK, where the data bits are represented as presence r absence of transmission. Therefore, a persistent jamming signal will only cause rrors when the transmission is supposed to be absent, which can be assumed to be alf of the time for iid data.

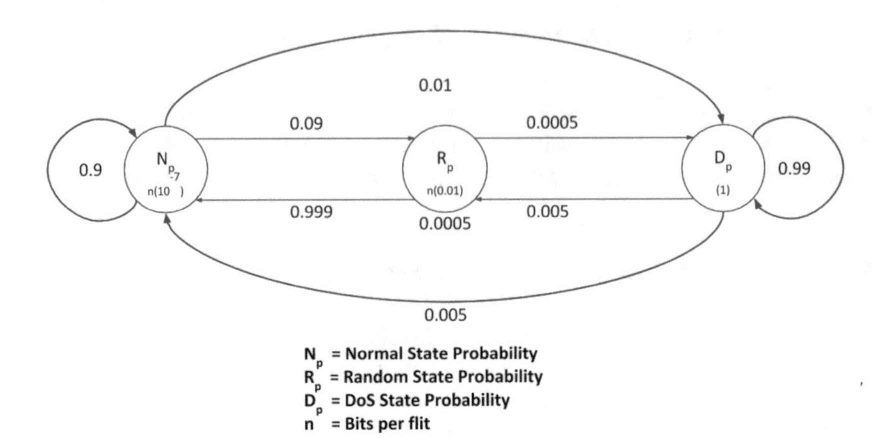

Fig. 16.4 Markov chain process with different operating states

The simulator is modeled to create flit errors based on these BER information which are then assumed to be detected by the BEU. The simulator is made t operate in one of the three modes dynamically by using a Markov Chain drive process, as shown in Fig. 16.4. The manifestation of the DoS attack is considered t result in the same kind of burst errors for both internal and external attackers. Th probability of staying in the attack mode, when already under attack is considere high, as a persistent jamming attack is effective only when it is sustained for a lon duration. The probability of staying in a random burst error mode when already in it is modeled as low as random burst errors are short-lived phenomena. The probabilit of transition into normal mode from a random burst error mode is therefore high The specific probability values can be altered to model any particular scenario This observed data (number of errors, flits transmitted and received) along with th operating mode as encountered in each WI is used to train the ML classifier at tha WI. As the duration of the individual mode are determined by the Markov Chai randomly, each specific instance of the mode has varying duration, resulting in diverse training data set.

For the inference, i.e., attack detection, the ML classifiers are fed runtim information such as whether a flit is received or not, and whether a burst error i detected or not to detect the mode of operation of system. Training of ML classifier is performed with a hundred thousand cycles of data.

16.4.4.2 Strengthening Attack Detection with Adversarial Learning

In order to craft the adversarial perturbations, we consider a functionally revers engineered ML classifier, i.e., a neural network with θ as the hyper-parameters, as the input to the model (communication information such as number of packet transmitted, packet errors), and y as the output for a given input x, and $L(\theta, x, y$

as the cost function used to train the neural network. Then the perturbation required to misclassify the ongoing communication is determined based on the cost function gradient of the neural network (in this case). The adversarial perturbation generated based on the gradient loss, similar to the FGSM [35], is given by

$$x^{adv} = x + \epsilon sign(\nabla_x L(\theta, x, y)) \tag{16.1}$$

where ϵ is a scaling constant ranging between 0.0 and 1.0 is set to be very small such that the variation in x (δx) is undetectable. In case of FGSM the input x is perturbed along each dimension in the direction of gradient by a perturbation magnitude of ϵ. Considering a small ϵ leads to well-disguised adversarial samples that successfully fool the machine learning model. In contrast to the images, where the number of features is large, the number of features in our environment, i.e., flit errors is limited. Thus the perturbations need to be crafted carefully and also ensured that they can be generated during runtime by the applications. For instance, a flit error higher than transmitted flits makes no sense and is impossible to implement. Hence, we include a lower bound on the adversary values that can be predicted.

Once the adversarial pattern is predicted or determined, the attacker crafts the attacks through induced errors or by spacing the attack in time so that the errors split over time as predicted. The attacker internal or external, is modeled to display the adversarial behavior as discussed above to create errors in the communicated flits only when the adversarial model allows rather than assuming constantly high BERs when in the attack state of the Markov Chain as in the previous subsection. Therefore, even when the simulator is in the attack stage, BERs may not be consistently high making the attack more sophisticated and decrease the likelihood of a detection. In order to defend against such threats, we incubate a hardener unit. The hardener unit predicts the adversarial samples, similar to the aforementioned attack and updates the ML classifier model through adversarial training [43]. The hardener is deployed off-chip (on a connected system), but it updates the weights of the ML classifier to robustify against the adversarial threats. One can argue that the adversarial training is inefficient in defending against wide range of crafted threats and large range of perturbations. However, in this given context, crafting too many vivid range of threats is not feasible due to the correlation between features. Further, large variations or perturbations can be easily caught, as large deviation in the errors clearly indicate the presence of anomaly.

16.4.4.3 Attack Detection Unit Operation

In this section, we discuss the logic block that is designed to distinguish an external attacker from an internal one in the proposed secure WiNoC, ensuring different defense mechanisms are activated. The detector takes as an input the signal from the ML classifier that detects the occurrence of a jamming-based DoS attack. On the detection of an attack, the ADU activates the probe mode, in which all the WIs operate according to the token based MAC mechanism controlled by the MAC-

LFSR. The MAC-LFSR is enabled when the ML classifiers of any of the WIs detects an attack. They send this single-bit signal to the MAC-LFSR. We consider the MAC-LFSR to be located in a secure part of the chip and it is reasonable to assume that it is not affected by the wireless jamming attack model assumed here. The MAC-LFSR then grants access to the wireless medium to each WI in a pseudo-random pattern. A probe-clock triggers the MAC-LFSR to generate the encoded grant signal which is decoded to create a one-hot signal which is sent over a pipelined link to the transmitters of all the WIs. A parallel-load shift register is used to serialize this one-hot signal. The token register in each WI is converted into a scan Flip-Flop. At each transmitter this signal is ANDed with the power supply routed from a secure Power Management Unit (PMU) [44], which is not vulnerable to the wireless attacks, to regulate the power supply to the transmitter. Thus, only one transmitter transmits data flits over the WI in one instance.

The very first signal is initialized as an all-zero signal to disable all WIs from transmitting. In this case, if any of the WIs still receives wireless transmission it implies that the jamming source is an external attacker as none of the internal transmitters are powered on. The probe mode is then terminated and the decision is sent to the defense block for appropriate action. However, if in this case, there is no RF transmissions received, the MAC-LFSR progresses to further probing by cycling through the MAC-LFSR, where only one transmitter is powered on in each cycle. In these cases, where the enabled WI is not the internal attacker, there will be interference in received flits at the WIs due to continuous jamming from the attacker. Only in the case where the MAC-LFSR enables the attacker there will be no interference and correct reception will be received at the WIs. The ID of this WI is then passed to the defense block. The data packet that each WI sends in the probe mode can be a pre-programmed pseudo-random data that can be distinguished from random bits when the WIs receive this packet. It could also be same or different for each sending WI. Alternatively, these can be generated by a local LFSR and compacted into signatures at the receiving WIs to match with a known signature. So, the algorithm declares the WI that is enabled by the MAC-LFSR in which case correct data packet or signature is received, as the internal attacker. Figure 16.5 depicts the ADU process.

16.4.4.4 WiNoC Defense Mechanism Against DoS Attack

The ADU passes the address of the WI that is determined to be the attacker to the DU. In case the attacker is an external agent, the address is an all-zero string. If the address received indicates an external attacker, the DU sends a signal to the secure PMU to shut down all the WIs and also update the routing tables of the WIs such that the wireless links are not used for data routing. These updates of the routing tables can be done without hardware overhead as these alternative values can be pre-computed for each WI for the alternative shortest path routing when the WIs are not available and stored in the OS. Therefore, in this case, all the WIs are disabled and data is routed via the wired links, eliminating the advantage of the wireless

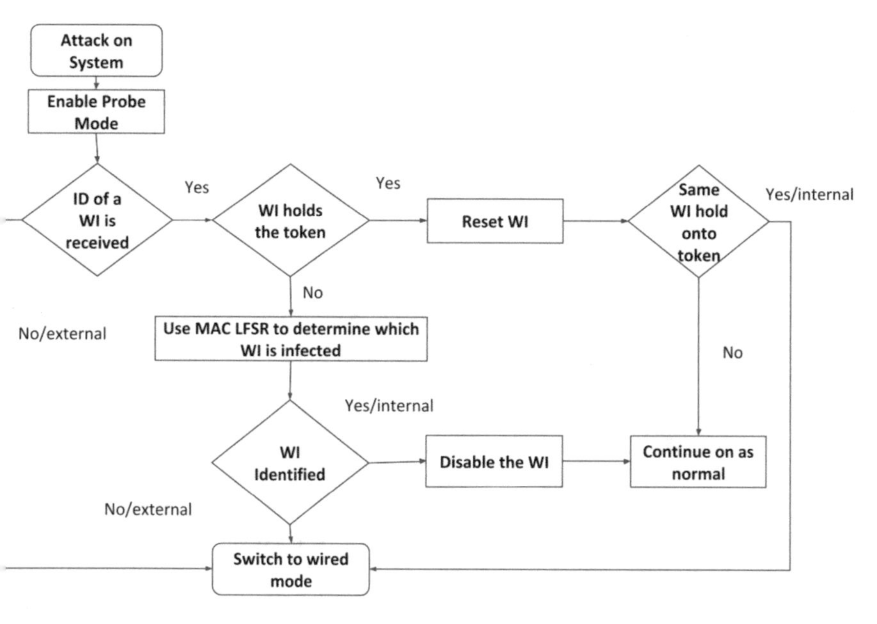

'ig. 16.5 Jamming-based DoS attack detection and defense mechanism flow diagram in WiNoC

nterconnections. In order to benefit from the wireless interconnection, the probe node is periodically activated by the ADU to check if the attack has stopped. In his case the use of the WIs can be resumed by using the PMU and by updating the outing tables.

If the address passed on to the DU indicates the address of an internal attacker, he DU sends a signal to disable only the power supply to the indicated WI and pdates the routing table of its NoC switch to not use the WI. In this way, only the rojan infected WI is disabled and the rest of the WIs continue to use the wireless nedium. Unlike the previous case, as the attacker is an internal hardware trojan, the ssociated WI may never be safe to use again and therefore will be permanently lisabled using the PMU and quarantined. The core or cores attached to the infected VI will continue to route their packets over wired links using the NoC switch as the IT does not influence the wired part of the NoC in the threat model that we have onsidered in this work.

16.4.5 *Defending WiNoC Against Eavesdropping*

n this section protection against passive eavesdropping has been discussed as it s relatively easy to launch and very difficult to detect. We discuss our protection trategy against both an internal and an external passive eavesdropping attack from single attacker.

16.4.5.1 Defense against External Eavesdropper

For a single agent external passive ED, it is complex and nearly impossible to detect with the available resources and capability of the WiNoC. This is because there may not be any change in the behavior of the overall system during such an attack. Furthermore, we assume that the wireless NoC communications will have enough power and/or the eavesdropping attacker is sensitive enough to pick up the transmission and decipher the information. This is in-line with real world eavesdropping attack scenarios. The attacker needs to be equipped with a wireless receiver tuned to the wireless channel used in the WiNoC and have basic depacketization functions which are extremely simple and low-overhead in NoC and therefore easy to instantiate.

In order to address this threat, we propose to deploy a simple XOR-based data scrambling approach. The header flits are not encoded to enable routing as in traditional networks. The rest of the flits, which are the body flits, are XORed with a code word from each WI and transmitted over the wireless channel. We propose to use the same length of the code word as that of each flit with parallel bitwise XOR gates to reduce, the delay in communication. Therefore, the bandwidth is not affected as the number of bits transmitted for a flit does not change. At the receiver the same code will be used to XOR the received flit to receive the uncoded data back. In general, unless an eavesdropper has the same code, it cannot decode the received flit. However, with enough time, an eavesdropper can determine the used code with brute-force trials. Therefore, such schemes continuously change the code used by each transmitter. In order to change the code periodically we generate the codes from an LFSR in each WI. The LFSR can be of the same length as that of the flit size (in number of bits). If a higher degree of pseudorandomness is desired then a larger LFSR can also be used. In this paper, we consider the LFSR to be of the same size as that of the flit. We refer to these LFSRs as code-LFSRs. The enable signals for these code-LFSRs to generate a new code can be routed from the Security Controller through the serializer in the normal mode of operation (not in the probe mode when a jamming attack has been detected by the ADU). The special all-"1" code can be used to signal all the code-LFSRs to change the code they are creating to the next pseudo-random code in all the transmitters. All transmitters have the same code-LFSR which is shared with the receivers collocated with the transmitter. This code is used by the security encoder (Sec-ENC) and the security decoder (Sec-DEC) in each WI as shown in Fig. 16.3. Therefore, the code used by all the WIs is same at all times. Each transmitter does not need to have a unique code-LFSR as this mechanism is for protection from an external eavesdropper and not an internal one.

16.4.5.2 Defense Against Internal Eavesdropper

We model the internal ED as follows: we assume that one of the WIs is an internal eavesdropper. The attack model is such that this WI is either always or intermittently processing data packets transmitted over the on-chip wireless medium which are not meant for it. Therefore, the attacker WI can receive and leak (to the outside) data that is not meant for it. This can be achieved with a HT which is embedded at the wireless input port which does not allow the port to drop a packet that is not addressed for the particular WI. As this is an internal attacker, we propose a mechanism to detect such an attacker and to protect the WiNoC once such an attack is detected.

In order to detect the internal eavesdropper, power consumption based detectors could be deployed. However, deploying such power measurement units incur additional silicon footprint as well as computational overheads even in the absence of attacks. Therefore, we propose equipping the input port of each WI with a low-complexity rule-based checker. Moreover, as a WI transmits a packet over the wireless medium it will embed the address(es) of the recipient WIs which may then pass the packet further downstream to the final destinations. The rule checker will match the WI address(es) of the header with the local address of the receiving WI. If there is no match, the WI should not pass this header to any downstream port and kill the packet to avoid packet duplication in the WiNoC. However, if this WI sends this packet to any outgoing port including the local port to the core, the checker raises a flag and this triggers an action in the secure PMU. The PMU then powers down the particular WI to prevent it from eavesdropping further.

The location of the checker is critical in order to reduce the overheads and delay in detecting such an eavesdropping. As the location of the checker should be downstream from the logic block that is supposed to flush out a packet not meant for the WI we propose to implement this checker after the input arbiter of the WI switch. In this way, an eavesdropped packet can be detected if it is not flushed out of the input buffers and progresses to the next step of routing. As during routing the destination address of the header flit will be parsed anyway, it can also be used in parallel to check for eavesdropping. This will minimize the additional overhead of this checking. Moreover, in this way we do not delay the routing of the header of all legitimate packets due to this checking. If the result is positive (detected ED) then the flag is simply sent to the PMU which will prevent further reception of packets at that WI. In addition, the flag is also sent to all the output ports of the WIs to flush out the current packet when routing is completed to prevent information leakage of that packet. This ensures quick reaction on detection of an internal eavesdropping. The flushing of the input or output buffers is achieved by activating the reset on the buffers without the need for any additional circuitry.

However, there are some exceptions to the proposed internal eavesdropper detection and defense. For instance, when the packets are broadcast to all the WIs or the eavesdropper WI happens to be one of the addressees in the packet header, the proposed defense mechanism will fail and will have to rely on mechanisms at higher layers of the system such as the application layer.

16.5 Experimental Results and Analysis

In this section we present the evaluations of the proposed secure WiNoC and the simulation tools used to evaluate it.

16.5.1 Simulation Setup

Simulation of wireless interconnection requires a combination of multiple simula-tion tools. We use ASIC design flows with Synopsys Design Compiler using 65 nm CMP standard cell libraries (https://mycmp.fr/) to model the digital parts of the WiNoC such as NoC switches and the WSU. The BEU encoder and decoder are implemented as two pipelined stages in the WIs to accommodate their delay [34 thereby maintaining the pipelined communication of the WiNoC. Each switch has three pipeline stages implementing backpressure flow control [45]. We conside each input and output port of a switch including those with the wireless transceiver to have 8 VCs with a buffer depth of 4 flits for all the architectures considered in this work. We consider a packet size of 64 flits with a flit size of 32 bits in our experiments. A uniform random traffic distribution is assumed with self-simila temporal behavior at maximum injection load of 1 flit/core/cycle to evaluate the NoCs under worst case traffic. All the digital components are driven by a 2.5 GH clock and 1 V power supply. The delay and energy dissipation on the wireline links is obtained through Cadence simulations considering the specific lengths of each link based on the NoC topology assuming a 20 mm×20 mm chip. The adopted wireless transceiver circuits consume 2.075 pJ/bit at 16 Gbps in 65 nm technology [41, 42].

The adopted antenna has a 3-dB bandwidth of 16 GHz [40]. The characteristic of the transceivers, routers, and wired links are annotated into a system-leve cycle-accurate simulator to evaluate the performance of the WiNoC in presence of DoS attacks and the proposed defense mechanism. The simulator monitors the progression of flits on a cycle-by-cycle basis accounting for all flits that move or are stalled. We evaluate the proposed system in terms of average packet latency, peak bandwidth per core and average packet energy. Average packet latency is defined a the number of cycles required for a packet to reach its final destination after being injected on an average. Peak bandwidth per core is defined as the number of bits received per core of the WiNoC per second with full injection load. Average packe energy is the average energy dissipated by a packet to be transferred to the fina destination over the WiNoC fabric through switches, wired and wireless links.

16.5.2 *ML Classifier Performance for DoS Attack Detection*

Table 16.1 presents the accuracy and robustness (in terms of precision, recall, and the area-under-curve (AUC) metrics) of different ML classifiers when deployed to detect the DoS attacks. Higher the value of accuracy and robustness metrics, better will be the performance.

One can observe from Table 16.1, among different classifiers, KNN achieves high attack detection accuracy of nearly 99.87%, which is higher than the other techniques. We anticipate this behavior, as no assumptions are made regarding the data during the training phase of KNN. For the KNN, a Euclidean distance function is employed with $k = 1$ due to its lower complexity. Therefore, KNN is deployed in this ADU for attack detection. Though SVM displayed high accuracy, it is observed in experiments that it is not able to detect sporadic variations such as spontaneous random errors, and is hence not the best option. For the neural network (MLP) a single hidden layer with 20 nodes is utilized. It can be argued that the hyper-parameters of the ML classifiers can be tuned to improve the performance, however, optimizing the ML classifiers is not the focus of this work. To compare the ML classifiers with a heuristic method, we consider a threshold-based approach. As shown in Table 16.1, the threshold-based mechanism is not as accurate as the chosen machine learning (KNN) approach. Despite having low latency, threshold-based approach has higher area and power consumption due to the involved floating point computations and comparisons, as shown in Table 16.3. In this threshold-based approach, two thresholds are necessary, to separate between the attack mode, burst error mode, and normal mode. The thresholds are computed based on the same data that was used to train the machine learning algorithms. The threshold between the attack mode and burst error mode is chosen to be equidistant from the average number of erroneous flits in burst errors and jamming induced errors. Likewise, the threshold to separate the burst error mode from the normal mode is chosen to be equidistant from the average number of flit errors in burst mode and normal mode. For all the employed classifiers, the inputs (flits received, flits at error, and flit error ratio) and output classes (normal, random error, DoS error) are same.

Table 16.1 ML classifiers' performance for attack detection

ML classifier	Accuracy (%)	Recall	F-score	AUC
MLP	47.86	0.48	0.65	0.47
SVM	98.96	0.98	0.98	0.99
KNN	99.87	0.99	0.99	0.99
DT	52.46	0.52	0.69	0.53
Thresh	94.55	0.92	0.92	0.95

16.5.3 Detection Accuracy with Adversarial Attacks

We also evaluate the impact of the crafted adversaries on the traditional ML-base threat detectors and the impact on the enhanced detector, i.e., hardener unit traine with adversarial samples. Table 16.2 presents the performance of the traditional an hardener detectors. As one can observe that under normal threat conditions, th ML classifier (KNN) is able to achieve an accuracy of 99.87%. However, under th adversarial scenarios, the accuracy of the same KNN drops to 85.67%. A simila degradation in terms of performance is observed in other metrics too. Subsequentl through the adversarial training an improvement in the accuracy to 95.95% i observed with a similar trend in other performance metrics. One can observe that th performance with adversarial training makes the classifier to have lower accurac compared to the normal classifier. However, it should be noted that in this case th system is under attack from a smarter attacker which has adversarial knowledge o the system and that without the adversarial training the WSU would be much les accurate.

In addition to performance benefits, ML classifiers also incur silicon and resourc overheads. Table 16.3 presents the incurred overhead in terms of area, power, an delay of the deployed ML Classifiers. The characteristics of the various classifier are obtained from post-synthesis RTL models are synthesized using 65 nm standar cell libraries, as mentioned earlier. As the KNN Classifier has the highest accurac and lowest area and power consumption, we adopt the KNN Classifier for th evaluation of overall system. Although, the delay of the KNN classifier is nc optimal, we choose KNN for attack detection, as the ML Classifier is not in th path of data transmission of the WiNoC, as shown in the proposed secure wireles architecture in Fig. 16.3. One can question the impact of false Negatives, i.e., Do not detected despite its presence. This scenario can lead to a DoS attack. Howeve for any classification technique, false negatives/positives are inevitable. Howeve the deployed classifier has shown robustness against such scenarios (0.13% fo employed KNN, smaller compared to others), which indicates a high detectio capability and low probability of misclassification.

Table 16.2 AML performance evaluation

	Accuracy (%)	Recall	F-score	Precisio
After attack	85.67	0.94	0.86	0.79
W Adv. Training	95.95	0.97	0.97	0.97

Table 16.3 Implementation overheads of ML classifiers

Classifier	Area (μm^2)	Power (μW)	Timing (ns)
MLP	34,448.79	6299.3	0.41
SVM	5412.01	8076.1	0.37
KNN	105.28	27.075	0.56
DT	127.32	41.12	0.23
Thresh	24,262.63	22,515.2	0.07

able 16.4 WSU implementation overheads

Metric	MAC LFSR	Decoder	Scan FF	ML Detector	BEU	Code-LFSR	Sec-ENC/DEC	Address Checker	Wireless Tx-Rx
Area (μm^2)	41.08	37.96	15.08	105.28	4357.5	326.4	26	219.49	200,000
Power (mW)	0.0594	0.0147	0.0247	0.027	0.0047	0.48	0.0361	0.859	36
Delay (ns)	0.16	0.09	0.07	0.56	0.80	0.16	0.07	0.24	0.0625

6.5.4 Performance of the WiNoC in Presence of DoS Attacks

his section describes the performance of the proposed WiNoC in presence of DoS ttacks from internal and external attackers. We consider a WiNoC with 64 cores in 20 mm × 20 mm die interconnected with a wired mesh and overlaid with 4 WIs at he central node of each subnet of 16 cores. The WiNoC with embedded security s also compared with an equivalent 64 core wired mesh in terms of performance. he characteristics of the individual blocks in the WIs of the secure WiNoC are hown in Table 16.4 and used in the simulation platform for the system-level valuations. From Table 16.5, it is clear that the WiNoC outperforms the wired nesh in terms of peak bandwidth, latency, and packet energy due to the low power vireless shortcuts between distant cores, which reduce the average path length and lso use a low-power wireless medium for communication. This performance can hange depending on the number of subnets and WIs deployed on the WiNoC [40], owever, that study or optimization is not within the scope of this work. The security neasures developed in this chapter will be effective irrespective of the number of VIs for the assumed attack model.

Next, it can be seen that in presence of an external DoS attack, the performance f the WiNoC is similar to that of the wired mesh. This is expected, as on detecting n external attacker, the WSU deactivates all the WIs leaving wired links as the nly medium of communication for the purpose of security. On the other hand, vhen the attacker is an internal agent, only the infected WI is disabled, retaining ne advantage due to the presence of the rest of the WIs. Thus, in the case of an nternal DoS attack, a degradation of $<3\%$ in communication bandwidth compared WiNoC without any attack is achieved. That is why the ADU is an important esign element to distinguish an internal attacker from an external attacker.

6.5.5 WiNoC Performance Against Eavesdropping

n this section, we evaluate the performance of the WiNoC in presence of eavesdrop-ing attacks. For external eavesdropping, we adopt the XOR-based data scrambling pproach. The overheads of the additional code-LFSR and Sec-ENC/DEC is shown n Table 16.3. Due to the parallel XOR gates scrambling all bits of the body flits

Table 16.5 Analysis and comparison of WiNoC performance under proposed attacks.

	Wired mesh	WiNoC with 4 WIs	WiNoC external DoS	WiNoC internal DoS	WiNoC external ED	WiNoC internal ED
Bandwidth/core (Gbps)	26.4	30.4	26.4	29.5	26.5	29.5
Latency (Cycles)	395.96	286.80	396.00	319.00	299	319.00
Packet energy (nJ)	100	61	101	78	63.06	78

in parallel, the delay of the encoder or decoder is very low, minimally affecting the packet latency. The delay of the code-LFSR is not in the path of the data, therefore, it does not impact the packet latency. Due to the adopted light-weight scrambling approach, the impact of a threat of external eavesdropping is negligible on the performance of the secure WiNoC.

In case of internal eavesdropping where the rule checker is able to detect the attack, it will disable the infected WI and therefore, that WI will neither be able to send nor receive packets over the wireless interconnections. Moreover, the checker will add additional overhead albeit really marginal, as shown in Table 16.4. Therefore, the performance will degrade compared to the system with no attack, as shown in Table 16.5. Due to the disabling of the infected WI the overall performance of this system is similar to the case of a detected internal DoS attack as in that case, the system disables the attacking WI as well.

16.6 Security Framework for Multichip Systems with Wireless Network-in-Package (WiNiP) Interconnect

In this section, we describe how we can expand the proposed on-chip WiNoC security framework for MCMC systems using mm-wave wireless interconnect for inter-chip communication. As discussed earlier, we refer MCMC systems using mm-wave wireless interconnect for off-chip communication as WiNiP. Like WiNoC, for WiNiP, we consider the same attack model, ML and AML-based detection mechanism for persistent jamming-based DoS attack. However, unlike WiNoC, WiNiP has multiple chips in the system and inter-chip communication happens only through WIs located in different chips. Therefore, under external jamming attack, especially for MCMC systems, it is non-trivial to synchronize and inform all other WIs located at different chips about the presence of an adversary and continue inter-chip communication in such attack conditions. Moreover, the existing chip-to-chip wireless communication mechanism needs to be reconfigured as the attacker has invaded the current carrier frequency and the wireless communication channel has been compromised which is the only possible way to communicate with other chips in a WiNiP architecture. To address this issue, we develop a MCMC wireless communication protocol along with a reconfigurable MAC that can ensure robust and secure communication under internal and external persistent jamming attack for WiNiPs. Therefore, in this section of the chapter, we mainly focus on designing the reconfigurable MAC and communication protocol to design a sustainable WiNiP infrastructure for MCMC communication during jamming-based DoS attack.

16.6.1 Multichip Topology

To meet the increasing memory demands for current and emerging applications and mimic real MCMC architectures, we consider an MCMC system with multicore processors and in-package memory modules. The memory modules are connected to the edge cores through wired interconnect. Each tile in the multicore chips is composed of a processing core, a switch, L1 private cache, and a distributed shared last level cache (LLC). Tiles in each chip are connected with each other through a regular wired mesh-based NoC. For inter-chip communication, in each chip, we equip two NoC switches with WIs as shown in Fig. 16.6. Keeping the number of WIs minimum for inter-chip communication helps to reduce the communication overhead during jamming attack for our proposed approach.

However, a minimum of two WIs are necessary for each chip to ensure connectivity and reliable communication with the rest of the system even if one of them is compromised by an internal HT. A higher number of HTs within a single chip are assumed to be unlikely as it will make HT detection easier. Typically the footprint of HTs is minimal by design and hence we assume a maximum of a single HT per chip in our analysis. Although inter-chip communication happens

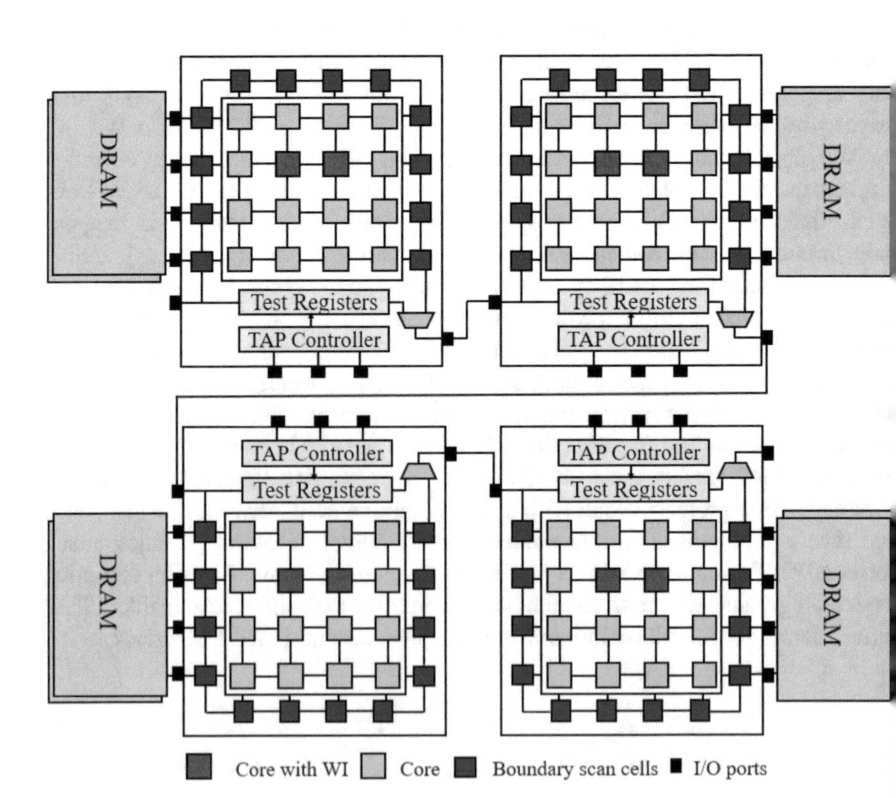

Fig. 16.6 A WiNiP topology in MCMC environment

only through the WIs in functional mode, the MCMC system is compliant to Joint Test Action Group (JTAG) test architecture where their boundary scans are daisy chained. We leverage this JTAG infrastructure for enhancing the security of MCMC system.

16.6.2 *Persistent Jamming-based DoS-Aware Reconfigurable MAC*

A wireless medium MAC mechanism enables a contention-free communication over the shared wireless channel among multiple transceivers. So far, no MAC has been proposed which is jamming-aware and can sustain communication in both normal and attack scenario. Therefore, we consider designing a reconfigurable MAC mechanism operating in two modes for sustainable communication even under persistent jamming attack. In the absence of persistent jamming attacks, we consider using a reservation-based MAC, termed as Normal MAC (NMAC) for MCMC communication. In NMAC, to get the channel access, each sender sends a non-overlapping reservation request to all the receivers encoded by a Common (C) code. Figure 16.8 shows the structure of the reservation packet and is discussed in details in the next subsection. As each receiver is equipped with same arbitration logic, each of them grants access to the same transmitter that gets the whole channel access at a time. In NMAC, as one sender gets the whole channel access, it ensures a contention free, high bandwidth off-chip communication.

The above mentioned mode of WiNiP communication is unaware of any persistent external jamming. Therefore, we switch the MAC to Pseudo-random Noise (PN) encoded Asynchronous Code Division Multiple Access (ACDMA) during external jamming attack and call it Attack MAC (AMAC). In this work, by ACDMA we only refer to using PN sequences and no other protocol overheads present in ACDMA communication in mobile cellular network. Data encoded with PN sequence is jamming and eavesdropping resistant because of the spread spectrum technology where the transmitted signals appear as noise to every receiver, except the one that has the PN code which was used to encode the data during its transmission. Therefore, any transmission not encoded with the same code appears as noise due to the weak cross-correlation, making this AMAC resilient to jamming. The PN codes used for ACDMA communication should have a strong auto-correlation and weak cross-correlation property. While maximal-length sequence (m-sequence) and Kasami sequence can be used to generate PN sequences, these sequences have worse cross-correlation property to Gold sequence [46]. Moreover, Gold sequence can also support more users than both Kasami and m-sequence. Therefore, we consider generating PN codes using Gold sequence.

We use the hybrid Transmitter-Common (TC) [47] PN code protocol to enable communication in AMAC mode where each transmitter has specific codes to encode packets they transmit and receivers have decoders for all channels to

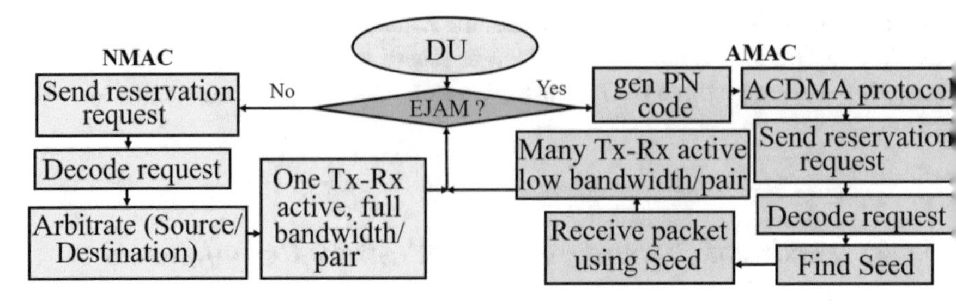

Fig. 16.7 Reconfigurable MAC flow diagram

be able to receive data simultaneously from multiple transmitters. The commo channel is used for arbitration and attack information propagation. We do not us AMAC in normal, attack free operation circumstances, as it reduces communicatio bandwidth of each link by its spreading factor. The focus of this paper is to ensur robust WiNiP communication in presence of persistent jamming attack on a high bandwidth WiNiP not to ensure high performance during such attack. Figure 16. shows the proposed reconfigurable MAC with the underlying operations.

16.6.3 Attack and Normal Mode Communication Protocol

Some of the key challenges of such jamming-aware hybrid MAC are to ensur proper switching and synchronous operation across MCMC system for both NMAC and AMAC modes with low overheads. In this section, we discuss our propose flow control that addresses these issues.

For high bandwidth off-chip communication during NMAC, each WI sends it reservation signal encoded by a fixed common PN code to all other WIs. The PN code being common to every WI increases the chance of corrupting the source destination addresses of the multiple simultaneous requests. Therefore, we propos a non-overlapping/non-interfering source-destination representation. As shown in Fig. 16.8a and b, each transmitter has its own slot to define its intended receivers The slots being non-overlapping and orthogonal do not create any interference with each other in their aggregate signal as shown by Fig. 16.8c. Hence, receivers ca arbitrate among multiple requests and grant the channel to a single transmitter in NMAC mode. The adopted arbitration logic considers channel access starvation fo WIs and provides priority to multi-cast traffic [48]. We re-use such non-overlapping signals to ensure synchronous operation even under jamming attack as discussed in the next paragraph.

When the MCMC system is under attack, all the WIs in the system change it MAC to ACDMA mode and continue their communication, but with a reduce bandwidth as the data is now encoded with PN sequence. We consider providing the highest priority to attack conditions which is indicated by the attack flag in

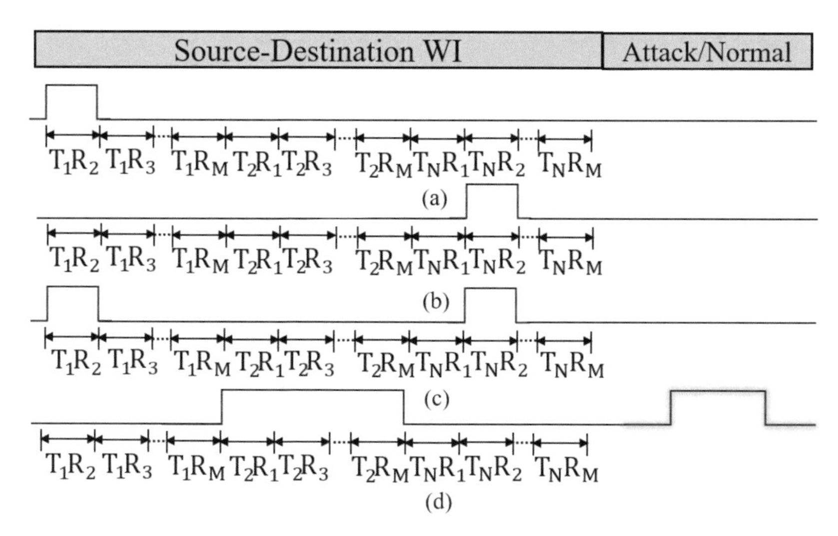

Fig. 16.8 Channel reservation and AMAC mode synchronization using attack flag

Fig. 16.8d. After detecting a potential external jamming attack as described in Sect. 16.4.4, a WI uses such signaling encoded by fixed PN code to inform other WIs during external jamming. All the other WIs in MCMC system after receiving the attack signal switches to AMAC mode simultaneously due to the priority in attack bit. The PN sequence generation and AMAC communication are described in the next subsections.

16.6.3.1 Selection and Generation of PN Code

The PN codes are binary sequences that appears to be random, but, they can be generated in a deterministic manner. However, to generate Gold sequence, two preferred m-sequences of the same length are required. In each of the transmitters, we configure two LFSRs according to the preferred polynomial pair and XOR their output to finally generate the desired Gold sequence. Figure 16.9 shows the LFSR configuration to generate a 32-bit gold code. Moreover, to generate a different PN sequence for each of the transmitter, we choose different seed values for each of the transmitters.

16.6.3.2 ACDMA Communication Mechanism in Attack Mode

During any persistent jamming attack, all the WIs in the multichip system change the MAC to ACDMA mode as discussed in Sect. 16.6.2. In ACDMA mode, the PN codes are managed using TC protocol. Before any transmission, similar to reservation assisted NMAC mode, the senders use a common PN code to send

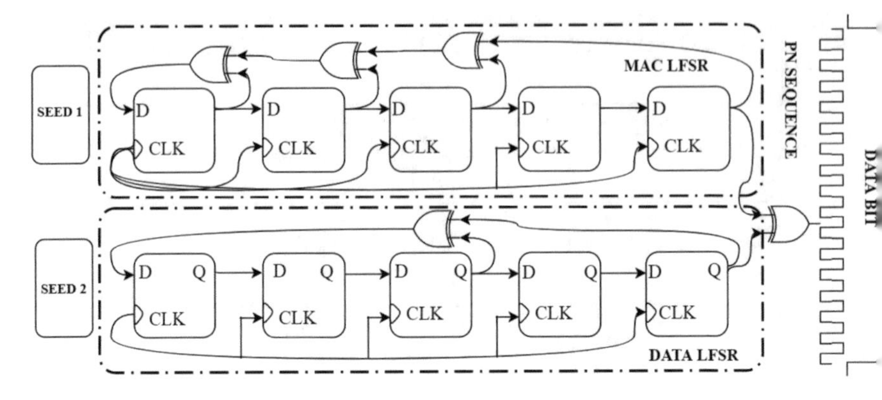

Fig. 16.9 PN code generation using Data and MAC-LFSR

non-overlapping send requests as shown in Fig. 16.8. However, based on the received requests, multiple receivers can grant access to multiple transmitters as now communication happens through different ACDMA channels. We consider the LFSR length to be 5 so that each PN sequence repeats after 32 cycles which is exactly the same time duration of a single bit of the baseband signal. Therefore each signal in a particular transmitter will be modulated by the same PN sequence. However, different transmitters use different codes of the same length because of having different seed values. Each receiver stores the seed values in a small tamper proof memory where the address of the seeds matches their transmitter address. Therefore, the receiver already knows which PN code to use for demodulation in a particular channel while granting the channel access through reservation requests. Hence, the additional delay for seed search does not have any impact on data transmission. To enhance security the seed values can be dynamically changed as commonly practiced in cellular networks [49]. The AMAC steps are also depicted in Fig. 16.7. The transmitter and receiver architecture will be discussed in the next section.

16.6.4 DoS Attack Detection and Defense for WiNiP

We utilize the same WSU unit and AML-based detection mechanism discussed in Sect. 16.4.4 for WiNoC. However, as WiNiP does not have any wired interconnection between multiple chips and chip-to-chip communication happens only through wireless channel, the defense mechanism for external jamming is completely different than WiNoC. Here, we describe the defense for the internal and external DoS attack in WiNiP.

DU implements different defensive measures based on the attack type. The ADU passes the address of the WI that is determined to be the attacker to the DU. If the address passed on to the DU indicates the address of an internal attacker, the DU sends the signal to disable only the power supply to the indicated WI and update ne routing table of its NoC switch to prevent the use of the WI equipped port. Moreover, as there are at least 2 WIs in each chip, the WI that is not compromised will inform other WIs in the MCMC system to update their routing table for the compromised WI. Now, all the incoming packets at the compromised WI will be diverted to the other WI on the chip via wired links. Hence, only the HT infected WI is disabled and other WIs continue to use the wireless medium.

In case the attacker is an external agent, the DU enables the detecting WI to send control signal as shown in Fig. 16.8d over the common reservation channel by setting the reservation flag on the transmitter side. The reservation channel like the other ACDMA channels is resilient to jamming. Moreover, as the signal has the attack flag set and is broadcast in nature, every WI in MCMC can switch the MAC mode to AMAC simultaneously and continue communication even under external persistent jamming attack. Moreover, for an external attack, the ADU periodically probes the system to restore the system to NMAC mode once the external jammer is no longer active.

16.7 Simulation Results

Here, we evaluate the performance of the proposed unified test and security of WiNiP under different attack scenarios. The section concludes with our study on the code length selection and system scaling.The simulation parameters are listed in Table 16.6.

We evaluate the proposed system in terms of average packet latency and average packet energy for application-specific traffic patterns from PARSEC and SPLASH2 benchmark suites. We consider a 4-chip system with 4 in-package memory modules. The core configurations in Table 16.6 have been used to extract the core-to-memory and cache coherency traffic for these applications when they are executed until completion using SynFull [51]. In order to map these traffic patterns to the MCMC environment, we consider multiple threads of the same application kernel running on the MCMC system where each processing core executes a single thread and

Table 16.6 Simulation parameters considered for WiNiP

Component	Configuration
System size	64 cores, Out-of-Order, 16cores/chip
Cache	32 KB (private L1), 512 KB (shared L2), MOESI
NoC router	3 stage pipelined 5 ports,0.078 pJ/bit(wired)
Total VC	4, each 8 flits deep, 32 bits/flit
Wired NoC links	32-bit flits, single cycle latency, 0.2 pJ/bit/mm
OOK transceiver	16 Gbps, 2.07 pJ/bit, 60 GHz, 2WIs/chip
CDMA	Encoder, decoder [50], 16 Gbps, 0.66 pJ/bit
HBM links	128 Gbps, 6.5 pj/bit
Technology	65 nm, 1 V supply, 1 GHz clock

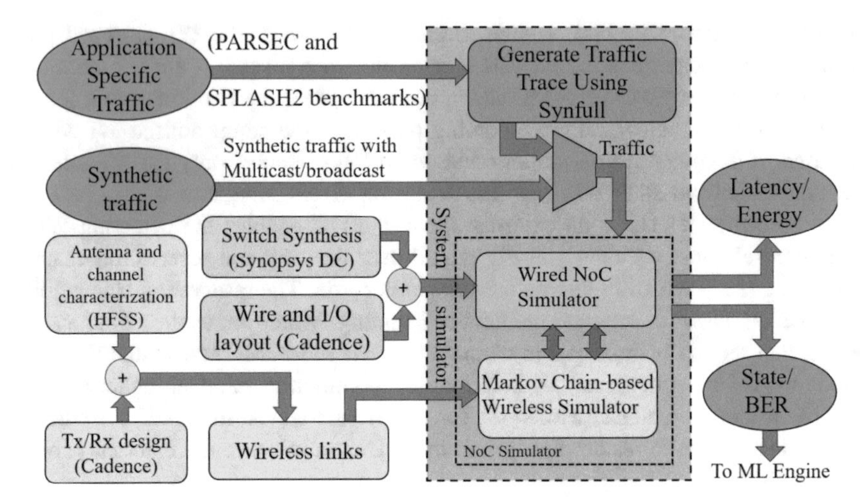

Fig. 16.10 Simulation process

the memory stacks are shared among threads. Figure 16.10 depicts the simulation process for the WiNiP architecture considered in this chapter. In the next section we discuss the performance of the WiNiP architecture in presence of internal and external jamming-based DoS attack.

16.7.1 Evaluation Under Persistent Jamming-Based DoS Attack

As the proposed architecture takes different defensive measures for internal and external attack, in this subsection, we study the impact of such measures on system energy and latency using application-specific traffic patterns.

16.7.1.1 Internal jamming

Disabling a compromised WI (CWI) in case of internal attack forces the incoming flits to change its route toward the remaining WI for chip-to-chip communication. Therefore, it introduces congestion for other WI nodes and increases latency as well as energy consumption. We consider three scenarios for our performance evaluation under internal attack. First, we consider MCMC system with one CWI for the entire (4 chip) system (1-CWI/sys). Second, we consider an MCMC system having one CWI per chip (1-CWI/chip). We compared these scenarios with a wired only MCMC system where the cores at the edges of each chip are connected to corresponding cores in the other chip with a mesh topology over high-speed I/Os. As we considered two WIs per chip, a system having more than one CWIs in a

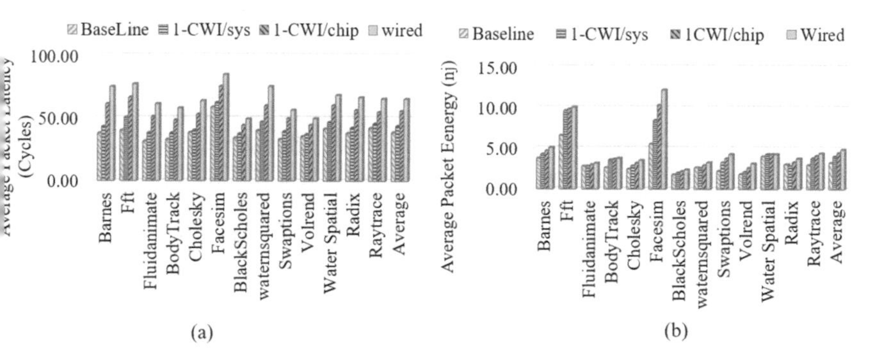

ig. 16.11 (**a**) Latency. (**b**) Energy under internal jamming attack for different MCMC systems

hip indicate a complete system failure and JTAG chain can be used for multichip ommunication with huge latency penalty. It can be observed from Fig. 16.11 that, lthough both the latency and energy consumption of the WiNiP increase with ncreasing number of compromised nodes, it is still lower than the wired MCMC ystem as each flit does not have to traverse through energy and latency-hungry NoC links and I/O modules. However, the average packet latency is 1.44× of the aseline system.

6.7.1.2 External Jamming

n the presence of an external persistent jamming attack, the MAC switches to ACDMA which ensures secure communication. However, it increases the average acket latency due to the encoding and decoding through PN sequence. Moreover, he runtime PN sequence generation through LSFRs and CDMA transceivers ntroduces additional energy overhead. The energy and latency overhead of the MCMC system increase with the PN Code Length (CL). The relative performance egradation of ACDMA communication under external persistent jamming with espect to the baseline NMAC mode communication for different PN CL in bits 16b, 32b, 64b) has been shown in Fig. 16.12. It can be observed from the figure hat, using a higher CL increases latency and energy consumption while providing igher security. It is also interesting to note here that, even with a PN CL of 32b, the ViNiP system under external attack outperforms the wired MCMC system. Only or a PN CL of 64b, the performance of the WiNiP drops below the wireline system.

6.7.2 *Optimum PN Code Length Selection*

n AMAC mode communication, the system performance and communication ecurity are heavily depended on PN code length. Figure 16.12 shows the effect

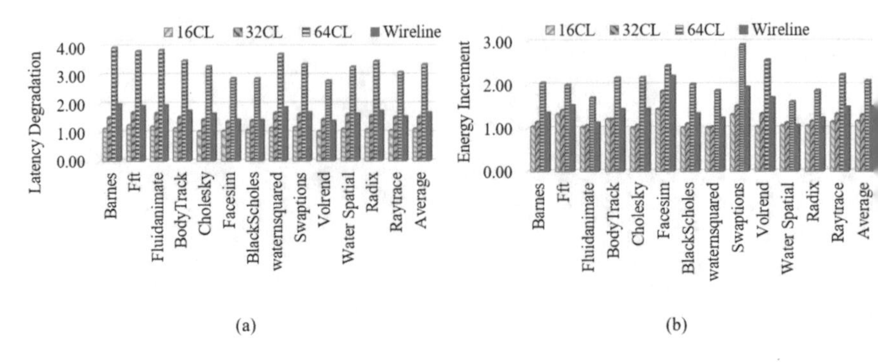

Fig. 16.12 Relative (**a**) Latency. (**b**) Energy degradation under external jamming attack for PN code length variation

of PN CL on system latency and energy. In this subsection we analyze the effect of CL on system security.

In ACDMA, all the simultaneous wireless transmissions appear as noise for a particular receiver. Moreover, the attacker can also introduce its interference noise and vary its output power to decrease the Signal to Noise plus Interference Ratio (SINR). Therefore, we determine the maximum power of the attacker that can be tolerated for a reliable communication for each of the CL considered above. We target an SINR of 15 db [8] that results in a BER of 10^{-15} which is comparable of wired link's BER. For each transmitter and receiver pair we adopt the transmitter power of −23.93 dBm, the noise floor of −69.43 dBm, and the path loss of 26.5 dB [50]. We consider one valid communicating WI pair and model other and attacker transmission as noise in the receiver side. Figure 16.13 shows the SINR variation for various PN CL (in bits) in any receiver after considering the auto and cross-correlation among PN codes. The 16b PN code results in lower SINR although it showed better latency and energy performance in Fig. 16.12. The 64b PN code though provides marginally better SINR than 32b PN, its latency and energy performance are worse than wireline interconnection architecture as shown in Fig. 16.12. From Fig. 16.12 it can be seen that the 32-bit PN sequence increases the average packet latency by 1.56×, and average packet energy by 1.31× compared to baseline while still outperforming the wired counterpart and therefore, we choose the 32b PN code for the best trade-off between performance and security.

16.7.3 Eavesdropping in WiNiP

Like WiNoC, we consider similar passive internal and external ED attacks as described in Sect. 16.3, where the attacker can listen to any intra or inter-chip wireless communication if it is tuned to the carrier frequency of the communicating WIs

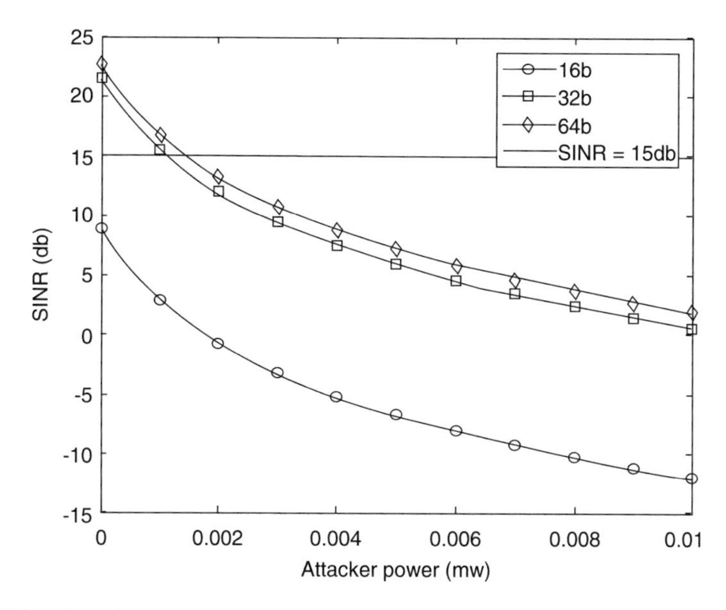

'ig. 16.13 Effect of PN code length variation on SINR for the proposed WiNiP architecture

n WiNiP, the PN encoded communication during AMAC mode prevent external aves dropping and we re-use the PMU managed power gating methodology for nternal ED attack as discussed in Sect. 16.5.5.

'6.7.4 Overhead Analysis

\s noted in Fig. 16.3, each WI is equipped with the BEU, ML Classifier, ADU,)efense Units, code-LFSR, sec-ENC/DEC, and the ED checker. The largest blocks .s shown in Table 16.4 are the BEU, KNN classifier, and the code-LFSR. The dopted KNN Classifier occupies an area of 105.3 μm^2. The BEU [34] occupies n area of 4357.5 μm^2. The code-LFSR occupies 326.4 μm^2 in each WI. The area f the ADU, Defense Unit, and Sec-ENC/DEC blocks are negligible. Therefore, the otal area overhead for each WI is 0.005 mm^2. The area of each wireless transceiver s 0.2 mm^2, making this overhead 2.5% per wireless transceiver in the system. The rea of the single MAC-LFSR and its decoder are 41 μm^2 and 38 μm^2. As can be een that the area overhead incurred by embedding proposed secure mechanism is mall when compared to the die size of 400 mm^2 considered in this work.

16.8 Conclusions

Though wireless interconnects can improve the performance of the on and off-chi communication through energy-efficient single hop links, they are also vulnerabl to various security threats like jamming-based DoS attack, eavesdropping. In thi chapter, we described a novel detection and defense mechanism for WiNoCs agains jamming-based DoS attacks and eavesdropping originating from either an interna HT or an external attacker. We also demonstrated how such security architecture an AML-based detection mechanism can be extended for multichip systems using in package mm-wave wireless interconnect, known as WiNiP, to ensure a sustainabl communication even under jamming attack. Unlike WiNoC, a novel reconfigurabl MAC and communication protocol were required to combat the jamming-base DoS attack in WiNiPs. In summary, with the proposed ML-based attack detectio and defense scheme, the security architecture considered in this chapter can detec both external and internal persistent jamming-based DoS attack with an accurac of 99.87%. Moreover, the proposed ML is also robust and shows an accuracy o 95.95% even in presence of adversaries. Most importantly, with the reconfigurabl MAC proposed in this paper, the MCMC system could sustain on and off-chi communication even under persistent jamming attack with an average latenc increment of 1.56× compared to baseline for a 32b PN code length while th WiNoC suffers only a bandwidth degradation of <3% . However, both WiNiP an WiNoC secure interconnection architectures outperformed the wired counterpar for internal as well as external persistent jamming attack with very minimal are overhead.

References

1. P.D.S. Manoj, J. Lin, S. Zhu, Y. Yin, X. Liu, X. Huang, C. Song, W. Zhang, M. Yan Z. Yu, H. Yu, A scalable network-on-chip microprocessor with 2.5D integrated memory an accelerator. IEEE Trans. Circ. Syst. I: Regul. Pap. **64**(6), 1432–1443 (2017
2. J. Lin, S. Zhu, Z. Yu, D. Xu, P.D.S. Manoj, H. Yu, A scalable and reconfigurable 2.5D integrated multicore processor on silicon interposer, in *IEEE Custom Integrated Circuit Conference* (2015)
3. U.Y. Ogras, R. Marculescu, It's a small world after all: NoC performance optimization vi long-range link insertion. IEEE Trans. on Very Large Scale Integr. Syst. **14**(7), 693–706 (2006
4. L.P. Carloni, P. Pande, Y. Xie, Networks-on-chip in emerging interconnect paradigms advantages and challenges, in *ACM/IEEE Int. Symp. on Networks-on-Chip* (2009)
5. P.D.S. Manoj, H. Yu, Y. Shang, C.S. Tan, S.K. Lim, Reliable 3-D clock-tree synthesi considering nonlinear capacitive TSV model with electrical-thermal-mechanical coupling IEEE Trans. Comput.-Aided Des. Integr. Circ. Syst. **32**(11), 1734–1747 (2013)
6. S.S. Wu, K. Wang, P.D.S. Manoj, T.Y. Ho, M. Yu, H. Yu, A thermal resilient integration o many-core microprocessors and main memory by 2.5D TSI I/Os, in *Design, Automation Tes in Europe Conference Exhibition (DATE)* (2014)
7. A. Ganguly, N. Mansoor, M.S. Shamim, M.M. Ahmed, R.S. Narde, A. Vashist, J. Venkatara man, Intra-chip wireless interconnect: the road ahead, in *International Workshop on Networ on Chip Architectures* (2017)

8. M.S. Shamim, N. Mansoor, R.S. Narde, V. Kothandapani, A. Ganguly, J. Venkataraman, A wireless interconnection framework for seamless inter and intra-chip communication in multichip systems. IEEE Trans. Comput. **66**(3), 389–402 (2017)
9. R. Mahajan, D. Mallik, R. Sankman, K. Radhakrishnan, C. Chiu, J. He, Advances and challenges in flip-chip packaging, in *IEEE Custom Integrated Circuits Conference* (2006), pp. 703–709
0. A. Ganguly, M.M. Ahmed, R. Singh Narde, A. Vashist, M.S. Shamim, N. Mansoor, T. Shinde, S. Subramaniam, S. Saxena, J. Venkataraman et al., The advances, challenges and future possibilities of millimeter-wave chip-to-chip interconnections for multi-chip systems. J. Low Power Electron. Appl. **8**(1), 5 (2018)
1. G. Kolhe, P.D.S. Manoj, S. Rafatirad, H. Mahmoodi, A. Sasan, H. Homayoun, Making LUT obfuscation a practical solution: breaking the design and security trade-offs using custom LUT-based obfuscation, in *ACM Great Lakes Symposium on VLSI* (2019)
2. B. Wu, J. Chen, J. Wu, M. Cardei, A survey of attacks and countermeasures in mobile ad hoc networks, in *Wireless Network Security*, ed. by Y. Xiao, X.S. Shen, D.Z. Du. Signals and Communication Technology (Springer, Boston, MA, 2007), pp. 103–135. https://doi.org/10.1007/978-0-387-33112-6_5
3. H. Sayadi, N. Patel, P.D.S. Manoj, A. Sasan, S. Rafatirad, H. Homayoun, Ensemble learning for effective run-time hardware-based malware detection: a comprehensive analysis and classification, in *ACM/EDAA/IEEE Design Automation Conference* (2018)
4. J. Sepúlveda, D. Flórez, M. Soeken, J. Diguet, G. Gogniat, Dynamic NoC buffer allocation for MPSoC timing side channel attack protection, in *IEEE Latin American Symposium on Circuits Systems* (2016)
5. P. Cotret, J. Crenne, G. Gogniat, J. Diguet, L. Gaspar, G. Duc, Distributed security for communications and memories in a multiprocessor architecture, in *IEEE International Symposium on Parallel and Distributed Processing Workshops* (2011)
6. C.H. Gebotys, R.J. Gebotys, A framework for security on NoC technologies, in *IEEE Computer Society Annual Symposium on VLSI* (2003)
7. A. Ganguly, M.Y. Ahmed, A. Vidapalapati, A denial-of-service resilient wireless NoC architecture, in *Great Lakes Symposium on VLSI* (2012)
8. F. Pereñíguez García, J.L. Abellán, Secure communications in wireless network-on-chips, in *International Workshop on Advanced Interconnect Solutions and Technologies for Emerging Computing Systems* (2017)
9. B. Lebiednik, S. Abadal, H. Kwon, T. Krishna, Architecting a secure wireless network-on-chip, in *IEEE/ACM International Symposium on Networks-on-Chip (NOCS)* (2018)
0. A. Vashist, A. Keats, S.M. Pudukotai Dinakarrao, A. Ganguly, Securing a wireless network-on-chip against jamming based denial-of-service and Eavesdropping attacks, in *IEEE Transactions on Very Large Scale Integration Systems (TVLSI)* (2019)
1. A. Moser, C. Kruegel, E. Kirda, Limits of static analysis for malware detection, in *Annual Computer Security Applications Conference* (2007)
2. M.A. Faizal, M.M. Zaki, S. Shahrin, Y. Robiah, S.S. Rahayu, B. Nazrulazhar, Threshold verification technique for network intrusion detection system. CoRR **abs/0906.3843** (2009)
3. W. Zhao, Y. Ha, M. Alioto, AES architectures for minimum-energy operation and silicon demonstration in 65nm with lowest energy per encryption, in *IEEE International Symposium on Circuits and Systems (ISCAS)* (2015)
4. B. Lebiednik, S. Abadal, H. Kwon, T. Krishna, Spoofing prevention via RF power profiling in wireless network-on-chip, in *Proceedings of the International Workshop on Advanced Interconnect Solutions and Technologies for Emerging Computing Systems* (2018), pp. 1–4
5. M.M. Ahmed, A. Ganguly, A. Vashist, S.M. Pudukotai Dinakarrao, AWARe-Wi: A jamming-aware reconfigurable wireless interconnection using adversarial learning for multichip systems. Sustainable Computing: Informatics and Systems, **29** (2021), p. 100470
6. V. Navda, A. Bohra, S. Ganguly, D. Rubenstein, Using channel hopping to increase 802.11 resilience to jamming attacks, in *IEEE INFOCOM* (IEEE, New York, 2007), pp. 2526–2530

27. E. Kakoulli, V. Soteriou, T. Theocharides, Intelligent hotspot prediction for network-on-chip based multicore systems. IEEE Trans. Comput.-Aided Des. Integr. Circ. Syst. **31**(3), 418–43 (2012)
28. M. Zekri, S.E. Kafhali, N. Aboutabit, Y. Saadi, DDoS attack detection using machine learnin techniques in cloud computing environments, in *International Conference of Cloud Computin Technologies and Applications (CloudTech)* (2017)
29. R. Karimazad, A. Faraahi, An anomaly-based method for DDoS attacks detection using RB neural networks, in *International Conference on Network and Electronics Engineering* (2011
30. P.A. Raj Kumar, S. Selvakumar, Distributed denial of service attack detection using a ensemble of neural classifier. Comput. Commun. **34**(11), 1328–1341 (2011)
31. R. Doshi, N. Apthorpe, N. Feamster, Machine learning DDoS detection for consumer interne of things devices. CoRR **abs/1804.04159** (2018)
32. M. Suresh, R. Anitha, Evaluating machine learning algorithms for detecting DDoS attacks, i *Advances in Network Security and Applications* (2011)
33. E.M. Rudd, A. Rozsa, M. Günther, T.E. Boult, A survey of stealth malware attacks, mitigatio measures, and steps toward autonomous open world solutions. IEEE Commun. Surv. Tuto **19**(2), 1145–1172 (2017)
34. B. Fu, P. Ampadu, Burst error detection hybrid ARQ with crosstalk-delay reduction for reliabl on-chip interconnects, in *IEEE International Symposium on Defect and Fault Tolerance i VLSI Systems* (2009)
35. I.J. Goodfellow, J. Shlens, C. Szegedy, Explaining and harnessing adversarial examples, i *International Conference on Learning Representations (ICLR)* (2015)
36. N. Papernot, P. McDaniel, S. Jha, M. Fredrikson, Z.B. Celik, A. Swami, The limitations o deep learning in adversarial settings, in *IEEE European Symposium on Security and Privac (Euro S&P)* (2016)
37. Y. Liu, X. Chen, C. Liu, D. Song, Delving into transferable adversarial examples and black-bo attacks, in *International Conference on Learning Representations (ICLR)* (2017)
38. K.N. Khasawneh, M. Ozsoy, C. Donovick, N. Abu-Ghazaleh, D. Ponomarev, EnsembleHMD accurate hardware malware detectors with specialized ensemble classifiers. IEEE Trans Depend. Secure Comput. **17**(3), 620–633 (2018)
39. P.D.S. Manoj, S. Amberkar, S. Bhat, A. Dhavlle, H. Sayadi, S. Rafatirad, H. Homayour Adversarial attack on microarchitectural events based malware detectors, in *Design Automatio Conference* (2019)
40. S. Deb, K. Chang, X. Yu, S.P. Sah, M. Cosic, A. Ganguly, P.P. Pande, B. Belzer, D. Heo, Desig of an energy-efficient CMOS-compatible NoC architecture with millimeter-wave wireles interconnects. IEEE Trans. Comput. **62**(12), 2382–2396 (2013)
41. X. Yu, S.P. Sah, H. Rashtian, S. Mirabbasi, P.P. Pande, D. Heo, A 1.2-pJ/bit 16-Gb/s 60 GHz OOK transmitter in 65-nm CMOS for wireless network-on-chip. IEEE Trans. Microwav Theory Tech. **62**(10), 2357–2369 (2014)
42. X. Yu, H. Rashtian, S. Mirabbasi, P.P. Pande, D. Heo, An 18.7-Gb/s 60-GHz OOK demodulato in 65-nm CMOS for wireless network-on-chip. IEEE Trans. Circ. Syst. I **62**(3), 799–806 (2015
43. U. Shaham, Y. Yamada, S. Negahban, Understanding adversarial training: increasing loca stability of neural nets through robust optimization (2015). ArXiv e-prints
44. R. JayashankaraShridevi, C. Rajamanikkam, K. Chakraborty, S. Roy, Catching the flu emerging threats from a third party power management unit, in *ACM/EDAC/IEEE Desig Automation Conference* (2016)
45. P.P. Pande, C. Grecu, M. Jones, A. Ivanov, R. Saleh, Performance evaluation and design trade offs for network-on-chip interconnect architectures. IEEE Trans. Comput. **54**(8), 1025–104 (2005)
46. M.A. Abu-Rgheff, *Introduction to CDMA Wireless Communications* (Academic, New York 2007)
47. X. Wang, T. Ahonen, J. Nurmi, Applying CDMA technique to network-on-chip. IEEE TVLS **15**(10), 1091–1100 (2007)

48. M.M. Ahmed et al., A one-to-many traffic aware wireless network-in-package for multi-chip computing platforms, in *IEEE SOCC* (2018)
49. B. Wu, J. Wu, E.B. Fernandez, S. Magliveras, Secure and efficient key management in mobile ad hoc networks, in *19th IEEE International Parallel and Distributed Processing Symposium* (IEEE, New York, 2005), pp. 8
50. V. Vijayakumaran et al., CDMA enabled wireless network-on-chip. ACM JETC **10**(4), 28 (2014)
51. M. Badr, N.E. Jerger, Synfull: synthetic traffic models capturing cache coherent behaviour, in *ACM SIGARCH Computer Architecture News*, vol. 42(3)(IEEE Press, Hoboken, 2014), pp. 109–120

Chapter 17
Securing 3D NoCs from Hardware Trojan Attacks

Venkata Yaswanth Raparti and Sudeep Pasricha

7.1 Introduction

With the rise in number of processing cores and growing parallelism in applications, ne communication traffic in a manycore processor has been increasing. Chip esigners and manufacturers are moving towards network-on-chip (NoC) as their e-facto intra-chip communication fabric [1, 2]. Typically, emerging manycore rocessors have tens to hundreds of components that are designed either by in-house ngineers or obtained from third-party vendors (3PIP), and then finally integrated ogether in a single global facility. With the growing complexity in NoC design, esigners are opting for third-party NoC IPs, e.g., [3], to connect the components in neir processors. This global trend of distributed design, validation, and fabrication as led to major challenges in ensuring secure execution of applications on nanycore platforms, in the presence of potentially untrusted hardware and software omponents.

3D integration has gained much attention recently as it brings numerous advanages over traditional 2D integration to overcome CMOS scalability bottleneck [4]. D ICs have demonstrated higher transistor density, lower power dissipation, and maller area footprint compared to 2D ICs. Additionally, 3D ICs deliver better pplication performance owing to their shorter distance between compute and nemory units that are connected using through-silicon-vias [5]. However, in 3D ntegration, dies are individually tested and stacked and often not subjected to xtensive examination of the inter-die integration defects before and after stacking. Thus, security of crucial components such as NoC in 3D ICs is compromised by nalicious agents at third party foundries or manufacturers.

V. Y. Raparti · S. Pasricha (✉)
Department of Electrical and Computer Engineering, Colorado State University, Fort Collins, CO, USA
-mail: sudeep@colostate.edu

. Mishra, S. Charles (eds.), *Network-on-Chip Security and Privacy*,
https://doi.org/10.1007/978-3-030-69131-8_17

Much work has been done to mitigate side-channel attacks on shared resources and to detect counterfeit ICs that compromise manycore chip performance in both 2D [6] and 3D NoCs [7]. This work focuses on an orthogonal attack scenario where an adversary can insert a hardware Trojan (HT) into the RTL or the netlist of a manycore processor to disrupt or alter the integrity of its behavior without being detected at the post silicon verification stage. HTs can be inserted by an intellectual property (IP) vendor, untrusted CAD tool/designer, or at the foundry via reverse engineering [8]. We focus on one such attack called a data-snooping attack where a malicious software and an HT work together to steal information from applications executing on manycore processors.

3D NoCs are ideal candidates for such attacks as they have a complex design that can be used to hide an HT which cannot be easily detected via functional verification. HTs can be placed in NoC links, routers, or network interfaces (NIs) to secretly snoop on the data or corrupt data passing through them. Typically, in data snooping attacks HTs create duplicate packets with modified headers and send them into the NoC for an accomplice thread to receive them [9]. Several works propose packet encoding/error correction mechanisms such as parity bits and ECC in NoC packets to detect faulty data packets at the receiver [10, 11]. Other works such as [9, 12–14] have also proposed data protection mechanisms in the presence of an HT in NoC components. However, there are three major shortcomings with the state-of-the-art: (1) these works assume the presence of HTs in 2D NoC routers or links which can be detected by physical inspection or functional verification, without employing costly security mechanisms; (2) the mechanisms proposed in prior works protect application data from snooping attacks but do not detect the attack and mitigate future attacks; and (3) most of the security enhancement mechanisms are costly to implement and increase NoC latency and power consumption which worsens the overall performance. *It is important to design and deploy lightweight mechanisms that can detect the operation of malicious HTs embedded in 3D NoCs and accomplice threads, and secure against their data-snooping attacks in emerging manycore processors.*

In this work we focus on security enhancement that do not notably increase performance and power overheads. We provide robust yet low-power mechanisms to detect the source of the attacks by utilizing controlled aging in circuits at runtime which is not easy to obfuscate or tamper with in the design and fabrication process. Our novel contributions in this work are as follows:

- We first design and demonstrate a data-snooping attack using an HT in the 3D NoC interface that duplicates packets and injects them into the 3D NoC with a minimal area and power footprint, making it difficult to detect by traditional functional verification mechanisms;
- We then protect against such data-snooping attacks by proposing a novel snooping invalidation module (*SIM*) that uses an encoding-based duplicate packet detection mechanism;

We further propose a novel data-snooping detection circuit called *THANOS* that uses threshold voltage degradation as a means to detect an on-going attack at runtime and blacklist the malicious software task that initiated the attack; Experimental analysis shows that *SIM* with *THANOS* provides security against HTs with minimal area and power overhead in 3D NoCs.

17.2 Related Work

Significant research has been done to increase robustness against attacks by HTs in NoCs by assuming that an HT tampers or snoops data passing through it. In [12], bit shuffling and Hamming ECC are used to reduce the effectiveness of HTs that corrupt data. In [13], security zones managed by a centralized security manager are proposed to protect sensitive information from being accessed by malicious agents. In [9] data scrambling, packet authentication, and node obfuscation are proposed to prevent data stealing by a compromised NoC. Data scrambling, and packet-authentication mechanisms use a one-time pad XOR cipher that can be broken by the malicious tasks when enough encrypted packets are accumulated. In [14], CRC and algebraic manipulation detection (AMD) are used to encode packet headers to safeguard from faults and snooping attacks. In [15], a novel wave-based scheduling mechanism for NoCs is proposed that eliminates the need for TDMA-based NoC resource sharing, hence providing non-interference between different domains of applications. In [16], a process variation-based packet encoding and decoding mechanism is proposed to prevent data-snooping in silicon photonic NoCs. *Most of these schemes that protect application data from NoC security attacks lack an efficient and low-power attack detection mechanism which makes them incomplete in providing security.*

A few works address HT detection in NoC components at design-time and runtime. At design time, techniques such as physical inspection [17], functional testing [18], and side channel analysis [19] have been proposed. But testing for HTs at design time is still in infancy, and the growing complexity of 3D NoC components make this even more difficult. Hence, designers are now exploring runtime detection methods. A key logic built-in self-test (LBIST) was proposed in [20] that uses test vectors generated by programmable keys to detect Trojans. However, LBIST requires that the chip operation should be paused while testing at regular and frequent intervals, which is not suitable for NoCs that should function seamlessly. A few other works such as [21, 22] propose in-situ HT detection modules that rely on verification units placed in NoC components to detect HTs. There are two limitations with all of these works: (1) the verification units used to detect HTs can also be reverse-engineered and tampered, (2) these mechanisms are used to detect only HT induced data-corruption attacks. Data-snooping attacks unlike data-corruption attacks attempt to leak critical application data to malicious software tasks. *None of the prior works have addressed the problem of detecting the software task that initiates data-snooping attacks to blacklist and prevent future attacks.*

In [23], a run-time technique called NoCAlert is proposed to detect failures in th control logic of NoC components. This technique is further enhanced by [24] tha proposes modules which alert the host system if the control logic in NoC router detects invariance violations caused by HTs placed in its control-path, e.g., logi for route computation (RC) or virtual channel allocation (VCA). However, thes techniques focus on NoC components that have substantial control logic, such a routers. They ignore the network interface (NI) which prohibits easy placement o model checkers to detect packet duplication. In [25] a novel snooping invalidatio module (*SIM*) in the NI that can mitigate snooping attacks is proposed. Further, low-overhead techniques to detect the source of data-snooping attacks in 2D NoC is also proposed in [25]. To the best of our knowledge, [25] is the first work tha mitigates snooping attacks in 2D NoCs, while also detecting the source of snoopin attack to protect against future attacks. In this chapter, we embrace the technique proposed in [25] to mitigate snooping attacks in 3D NoCs that face the similar threa from data snooping attacks.

17.3 Background and Attack Model

17.3.1 Background

In this section we discuss our assumed baseline NoC design. We consider traditional 3D mesh based NoC with processing elements (PE) connected to the No via a network interface (NI). The packets entering the 3D NoC are routed toward their destination by routers that use a hop-by-hop, turn-based distributed deadloc free 4NP-first routing algorithm [26]. Figure 17.1 shows the schematic of the 2I mesh NoC with an NI and PE connected to routers. In a 3D mesh NoC, a route has two additional I/O ports {*up, down*} and their associated channels along wit the four I/O ports {*north, east, west, south*}of a 2D NoC router. We use traditiona 3-stage {*buffer write, RC* + *VCA* + *SA, LT*} pipelined routers in the 3D NoC wit wormhole switching and 4-VC buffers at each input port. PEs communicate usin messages that are passed to the NI which packetizes them before sending them t the 3D NoC. The packets received by the NI from the 3D NoC are de-packetized an

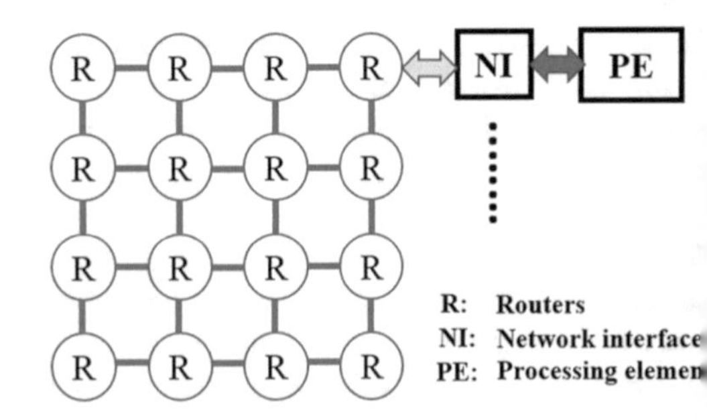

Fig. 17.1 Baseline NoC architecture with example routers, a PE and an NI as shown in [25]

sent to the connected PE. We consider ARM Cortex-A73 cores in our PEs that use the AXI interface for communication. Each PE has a private L1 cache and a shared distributed L2 cache that uses a scalable directory-based cache coherence protocol to send messages in the form of NoC packets.

17.3.2 *Attack Model*

Prior works [9, 12–14] assumed data-snooping attacks to be carried out by HTs embedded in NoC routers or by compromised links that enable HTs to modify the packet headers. These HTs, once activated by a flit with a special activation sequence, make copies of packets passing through the router and transmit them to the PE that has a malicious accomplice task running on it. Once an HT is activated in a router, it generates new packets, or diverts an existing packet to the PE running the accomplice task. This type of HT, that has a high 4% area overhead [9], may be noticed by testers while conducting physical inspection or side channel analysis. Moreover, this type of attack can lead to illegal utilization of router resources such as buffers, VCs, and switch allocators, which cause control logic violations that can be detected by secure model checkers [24]. We thus focus on a harder-to-detect attack with an HT embedded in the NI where packets are generated, and hence packets can be duplicated with relatively simpler logic without interfering with the basic NI functionality.

Figure 17.2a shows an overview of an on-going data-snooping attack taking place in a single layer of a 3D NoC based manycore processor with multiple HTs activated in the 3D NoC NI modules shown in red, and a malicious task running on a PE connected to the yellow router and NI. The HTs in NIs make duplicate copies of packets that are sent to the malicious task.

17.3.3 *Design Details: Network Interface with a Hardware Trojan*

Figure 17.2b shows the microarchitecture of an NI with an embedded HT in the packetizer module. The NI receives messages from the PE via the AXI interface that are then stored in its buffers. The messages usually are read/write commands with address and data fields. The packetizer module appends source ID, destination ID, and virtual channel ID information to the commands and creates packets. A packet is further divided into flits, with the header flit containing the 3D NoC routing related information. The packet flits are then injected into the circular flit queue that is accessed via head and tail pointers. After the packetizer injects a flit, the tail pointer of the queue is incremented. After a flit is transmitted to a router, the head pointer is incremented to transmit the next flit.

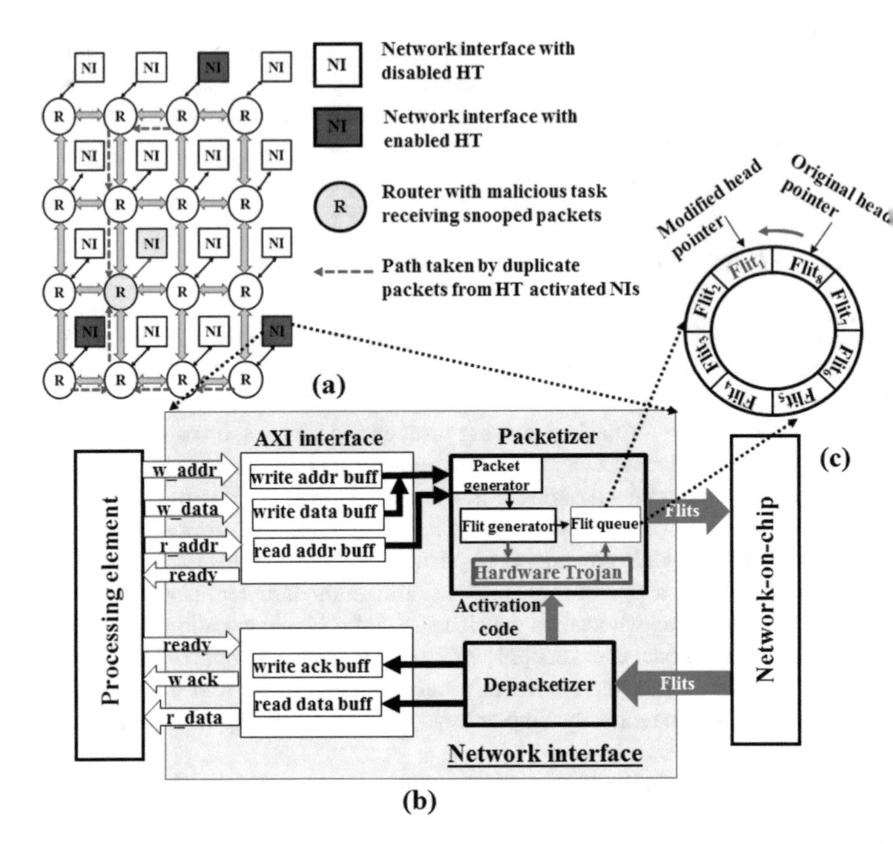

Fig. 17.2 (**a**) Overview of attack model on a NoC with a malicious software task coordinating the data-snooping attack [25], (**b**) microarchitecture of network interface (NI) with a hardware Trojan embedded in packetizer module [25], (**c**) FIFO queue modification by hardware Trojan [25]

An HT can potentially tamper with the pointer values to re-send duplicate packets intelligently. Once a flit has been transmitted from NI to the router, it stays in the cyclic queue until a new flit is overwritten on that location. The HT can keep track of these locations to read a header flit that has already been transmitted to the router, append it with a duplicate destination ID of the malicious node, and *update the head-pointer*. Figure 17.2c shows how the HT modifies the head pointer. By moving the head pointer at regular intervals, the HT can send duplicate packet flits without having to store them externally. The duplicate packet is re-sent to the router for transmission. If the flit queue is full (head pointer = tail pointer), both the HT and packetizer do not inject new flits into the queue, and do not accept any more incoming data from the PE until the outstanding flits are transmitted. The HT does not interfere with the control logic which is mostly present in the AXI interface, and an attacker can snoop on data using this HT in NIs between two PEs, or between a PE and a memory controller that is connected to main memory channels.

Table 17.1 FPGA Implementation of NI packetizer with and without Hardware Trojan (HT) [25]

	Timing (ns)	Number of FFs	Number of LUTs
NI without HT	3.45	258	535
NI with HT	3.45	273	549

We now perform an overhead analysis of this HT. The proposed HT requires an nternal memory to save the head flit to modify its destination ID, save the header ointer of the queue, and save the current state of the HT (~72 bits). We designed ne NI shown in Fig. 17.2b by modifying the CONNECT open-source NoC model 27] and used Xilinx's Vivado HLS [28] tool to analyze the overheads. Table 17.1 hows the clock cycle period, number of flipflops and LUTs used for an FPGA nplementation of the packetizer. The optimized design indicates that the NI with an IT requires an additional ~5% FFs and ~ 1% LUTs (1.3% area overhead) without ncurring additional timing latency. This low overhead HT can be inserted at the TL level, or by reverse engineering and changing the netlist at the place and routing tage [8, 17]. The small size of the HT makes it hard to detect by physical inspection r by side-channel analysis. Also, the run-time secure model checkers from [23, 24] re not able to check the validity of flits in the NI as it does not interfere with the ontrol logic. *Hence, there is a need to design a low-overhead flit validation module n NIs to check flit validity before injecting them into the* 3D *NoC.*

7.3.4 Hardware Trojan Attack Model in 3D NoCs

n 3D ICs, a malicious agent can be placed across different dies. Attackers may plit the implementation of a Trojan across multiple tiers. This kind of split nplementation further reduces the area and the power footprint of HTs hidden n NoC components. For example, as shown in Fig. 17.3, the activation module or a Trojan may be placed in the bottom most layer, the payload is created in the niddle layer, and the accomplice software thread may listen to the snooped data n the topmost layer. Traditional testing mechanisms are not effective in detecting nese types of hardware Trojans as the testers evaluate each die (for each tier) ndividually. If the dies pass those tests, they are stacked vertically. Beyond testing or die alignment errors, no further testing is typically performed to detect multi-ayer Trojan implementations. These Trojans take much lower footprint per die than f they were implemented on a single die, and are thus also harder to detect. The rojans are also easy to activate using mechanisms such as thermal trigger circuits 29] or external probe sensors [30].

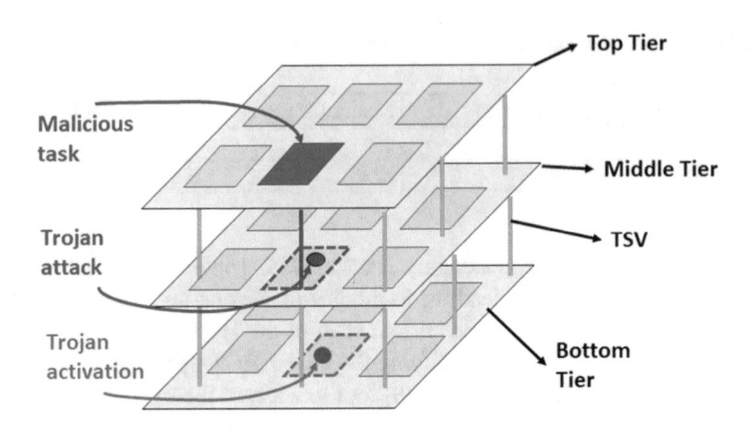

Fig. 17.3 Hardware Trojan implemented across different tiers to carry out attacks

17.4 Mitigation of 3D NoC Snooping Attacks

We propose a novel framework that integrates two mechanisms to mitigate data snooping attacks from taking place, as well as to detect the source of an on-going attack, and protect against future snooping attacks. We rule-out data corruption attacks as they can be detected and corrected using ECC codes such as in [14]. Our proposed framework consists of two security mechanisms, *(1) a snooping invalidator module at the NI output queue to discard duplicate packets, (2) detection of data-snooping attacks at the PE where an accomplice thread is executing.* This comprehensive protection framework ensures that we proactively mitigate future attacks and safeguard the application data for the entire lifetime of the processor. We have designed our security mechanism to be hard to be tampered by adversaries that use reverse engineering techniques to insert HTs in the netlist. Our approach also works irrespective of the HT triggering process to start snooping attacks such as special flit data, circuit aging, or temperature [31]. The following sections discuss our two security mechanisms.

17.4.1 Security Enhanced NI: Preventing Data-Snooping Attack

The first security enhancement mechanism is to prevent a snooping attack with the help of a snooping invalidator module (*SIM*) at the NI. Using *SIM*, we aim to discard packets with invalid header flits from being injected into the 3D NoC. Unlike traditional ECC-based security enhancement mechanisms, *SIM* incurs low power and low latency overheads because of its lightweight computations that are designed solely to mitigate snooping attacks. Figure 17.4a shows an overview of the security enhancement in NI using *SIM*.

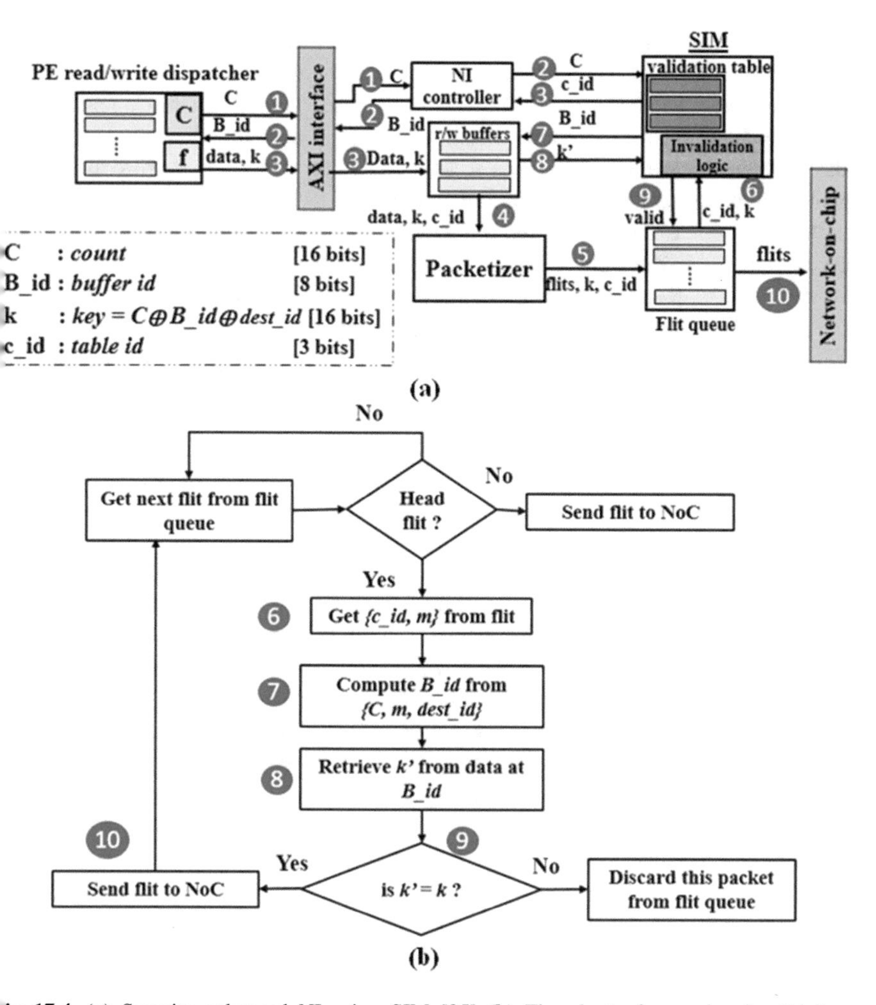

Fig. 17.4 (**a**) Security enhanced NI using SIM [25] (**b**) Flowchart of snooping invalidation mechanism in NI [25]

We divide the implementation of *SIM* across the PE and NI to prohibit 3PIP NoC designers/testers to reverse engineer or tamper with the secret encoding/decoding information at runtime. The PE and NI communicate using the standard AXI hand-shake protocol (ready, valid, and valid ready signals). A typical NI receives messages from the PE to be packetized and sent to the 3D NoC and vice-versa. In the security enhanced NI, additional encoding information (*key*) is attached with the data received from the PE to validate the uniqueness of data packets. The numbered sequence of steps shown in Fig. 17.4 describe how a packet is validated using *SIM*. These steps are discussed next.

In step 1, the PE data dispatcher sends a *count* (*C*: increments with each outgoing data) value to the NI controller along with the AXI ready signal. In step 2, the

NI controller sends a buffer id (*B_id*) that is reserved to store the incoming dat along with the AXI valid signal to the PE data dispatcher. The NI controlle simultaneously sends *C* to *SIM* that stores it in a "validation table". The PE use an XOR function *f* to generate an encoded key *k* as a function of *C*, *B_id*, an destination ID (*dest_id*) of the packet, as shown in eq. (17.1) below. In step 3, the P sends a message {*data, k*} combination to the data buffers and toggles the valid read signal to high. At the same time, the validation table sends a *c_id* (location wher *count* is stored in the table) to the NI controller. This is stored with the message sen by the PE. The {*data, k, c_id*} combination is stored as a unit in read/write buffer till the packet is sent out of the NI. While the size of payload *data* varies from 8B t 128B depending on message type, the values of *k, C* and *c_id* require only few bit of storage (see legend of Fig. 17.4a).

$$k = C \wedge B\_id \wedge dest\_id \tag{17.1}$$

$$B\_id = C \wedge k \wedge dest\_id \tag{17.2}$$

In step 4, the {*data, k, c_id*} combination is sent to the packetizer to generat packets. In step 5, flits are generated with *k* and *c_id* copied into the header fli We save *k* and *c_id* in the 24-bits reserved in the header flit to store destinations o source-routing path [32] which are unused as we adopt distributed routing for our 3 NoC. The flits are then saved in the output flit queue. *Steps 6–9 are part of snoopin invalidation flow explained in more detail in* Fig. 17.4b. *SIM* tries to retrieve th encoded key *k* from the *C* entry in the validation table as a part of packet validation In step 6, *SIM* reads *dest_id, k, and c_id* bits of the header flit, and performs decoding operation shown in eq. (17.2) to obtain *B_id* of the buffers that stored th corresponding packet data, and *k* sent by the PE (step 7). In steps 8 and 9, *SI* retrieves the value of *k'* stored in the buffer located at *B_id* and compares with *k* tha is read from the header flit. If *k* = *k'*, *SIM* sends a valid signal that the header flit i valid, and it is injected into the 3D NoC. If *SIM* sends an invalid signal, the flit queu discards all the flits corresponding to the duplicate packet. *SIM* efficiently detect duplicate packets because, if the value of *dest_id* is modified by the HT, eq. (17.2 leads to an incorrect value of *B_id* that does not retrieve the *k* value corresponding t the data of packet sent in step 3. Note that for broadcast/multicast packets, multipl keys are generated for each *dest_id* value and key verification steps 8 and 9 ar performed on each of them separately. After a packet is sent out, the correspondin read/write data buffer and validation table entries are reused for new data. *This low overhead SIM module with minor modifications can also be used to curb potenti data duplication at router-link interfaces or within a router.*

17.4.1.1 Overhead Analysis

Several steps in the *SIM* module can be performed in parallel. The existing communication data channel between the PE and the NI that is established by AXI interface is used to communicate both packet data and *SIM* metadata (*C, k, B_id* in steps 1 and 2). Hence, no additional wires are needed to transmit *SIM* metadata. Steps 1 and 2 are performed in parallel with AXI interface's ready and valid signal exchange to minimize the latency overhead. Also, there is no additional overhead involved in steps 3 to 5. Steps 6 to 9 take one cycle in a 3D NoC that is clocked at 1 GHz frequency which was verified via FPGA synthesis of the modified NI [28]. This increases the number of pipeline stages of the NI microarchitecture. *SIM* takes additional memory to maintain a validation table, and additional logic to perform XOR and comparison operations. *SIM* incurs ~5.5% more power and ~ 2.15% more area overhead compared to the baseline NI with a buffer capacity of 16 packets, at the 22 nm technology node.

17.4.2 Detecting the Source of a Data-Snooping Attack

Using our security enhanced NI with the integrated *SIM*, we can curb packet duplication at the NI. However, the malicious task that is the source of the attack is still not detected that could initiate attacks from compromised routers or links [14]. In this section, we propose a module called *THANOS,* a novel threshold activated snooping attack detector that is implemented at the interface between an NI and a PE, as shown in Fig. 17.5a to detect the source of the snooping attack.

A PE sends and receives various types of messages into the 3D NoC that can be broadly classified into two types: (1) direct messages between cores for

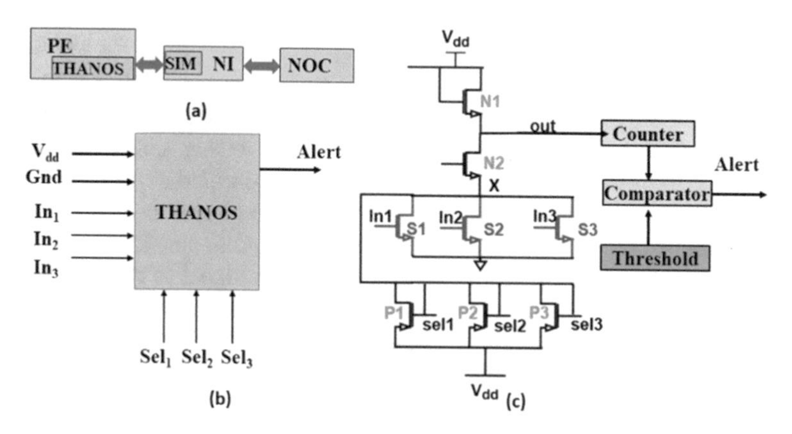

Fig. 17.5 (**a**) Overview of THANOS [25] (**b**) block diagram of THANOS showing inputs and outputs [25] (**c**) snooping detecting circuit used in THANOS [25]

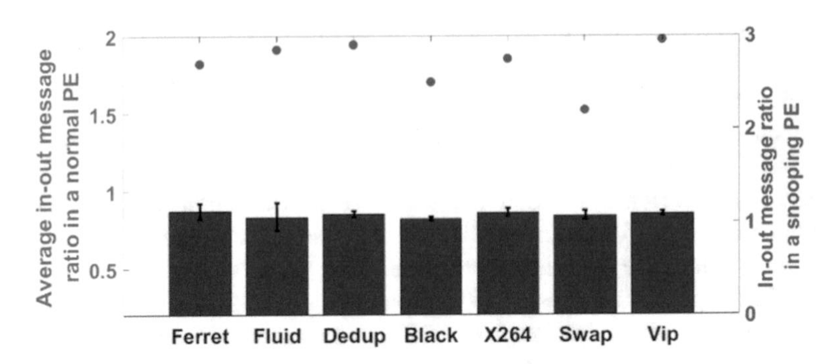

Fig. 17.6 Average incoming-outgoing message ratio at normal PE (left), and at snooping P (right) across different applications [25]

inter-core communication, and (2) cache-coherence messages between a PE an directory table. Figure 17.6 (blue bars) shows the average incoming-outgoin message ratio sent over 64 cores in a 3D NoC by different PARSECv2.1 [33 benchmark applications with 64 tasks each. The error-bars in Fig. 17.6 represen variance across 3D NoC nodes. Figure 17.6 shows that the ratio is less than 1 ove all the benchmarks with each node receiving a smaller number of messages tha the messages it sends out (number of "packets" in a "message" can vary based o message type). Another important observation from Fig. 17.6 is that the incoming outgoing message ratio is much greater than 1 (red points) when a data-snoopin attack takes place, because a PE receives significantly higher number of message (and packets) than it sends out. This phenomenon can be easily detected in the sho term by placing a counter in the NI and observing the number of incoming an outgoing messages over an epoch of time. However, observing messages in the sho term can lead to false positives, e.g., due to periodic bursts of messages from a tas that requires higher volumes of input data. Also, a message counter is not the mos secure way to detect a snooping attack, given the reverse engineering technique available to tamper digital logic [17].

In *THANOS* we devise a mechanism that observes the ratio of incoming an outgoing messages over a period of few hours and identifies the source of a snoopin attack. *THANOS* is designed using a combination of analog and digital logic t detect if a PE is snooping on messages over a duration of time. As *THANOS* i not entirely a digital logic implementation, it is hard to reverse engineer or tampe with, and can be used as a final frontier to mitigate data-snooping attacks. *THANO* receives inputs from the PE and sends a security alert signal to the PE as show in Fig. 17.5b. The PE then identifies the source of data-snooping attacks and take preventive steps to mitigate future attacks. *THANOS* is designed as a standalon module that can also be used with prior data protection Schemes [9, 12–14] to detec the source of attack.

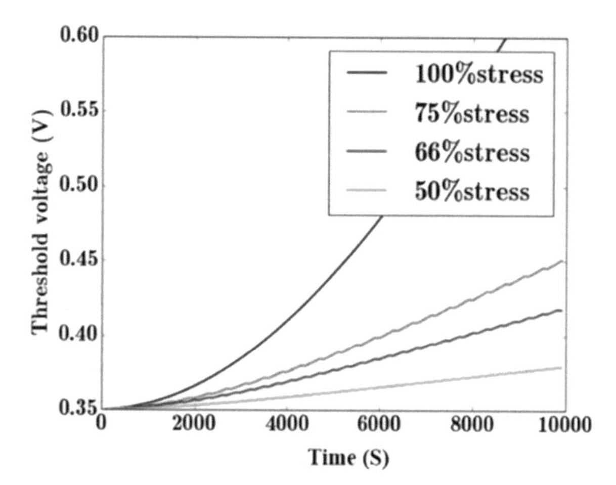

ig. 17.7 Threshold voltage egradation observed across ifferent stress-recovery atios in a NMOS transistor at 2 nm technology node [25]

7.4.2.1 Overview of Snooping Detection Circuit

Ve take inspiration from a controlled aging module [6] that uses threshold degra- ation of NMOS transistors due to aging phenomenon such as bias temperature nstability (BTI) and hot carrier injection (HCI) to detect chip usage, which elps identifying counterfeit ICs. In *THANOS* we use NMOS threshold voltage egradation to detect a PE that is receiving duplicate packets injected by multiple IT activated NIs in the 3D NoC. NMOS transistors undergo stress-recovery periods n their ON and OFF operations that leads to threshold voltage (V_{th}) degradation 34]. Figure 17.7 shows the V_{th} degradation observed across different ratios of stress nd recovery in an NMOS transistor at 22 nm using the long-term aging model roposed in [34]. At 100% stress (no recovery) the V_{th} of a transistor increases by 100 mV in about 2 hours duration. We use this phenomenon to detect snooping ttacks.

7.4.2.2 Operation of Snooping Detection Circuit

n our snooping detecting circuit shown in Fig. 17.5c, there are 2 transistors; N_1 that cts as a diode connected load and N_2 that acts as gate-source voltage (V_{gs}) sensor. $_1$, P_2, P_3 are diode-connected PMOS transistors that pull the drain voltages of S_1, $_2$, S_3 to high. Transistors S_1, S_2, S_3 are driven using low over-drive voltages In_1, In_2, $_3$ that barely switch them ON. *We artificially induce stress in a selected transistor mong S_1/S_2/S_3 when a message is received and induce recovery when a message sent out. Hence, we call them stress-transistors.* At any point only one of S_1,S_2,S_3 re connected to the circuit (using *In* and *sel* signals). When S_1/S_2/S_3 is turned ON, e source (V_x) of N_2 is pulled low, which turns ON N_2, leading to a "low" *out* state. ut, when a stressed transistor (S_1/S_2/S_3) undergoes V_{th} degradation, its over-drive oltage ($In = V_{gs}$-V_{th}) is not high enough to turn ON the stress-transistor and hence

Table 17.2 State transition of snooping detection circuit

Stress-transistors ($S_1/S_2/S_3$)	V_x	N_2	out
Saturated	Low	ON	Low
Triode	High	OFF	High

drives it into the triode region. When $S_1/S_2/S_3$ is in the triode-region, the sourc voltage (V_x) of N_1 is not pulled low and the *out* signal is set to "high".

Table 17.2 gives the states of different transistors and the corresponding change in *out* signal state. When a PE is not receiving snooped packets, its incoming outgoing message ratio is less than 1 as shown in Fig. 17.6. Hence, for normal 3I NoC traffic the stress-recovery ratio of stress-transistors ($S_1/S_2/S_3$) is less than 40% Generally, BTI and HCI are slow wear-out phenomenon in logic circuits. But, w input low over-drive (V_{gs}-V_{th}) voltage of ~100 mV to the stress-transistors throug input signals $In_1/In_2/In_3$. Hence, the circuit would set the *out* signal to high state i a duration of 2–3 days. However, when a malicious task on a PE is snooping with u to four HT activated in NIs, its incoming-outgoing message ratio is 3× the averag ratio (shown in Fig. 17.6). As a result, the stress-transistors in *THANOS* underg 80–90% more stress than recovery when there is a snooping attack. From Fig. 17.7 when a stress-transistor receives ~90% stress, its threshold voltage increases over shorter duration (~3–4 hours). Hence the snooping detection circuit toggles the *ou* signal to "high" state quicker when PE receives snooped packets.

In *THANOS*, we use a counter to track the time taken for the *out* signal t change its state and compare it with a threshold time that is configured by a truste PE firmware, as shown in Fig. 17.5c. *THANOS* sends an ALERT signal whe the time taken by the *out* signal to switch the state is less than the threshold *Overall, THANOS sends a notification about a potential malicious task anywher from ~ 2 hours to ~ 2 days based on the number of HTs that are active.* The truste PE firmware then alerts the OS about the malicious application task executing o the PE, so that preventive measures can be taken.

The snooping detection circuit should last for the lifetime of the processor t detect snooping attacks. However, due to artificially induced stress and recover cycles, the stress-transistors (S_1, S_2, S_3) wear-out much more rapidly than the res of the chip. To increase the lifetime of *THANOS* we take two measures: (1) We inpu low over-drive voltage (In-$V_{th} \approx 100$ mV) and high V_{dd} using separate power line for stress-transistors; after every state change of the *out* signal, we increment the *I* signal by ~100 mV until we satisfy the MOS saturation condition (In-$V_{th} < V_{dd}$) (2) The stress-transistors are over-provisioned; we use only one stress-transistor a any time to detect an attack and when an *In* voltage of a stress-transistor can n longer be incremented without violating the saturation condition, *THANOS* switche to the next stress-transistor using the *sel* signal. Using three stress-transistors an $V_{dd} = 3$ V, *THANOS* can seamlessly detect snooping attacks for up to 1.5 years. Th number of stress-transistors in *THANOS* is hence left to the decision of the designe The overhead of *THANOS* is negligible in power (~50 μW) and area (~0.9 μm^2 compared to the PE (~1 W, ~318 mm^2) at 22 nm technology node as it requires jus 8 MOSFETs, a counter, a comparator, and a simple control logic block to send inpu signals.

17.5 Experiments

We target a 60-core manycore chip with low power ARM cortex-A73 cores and a 3D mesh NoC with 5 × 4 × 3 dimension to test the performance, latency, energy, and area overheads of the proposed lightweight snooping invalidation module *(SIM)* and snooping detection circuit *(THANOS)* compared to the state-of-the-art. For simulations, we modeled the behavior of *SIM* and *THANOS* as part of the cycle-accurate NoC simulator Noxim [35]. We obtained the power and area overheads of *SIM* and *THANOS* modules from post-synthesis vectorless estimation in Vivado [28], and Cadence Virtuoso [36], at 22 nm. We integrate the latency and energy overheads of *SIM* and *THANOS* with Noxim for our simulations. We tested our framework using PARSECv2.1 benchmark NoC traces generated by netrace [37] to capture the request-response dependencies to accurately simulate parallel application performance.

We compare our work with a baseline 3D NoC (with a configuration that is described in Sect. 3.1) with no security mechanism employed, and with two prior works, FortNoCs [9], and P-Sec [14]. In [9], only data obfuscation and data scrambling techniques are implemented for a fair comparison in 3D NoC. In [14] end-to-end algebraic manipulation detection (AMD) and cyclic redundancy codes (CRC) are appended to the header flit for reliability against faults and HT attacks. We set the threshold time in the snooping detection circuit of *THANOS* as ~2.5 days to get a security violation alert. We first present results of application performance, 3D NoC latency and 3D NoC energy consumption for 4 actively snooping HTs that are randomly placed in the 3D NoC. Subsequently we present results for scenarios with 1 and 2 HTs operating in the 3D NoC.

Figure 17.8a shows the comparison of application execution time across different NoC security mechanisms. P-Sec and FortNoCs cannot prevent the injection of duplicate packets at the NI, and only discard faulty packets at the receiver, which leads to higher 3D NoC traffic. Moreover, P-Sec takes two extra cycles for CRC + AMD encoding/decoding, and FortNoCs takes at least four extra cycles at the NI for node obfuscation and data scrambling techniques on the entire packet. This leads to poor application performance with P-Sec and FortNoCs compared to the baseline. *SIM* mitigates duplicate data packets near the source, resulting in less 3D NoC congestion, thereby actually achieving ~49% average improvement in application execution time, compared to the baseline.

A similar trend is observed for network latency, shown in Fig. 17.8b. The network latency of FortNoCs is higher due to the packet scrambling mechanism that encrypts/decrypts the entire packet using XOR operation, which is time consuming for packets with high payload size. FortNoCs incurs additional overhead due to packet authentication as an additional security mechanism. *SIM*, in comparison, takes one cycle only at the sending NI to detect duplicate packets, and *THANOS* has no latency overhead. In the absence of duplicate packets in the 3D NoC, *SIM* has the lowest 3D NoC latency, and achieves an average of ~69%, ~80% and ~ 70% latency reduction compared to the baseline, FortNoCs, and P-Sec in the presence of active data-snooping HTs.

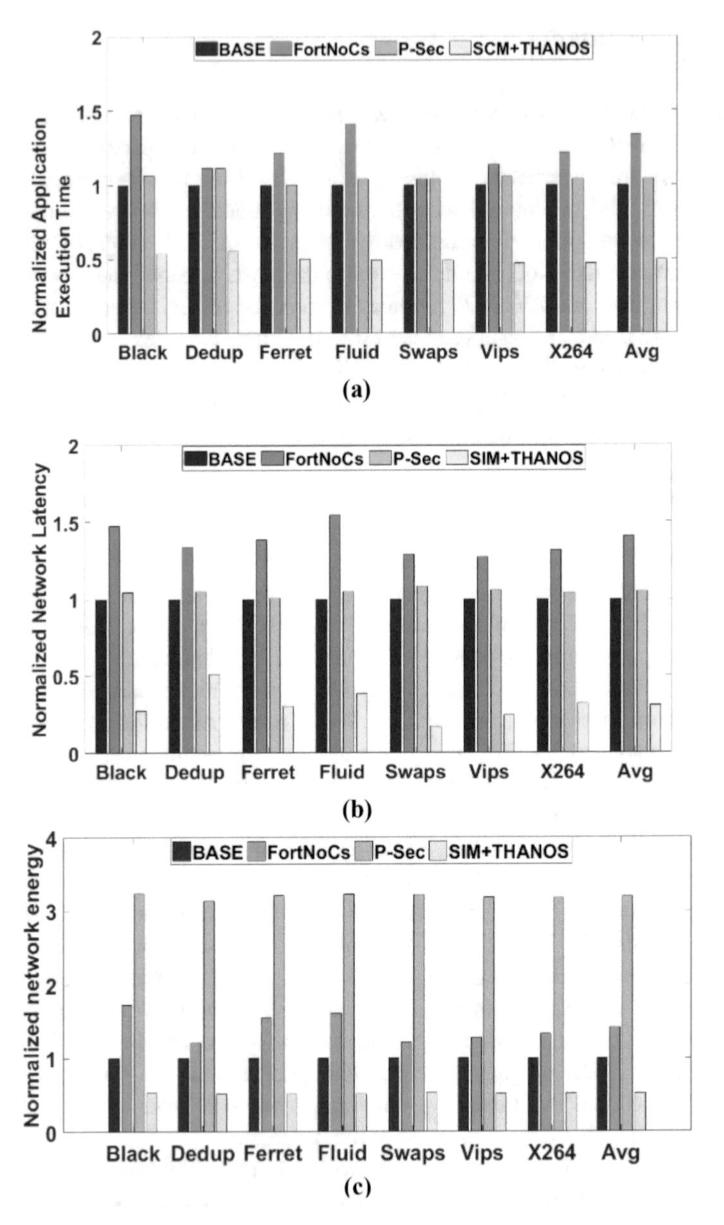

Fig. 17.8 (**a**) Normalized application execution time, (**b**) normalized network latency, (**c**) normalized 3D NoC energy consumption, across 3D NoCs with different security mechanisms in the presence of 4 active HTs attempting to inject duplicate packets to an accomplice thread

Next, we analyze 3D NoC energy consumption. Although *SIM + THANO* consumes ~5.5% additional NI power, its energy consumption is ~50% lowe compared to baseline on average due to the lower application execution time a

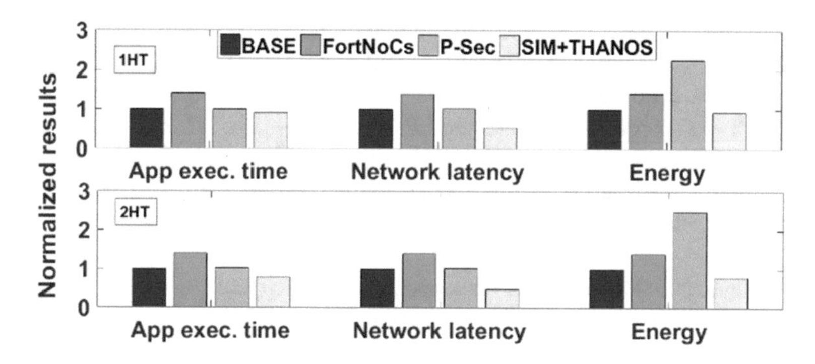

ig. 17.9 Normalized average values of application execution time, network latency, and 3D NoC nergy consumption across different security mechanisms with 1 HT (top), 2 HTs (bottom)

able 17.3 Area footprint of different 3D NoC security enhancement mechanisms

SIM + THANOS	FortNoCs	P-Sec
2.2 μm^2	4.9 μm^2	500 μm^2

hown in Fig. 17.8c. FortNoCs consumes around ~42% additional energy compared the baseline due to increased execution time and the overheads incurred to employ OR encryption/ decryption logic in the NI. P-Sec consumes up to 200% more nergy compared to the baseline due to its costly AMD, and CRC codec engines resent in NIs and 3D NoC routers. P-Sec is thus much more expensive, although it rovides combined safety against faults and snooping attacks.

We observe similar trends in application execution time, 3D NoC energy, and atency even when fewer number of HTs are active as shown in Fig. 17.9, with *IM + THANOS* performing better than the baseline unlike FortNoCs and P-ec. *This shows that our proposed snooping invalidation and snooping detection echanisms, SIM + THANOS, does not trade-off 3D NoC performance and 3D oC energy consumption to provide security.* Lastly, we compare area footprint of *IM + THANOS* with other schemes. As shown in Table 17.3, *SIM + THANOS* has e lowest area footprint amongst the three security mechanisms. *SIM + THANOS* echanism consumes only 2.15% additional area in the NI to implement the packet alidation mechanism.

7.6 Conclusions

this chapter we proposed a low-overhead mechanism called *SIM* to prevent data-nooping attacks that are initiated by HTs embedded in 3D NoC network interfaces. e also proposed a lightweight standalone snooping-attack detection mechanism alled *THANOS* that uses controlled circuit aging to detect the source of attacks that an help processors take preventive steps to mitigate future attacks. In FortNoCs and P-Sec it is impossible to detect the source of the attack, which can be addresse by using *SIM* + *THANOS*. Experimental results show that *SIM* + *THANOS* reduce application execution time by ~63% and ~ 49% and energy consumption by ~65% and ~ 85% compared to FortNoCs and P-Sec. *SIM* + *THANOS* incurs a minima additional 5.5% power and 2.15% area overhead, compared to the baseline, muc lower than the overhead for FortNoCs and P-Sec. Thus *SIM* + *THANOS* represent a promising solution to enhance 3D NoC security in manycore processors.

Acknowledgement This material is based upon the work supported by the National Scienc Foundation under grants CCF-1813370 and CCF-2006788.

References

1. V.Y. Raparti, S. Pasricha, RAPID: memory-aware NoC for latency optimized GPGPU archi tectures. IEEE Trans. Multi Scale Comput. Syst. **4**(4), 874–887 (2018)
2. V.Y.Raparti, S. Pasricha, DAPPER: data aware approximate NoC for GPGPU architecture Proceedings of IEEE/ACM NOCS, (2018)
3. Arteris, http://www.arteris.com/
4. S.S. Iyer, Three-dimensional integration: An industry perspective. MRS Bull. **40**(3), 225–23 (2015)
5. D. H. Kim et al., 3D-MAPS: 3D massively parallel processor with stacked memory. Proceed ings of International. Solid-State Circuit Conference, (2012)
6. N.E.C. Akkaya, et al. Secure chip odometers using intentional controlled aging. Proceeding of IEEE HOST, (2018)
7. S. Patnaik, et al., Best of both worlds: Integration of split manufacturing and camouflaging int a security-driven CAD flow for 3D ICs. Proceedings of IEEE ICCAD, (2018)
8. S. Bhunia et al., Hardware trojan attacks: threat analysis and countermeasures. Proc. IEE **102**(8), 1229–1247 (2014)
9. D. Ancajas, et al., Fort-nocs: mitigating the threat of a compromised NoC. Proceedings c ACM design automation conference, (2014)
10. D. Park et al. Exploring fault-tolerant network-on-chip architectures. Proceedings of th international conference on dependable systems and networks (DSN), (2006)
11. S. Shamshiri, et al., End-to-end error correction and online diagnosis for on-chip network Proceedings of the international test conference (ITC), (2011)
12. JYV Manoj Kumar, et al., Run time mitigation of performance degradation hardware troja attacks in network on chip. Proceedings of IEEE computer society annual symposium on VLS (ISVLSI), (2018)
13. J. Sepúlveda, et al., Reconfigurable security architecture for disrupted protection zones in NoC based MPSoCs. Proceedings of IEEE ReCoSoC, (2015)
14. T. Boraten, et al., Packet security with path sensitization for NoCs. Proceedings of IEEE DATE (2016)
15. H.M. Wassel et al., SurfNoC: a low latency and provably non-interfering approach to secu networks-on-chip. Proc. ACM SIGARCH Comput. Architecture News **41**(3), 583–594
16. S.V.R Chittamuru, I. Thakkar, S. Pasricha, SOTERIA: exploiting process variations to enhanc hardware security with photonic NoC architectures. Proceedings ACM//IEEE design automa tion conference (DAC), (2018)
17. S. Skorobogatov, Physical attacks and tamper resistance, in *Introduction to Hardware Securi and Trust*, (Springer, Berlin/Heidelberg, 2011)

8. E. Dubrova, et al., Secure and efficient LBIST for feedback shift register-based cryptographic systems. Proceedings of international test conf. (ITC), (2014)
9. P. Kocher, et al., Differential Power Analysis. (Springer- Verlag, 1999)
0. E. Dubrova, et al., Keyed logic BIST for Trojan detection in SoC. Proceedings of IEEE international symposium on system-on-chip (SoC), (2014)
1. M. Oya, In-situ Trojan authentication for invalidating hardware-Trojan functions. Proceedings of IEEE ISQED, (2016)
2. M. Hussain, et al., EETD: an energy efficient design for runtime hardware Trojan detection in untrusted network-on-chip. Proceedings of IEEE computer society annual symposium on VLSI (ISVLSI), (2018)
3. A. Prodromou, et al., Nocalert: An on-line and real-time fault detection mechanism for network-on-chip architectures. Proceedings of MICRO, (2012)
4. T. Boraten, et al., Secure model checkers for Network-on-Chip (NoC) architectures. Proceedings of IEEE Great Lakes Symposium on VLSI, (2016)
5. V.Y. Raparti, S. Pasricha, Lightweight mitigation of hardware trojan attacks in NoC-based manycore computing. Proceedings of IEEE/ACM DAC, (2019)
6. S. Pasricha, et al., A low overhead fault tolerant routing scheme for 3D Networks-on-Chip. Proceedings of IEEE ISQED, (2011)
7. M.K. Papamichael, et al., CONNECT: re-examining conventional wisdom for designing nocs in the context of FPGAs. Proceedings of FPGA, (2012)
8. Vivado HLS tool, Xilinx. https://www.xilinx.com/products/design-tools/vivado/integration/esl-design.html
9. S.R. Hasan, et al., Tenacious hardware trojans due to high temperature in middle tiers of 3-d ics. Proceedings of IEEE MWSCAS, (2015)
0. X.T. Ngo, et al., Integrated sensor: a backdoor for hardware trojan insertions?, Proceedings of euromicro conference on digital system design, (2015)
1. S.F. Mossa et al., Self-triggering hardware Trojan: due to NBTI related aging in 3-D ICs. Integr. VLSI J. **58**, 116–124 (2016)
2. S.R. Vangal et al., An 80-tile sub-100-w teraflops processor in 65-nm cmos. IEEE J. Sol. St. Circ. **43**(1), 29–41 (2008)
3. M. Gebhart, et al. Running PARSEC 2.1 on M5, Technical Report TR-09-32, UT Austin, Department of Computer Science, (2009)
4. S. Bhardwaj, et al., Predictive modeling of the NBTI effect for reliable design. Proceedings of IEEE custom integrated circuits conference, (2006)
5. V. Catania, et al., Noxim: An open, extensible and cycle-accurate network on chip simulator. Proceedings IEEE ASAP, (2015)
6. Cadence, https://www.cadence.com/content/cadence-www/tools/custom-ic-analog-rf-design/layout-design/virtuoso-layout-suite.html
7. J. Hestness, et al., Netrace: dependency-driven trace-based network-on-chip simulation. Proceedings of ACM NocArch, (2010)

Part VI
Conclusion and Future Directions

Chapter 18
The Future of Secure and Trustworthy Network-on-Chip Architectures

Prabhat Mishra and Subodha Charles

18.1 Summary

Given the widespread acceptance of Network-on-Chip (NoC) architectures in designing System-on-Chip (SoC) based devices, it is critical to ensure the security and trustworthiness of NoC-based SoCs [6–12]. This book provides a comprehensive reference for SoC designers, security engineers as well as researchers interested in designing secure on-chip communication architectures. This book contains contributions from NoC security and privacy experts. Different chapters cover a wide variety of security attacks and state-of-the-art countermeasures. The topics covered in this book can be broadly divided into the following categories.

18.1.1 NoC-Based SoC Design Methodology

The first three chapters introduced the readers to the NoC-based SoC design methodology. Specifically, it outlined various challenges associated with designing secure and energy-efficient on-chip communication architectures including discussions on *security vulnerabilities in NoC-based SoCs* (Chap. 1), *modeling of NoC architectures* (Chap. 2), and *energy-efficient NoC design* (Chap. 3).

P. Mishra (✉)
University of Florida, Gainesville, FL, USA
e-mail: prabhat@ufl.edu

S. Charles
University of Moratuwa, Colombo, Sri Lanka
e-mail: s.charles@ieee.org

P. Mishra, S. Charles (eds.), *Network-on-Chip Security and Privacy*,
https://doi.org/10.1007/978-3-030-69131-8_18

18.1.2 Design-for-Security Solutions

The next four chapters described efficient design-time solutions for securing NoC architectures. The goal of these approaches is to discuss lightweight security solutions for designing trustworthy NoCs including *lightweight encryption using incremental cryptography* (Chap. 4), *trust-aware routing* (Chap. 5), *lightweight anonymous routing* (Chap. 6), and *secure cryptography integration* (Chap. 7).

18.1.3 Runtime Monitoring Techniques

The next four chapters deal with security solutions for monitoring runtime vulnerabilities. Specifically, it explores several runtime security monitoring techniques including *detection of denial-of-service attacks* (Chap. 8), *securing communication using digital watermarking* (Chap. 9), *NoC attack detection using machine learning* (Chap. 10), and *Trojan-aware NoC routing* (Chap. 11).

18.1.4 NoC Validation and Verification

The next three chapters of the book look at NoC validation and verification techniques. Specifically, it presents efficient techniques for verifying functional correctness as well as security vulnerabilities including *NoC security and trust validation* (Chap. 12), *post-silicon validation and debug* (Chap. 13), and *design of reliable NoC architectures* (Chap. 14).

18.1.5 Emerging NoC Technologies

The next three chapters survey security implications in emerging NoC technologies. Specifically, it looks at security vulnerabilities and countermeasures for optical (Chap. 15), wireless (Chap. 16) as well as 3D NoCs (Chap. 17).

18.2 Future Directions

This book covered security challenges in NoC-based SoC architectures. The future giga and tera-scale architectures can impose new challenges and opportunities. The introduction of emerging NoC technologies such as wireless and optical have already shown promising results. However, it is a major challenge to develop low-ost and flexible security solutions with minimal impact on area, performance, and nergy. We briefly outline some of the challenges ahead in designing secure and rustworthy NoC architectures.

18.2.1 *Confluence of Functional Validation and Security Verification*

)rastic increase in SoC complexity has led to a significant increase in SoC design nd validation complexity [2, 13, 17, 19, 24, 27, 31, 33, 35]. Therefore, it is crucial o verify both functional correctness and security guarantees of NoC-based SoCs.)ne promising direction is to utilize the existing functional validation methodology o perform NoC security verification. Specifically, assertion-based validation is videly used for functional validation of NoC-based SoCs. Verification engineers an develop security assertions for monitoring security vulnerabilities [32]. Similar o functional assertions, security assertions can be used to check for any pre-silicon ecurity vulnerabilities. They can also be synthesized as security checkers (coverage nonitors) for post-silicon security validation. Similar to activating functional ssertions, verification engineers can utilize the same test generation framework to enerate tests for activating security assertions. In the future, verification engineers an seamlessly integrate assertion-based functional validation with assertion-based ecurity validation to design secure and trustworthy NoC-based SoCs.

18.2.2 *Security of Emerging NoC Architectures*

'he increased usage of emerging NoC technologies have motivated researchers to xplore security in optical, wireless, and 3D NoC architectures. The applicability f the proposed ideas to emerging NoC technologies is a promising avenue for uture exploration. The inherent characteristics of emerging NoC technologies can reate unique security vulnerabilities as well. For example, wireless NoCs inherntly use broadcast message to communicate between nodes. In such a scenario, avesdropping and spoofing attacks can become more prominent. Therefore, future esearch can explore required modifications to the proposed approaches to fit the haracteristics of emerging NoC architectures.

18.2.3 *Seamless Integration of NoC Security Mechanisms*

Vhile existing literature has discussed different threat models, it is naive to think hat mitigating one particular type of threat will secure the SoC. For example, defending against eavesdropping attacks does not guarantee that eavesdropping is the only possible attack in that particular architecture. Developing security mechanisms for different threat models is a promising starting point. However seamless integration of a suite of security mechanisms is required to secure the hardware root of trust. For example, Intel SGX (Software Guard Extensions) [14 provides hardware based software protection techniques. Future research needs to explore how to integrate several NoC security mechanisms and ensure thei interoperability in hardware, firmware, and software layers in order to enable a truly secure cyberspace.

18.2.4 NoC Security versus Interoperability Constraints

NoC-based SoCs are widely used today in resource-constrained IoT devices Designers employ a wide variety of techniques to improve energy efficiency in NoC based SoCs [3–5, 20, 23, 39]. IoT applications bring three important considerations long application life, conflicting design constraints, and dynamic use-case scenarios For example, a car equipped with state-of-the-art security would be vulnerable in a few years since it was not designed to defend against future attacks. Similarly it may be acceptable for a smart watch to trust the wireless network at home and impose a light-weight security requirement in favor of a lower energy profile However, a stronger defense mechanism is necessary when communicating with the untrusted network in a coffee shop at the cost of power and performance. Ther is a critical need for IoT devices to dynamically adapt to the environment ove the lifetime based on four-way interoperability constraints consisting of security energy, connectivity, and intelligence. The future NoC-based SoCs need to utilize a reconfigurable security engine that can be tailored during execution based on the use-case scenarios as well as interoperability constraints [10].

18.2.5 Comprehensive NoC Security Vulnerability Analysis

Given the increasing design complexity coupled with diversity of attacks, it i critical to verify NoC-based SoCs to ensure that there are no vulnerabilities. In the future, there will be a comprehensive vulnerability analysis framework tha utilizes a wide variety of analysis/validation techniques across design stages. Whil design-time security validation techniques can detect certain types of vulnerabilitie [1, 15, 16, 18, 29, 30, 34, 36], it is infeasible to remove all possible vulnerabilitie during pre-silicon security validation [35]. Therefore, verification engineers need to utilize both pre-silicon and post-silicon security validation [16, 35]. There are thre major approaches for NoC security validation: simulation-based validation, forma verification, and side-channel analysis. While formal methods can provide securit guarantees, the complexity of NoC designs make the exploration space grov exponentially. Simulation-based techniques are scalable but they cannot provide 100% security guarantees due to input space complexity. Side-channel analysis is a promising alternative that relies on side-channel signatures (such as power, delay, electromagnetic emanation, etc.) to detect vulnerabilities [21, 22, 25, 26, 28, 37]. The future of NoC security validation needs to utilize an effective combination of simulation, formal methods, and side-channel analysis.

18.2.6 NoC Security and Privacy Analytics Using Machine Learning

The intersection of machine learning and security has not been given adequate attention in an NoC context. Apart from a few runtime monitoring techniques that uses machine learning concepts, this area is still in its infancy. Tools such as *Cisco Encrypted Traffic Analytics* [38] utilize machine learning to detect threats by observing traffic behavior and unencrypted packet header information. It has shown promising results in the computer networks domain. Models that can be trained offline and detect threats during runtime has the potential to provide security guarantees, especially for real-time and safety-critical applications. While this book covered a wide variety of NoC security vulnerabilities known today, it is expected that future attacks will exploit new vulnerabilities. Therefore, a synergistic integration of security validation and machine learning will be crucial to design emerging NoC-based SoCs.

Acknowledgments This work was partially supported by the National Science Foundation (NSF) grant SaTC-1936040.

References

1. A. Ahmed, F. Farahmandi, Y. Iskander, P. Mishra, Scalable hardware trojan activation by interleaving concrete simulation and symbolic execution, in *2018 IEEE International Test Conference (ITC)*, pp. 1–10 (IEEE, 2018)
2. A. Ahmed, F. Farahmandi, P. Mishra, Directed test generation using concolic testing on RTL models, in *2018 Design, Automation & Test in Europe Conference & Exhibition (DATE)*, pp. 1538–1543 (2018)
3. A. Ahmed, Y. Huang, P. Mishra, Cache reconfiguration using machine learning for vulnerability-aware energy optimization. ACM Trans. Embed. Comput. Syst. (TECS) **18**(2), 1–24 (2019)
4. S. Charles, A. Ahmed, U.Y. Ogras, P. Mishra, Efficient cache reconfiguration using machine learning in NoC-based many-core CMPs. ACM Trans. Des. Autom. Electron. Syst. (TODAES) **24**(6), 1–23 (2019)
5. S. Charles, H. Hajimiri, P. Mishra, Proactive thermal management using memory-based computing in multicore architectures, in *International Green and Sustainable Computing Conference (IGSC)*, pp. 1–8 (2018)

6. S. Charles, M. Logan, P. Mishra, Lightweight anonymous routing in NoC based SoCs, in *Design Automation & Test in Europe (DATE)* (2020)
7. S. Charles, Y. Lyu, P. Mishra, Real-time detection and localization of dos attacks in NoC based socs, in *Design Automation & Test in Europe (DATE)*, pp. 1160–1165 (2019)
8. S. Charles, Y. Lyu, P. Mishra, Real-time detection and localization of distributed dos attacks in NoC based socs. IEEE Trans. Comput. Aided Des. Integr. Circuits Syst. **39**(12), 4510–452 (2020)
9. S. Charles, P. Mishra, Lightweight and trust-aware routing in NoC based socs, in *IEEE Computer Society Annual Symposium on VLSI (ISVLSI)* (2020)
10. S. Charles, P. Mishra, Reconfigurable network-on-chip security architecture. ACM Trans. Des Autom. Electron. Syst. (TODAES) **25**(6), 1–25 (2020)
11. S. Charles, P. Mishra, Securing network-on-chip using incremental cryptography, in *IEEE Computer Society Annual Symposium on VLSI (ISVLSI)* (2020)
12. S. Charles, C.A. Patil, U.Y. Ogras, P. Mishra, Exploration of memory and cluster modes in directory-based many-core CMPs, in *IEEE/ACM International Symposium on Networks-on Chip (NOCS)*, pp. 1–8 (2018)
13. M. Chen, X. Qin, H.-M. Koo, P. Mishra, *System-Level Validation: High-Level Modeling and Directed Test Generation Techniques* (Springer, 2012)
14. V. Costan, S. Devadas, Intel SGX explained. IACR Cryptology ePrint Archive **2016**(086) 1–118 (2016)
15. F. Farahmandi, Y. Huang, P. Mishra, Trojan localization using symbolic algebra, in *2017 22nd Asia and South Pacific Design Automation Conference (ASPDAC)*, pp. 591–597 (IEEE, 2017)
16. F. Farahmandi, Y. Huang, P. Mishra, *System-on-Chip Security: Validation and Verification* (Springer Nature, 2019)
17. F. Farahmandi, P. Mishra, Automated debugging of arithmetic circuits using incremental gröbner basis reduction, in *2017 IEEE International Conference on Computer Design (ICCD)*, pp. 193–200 (IEEE, 2017)
18. F. Farahmandi, P. Mishra, FSM anomaly detection using formal analysis, in *2017 IEEE International Conference on Computer Design (ICCD)*, pp. 313–320 (IEEE, 2017)
19. F. Farahmandi, P. Mishra, Automated test generation for debugging multiple bugs in arithmetic circuits. IEEE Trans. Comput. **68**(2), 182–197 (2018)
20. U. Gupta, C.A. Patil, G. Bhat, P. Mishra, U.Y. Ogras, DyPO: Dynamic pareto-optimal configuration selection for heterogeneous MpSoCs. ACM Trans. Embed. Comput. Syst (TECS) **16**(5s), 1–20 (2017)
21. Y. Huang, S. Bhunia, P. Mishra, MERS: statistical test generation for side-channel analysis based trojan detection, in *ACM SIGSAC Conference on Computer and Communication Security (CCS)*, pp. 130–141 (2016)
22. Y. Huang, S. Bhunia, P. Mishra, Scalable test generation for trojan detection using side channel analysis. IEEE Trans. Inf. Forensics Secur. (TIFS) **13**(11), 2746–2760 (2018)
23. Y. Huang, P. Mishra, Vulnerability-aware energy optimization for reconfigurable caches in multitasking systems. IEEE Trans. Comput. Aided Des. Integr. Circuits Syst. **38**(5), 809–82 (2019)
24. Y. Lyu, A. Ahmed, P. Mishra, Automated activation of multiple targets in RTL models using concolic testing, in *2019 Design, Automation & Test in Europe Conference & Exhibition (DATE)*, pp. 354–359 (IEEE, 2019)
25. Y. Lyu, P. Mishra, A survey of side-channel attacks on caches and countermeasures. J Hardware Syst. Secur. **2**(1), 33–50 (2018)
26. Y. Lyu, P. Mishra, Efficient test generation for trojan detection using side channel analysis, in *Design Automation & Test in Europe Conference (DATE)*, pp. 408–413 (2019)
27. Y. Lyu, P. Mishra, Automated test generation for activation of assertions in RTL models, in *2020 25th Asia and South Pacific Design Automation Conference (ASPDAC)*, pp. 223–22 (IEEE, 2020)
28. Y. Lyu, P. Mishra, Automated test generation for trojan detection using delay-based side channel analysis, in *2020 Design, Automation & Test in Europe Conference & Exhibition (DATE)*, pp. 1031–1036 (2020)

29. Y. Lyu, P. Mishra, Automated trigger activation by repeated maximal clique sampling, in *Asia and South Pacific Design Automation Conference (ASPDAC)*, pp. 482–487 (2020)
30. Y. Lyu, P. Mishra, Scalable activation of rare triggers in hardware trojans by repeated maximal clique sampling. IEEE Trans. Comput. Aided Des. Integr. Circuits Syst. (2020)
31. Y. Lyu, P. Mishra, Scalable concolic testing of RTL models. IEEE Trans. Comput. (2020). https://doi.org/10.1109/TC.2020.2997644
32. Y. Lyu, P. Mishra, System-on-chip security assertions. Preprint (2020). arXiv:2001.06719
33. Y. Lyu, X. Qin, M. Chen, P. Mishra, Directed test generation for validation of cache coherence protocols. IEEE Trans. Comput. Aided Des. Integr. Circuits Syst. **38**(1), 163–176 (2018)
34. P. Mishra, S. Bhunia, M. Tehranipoor, *Hardware IP Security and Trust* (Springer, 2017)
35. P. Mishra, F. Farahmandi, *Post-Silicon Validation and Debug* (Springer, 2019)
36. Z. Pan, P. Mishra, Automated test generation for hardware trojan detection using reinforcement learning, in *Asia and South Pacific Design Automation Conference (ASPDAC)* (2021)
37. Z. Pan, J. Sheldon, P. Mishra, Test generation using reinforcement learning for delay-based side channel analysis, in *IEEE/ACM International Conference on Computer-Aided Design (ICCAD)* (2020)
38. S.R. Patil, G.B. Pularikkal, D. McGrew, B.H. Anderson, M. Nanjanagud, Encrypted traffic analytics over a multi-path TCP connection, August 8 2019. US Patent App. 15/891,708
39. W. Wang, P. Mishra, S. Ranka, *Dynamic Reconfiguration in Real-Time Systems* (Springer, 2012)

Index

A

B

C

P. Mishra, S. Charles (eds.), *Network-on-Chip Security and Privacy*,
https://doi.org/10.1007/978-3-030-69131-8

Printed in the United States
by Baker & Taylor Publisher Services